# Microwave solid state devices and applications

Edited by

D.V. MORGAN and M.J. HOWES
Department of Electrical and Electronic Engineering
University of Leeds

PETER PEREGRINUS LTD

Published by Peter Peregrinus Ltd., Stevenage, UK, and New York

ISBN: 0 906048 39 7



Printed in Great Britain by A. Wheaton & Co. Ltd., Exeter

# PREFACE

This book contains extended versions of the lectures which formed the basis of the 6th IEE Microwave Summer School on "Microwave Solid-State Devices and Applications". The summer school, held at Bodington Hall, University of Leeds in July 1980, reflected the current interests and developments in the general area of solid-state device applications at microwave frequencies, and was particularly concerned with sub-system design. The school and the present text were intended to provide a compact, high-level introduction to microwave amplifiers, osicllators, and mixers, and was particularly aimed at the University and industrial research workers wishing to gain expertise in this area. Although no prerequisites were required for the course, it was assumed that students would have a degree in physics or electronic engineering in order to cope with the technical content.

The text is divided into fourteen chapters, and each chapter embodies up to four lectures written by acknowledged academic or industrial research workers who have been closely associated with particular aspects of solid-state microwave engineering. The contents of the book divide naturally into four basic categories; (i) the physics and technology of solid-state microwave devices (Chapters 1-4), (ii) circuit and device characterisation (Chapters 5-7) (iii) The design of oscillator and amplifier modules (Chapters 8-13), and (iv) the design of mixers and detectors (Chapter 14). Clearly, there is much overlap between these topics and indeed one of the philosophys which the school organising committee (and a feature of previous schools) wished to emphasise in the lectures was the close interaction between the device physics and the circuit in which the device is embedded.

We are very grateful to all those who have contributed to the present and past microwave summer schools and in particular to the authors of the present text, the school organising committee, and the staff of the IEE. We are also indebted to our Leeds colleague, Roger Pollard, for his critical comments, and advice on the contents and balance of this text.

School Co-Chairmen and Editors:
David V Morgan and Michael J Howes

University of Leeds
January 1980

# CONTRIBUTORS


M J ARNOLD — *Allen Clark Research Centre, Plessey Company, Towcester, England*
P BROOK — *GEC Hirst Research Centre, Wembley, England*
S CRIPPS — *Watkins Johnson (UK) Ltd, Windsor, England*
M J HOWES — *Dept of Electrical and Electronic Engineering, University of Leeds, England*
P HUISH — *Post Office Research Centre, Ipswich, England*
E HUGHES — *GEC Hirst Research Centre, Wembley, England*
N KENYON — *Post Office Research Centre, Ipswich, England*
J S LAMMING — *GEC Hirst Research Centre, Wembley, England*
B MAZONAS — *Microwave Associates Ltd, Dunstable, England*
D V MORGAN — *Dept of Electrical and Electronic Engineering, University of Leeds, England*
F MYERS — *Allen Clark Research Centre, Plessey Company, Towcester, England*
T OXLEY — *GEC Hirst Research Centre, Wembley, England*
J PATTISON — *RSRE Malvern, England*
R PENGELLY — *Allen Clark Research Centre, Plessey Company, Towcester, England*
R D POLLARD — *Dept of Electrical and Electronic Engineering*
R RANSON — *MEL Ltd, Crawley, England*
J D RHODES — *Dept of Electrical and Electronic Engineering*
J O SCANLAN — *Dept of Electrical and Electronic Engineering, University College, Dublin, Eire*
J A STEWART — *Dept of Electrical Engineering, Queen's University, Belfast, N Ireland*
A T M VAUGHAN — *RSRE Malvern, England*

# CONTENTS

# CHAPTER 1

## Passive microwave devices

Dr. D.V. Morgan

# PASSIVE MICROWAVE DEVICES

D V Morgan

University of Leeds, UK

## 1. INTRODUCTION

This course of lectures will consider the basic physical principles underlying the operation of solid state microwave devices. Of particular importance to the device/circuit engineer, is the relationship between the physical structure and the equivalent circuit model for the device. This allows the device to be designed into a practical microwave system. As a starting point to these lectures the basic foundation of semiconductors will be assumed, consequently, only a brief resume of this background will be presented here. In the 5 lectures under the title of microwave devices, the following will consider i) Simple PN junctions, ii) Schottky barrier diodes, iii) transferred electronic devices, iv) Impatt diodes, and v) Bipolar and field effect transistors.

## 2. RESUME OF SEMICONDUCTOR PHYSICS

Effective mass in crystal $m^* = \hbar^2/[d^2E/dk^2]$ (1)

Intrinsic carrier density

$$n_i = p_i = (N_V N_C)^{\frac{1}{2}} \exp\left[\frac{-(E_c - E_v)}{2kT}\right] \qquad (2)$$

$$N_c = 2\left[\frac{2\pi m_e^* kT}{\hbar^2}\right]^{3/2} \qquad N_v = 2\left[\frac{2\pi m_h^* kT}{\hbar^2}\right]^{3/2} \qquad (3)$$

Position of Fermi level in intrinsic semiconductor

$$E_F = \left\{\frac{E_c + E_v}{2}\right\} + kT\ln\left(\frac{N_c}{N_v}\right) \qquad (4)$$

Extrinsic semiconductors

n-type $N_d$ $\quad p_{no} = n_i^2/N_d$ (5)

p-type $N_a$ $\quad n_{po} = n_i^2/N_a$ (6)

Mobility

$$\mu_n = \frac{q\tau}{2m_e^*}, \quad \mu_p = \frac{q\tau}{2m_h^*} \qquad (7)$$

Current density $J = (nq\mu_n + pq\mu_p)\bar{F}$ (8)

Conductivity $\sigma = (nq\mu_n + pq\mu_p) = (J/F)$ (9)

Drift and diffusion

$$\left.\begin{aligned} J_n &= nq\mu_n + qD_n\left(\frac{dn}{dx}\right) \\ J_p &= pq\mu_p - qD_p\left(\frac{dp}{dx}\right) \end{aligned}\right\} \qquad (10)$$

Continuity equations:

$$\left.\begin{aligned} \frac{\partial n}{\partial t} &= g - U + \mu_n\nabla\cdot(n\bar{F}) + D_n\ \nabla\cdot\nabla n \\ \frac{\partial p}{\partial t} &= g - U - \mu_p\nabla\cdot(p\bar{F}) - D_p\ \nabla\cdot\nabla n \end{aligned}\right\} \qquad (11)$$

Where g is the generation rate and U the recombination rate

Poisson's equation

$$\rho = \varepsilon\nabla.F \qquad (12)$$

## 3. P.N. JUNCTION DIODES

Semiconductor p-n junctions are very commonplace in present-day electronic circuits. They are two-terminal devices and are extensively employed as polarity-sensitive switches with wide applications in the field of a.c. to d.c. conversion. p-n junctions, however, are more versatile than this simple picture would suggest - they have associated with them an a.c. impedance which is voltage dependent and which can exhibit either capacitive or inductive effects. In order to understand the terminal behaviour of p-n junctions it is necessary to consider the physical processes taking place inside the semiconductor and, from the basic physics, derive the d.c. and small-signal a.c. circuit models.

We can define a p-n junction (1) as a semiconductor device comprising a length of semiconducting material where the net doping $N(x) = (N_a - N_d)$ changes from p-type ($N_a > N_d$) to n-type ($N_a < N_d$). The point where $N_a = N_d$ is defined as the metallurgical junction. The detailed nature of the terminal electrical characteristics are very dependent on the shape of the doping profile N(x).

The behaviour of electrons and holes in the junction under the influence of an externally applied electric field is described by the transport equations which relate the electron and hole concentrations (p and n) and the current densities ($J_n$ and $J_p$) to the electric field, F.

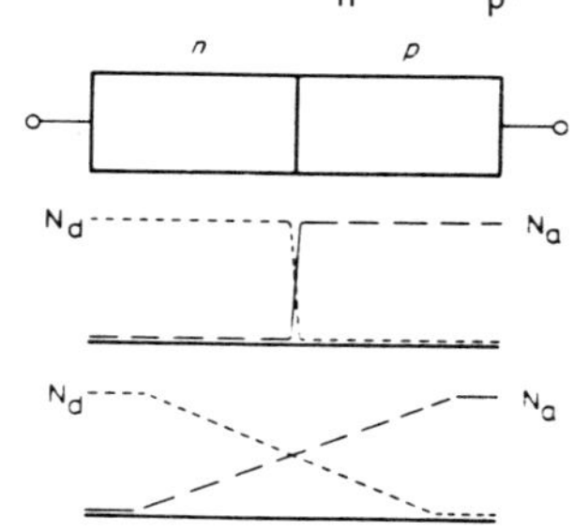


Figure 1. Illustrating the doping profile for an abrupt and linearly graded junction.

Equation 11 is the continuity equation, Equation 10 the current density equations, and Equation 12 is Poisson's equation. By inserting (10) into (11), a set of three equations is obtained which must be solved for the three unknowns n, p and F, to completely specify the state of the semiconductor sample at a particular point. In this general form the problem has no simple analytical solution, and one must resort to numerical techniques to effect a solution. In order to gain some insight into the basic physical processes we will consider two approximate models which can be solved analytically. These are the abrupt junction and the linearly-graded junction, shown in Figure 1. In the former, the doping undergoes an abrupt (step function) change at x = 0 from p-type to n-type. On the other hand, the linearly graded junction changes from $N_a$ at $-x_a$ to $N_d$ at $+x_d$ according to the equation $(N_d - N_a) = ax$, where the constant a defines the grading of the junction. Apart from the mathematical convenience of these two approximations, real junctions can be fabricated which closely approximate to one or other of these two ideal cases.

### 3.1 Static Charge, Field, and Potential Distribution

In order to simplify the discussion, consider what happens at an abrupt p-n junction. Electrons will diffuse across the interface from the n-type region to the p-type leaving a region of semiconductor where the donors are not compensated. This gives rise to a region of positive space charge. A similar situation exists in the p-type region; holes will diffuse across the junction leaving a negative space charge. This space charge will set up an electric field, which inhibits further diffusion of free carriers across the junction. In other words the Fermi levels will balance, as shown in Figure 2.

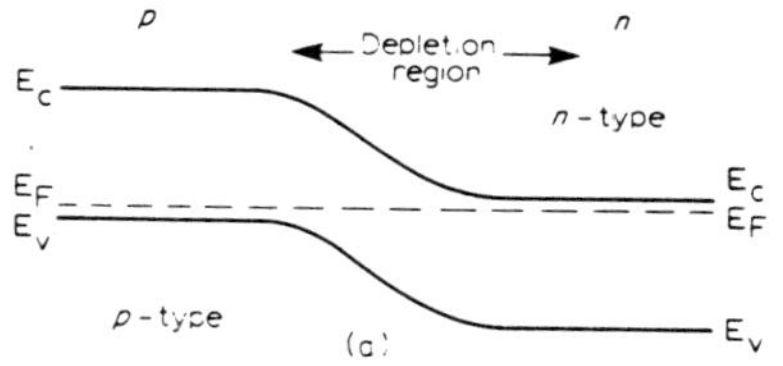


Figure 2 Illustrating how the Fermi levels match up in a p-n junction in equilibrium.

In our model the following simplifying assumptions will be made:

(i) The diode junction region is fully depleted of carriers, and the boundaries of the depletion region are sharp.

(ii) The values of the carrier concentrations at the depletion boundaries and the external contacts are known.

(iii) Recombination of carriers is described by a linear recombination law.

We shall now calculate the potential difference, using Poisson's equation. To do this we need to write down the total charge density inside the semiconductor. There are four contributions: (i) free electrons, (ii) holes, (iii) ionized donor atoms, (iv) ionized acceptor atoms. Thus the charge density is

$$\rho(x) = q(p(x) - n(x) + N_d(x) - N_a(x)).$$

In general these four terms are functions of position. In our ideal diodes, we reduce the problem to one dimension only. Poisson's equation becomes

$$\frac{d^2\psi}{dx^2} = \frac{q}{\varepsilon}\left[ p(x) - n(x) + N_d(x) - N_a(x) \right] \qquad (14)$$

#### Complete depletion approximation

(i) Abrupt junction

The simplest solution to Poisson's equation is achieved by assuming that the junction is completely depleted of mobile charge carriers in the region $-x_p$ to $x_n$. We can then solve Equation 14 in the p and n regions separately, subject to the boundary conditions.

$$\frac{d\psi}{dx} = 0 \text{ at } x = x_n \text{(n-region) and } x = -x_p \text{(p-region)}$$

Thus

$$\left[F = \frac{q}{\varepsilon} N_d(x_n - x),\ x>0\right] \qquad \left[F = \frac{q}{\varepsilon} N_a(x_p + x),\ x<0.\right] \qquad (15)$$

This is illustrated in Figure 3. The maximum field $F_o$ occurs at the metallurgical junction (x=0), where

$$F_o = \frac{q}{\varepsilon} N_d x_n = \frac{q}{\varepsilon} N_a x_p. \qquad (16)$$

Integrating a second time and using the boundary condition $\psi(0) = 0$

$$\psi(-x_p) = \frac{qN_a x_p^2}{2\varepsilon},\ -\psi(x_n) = \frac{qN_d x_n^2}{2\varepsilon} \qquad (17)$$

Summing the two contributions yields the total potential

$$V_o = |\psi(-x_p)| + |\psi(x_n)|$$

$$= q\left[N_a x_p^2 + N_d x_n^2\right]/2\varepsilon \qquad (18)$$

This is the built-in potential and its polarity is such as to oppose the flow of majority carriers across the junction. Now, since $x_n$ and $x_p$ are related by Equation 16 and $V_o$ can be calculated using (18), the depletion widths $x_n$ and $x_p$ can be calculated. The total depletion layer width ($X_D = x_n + x_p$) is given by

$$X_D = \left(\frac{2\varepsilon V_o}{q(N_a + N_d)}\right)^{1/2}\left[\left(\frac{N_a}{N_d}\right)^{1/2} + \left(\frac{N_d}{N_a}\right)^{1/2}\right] \qquad (19)$$

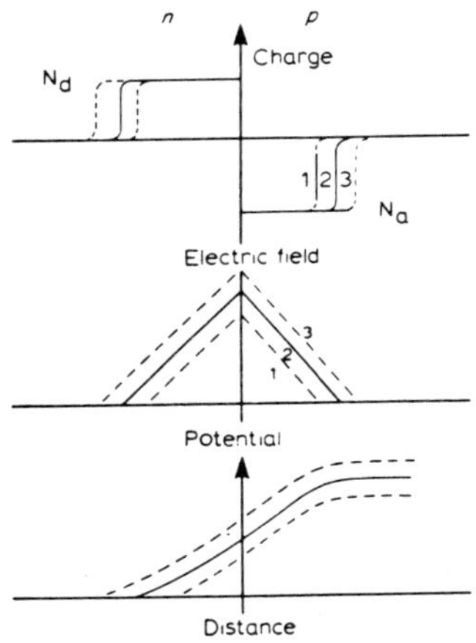


Figure 3. The charge, electric field, and potential distributions across an abrupt p-n junction.

For asymmetrical junctions we have

| $p^+n$ | $n^+p$ |
|---|---|
| $N_a >> N_d$ | $N_d >> N_a$ |
| $X_D \simeq x_n = \left(\frac{2\varepsilon V_o}{2N_d}\right)^{1/2}$ | $X_D \simeq x_p = \left(\frac{2\varepsilon V_o}{2N_a}\right)^{1/2}$ |

(20)

Thus, in an asymmetrical junction, the depletion layer extends mainly into the lightly doped semiconductor.

The effect of an externally-applied bias is to assist (reverse bias) or oppose (forward bias) the contact potential. Consequently, the variation of depletion width and stored charge with applied voltage V can be obtained by replacing $V_o$ with $V_T = V_o - V$ where V is positive for forward bias and negative for reverse bias. These results are summarized in Figure 3.

(ii) Linearly graded junction

The situation in a linearly graded junction is depicted in Figure 4 where

$$N_d - N_a = ax$$

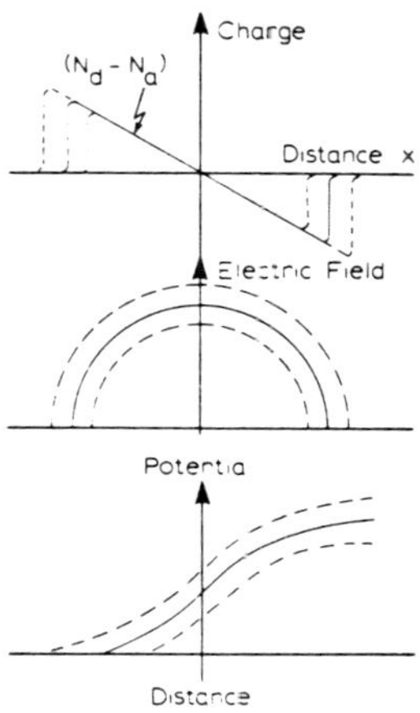


Figure 4. The charge, electric field, and potential distributions across a lineraly graded junction.

Hence Poisson's equation becomes

$$\frac{-dF}{dx} = \frac{d^2\psi}{dx^2} = \frac{q}{\varepsilon}\,ax$$

We note that $x_n = x_p = \frac{1}{2}X_D$ at all times, because of the symmetry of the junction. Integrating as before gives

$$F(x) = \frac{-qa}{2\varepsilon}\left(\tfrac{1}{4}X_D{}^2 - x^2\right)$$

$$F_o = \frac{qaX_D{}^2}{8\varepsilon} \quad (\text{at } x = 0)$$

$$V_o = \frac{qaX_D{}^3}{12\varepsilon} \tag{21}$$

For an externally applied voltage V, Equation 21 becomes

$$(V_o - V) = \frac{qa}{12\varepsilon}\,X_D{}^3 \tag{22}$$

## Current-Voltage Characteristics

### Diffusion current

Up to this point we have discussed the static properties of the junction depletion region. We now wish to derive the relationship between the terminal current and the externally applied voltage. Before proceeding let us remind ourselves of the role of the externally applied voltage.

It opposes (forward bias) or assists (reverse bias) the built-in voltage. In the former case the Fermi level in the n-side is raised with respect to that in the p-side, causing the depletion region to decrease in size and majority carriers from each side to be injected across the depletion region. The behaviour of the majority carriers injected across the junction to become minority carriers is governed by Equation 11 Once these carriers pass through the depletion regions they enter a field-free region, where diffusion takes place in the direction of the two respective contacts Gradually, minority carriers will be lost by recombination with the abundant majority carriers. These lost carriers must be replaced by a current drawn from the terminals. In reverse bias the applied field assists the internal field, reducing the injection of majority carriers but at the same time encouraging the extraction of minority carriers from the regions neighbouring the junction. This loss of minority carriers must be fed by the diffusion of minority carriers from the bulk. The three situations are depicted in Figure 5. From this qualitative picture we can isolate three steps which lead to a solution to the problem. (i) knowing the magnitude of $p_n(o)$ and $n_p(o)$ one may calculate the variation in the minority carrier density on either side of the junction as the applied voltage is varied. (ii) The detailed shape of the profile of the injected carriers as they diffuse into the bulk of the semiconductor may be calculated by solving the continuity equations Equations (iii) The current, which is a measure of the injected carrier concentration, can be calculated by applying Equations10 at the edge of the depletion region. That is,

$$J_n = qD_n\left(\frac{dn}{dx}\right)_{x=x_n} \qquad J_p = qD_p\left(\frac{dp}{dx}\right)_{x=x_p} \tag{23}$$

Figure 5 illustrates the carrier distribution in the vicinity of the depletion region. Our problem is to solve the continuity equation for the minority carriers to enable us to predict the shape of these injected carrier profiles. We will only consider the electrons injected from the n-type into the p-type material. Because of the symmetry, the argument for the holes injected from the p to the n regions will be identical. For all values of $x > x_n$, $-x < -x_p$ the electric field (F) is assumed zero. A second assumption is that our solution is restricted to low injection levels (i.e. $p_n \ll n_n \doteq n_{no}$ and $n_p \ll p_p \doteq p_{po}$). Under conditions of high injection the minority and majority carrier continuity equations must be solved in both regions. The third assumption is that generation and recombination of carriers in the depletion region is negligible.

With these approximations we can reduce the minority carrier continuity equations to

$$\frac{\partial^2 p_n}{\partial x^2} - \frac{p_n - p_{no}}{D_p\tau_p} = 0 \qquad \frac{\partial^2 n_p}{\partial x^2} - \frac{n_p - n_{po}}{D_n\tau_n} = 0. \tag{24}$$

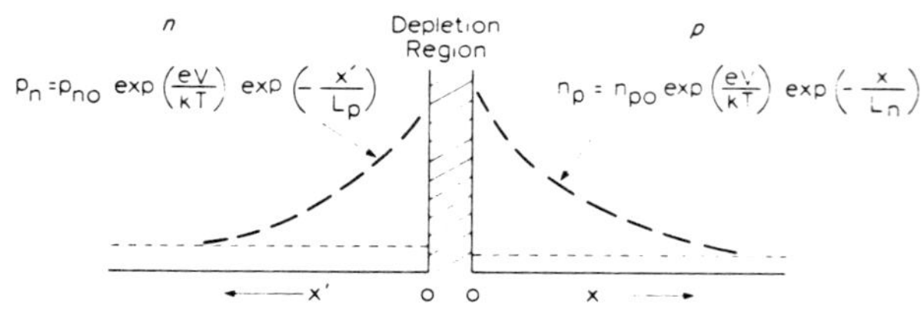


Figure 5. The minority carrier dsitributions for a forward biased p-n junction.

These eouations must be solved, subject to the following boundary conditions: (i) at $x = x_{no}$ and $x_{po}$ the minority carrier concentrations are defined by the injection values, (ii) at $x = \infty$ (n-region), $p_n \rightarrow p_{no}$ and at $x = \infty$, $n_p \rightarrow n_{po}$ (p-region). Hence

$$\Delta p_n = p_n - p_{no} = p_{no}\left[\exp\left(\frac{qV}{kT}\right) - 1\right]\exp\left[\frac{(x - x_{po})}{L_p}\right] \tag{25}$$

$$\Delta p_n = p_n - n_{po} \left[\exp\left(\frac{qV}{kT}\right) - 1\right] \exp\left[\frac{(x - x_{po})}{L_n}\right] \qquad (25)$$

where

$$L_p = \sqrt{(D_p \tau_p)}$$

$$L_n = \sqrt{(D_n \tau_n)}$$

are the diffusion lengths. This situation is depicted in Figure 5 for forward bias. Finally, the injected current components can be obtained by using Equation 23.

$$J_p = \frac{qD_p p_{no}}{L_p} \left[\exp\left(\frac{qV}{kT}\right) - 1\right] \text{ and } J_n = \frac{qD_n n_{po}}{L_n} \left[\exp\left(\frac{qV}{kT}\right) - 1\right] \qquad (26)$$

The total current density which has to be supplied at the terminals is the sum of these two components:

$$J = J_n + J_p$$

$$J = \left(\frac{qD_p p_{no}}{L_p} + \frac{qD_n n_{po}}{L_n}\right) \left[\exp\left(\frac{qV}{kT}\right) - 1\right] \qquad (27)$$

$$= J_s \left|\exp\left(\frac{qV}{kT}\right) - 1\right| \qquad (28)$$

Reverse bias (V<0)

In these circumstances exp(qV/kT)<1 and when qV>few kT, exp(qV/kT)<<1 so that Equation 27 becomes

$$J = -\left[\frac{qD_p p_{no}}{L_p} + \frac{qD_n n_{po}}{L_n}\right] = -J_s \qquad (29)$$

The full terminal characteristics are shown in Figure 6

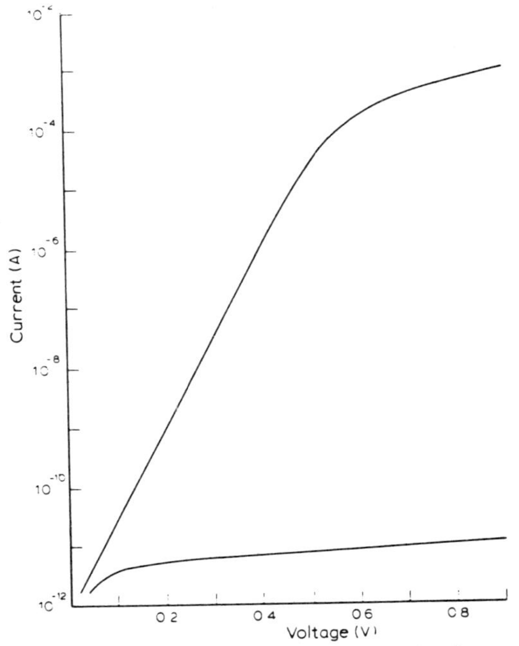


Figure 6. Current-voltage characteristic for a real p-n junction plotted on a semi-logarithmic scale.

Forward bias (V>0)

Rearranging Equation 28 we obtain

$$\ln\left[\frac{J + J_s}{J_s}\right] = \left(\frac{q}{kT}\right) V$$

or for $qV \geq$ few kT

$$\ln(J) - \ln(J_s) = \left(\frac{q}{kT}\right) V.$$

Thus a plot of ln(J) versus V will yield a straight line with a slope of (q/kT) and intercept $J_s$.

Incremental resistance

From the terminal characteristics we can define an incremental resistance (quasi-static)

$$r_{ed} = \frac{dV}{dI} = \frac{kT/q}{AJ_s\left[\exp(qV/kT)\right]} \qquad (30)$$

where A is the diode area.

$$r_{ed} = \frac{kT}{q(J + J_s)} \text{ per unit area.} \qquad (31)$$

In reverse bias $r_{ed}$ tends to infinity (i.e. the diode is open circuit to small-signal a.c.).

Generation/recombination current

In order to illustrate the nature and magnitude of the generation/recombination current, consider the following idealized case. A simple abrupt junction is forward biased with a small voltage. In this case, however, the depletion region contains a finite density of some efficient acceptor recombination centre with a mid-band energy level. Consider the situation at low injection levels (small applied voltage). An electron travelling from the n region to the p region has a finite chance of being trapped by the recombination centre. If, subsequently, this centre makes a further capture of a hole travelling from p→n before the captured electron is emitted, then an electron→hole pair has been lost, resulting in a decrease in the terminal current. As the forward bias is increased the size of the depletion region decreases in the recombination volume. At the same time, of course, the diffusion current, described earlier, increases very rapidly with applied voltage (i.e. exponentially) and eventually dominates the I-V characteristic at moderate applied voltages. In reverse bias a similar situation holds, but here the centre acts in reverse: the emission of a hole followed by the emission of an electron results in the generation of an electron-hole pair and an increase in the terminal current. The field in the depletion

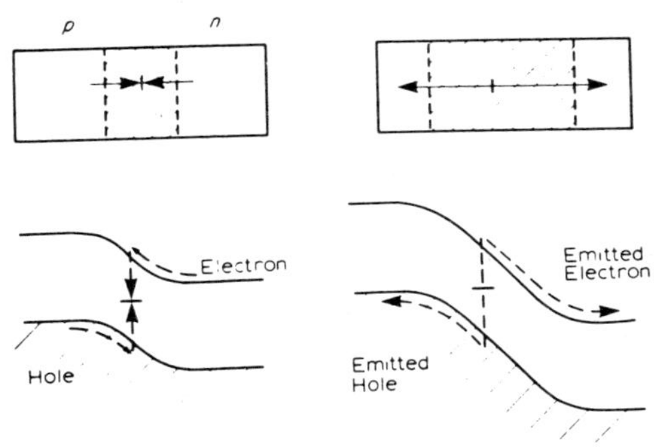


Figure 7. Illustrating the process of recombination and generation in the depletion region.

region ensures that the extracted minority carriers (electrons from the p region and holes from the n region) are moved swiftly through the depletion region thereby reducing the probability of recombination. Since the current generated is dependent on the number of centres, and the depletion region volume increases with reverse bias, the generation current increases with reverse bias. The result of this is to cause the reverse saturation current predicted from the diffusion model to have a slow dependence on the applied voltage. The process is illustrated in Figure 7.

The role of generation/recombination on the terminal characteristics was put onto a quantitative footing by the theoretical model of Sah Noyce and Shockley (3) which yields the following ecuations for the terminal I-V characteristics: In forward bias we have

$$I_{gr} \doteq - \left\{ \frac{A}{2} \frac{q\, ni\, X_D}{\tau_o} \right\} \exp \left[ \frac{qV}{2kT} \right] \quad (32)$$

where $\tau_o = 1/|\sigma v_{th} N_t|$ and $v_{th}$ the carrier thermal velocity and $N_t$ the recombination trap density.

Note that in this case

$$\ln(I_{gr}) = \ln(\text{const}) + \left\{ \frac{qV}{2kT} \right\}$$

and thus a plot of $\ln(I_{gr})$ versus V will yield a straight line with a slope (q/2kT), one half that obtained with the diffusion current. Generation recombination current is dominant over the diffusion current at low forward voltages and in wide bandgap semiconductors, this is illustrated by the set of curves for Ge, Si and GaAs diodes shown in Figure 8.

In reverse bias the corresponding current is

$$I_{gr} = - \left\{ \frac{q\, n_i A\, X_D}{2\tau_o} \right\} \quad (33)$$

This result indicates that the reverse current will not saturate, but will continue to increase with reverse bias since the depletion width $X_D$ increases. The relationship $X_D(V)$ is known for ideal diodes and therefore for a linearly graded junction.

$$I_{gr} \propto V^{1/3}$$

and an abrupt junction $I_{gr} \propto V^{\frac{1}{2}}$

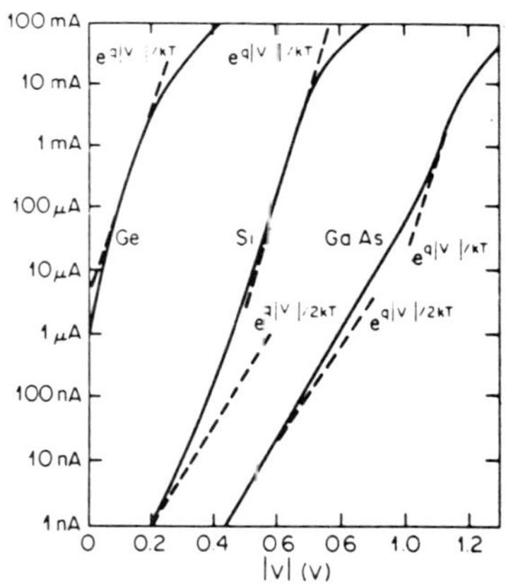


Figure 8. Comparison of forward bias current-voltage characteristics for Ge, Si, and GaAs diodes at 25°C (after Grove).

## Incremental resistance

Generation/recombination current dominates over the diffusion current for low values of forward bias and for reverse bias. In these circumstances the device conductance is no longer given by Equation 31 but has to be derived from Equation 32 for forward bias, and 33 for reverse bias. Thus we may write for forward bias

$$r_{eg} = \left( \frac{dV}{dI_{gr}} \right) = \frac{2kT}{qAJ} \quad (34)$$

or in the generalised form

$$r_{eg} = \left( \frac{mkT}{qAJ} \right) \quad . \quad (35)$$

In reverse bias the small signal resistance is no longer infinite. From Equation 33 we have

$$I_{gr} = \frac{-qn_i A}{2\tau_o} X_D(V)$$

where the dependence of $X_D$ on applied voltage is determined by the doping profile:for an abrupt junction $X_D = \alpha V^{\frac{1}{2}}$ where the constant $\alpha$ is defined by Equation 19. Hence

$$r_{eg} = \left( \frac{dV}{dI_{gr}} \right) = \frac{4\tau_o V^{\frac{1}{2}}}{qn_i A\alpha} = \frac{2V}{I_{gr}} \quad (36)$$

A similar equation may be derived for a linearly graded junction.

## Junction Capacitance

### Depletion capacitance

We saw earlier that the semiconductor junction has associated with it a dipole of static charge enclosed in the depletion region ($X_D = x_n + x_p$) and that the total enclosed charge is a function of the applied terminal voltage. When the applied voltage changes, charge must move into or out of the depletion region, and since this charge has to be supplied through the terminals it gives rise to the depletion capacitance $C_d$ since the stored charge Q=f(V),

$$C_d = \left( \frac{dQ}{dV} \right) .$$

The physical basis of this capacitance is the finite time required to inject or withdraw the mobile carriers required to contract or expand the size of the depletion region.

The charge per unit area in one half of the depletion region is

$$Q = q \int_{-x_p}^{0} N_a(x)dx = q \int_{0}^{x_n} N_d(x)dx. \quad (37)$$

In the abrupt junction $N_a$ and $N_d$ are constants, and hence

$$Q = qN_a x_p = qN_d x_n .$$

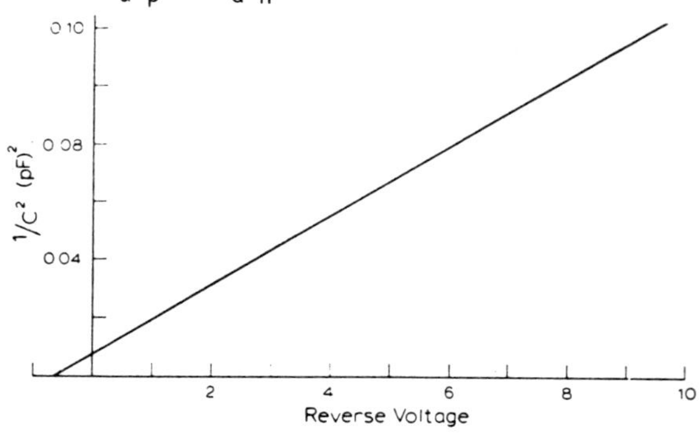


Figure 9. $(1/C^2)$ versus applied voltage V for a silicon $p^+n$ diode

Substituting for $x_n$ or $x_p$ from Equation 19

$$Q = (V_o - V)^{\frac{1}{2}} \left[ 2\varepsilon q \frac{N_a N_d}{(N_a + N_d)} \right]^{\frac{1}{2}}$$

or

$$Q = (V_o - V)^{\frac{1}{2}} B$$

Therefore

$$C_d = \frac{dQ}{dV} = \tfrac{1}{2} B (V_o - V)^{-\frac{1}{2}} \tag{38}$$

or

$$V_o - V = \left(\frac{1}{C_d^2}\right) \left(\frac{B^2}{4}\right) \tag{39}$$

A plot of $(1/C_d^2)$ versus applied voltage V yields a straight line with a gradient of $(B^2/4)^{-1}$ and an intercept $(4V_o/B^2)$ Figure 9 shows such a plot for an abrupt $p^+n$ Si diode.

For a linearly graded junction we must substitute for $X_D$ from Equation 22..

$$C_d = \left[ \frac{qa\varepsilon^2}{12(V_o - V)} \right]^{1/3} \tag{40}$$

$$V_o - V = \frac{1}{C_d^3} \left(\frac{qa\varepsilon^2}{12}\right). \tag{41}$$

A plot of $(1/C_d^3)$ versus applied voltage V will yield a straight line with a gradient of $(qa\varepsilon^2/12)^{-1}$ from which the impurity gradient may be calculated.

Diffusion capacitance

In forward bias, minority carriers are injected into the neutral regions beyond the edges of the depletion region. The excess carrier densities $\Delta n_p$, $\Delta p_n$, equation 25 represent a stored charge $Q_m$, which is a function of the applied voltage V. Consequently, we can derive a diffusion capacitance.

$$C_{diff} = \frac{dQ_m(V)}{dV}$$

We can write $Q_m = Q_1 + Q_2$, and hence limits are

$$Q_m = q \int_{x_n}^{W_n} \Delta p_n dx + q \int_{-W_p}^{-x_p} \Delta n_p \, dx \tag{42}$$

Substituting for $\Delta p_n(x)$ and $\Delta n_p(x)$ and integrating yields the result

$$Q_m = \left\{ \exp\left(\frac{qV}{kT}\right) - 1 \right\} q[p_{no}L_p + n_{po}L_r] \tag{43}$$

and hence

$$C_{diff} = \frac{q^2}{kT} (p_{no}L_p + n_{po}L_n) \exp\left(\frac{qV}{kT}\right) \tag{44}$$

which defines the low-frequency or quasi-static limit of the diffusion capacitance.

In order to obtain the full expression for the diffusion capacitance, incorporating the frequency dependence, we must use the rigorous procedure of calculating the diode admittance (6,2) and taking the imaginary part

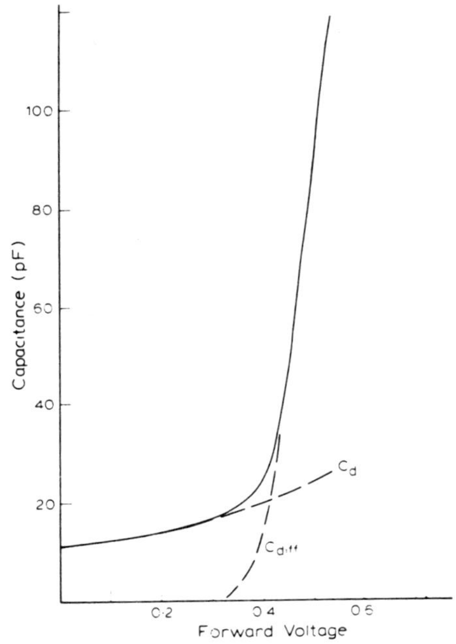


Figure 10. Capacitance-voltage curve for a forward bias $p^+n$ diode.

## 6. JUNCTION BREAKDOWN

We will now consider junction breakdown phenomena. The depletion region, as its name suggests, is depleted of mobile carriers and is hence an insulating region. In reverse bias, the majority of the voltage applied to the diode appears across this region giving rise to an intense electric field (Figure 3). As a result of this field two types of breakdown phenomenon can occur: (i) Zener or tunnelling breakdown, and (ii) avalanche breakdown. These phenomena are not restricted to p-n junctions; indeed a large body of literature dealing with these effects in thin insulating films predates the junction studies. We will now consider each of these effects in turn.

The Zener or tunnelling effect

Tunnelling is a quantum mechanical effect and is not predicted by classical mechanics. It is the ability of a particle, with insufficient energy to pass over a potential barrier, to pass through (or tunnel) the barrier. The importance of this effect in the present context is illustrated in Figure 11. When a junction is reverse biased such that there are vacant energy levels in the conduction band, having the same energy as the full states in the valence band, the possibility exists for an electron to tunnel from the valence band into the conduction band, resulting in the production of an electron-hole pair. These two new carriers will be swept quickly by the junction field out of the depletion region, thus increasing the terminal current.

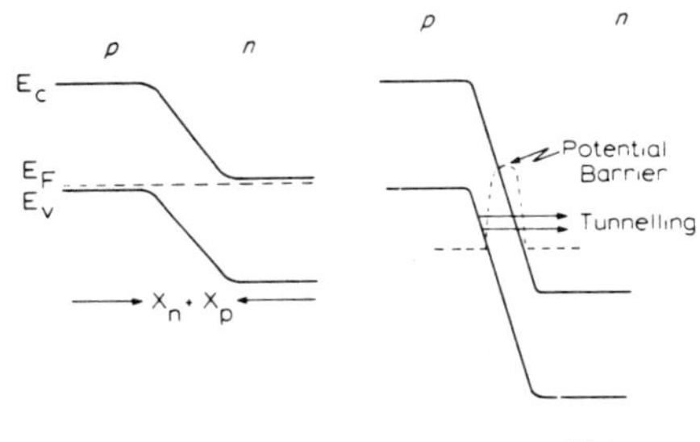


Figure 11. Illustrating the Zener tunnelling mechanism.

A mathematical treatment of the p-n junction tunnelling yields the following current density: (6).

$$J_t = \frac{\sqrt{(2m^*)}q^3}{4\pi^2\hbar^2 E_g^{\frac{1}{2}}} FV \exp\left(-\frac{4\sqrt{(m^*)}E_g^{3/2}}{3qF\hbar}\right). \quad (45)$$

When the junction field F approaches $10^6$V $cm^{-1}$ in Si and Ge diodes, the tunnelling current becomes significant. High junction fields of this magnitude are obtained by doping the p and n sides heavily.

Avalanche breakdown

Avalanche multiplication of the electrons and holes injected into the depletion region of p-n junctions was established for both Ge and Si by the early work of McKay (7) and Miller (8). The model envisaged for the avalanche process is the solid state equivalent to the well established gaseous avalanche breakdown and consequently much of the terminology and analysis is derived from this early source. There is one important difference in the solid state: multiplication occurs not only by the electrons, but also by the holes. This dual ionisation process provides a positive feedback to the avalanche process. Electrons and holes injected into the depletion region from the p and n regions respectively will be accelerated by the field and, when they have sufficient energy, may create electron-hole pairs by colliding with valence electrons. The collision products are then accelerated until they produce further pairs, and so on.

A mathematical treatment of the ionisation process yields the following equation for the breakdown voltage: (2)

$$V_B = \frac{F_o X_D}{2} = \left(\frac{\varepsilon F_o^2}{2q}\right) N_d \quad (46)$$

for an abrupt $p^+n$ junction and

$$V_B = \frac{2F_o X_D}{3} = \frac{4F_o^{3/2}}{3} \left(\frac{2\varepsilon}{aq}\right)^{\frac{1}{2}} \quad (47)$$

for a linearly graded junction.

## 4. METAL SEMICONDUCTOR CONTACTS

A second and very important class of rectifying diode consists of a metal semiocnductor (Schottky) contact. As in the case of the pn junction discussed in section 3, the rectifying properties of the Schottky barrier contact arise from the presence of an electrostatic barrier between the metal and the semiconductor. In order to understand the nature of the electrostatic barrier we will once again consider the application of Poisson's equation.

Space Charge Field and Potential Distribution

If, for the moment, we ignore the surface states present in the semiconductor, then figure 12 shows the simplified band diagrams for a metal of work function $\phi_m$ brought into contact with an n-type semiconductor with a work function $\phi_s$. In equilibrium the Fermi level must be continuous across the interface and hence if $\phi_m > \phi_s$, electrons will leave the semiconductor to move into the metal. The metal will gain a negative surface charge and the semiconductor a net positive depletion charge. A potential barrier will be set up which is sufficient to equate the Fermi levels and thus reduce the net flow of electrons to zero. The barrier height $\phi_b$ is controlled simply by the two work functions, thus:

$$\phi_b = (\phi_m - \chi_s) \quad (48)$$

and the diffusion potential $V_d$

$$qV_d = (\phi_m - \phi_s) \quad (49)$$

Clearly, from equation (54) we see that $\phi_b \simeq o$ when $\phi_m \leq \psi_s$. In these circumstances there will be no electrostatic barrier and the contact is non-rectifying. Similar arguments apply to p-type semiconductors, here however, the condition for rectification is $\phi_m < \phi_s$.

The linear dependence of barrier height on metal work function predicted by equation (48) is rarely observed in practice. In order to understand this problem, it is first necessary to discuss the near surface properties of semiconductors. Consider an n-type semi-conductor as shown in Figure 13. The surface will have associated with it a series of surface states whose origin can be either dangling bonds from the surface atoms or, alternatively, they may result from some extrinsic surface contaminant film. With such surface states we can define a neutral level $\phi_o$, such that the surface will be neutral if all the states up to $\phi_o$ are filled. Figure 13 shows how the free surface relaxes in equilibrium. Electrons in the conduction band near to the surface will occupy some of the surface states above $\phi_o$, giving rise to a negative surface charge. At the same time the loss of free electrons near the surface will provide the positive space charge required to compensate the surface charge. The corresponding potential, charge and field distributions are shown in Figure 13 . Both $\phi_o$ and the distribution in energy of the surface states depend very critically on the nature of the semiconductor surface and in the limiting case where the density of surface states is very high, the free surface band bending will be pinned near to $\phi_o$ and hence $\phi_b \simeq (E_g - \phi_o)$. Consider now, what happens when a metal of work function $\phi_m$ is brought into contact with an n-type semiconductor, as is shown in Figure 14. Here we assume that $\phi_m > \phi_{RS}$ and that the semiconductor is covered with a thin oxide film of thickness $\delta$. As soon as the two surfaces are brought into intimate contact the Fermi levels of the metal and the semiconductor must be equated as is shown in Figure 14 . The existence of surface states means that charge can be stored at the semiconductor surface. Consequently, part of the potential required to equate the Fermi levels can be accommodated by a potential drop across the oxide film resulting from a transfer of charge between the metal and the surface states. A change in the surface charge will also result in a readjustment in the depletion charge. However, in situations where the surface state density is high, considerable charge can be injected into, or removed from the surface states without the surface Fermi level deviating greatly from $\phi_o$. In these circumstances the semiconductor barrier height $\phi_b$ will be pinned by the surface states. The converse will be true for materials with low surface state densities. In these circumstances the surface will have no capacity for storing charge and the limiting case discussed earlier will occur.

The model described above, the Bardeen model (9), yields the following expression for the barrier height in an n-type semiconductor.

$$\phi_b = \gamma(\phi_m - \chi_s) + (1-\gamma)(E_g - \phi_o) \quad (50)$$

$$\text{where } \gamma = \frac{\varepsilon_i}{\varepsilon_i + q D_s} \quad (51)$$

and $D_s$ is the density of interface states in $eV^{-1}m^{-2}$

Clearly, as $D_s \to o$, $\gamma \to 1$ and $\phi_b \to (\phi_m - \chi_s)$, the correct result.

Conversely as $D_s \to \infty$ $\gamma \to o$ and $\phi_b = (E_g - \phi_o)$

(i.e. the surface states pin the barrier height).

The width of the depletion region is related to the barrier height. Applying Poisson's equation to the depletion region we have;

$$\frac{d^2\psi}{dx^2} = \frac{q}{\varepsilon} N_d$$

where the boundary conditions are

$\psi = o$ for $x = o$

$\psi = V_d - V$ for $x \geq x_n$

$\frac{d\psi}{dx} = o$ for $x \geq x_n$

hence $F = \frac{d\psi}{dx} = \frac{q}{\varepsilon} N_d(x_n - x)$

thus

$$F_o = \frac{q\,N_d\,x_n}{\varepsilon} \qquad (52)$$

Integrating a second time we obtain

$$\psi = +\frac{q\,N_d\,(x_n x - x^2/2)}{\varepsilon}$$

hence

$$(V_d - V) = \frac{q\,N_d\,x^2_n}{2\varepsilon} \qquad (53)$$

and

$$x_n = \left[\frac{2\varepsilon(V_d - V)}{q\,N_d}\right]^{\frac{1}{2}} \qquad (54)$$

Capacitance Voltage Relationship

The Schottky barrier resembles very closely an asymetrical abrupt pn junction (i.e. $p^+n$ or $n^+p$) and indeed exhibits a similar capacitance voltage relationship in reverse bias. In these circumstances the capacitance arises from the depletion region. If $Q_d$ is the total charge due to the uncompensated donors (n-type material) and $V_d$ the diffusion potential then we may write for an n-type semiconductor

$$Q_d = q\,N_d\,x_n$$

and since

$$x_n = \left\{\frac{2(V_d - V)\varepsilon}{q\,N_d}\right\}^{\frac{1}{2}}$$

we have

$$Q_d = \left\{2q\,N_d\varepsilon\,(V_d - V)\right\}^{\frac{1}{2}}$$

and hence the depletion capacitance $C_d$ per unit area.

$$C_d = \frac{dQ_d}{dV} = \left\{\frac{q\varepsilon\,N_d}{2(V_d - V)}\right\}^{\frac{1}{2}} = \left\{\frac{\varepsilon}{x_n}\right\} \qquad (55)$$

Thus

$$\left[\frac{1}{C_d^2}\right] = \frac{2}{q\varepsilon N_d}\,[V_d - V]$$

a straight line which extrapolated to the point where $(1/C_d^2) = o$ gives the diffusion potential $V_d$. It is instructive to compare this result with that derived for an abrupt $p^+n$ junction.

In terms of terminal capacitance, an important difference between the pn junction and the Schottky barrier is that the latter is a majority carrier device. In the ideal case, there is no minority carrier injection and hence there is no diffusion capacitance in forward bias.

4.2 Current Voltage Characteristics (Thermionic Emission Model)

The basic physical processes which allow current to be transported through a metal semiconductor contact have been discussed in detail in reference 9. In this lecture attention will be directed to one of these processes, thermionic emission of electrons over the top of the barrier, since, this is by far the most important process. According to this theory the terminal current voltage relationship may be written as

$$J = A^*T^2 \exp\left[\frac{-\phi_b}{kT}\right]\left\{\exp\left[\frac{qV}{kT}\right] - 1\right\} \qquad (56)$$

where $A^* = \frac{4\pi m^* qk^2}{h^3}$ is the Richardson constant.

This is the rectifier equation which corresponds to equation (28) in the pn junction. Deviation from this relationship may occur for a number of reasons which include: (i) Tunnelling through the barrier (highly doped semiconductors), (ii) generation/recombination in the space charge region (as discussed for pn junctions in section 3.2) and (iii) hole injection. Here again we will cite reference 9 for a detailed discussion of these points.

As a refinement of the theory we can take into account the image force on the electrons. This leads to a dependence of the barrier height on the applied bias as given by

$$J_s = A^*T^2 \exp\left\{\frac{-\phi_b}{kT}\right\} \exp\left\{\frac{0.44\sqrt{F_o/\varepsilon}}{T}\right\} \qquad (57)$$

and resulting in a non saturating reverse characteristic.

Although we have considered a metal/n-type semiconductor system, the theory for a p-type semiconductor can be developed along similar lines; the only difference is that the current across the barrier is carried by holes in that case.

As was mentioned earlier, deviation from these equations occur as the semiconductor doping is increased. Increasing $N_d$ (or $N_a$) results in a reduction in the depletion width and a corresponding thinning of the barrier. Initially, electrons below the top of the barrier are able to tunnel through resulting in an increase in the value of $J_s$. In these circumstances the transport is a combination of Thermoinic and field emission. At very high doping densities electrons near to the Fermi level can participate in tunnelling and the transport is almost entirely field emission. In the limiting case rectification will disappear and the contact will be ohmic. Indeed, this is a practical way of producing ohmic contacts.

Diode Noise

The noise process in p-n junctions and Schottky barriers can be represented as an emission across barriers and, since the passage of carriers across barriers constitutes a series of independent random events, we can treat the process at full shot noise.

(That is, the statistical nature of the current pulses is caused by the discrete charge on the electron). Fourier analysis yields a mean squared noise current with a white spectrum given by

$$\overline{i_n^2} = 2qIB \qquad (58)$$

where I is the d.c. current and B the bandwidth.

In order to keep our discussion on a simple basis we will consider the specific case of a $p^+n$ diode at low and intermediate frequencies and where the d.c. characteristic is dominated by the diffusion component of current (Section 1.3.1). In these circumstances the bulk of the current is carried by the two components of holes: (i) majority carriers from the $p^+$ region injected into the n region and recombining, and (ii) holes generated in the n region and flowing down a potential gradient into the p region. Thus the full shot noise of the currents $(I + I_s)$ due to hole of group (i), and $-I_s$ due to holes of group (ii), is

$$\frac{\overline{i_n^2}}{B} = 2q\left[(I + I_s) + I_s\right] \qquad (59)$$

$$= 2q(I + 2I_s) \qquad (60)$$

The small-signal conductance given in Equation (31) is

$$g = \frac{q(I + I_s)}{kT}$$

and hence we may write

$$\overline{i_n^2} = 2kTg\left(\frac{I + 2I_s}{I + I_s}\right) \qquad (61)$$

corresponding to full thermal noise 4kTBg at zero current (I=0) and to half thermal noise at large currents ($I \gg I_s$). In addition to shot noise from the junction, the series resistance $R_s$ corresponding to the semiconductor between the depletion region and the contacts will exhibit full thermal noise given by

$$\overline{i_n^2} = \left(\frac{4kTB}{R_s}\right) \qquad (52)$$

The full noise current is then given by the sum of these two components.

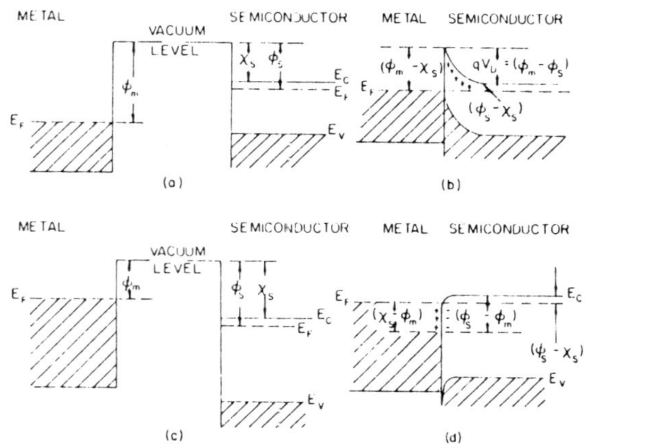


Figure 12 Energy band diagrams of metal contacts to an n-type semiconductor. (a) and (b) correspond to $\phi_m > \phi_s$ and (c) and (d) to $\phi_m \leq \phi_s$.

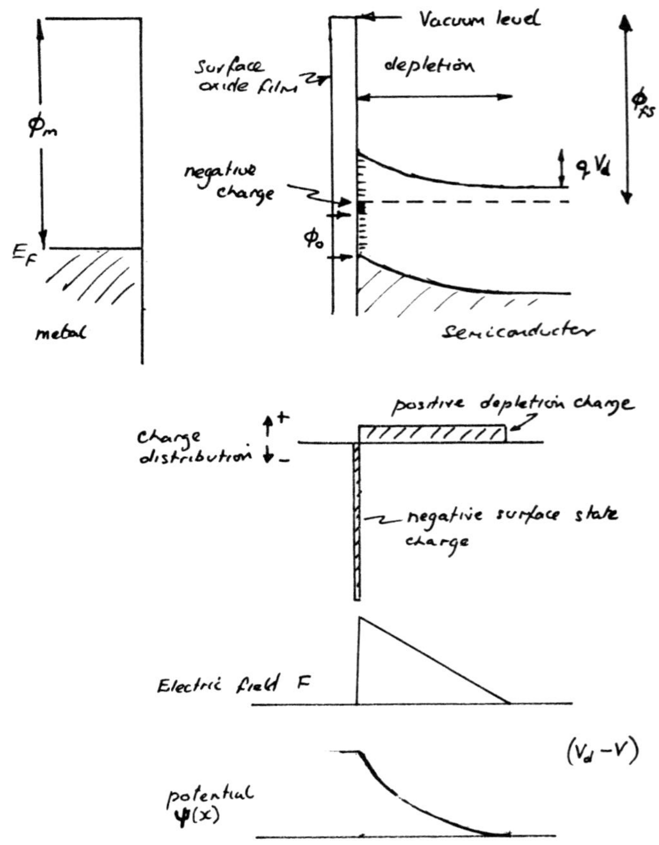


Figure 13 Free surface band bending for a semiconductor with a high surface state density.

Figure 14 Over page
A metal semiconductor contact taking into account surface states.

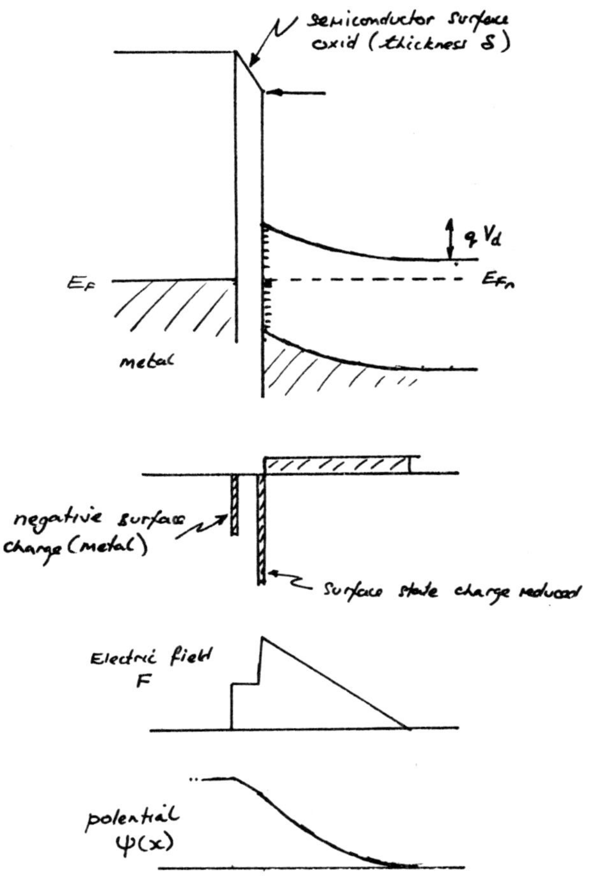


REFERENCES


1. Shockley, W., "The theory of p-n junctions in semiconductor and p-n junction transistors", Bell System Technical Journal, 28, 435-474 (1949).
2. Sze, S. M., "Physics of Semiconductor Devices", Wiley, New York, 1969.
3. Sah, C.T., Noyce, R.N., and Shockley, W., "Carrier generation and recombination in p-n junctions and p-n junction characteristics", Proc. IRE, 45, 1228 (1957).
4. Grove, A. S., "Physics and Technology of Semiconductor Devices", Wiley, London, 1967.
5. Ashburn, P., and Morgan, D.V., "The role of radiation damage on the current-voltage characteristics of p-n junctions", Solid State Electronics, 17, 689 (1974).
6. Moll, J.L., "Physics of semiconductors", McGraw-Hill, New York, 1964.
7. McKay, K.G., "Avalanche breakdown in silicon", Phys. Rev., 94, 4, 877 (1954).
8. Miller, S.L., "Avalanche breakdown in germanium", Phys. Rev., 99, 4. 1234 (1958).
9. Rhoderick, E.H., "Metal Semiconductor contacts", Clarendon Press, Oxford (1977).
10. Van der Ziel, A., "Noise sources, Characterisation and Measurements", Prentice Hall, Englewood Cliffs 1970.


Tutorial Question: Chapter 1

1. A small diode has a capacitance of 50pF at a reverse bias of 0.1V, 44.7pF at 0.5V and 41.1pF at 0.9V. Show that these results are consistent with a linearly graded impurity profile. Is it possible to tell from the measurements whether the diode is made of Si, Ge or GaAs?

2. A germanium pn junction has a bulk resistivity of $42 \times 10^{-4}$ Ωm and $2.08 \times 10^{-2}$ Ωm for the p and n type material respectively. If $\mu_n = 3 \times 10^{-1}$, $\mu_p = 1.5 \times 10^{-1} m^2 V$ for this material, and $n_i = 2.5 \times 10^{19} m^{-3}$ at room temperature, calculate the height of the potential barrier at the junction.

3. In problem 2, the lifetimes of the p and n-type material are 75 and 150μs respectively, and the sample has a cross-section of $10^{-6} m^2$. Assuming that there is a large thickness of semiconductor on each side of the junction, calculate the saturation current $I_s$. What proportion of the current is carried by holes?

   Note the following:

   $$D_n = \frac{kT}{q} \times \mu_n \qquad D_p = \frac{kT}{q} \times \mu_p$$

   at 300°K $\quad \frac{kT}{q} = 0.025V$

4. A p-n diode has the same impurity concentration in both regions, each of which is narrow compared with the diffusion lengths and the same width. Show that the ratio of the electron and hole current in the forward direction is

   $$(I_p/I_n) = (D_p/D_n)$$

   Show also that if the two regions are very wide this becomes

   $$(I_p/I_n) = (D_n\tau_p/D_p\tau_n)^{\frac{1}{2}}$$

   (Hint. Start by assuming values for the excess injected minority carrier density given by equations 25a) and b).

5. Tables 1 and 2 give the experimental results of the current-voltage and capacitance-voltage characteristics of a certain semi-conductor device. Discuss in detail why the current-voltage characteristic deviates from the theoretical relationship

   $$I = I_s \left[ \exp\left(\frac{qV}{kT}\right) - 1 \right]$$

   What information can you determine on the following points: (i) the impurity profile at the junction, (ii) the contact potential $V_o$, (iii) the semi-conductor material? Discuss any approximations used.

Table 1

| Forward bias | | Reverse bias | |
|---|---|---|---|
| V | I | V | I(pA) |
| | | 0.5 | 16 |
| 0.01 | $3 \times 10^{-12}$ | 1.0 | 20 |
| 0.05 | $3 \times 10^{-11}$ | 2.0 | 26 |
| 0.1 | $6 \times 10^{-10}$ | 4.0 | 40 |
| 0.3 | $1 \times 10^{-7}$ | 6.0 | 53 |
| 0.4 | $1 \times 10^{-5}$ | 8.0 | 66 |
| 0.5 | $8 \times 10^{-5}$ | 10.0 | 80 |
| 0.6 | $3 \times 10^{-4}$ | 10.2 | 100 |
| 0.7 | $6 \times 10^{-4}$ | 10.4 | 200 |

Table 2

| C(pF) | V |
|---|---|
| 50 | 0.1 |
| 44.7 | 0.5 |
| 41.1 | 0.9 |

6. A Schottky diode of area $10^{-8}m^2$ is operated at reverse bias of 10 volts. The measured capacitance at that operating point is 1.5pF. Assuming that $\varepsilon = 9$ $V_d = 1V$, calculate (a) the depletion layer thickness, (b) the (uniform) doping density $N_d$.

7. Claculate the avalanche breakdown voltage of a germanium abrupt pn junction in which $N_a = N_d = 10^{21}/m^3$. You may take $\varepsilon = 16$ and the critical field strength to be $2 \times 10^7 Vm^{-1}$. Repeat the calculation for an assymetric $n^+p$ diode, where $N_d >> N_a = 10^{21}M^{-3}$.

# CHAPTER 2

# Active microwave devices—TEDS

J. Pattison

# CHAPTER 2 - ACTIVE MICROWAVE DEVICES - TEDs

J. Pattison

Royal Signals and Radar Establishment, Malvern, Worcs

## INTRODUCTION

Transferred electron devices (TED) are a class of active semiconductor devices used mainly for generating power in the microwave and, more recently, in the millimetre wave part of the electromagnetic spectrum. They are used in a wide range of military and commercial systems for example in intruder alarms, local oscillators and low power radar transmitters.

The possibility of obtaining a negative resistance in a semiconductor by electron transfer was first proposed by two independent groups working in the UK in the early 1960's. Ridley and Watkins and Hilsum suggested that in certain semiconductors with an appropriate conduction band structure as shown in Fig 1 it might be possible to transfer electrons from the lowest energy band (i) to the higher energy band (ii) by the application of a high electric field. If band (ii) had a lower mobility then the electrons would slow down. That is increasing the electric field (α voltage) would cause a reduction in electron velocity (α current) and a negative resistance would result. GaAs was suggested as a suitable material for this effect by Hilsum and experimental support for these theories was given by the work of Gunn in his investigation of the high electric field properties of GaAs. In this study samples between 20 and 200 μm in thickness with ohmic contacts at either end were subjected to short (~ 10 ns) voltage pulses. Microwave oscillations were observed in all samples in the frequency range 0.47 to 6.5 GHz. Other major conclusions from this work were that the frequency of operation was roughly proportional to the inverse sample length and there was a well defined threshold field below which no oscillations occurred.

## DEVICE PHYSICS

### The Transferred Electron Effect in GaAs

The band structure of GaAs is shown in some detail in Fig 2. It is a direct gap material with an energy gap of 1.4 eV. The lowest energy conduction band minima is at the zone centre [Γ, <000>] and has a high electron mobility (~ 8000 $cm^2v^{-1}s^{-1}$), a low effective mass (~ 0.07 $m_o$) and a low density of states. About 0.36 eV higher in energy are the next highest conduction band minimas, six of them, in the <111> crystal direction. Here the mobility is lower (~ 150 $cm^2v^{-1}s^{-1}$), the effective mass higher (~ 0.4 $m_o$) and the density of states is higher than in the central valley, At room temperature all the electrons are in the central lowest energy valley. If an electric field is applied to a sample of GaAs, initially the electrons are able to lose almost all the energy gained by the electric field to the lattice by polar optic phonon scattering. The resulting electron velocity, electric field characteristics is linear with a slope given by the mobility. As the electric field is increased the electrons can no longer transfer their energy to the lattice and the energy increases rapidly with field as shown in Fig 3. Two possibilities exist. The material will either suffer dielectric breakdown and be destroyed or, as in the case of GaAs and a few other materials, electrons which have gained sufficient energy will transfer into the higher set of conduction band minima. In the case of GaAs the gain in energy is 0.36 eV, and as the density of states in the higher valleys is larger, a substantial fraction of the electrons will transfer. However, the mobility is low and effective mass is high in these valleys and as a result the mean velocity of electrons will fall. This results in the well known velocity field curve for GaAs as shown in Fig 4. There is a peak in the drift velocity at an electric field of 3500 $Vcm^{-1}$, beyond that the velocity falls to the valley velocity and there exists a region of negative differential mobility or expressed more simply a negative resistance. Any negative resistance is inherently unstable and it would not be surprising if a sample of GaAs biassed into this regime produced oscillations. In detail the onset of electron transfer causes a redistribution of the electron population in the material which is dependent upon the doping of the material, the r.f. circuit the sample is in and the type of electrical contact which is made to the material.

### Modes of Operation in Transferred Electron Devices

Domain Mode. In a long sample (~ 50 μm) of GaAs the ultimate form of the instability resulting from the negative resistance of the material is shown in Fig 5(a). The result is a dipole domain consisting of an accumulation of charge close to the cathode with a depletion of charge to the positive side. This instability will drift towards the anode at a velocity close to the valley velocity of the material. When the instability reaches the anode a pulse of current will be fed to the external circuit. This type of operation explains the observations made by Gunn in his initial experiments but is not observed in most transferred electron devices (TED's) which are used commercially.

Accumulation Mode. Most TED's devices have active lengths of 10 μ or less and have ohmic anode and cathode contacts. The mode which is appropriate in this device is an accumulation layer. It should be noted that the anode field in this device is high and this effects the maximum voltage which can be applied to the device and therefore limits the output power. Additionally, and of more significance, electrons entering the device at the cathode have no excess kinetic energy and therefore need to gain energy from the applied electric field before they can transfer. This results in the portion of the field profile in Fig 5(b), shown shaded, where the field close to the cathode is low. This dead space prevents the device from switching fully into a high voltage state and so reduces efficiency. This point will be referred to again in the next section on InP TED's.

LSA (Limited Space Charge Accumulation) Mode. This mode exploits the bulk negative resistance of the material in a way which is not transit time limited. The sample thickness must be significantly longer than the transit time thickness at the desired frequency of operation. On the application of bias an instability will initially form and the voltage will rise above threshold. If there is an appropriate resonant circuit around the device then the voltage will be forced down below threshold and the instability will be quenched. This is illustrated in Fig 6. Thus the frequency is determined totally by the circuit rather than the sample thickness. Very high peak powers have been achieved by this technique with high efficiencies. However because the sample is thick the thermal characteristics are poor and high mean power or continuous operation is very difficult to achieve.

## The Transferred Electron Effect in Indium Phosphide

There exists several materials other than GaAs which exhibit the transferred electron effect, for example CdTe, ZnSe and InP. Of these alternative materials InP has certain advantages which merit consideration. The principal advantage is that the peak to valley velocity ratio for this material is higher than for GaAs (3.5 compared with 2.2). For the simplest device which switches from a high voltage, low (valley) current state to a low voltage, high (peak) current state the ratio of peak to valley velocity will have a significant effect on efficiency. The larger this ratio the higher the efficiency. It is therefore expected that the limiting efficiency of an InP device will be in the region of 40% compared with about 25% for GaAs. It should be noted that these are theoretical predictions and do not take into account circuits and other limitations.

The technology of InP lags behind that of GaAs but recent work has shown that the predicted advantages of the material can be realised. However the type of device which fully exploits the potential of InP differs in principal for any device made in GaAs in that serious attention has been paid to the contacts.

As the intervalley separation, $\Delta E$ in Fig 2 is larger for InP (0.55 eV) than for GaAs (0.36 eV) the dead space shown in Fig 5(b) will be a larger fraction of the device length and the efficiency will be reduced. This device has ohmic contacts and so injected electrons have no excess kinetic energy. Cathode contacts have been developed for InP devices with non ohmic characteristics. In this case the injected electrons enter the device with some excess kinetic energy and there is no dead space as shown in Fig 7. In addition the anode field is reduced compared with ohmic contacts and this allows a greater voltage to be applied to the device and therefore more power. In practice device efficiencies up to 24% and peak powers up to 37 Watts have been achieved in the frequency band 12-18 GHz.

## DEVICE TECHNOLOGY

To exploit the mechanisms in previous sections requires considerable ingenuity in a number of areas of materials and device technology.

Materials of very high purity are necessary, complicated fabrication techniques to optimise the performance of the material are required and complex circuits need to be engineered.

### Materials

To achieve electron transfer in either GaAs or InP the principal low field scattering mechanism must be optic phonon scattering rather than impurity scattering, this puts an upper limit on the doping of the material which may be used. The lower limit is set by the type of mode which exists in the device. Typical carrier densities are in the range $n_D - n_A = 1 - 5 \times 10^{15}\ cm^{-3}$. This is a purity of 1 part in $10^8$. The frequency is broadly set by the thickness being in the range 2-15 μm for most devices. The technique usually employed is to grow the active device material epitaxially onto a conducting substrate of the same material. Many epitaxial techniques can be used but Vapour Phase Epitaxy and Liquid Phase Epitaxy are the more usual. Of these vapour phase epitaxy has the advantage that doping can be changed very rapidly by adding suitable impurities during growth.

### Device Fabrication

This is a multifarious subject and I shall discuss only two areas; namely contacts and thermal engineering.

A cross section of a slice of material suitable for devices is shown in Fig 8. The simplest device to fabricate is one in which the material is metallised on both sides and then cut into small square chips for mounting into an appropriate package. Most contacts to GaAs and some to InP are required to be ohmic; my definition of ohmic is that the current is controlled by the bulk properties of the semiconductor rather than the contact. The usual way is to introduce a dopant at very high doping density close to the surface of the material. If this is achieved the current from the metal into the semiconductor consists of tunnelling electrons. The most elegant way of achieving this high surface doping is Ion Implantation in which a beam of high energy ions is directed at the surface. This technique has given surface concentrations as high as $5 \times 10^{19}\ cm^{-3}$ which is adequate for good ohmic contacts. The more usual way to achieve an ohmic contact is to alloy a combination of metals usually silver/tin or nickle/germanium/gold at temperatures of a few hundred $^oC$. The exact proportions of metals and alloying temperatures vary from laboratory to laboratory but the result is generally identical. The detailed metallurgy is complex and has been the subject of much interesting research in recent years. Current limiting contacts, used extensively for InP, can be considered as modified forms of Schottky Barrier contacts. In this case the current is controlled by the height of the potential barrier between the metal and the semiconductor. Silver is the metal used for Indium Phosphide current limiting contacts and in this case the height of the barrier can be reduced by heat treatment to attain the appropriate degree of current limiting. A most important consideration is how the metal contact will influence the properties of the semiconductor, particularly if the device is required to operate at high temperatures. This consideration often means that anti-diffusion barriers are introduced between the contact metallisation and the metal films required for bonding or heat sinking purposes.

Transferred electron devices operate at power densities which can be as high as 200 kW $cm^{-3}$, and although the area is small, in the range $5 \times 10^{-5}$ to $1.5 \times 10^{-3}\ cm^{-2}$, the power produced as heat is considerable, even if the efficiency is 20%. To allow devices to operate continuously or at a high duty cycle when pulsed, it is necessary to remove the heat from the device. This technique usually employed is to place the device in intimate contact with a good thermal conductor, for example, Gold, Silver or Diamond. If the heat sink can be assumed semi infinite then

$$T_{max} = \left[ T_o + \frac{Q\pi d}{8K_s} \right] \exp \left\{ \frac{Q\ell}{2K_{sc}\ T_{max}} \right\}$$

where $T_{max}$ = Maximum device Temperature in $^oK$
$T_o$ = Ambient Temperature in $^oK$
$Q$ = Power Density in $Wcm^{-2}$
$d$ = Device diameter
$K_{HS}$ = Thermal conductivity for heat sink
$K_{SC}$ = Thermal conductivity of the semiconductor and is $\frac{150}{T_{max}}$ for GaAs

This calculation assumes that the heat is generated uniformly in the active low doped semiconductor. For most practical devices the maximum temperature is dominated by the spreading resistance term and clearly the larger the thermal conductivity of heat sink the lower the maximum temperature.

The most universal fabrication technique for heat sinking is the Integral Heat Sink (IHS) approach in which the epitaxial layer is electroplated, after contact metallisation, with a good thermal conductor to a thickness of about 70 µm. The semiconductor is then thinned from the substrate side and the device defined photolithographically. The final process is to mesa etch the semiconductor. A schematic of the final device is shown in Fig 9. It is useful to have a measure of the heatsinking achieved in a final device. The parameter which is usually used is the Thermal Impedance which is defined as the rise in temperature per unit dc input power ie

$$R_{th} = \frac{\Delta T}{P_{DC}} \quad \text{in units of } {}^oC/W$$

Measured values of thermal impedance for the device shown in Fig 9 are in the range 10 - 15$^oC/W$ for a device of area 1 x $10^{-3}cm^2$.

DEVICE PERFORMANCE

To obtain power from a transferred electron device it is necessary to place the device in a resonant circuit. Basically the device can be considered as a negative resistance in parallel with a capacitor. The circuit therefore must be a resistor in parallel with an inductance. In practice the situation is more complex and will be dealt with elsewhere.

Power can be obtained from transferred electron devices in the frequency range 1 to 110 GHz from either GaAs or InP, with InP preferred as the material for higher frequencies. Most attention has been given to continuously working (cw) GaAs devices with power levels of a few 10's of milliwatts. Devices of this type are used in applications such as transmitters for intruder alarms and local oscillators. The major attraction for these types of system is the low noise characteristics of transferred electron devices compared with other sources. Devices with high powers have also been developed with a higher cw power of 2.2 W at 8 GHz for a single 'chip' mounted on a diamond heat sink. The efficiency of these devices is in the region of 12 to 2% for the frequency range 10 to 75 GHz.

Pulsed devices are also required, for use in such systems as pulsed Doppler radars. GaAs devices give up to about 30 Watts with 10% efficiency in X band but at low duty cycle, usually about 0.1%. InP devices have given efficiencies up to 24% in J band with peak powers up to 37 Watts from a single chip. High mean powers are also possible with duty cycles up to 10%.

Transferred electron devices can also be used as reflection amplifiers. It is usual to use InP for these devices because of the lower noise figures which have been achieved, about 8 dB in X band. However with the success of GaAs FET's as low noise amplifiers the future of transferred electron amplifiers is not clear.

REFERENCES

1. Hobson, G. S., 1974, "The Gunn Effect", Monographs in Electrical and Electronic Engineering, Clarendon Press, Oxford.

2. Shaw, M. P., Grubin, H. L., Solomon, P. R., 1979, "The Gunn-Hilsum Effect", Academic Press.

3. Rees, H. D., Gray, K. W., September 1976, "Indium Phosphide: a Semiconductor for Microwave Devices", Solid State and Electron Devices, Vol 1, No 1, pp 1-8.

4. Eastman, L. F., 1972, "Gallium Arsenide Microwave Bulk and Transit-Time Devices", Artech House.

QUESTIONS

1. A sample of Indium Phosphide has a low field mobility of 5000 $cm^2V^{-1}S^{-1}$ and is 100 µm long with an area of $10^{-4}$ $cm^2$. If the low field resistance is 125 Ω and the peak current is 0.4 A calculate the peak velocity. If this sample was operated as a simple transit time oscillator at what frequency would it operate? (The peak to valley ratio for InP is 3.5:1 approximately.)

2. Calculate, for GaAs the central valley lattice temperature for a 1% electron occupation of the satellite valleys. (The density of states is proportional to $(M^*)^{3/2}$.)

3. A GaAs transferred electron device of length 6 µm and of diameter 50 µm is bonded onto a semi infinite silver heat sink. Given that the continuous output power is 100 mW calculate the maximum device temperature assuming an efficiency of 5%. Also calculated the operating current and voltage if the device is biassed at twice the threshold voltage. (The thermal conductivity of silver is 4.2 $W^oC^{-1}cm^{-1}$.

4. Comment on the upper frequency limit for TEDs.



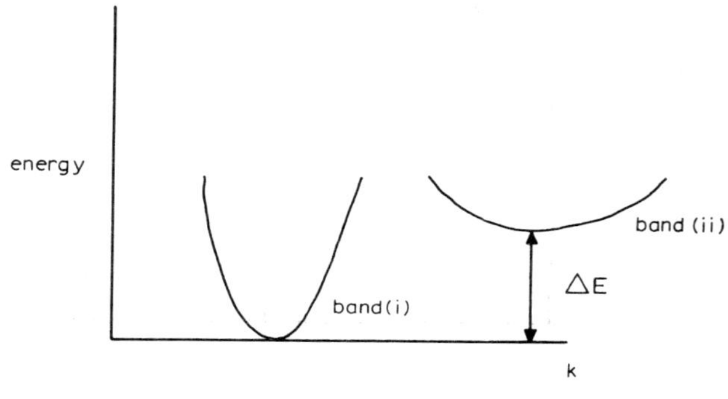


Figure 1 Schematic band structure necessary for electron transfer

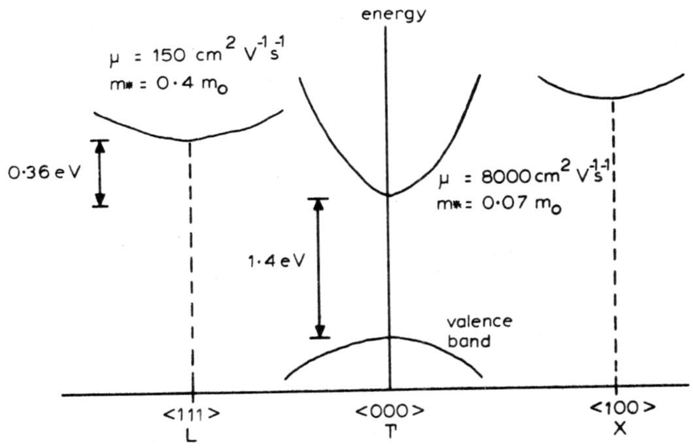


Figure 2 Band structure for GaAs

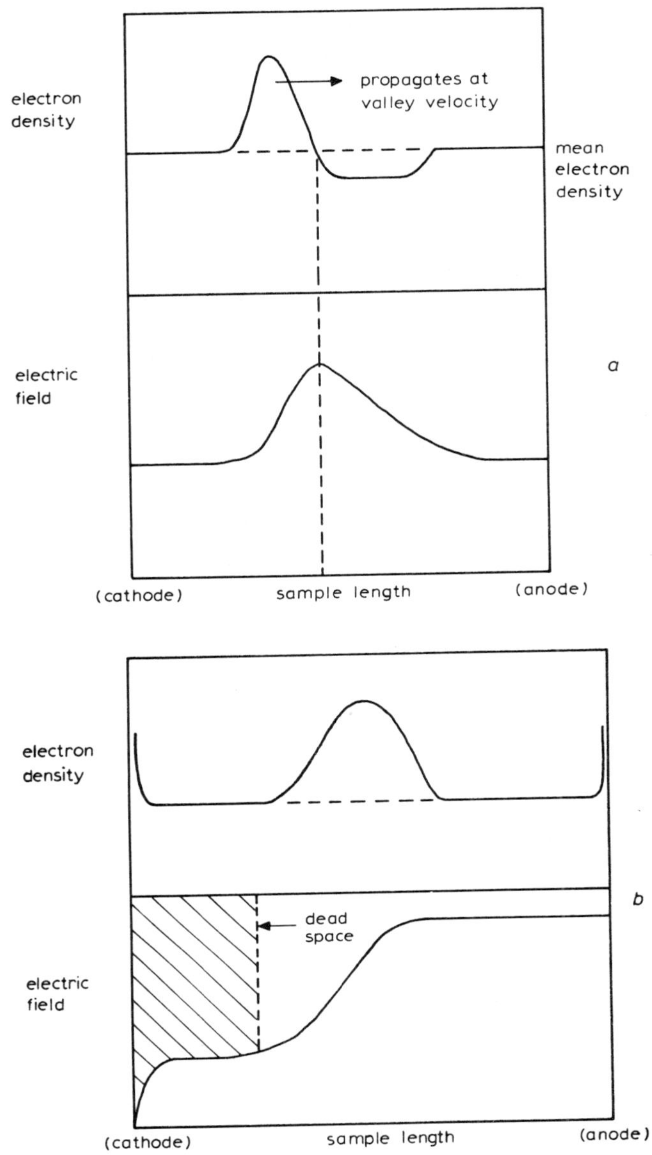


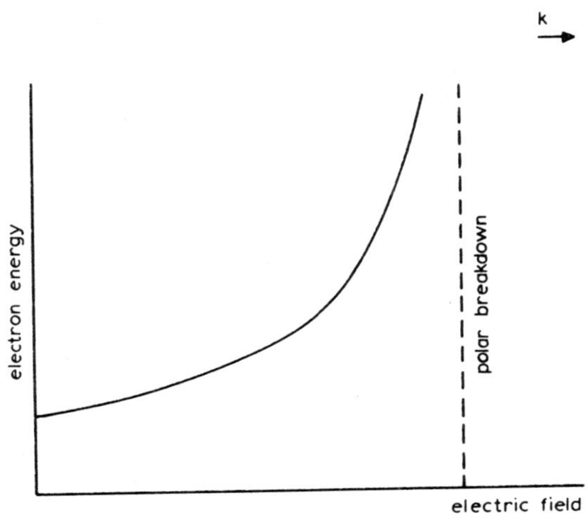


Figure 3 the relationship between mean electron energy and electric field

Figure 5 (a) Electron density and field profile for stable dipole domain

(b) Electron density and field profile for an accumulation layer.

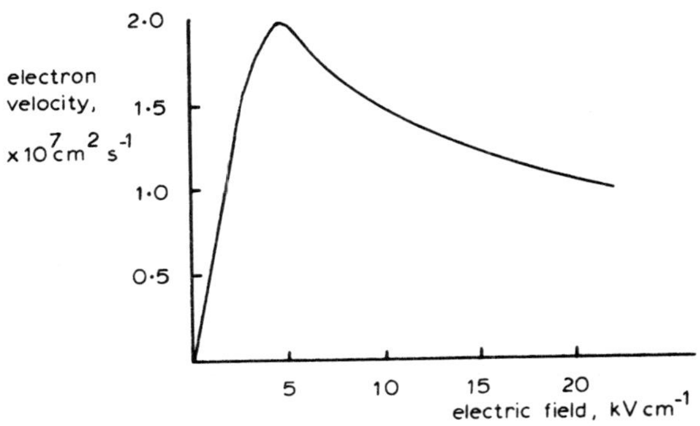


Figure 4 Velocity/field curve for GaAs

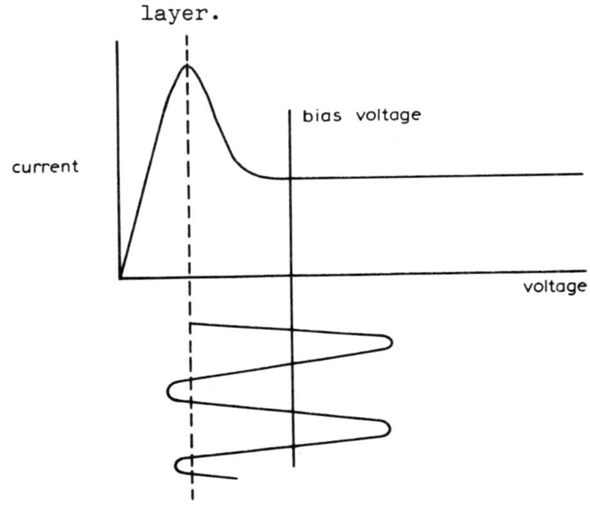


Figure 6 Schematic of LSA mode

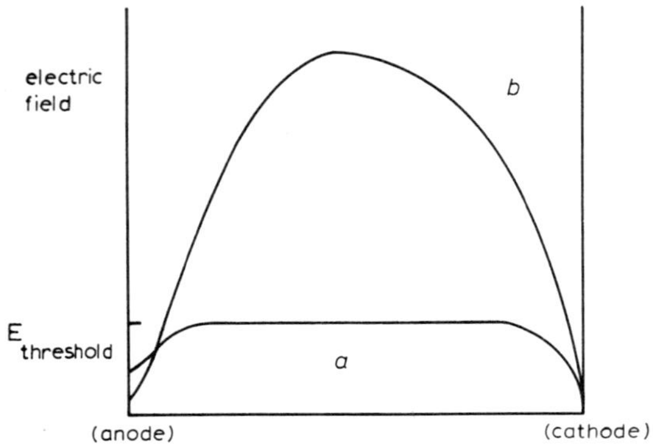


Figure 7 Electric field profiles for 'hot' electron injection cathodes shown (a) for the high current and (b) for the high voltage part of the r.f. cycle (note that the cathode field is low for the high voltage part of the r.f. cycle)

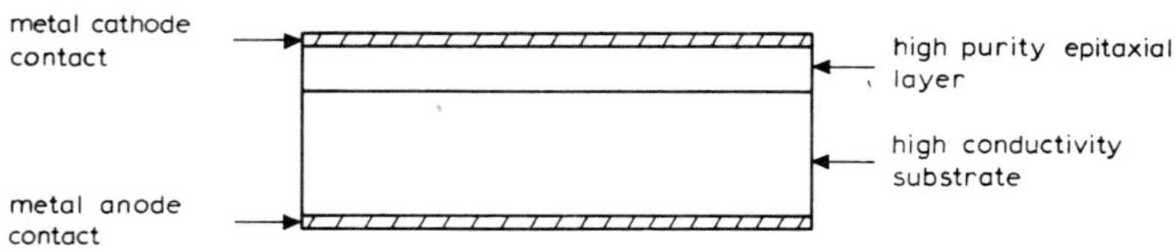


Figure 8 Schematic cross section of device

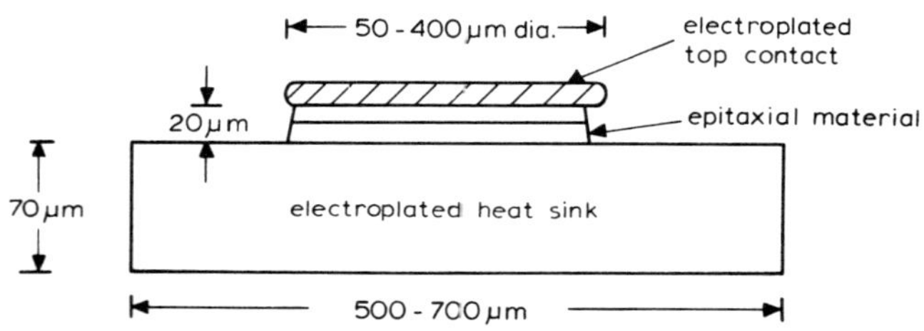


Figure 9 Schematic of integral heat sink device

# CHAPTER 3

## Active microwave devices-IMPATTS

Dr. P. Brook

# CHAPTER 3 - IMPATT DIODES

P Brook

Royal Signals and Radar Establishment, Malvern, Worcs

## INTRODUCTION

The term IMPATT diode (IMPact Avalanche and Transit Time) has come to describe a class of devices with a wide number of variants: single and double drift devices, high efficiency structures and devices in gallium arsenide and silicon. These devices are in widespread use in applications which call for solid state amplifiers and oscillators with high CW and mean power (around 1 to 10 watts at X Band), moderate efficiency (5 to 25% depending on structure) and, increasingly, for frequencies in the high microwave and millimetre wave region, where they maintain a useful performance. The detailed physics of IMPATT diodes differs significantly from one structure to another, but the essential features are common. This paper describes the theory in some detail for the simplest such device, the abrupt p-n junction single-drift diode, and then discusses the way in which performance is enhanced in the more complex structures.

## BASIC PRINCIPLES

All IMPATT devices are reverse-biased diodes, either Schottky or p-n junctions, operating in the voltage region beyond avalanche breakdown. Under such conditions, several aspects of hot-electron theory apply which are fundamental to operation.

### Impact Ionisation

When electrons and holes are accelerated to high velocities in an electric field they may gain sufficient energy to ionise the lattice atoms on collision, that is create new electron-hole pairs. Such a process can clearly lead to a chain reaction, since electrons and holes so created may themselves gain the necessary energy, around 1.5 times the semiconductor band-gap, to generate new pairs, thus leading to a large increase in device current and the familiar phenomenon of a breakdown voltage. The process can be described on the microscopic level by assigning to both electrons and holes a parameter, $\alpha$, the ionisation coefficient, defined as the probability per unit length that a carrier will undergo an ionising collision. The parameter can be considered a function only of local electric field and is in general different for electrons and holes, though for simplicity we will assume equality here. For a simple abrupt p-n junction, as shown in Figure 1, the electric field in the depletion region is non-uniform, having a peak value Ep at the junction and a slope given by solution of Poisson's equation as:

$$\frac{dE}{dx} = \frac{q}{\varepsilon} Nx, \quad \text{so} \quad Ep = \frac{q}{\varepsilon}NW \qquad (1)$$

where q is the electronic charge, $\varepsilon$ the dielectric constant and N the n-type doping concentration and W the depletion width. The condition for avalanche breakdown may be written:

$$\int_o^W \alpha(E)\, dx = \int_o^W \alpha(Ep - \frac{q}{\varepsilon} Nx)dx = 1 \qquad (2)$$

at which point carriers generated thermally at the depletion edges have unit probability of ionisation. For a given dependence of $\alpha$ on E, equation 2 may be solved to give Ep and hence the breakdown voltage $V_B$, as the area under the electric field curve in Fig 1. Graphs of $V_B$ and Ep as functions of N are given in (4) for the main semiconductor materials. It has been found experimentally that $\alpha$ is a very strong function of electric field, which has two consequences. Firstly the peak field at breakdown is relatively insensitive to doping concentration, being in the range $3\text{-}6\text{x}10^5$ volts/cm in silicon over practical ranges, so that breakdown voltage and depletion width at breakdown are approximately inversely proportional to doping. Secondly, the current multiplication is localised to a region of width $W_A$ close to Ep, as shown in Fig 1, which we may call the avalanche region. The rest of the depletion region, of width $W_D$, is called the drift region for reasons apparent below.

### Velocity Saturation

At low electric fields, electron and hole velocities are proportional to applied electric field, the proportionality constant being called the mobility. As the field increases the carriers interact more strongly with the lattice, especially in emitting optical phonons, so that the average velocity in the direction of the field falls much below the projection from low-field values. Eventually, for fields above about $10^4$ volts/cm, the drift velocity approaches a limiting value independent of field, called the scatter-limited or saturated velocity vs. Because of the electric field values encountered in avalanching junctions, we may consider the drift velocity essentially constant throughout the depletion region. Typical values of vs at device operating temperatures of 200°C are around 8 x $10^6$ cm/sec in silicon and 6 x $10^6$ cm/sec in gallium arsenide.

### Induced Current from a Moving Charge

It can be shown from simple electrostatics that if a carrier with charge q travels at a velocity v between the plates of a capacitor spaced at a distance W, a current I is induced in the external circuit which is given by:

$$I = qv/W \qquad (3)$$

This is the Ramo-Shockley theorem and may be applied to the case of carriers drifting in the depletion region of a semiconductor junction.

## THE IMPATT MECHANISM

The above principles are sufficient to describe the origin of microwave negative resistance in an avalanching p-n junction, using the qualitative treatment given by Read (2) in his original proposal for such devices. Consider a junction biased to the breakdown voltage, with a superimposed high frequency sinusoidal voltage Va. This is for the time being a theoretical convenience, but might arise in practice by embedding a device in a single-tuned circuit. The resulting conduction current density Ja flowing in the avalanche region is as shown in Fig 2(b) and (c). The current increases sharply while the voltage is above $V_B$ and starts to fall as the voltage passes through $V_B$ after one half-cycle; plotted linearly the avalanche current approximates to a series of sharp pulses.

At this point it is reasonable to question how such a process can be sustained, since it might be argued that once a device is biased above breakdown the current will multiply to infinite values leading to catastrophe. The answer lies in two areas. Firstly we are considering modulating the device voltage at microwave frequencies, at which the time-constant of the avalanche build-up, which is finite, is comparable with the period of the imposed waveform, so that the current growth is checked before runaway occurs. Secondly, the device is usually operated from a fast acting constant current supply which limits the average current drawn and hence damps the tendency for instability. (The consequences of using poorly stabilised bias supplies will be familiar to those who have had the misfortune to reduce an IMPATT diode to a mixture of melted semiconductor and fused bond-wire).

The current pulses produced in the avalanche region provide bunches of electrons which are injected into the drift region and holes which travel back to the junction. While the electrons traverse across the drift region at saturated velocity a conduction current is induced in the external circuit, as shown in Fig 2(d), which persists for the time it takes the electrons to reach the low-field end of the depletion region. By suitably designing the width of the structure it can be arranged that this transit time takes approximately half of the RF period. The fundamental component of the conduction current is then in antiphase with the impressed voltage, which implies an RF negative resistance. If we define a drift-region transit angle, θ, such that

$$\theta = \omega W_D/v_s \tag{4}$$

then the condition for complete antiphase occurs at a frequency fr, the Read frequency, given by:

$$fr = v_s/2W_D \tag{5}$$

The negative resistance so produced may be used as the basis either for a reflection amplifier or an oscillator, in which case the device is embedded in a circuit which presents a conjugate match to the diode impedance and power is delivered to a load.

To a first approximation, device performance peaks at the Read frequency (a more accurate relationship will be given below), so that, by way of example, we may see that for operation in silicon at 10 GHz, where $v_s = 8 \times 10^6$ cm/sec, we require a drift region width of around 4 micrometres.

It is important to note that the ideal phase depends on the initial $90^o$ delay between voltage and avalanche current, which may be viewed as an avalanche inductance. Other injection processes, such as tunnelling, which produce currents in phase with the applied voltage would lead to a positive terminal current during the time that RF voltage is positive and would thus produce less net negative resistance.

## READ'S EQUATION

A more quantitative treatment of the IMPATT diode, which leads to simple expressions for small and large signal device parameters, can be developed from Read's Equation, which describes the time-dependence of the avalanche region current density, Ja:

$$\frac{dJa(t)}{dt} = \frac{3Ja(t)}{\tau_A}\left(\int_o^W \alpha(E)dx - 1\right) \tag{6}$$

where $\tau_A$ is the electron transit-time through the avalanche region, ie $\tau_A = W_A/v_s$. This form of Read's Equation assumes negligible junction saturation current and includes the factor 3 instead of Read's value of 2 to include subsequent refinements to account for the effect of space-charge in the avalanche region.

## Small Signal Solution

For small sinusoidal perturbations of the peak electric field Ep, solution of equation 6 gives an expression for the effective inductance of the avalanche process:

$$L_A = W_A/A\, J_{DC}\, k \tag{7}$$

where A is the device area, $J_{DC}$ the mean current density and k is given by:

$$k = \frac{\partial}{\partial Ep}\left[\frac{3}{\tau_A}\left(\int_o^W \alpha(E)dx - 1\right)\right] \tag{8}$$

For a simple p-n junction:

$$k = \frac{3}{\tau_A}\frac{\varepsilon}{qN}\alpha(Ep) \tag{9}$$

The avalanche region also sustains a displacement current due to its capacitance $C_A$, where

$$C_A = \varepsilon A/W_A \tag{10}$$

These components will be shunt-resonant at a frequency fa, known as the avalanche resonance frequency, and given by:

$$(2\pi fa)^2 = \omega_a^2 = k\, J_{DC}/\varepsilon \tag{11}$$

A complete solution for conduction and displacement currents throughout the whole depletion region gives an expression for the device small signal impedance $Z_D$:

$$Z_D = \frac{W_D^2}{2v_s\varepsilon A}\,\frac{1}{\left(1-\frac{\omega^2}{\omega_a^2}\right)}\,\frac{1-\cos\theta}{\theta^2/2} + \frac{1}{j\omega C_D}\left\{1-\frac{\sin\theta}{\theta}+\left(\frac{\sin\theta}{\theta}+\frac{W_A}{W_D}\right)/\left(1-\frac{\omega_a^2}{\omega^2}\right)\right\} \tag{12}$$

This expression, originally derived by Gilden and Hines (3), show several points:

The real part of device impedance changes from positive to negative values as the frequency exceeds the avalanche resonance frequency.

The imaginary part of device impedance changes from inductive to capacitive at resonance and at high frequencies ($\omega^2 >> \omega_a^2$) is simply the capacitive resistance of the total depletion capacitance $C_T$ ($= \varepsilon A/W$) in series with the negative resistance.

At low frequencies the device resistance is real and equal to the space-charge resistance

$$R_{SC} = W_D^2/2v_s\,\varepsilon A \tag{13}$$

which is the DC resistance exhibited by a p-n junction above avalanche breakdown.

The full expression has been plotted schematically in admittance co-ordinates as a function of frequency for a typical X-band device as the curve marked "small signal" in Fig 3.

## Large Signal Solution

An approximate solution for large signals due to Tager (4) may be obtained by considering a finite sinusoidal variation in Ep, of amplitude Ea, so that equation 6 becomes:

$$\frac{dJa(t)}{dt} = Ja(t)\, k\, Ea \sin\omega t \tag{14}$$

This has a solution in modified Bessel functions, $I_o(x)$ and $I_1(x)$, where $x = kEa/\omega$. The result is an expression identical to equation 12, but with $\omega_a$ replaced by an effective value $\omega_a^1$, where $\omega_a^1 = F\,\omega_a$ and

$$F = \frac{2}{x}\,\frac{I_1(x)}{I_o(x)} \qquad (15)$$

the behaviour of the factor F is plotted in Fig 4 as a function of x, which can be considered as a normalised signal level. We see that F tends to unity at small signals and falls as signal level is increased, thus lowering the effective avalanche resonance frequency and leading to the family of large-signal admittance curves in Fig 3. The effect of large signals on admittance is solely through $\omega_a$ and is a contrary direction to the effect of current density. Thus the family of curves shown to represent increasing signal level could equally well represent decreasing current densities. The curvature of the constant-frequency contours, especially close to avalanche resonance, is a consequence of the electronic admittance of the avalanche region and accounts for many of the instabilities and non-linearities of IMPATT devices. In general, operation in this region is avoided.

It is worth noting that the ratio $I_1(x)/I_o(x)$ also describes the extent to which the avalanche current in a real device approaches the ideally sharp pulse described in Figure 2. The value 1 represents a "delta function" pulse and as can be seen from Fig 4 is approached for values of normalised signal level x of greater than about 6. For lower values of x, as will be seen below, the broadening of the avalanche current pulse leads to a reduction in impedance power and efficiency proportional to $I_1(x)/I_o(x)$. Apart from large signal levels, factors leading to a large value of x are low frequency and a narrow avalanche region. The latter factor constantly emerges as a desirable factor in IMPATT devices and provides much of the motivation for fabricating the more complex structures.

## Large Signal Impedance Power And Efficiency

Under conditions where $\omega^2 >> \omega_a^2$, equation 12, with large signal correction, may be simplified to:

$$Z_D = R_D + jX_D = -\frac{W_D^2}{2v_s\varepsilon A}\left(\frac{\omega_a^{1\,2}}{\omega^2}\right)\left(\frac{1-\cos\theta}{\theta^2/2}\right) + \frac{1}{j\omega C_T} \qquad (16)$$

Substituting for $\omega_a^1$ and making the assumption that the RF voltage, Va, is simply Ea x W, which is reasonable for frequencies well above avalanche resonance when the electronic admittance is small compared with the capacitive susceptance, we have for $R_D$:

$$R_D = -\frac{2}{\omega^2 C_D^2}\cdot\frac{W}{W_D}\cdot\frac{I_{DC}}{V\varepsilon}\cdot\frac{I_1(x)}{I_o(x)}\cdot\left(\frac{1-\cos\theta}{\theta}\right) \qquad (17)$$

where $I_{DC}$ is the DC current density through the device, and $C_D$ the capacitance of the drift region.

For practical devices $1/\omega^2 C_T^2 >> |R_D|^2$, so we may write for the device conductance, $G_D$:

$$G_D \approx R_D\omega^2 C_T^2 = -2\left(1 - \frac{W_A}{W}\right)\frac{I_{DC}}{Va}\cdot\frac{I_1(x)}{I_o(x)}\cdot\left(\frac{1-\cos\theta}{\theta}\right) \qquad (18)$$

Since by definition $-G_D = I_{ac}/Va$, we may introduce the current modulation factor M which is given by:

$$M = \frac{I_{ac}}{I_{DC}} = 2\left(1 - \frac{W_A}{W}\right)\cdot\frac{I_1(x)}{I_o(x)}\cdot\left(\frac{1-\cos\theta}{\theta}\right) \qquad (19)$$

For a device embedded in a lossless microwave circuit the power $P_D$, delivered to a matched load will be:

$$P_D = \frac{1}{2}|G_D|Va^2 = M.I_{DC}\,Va \qquad (20)$$

and the DC to RF conversion efficiency:

$$\eta = \frac{P_D}{V_B\,I_{DC}} = \frac{Va}{V_B}\,.\,M. \qquad (21)$$

The main implication of these expressions are:

For a given current and signal level, the frequency for maximum power and efficiency is largely determined by the term in $\theta$. This has a maximum value of 0.72 when $\theta = .75\pi$, ie at .75 of the Read frequency.

Power and efficiency are increased by a reduction in avalanche region width, both through the term in $(1 - W_A/W)$ and the function $I_1(x)/I_o(x)$.

Power and efficiency both increase with increasing signal level, the limits being discussed below.

Device negative conductance and resistance are proportional to bias current and decrease with signal level - the latter is a fundamental requirement for self-starting and stability as an oscillator. The form of this decrease is discussed in problem 3, and is the source of gain compression in IMPATT amplifiers.

If a total resistive loss of magnitude Rs is present in the device and circuit then the power and efficiency will be multiplied by a degradation factor $(1-Rs/|R_D|)$

## PHYSICAL LIMITS ON DEVICE PERFORMANCE

We have seen that device output power, in particular, increases with current and signal level and implicitly with device area. However limits exist on these parameters.

The modulation of current in the avalanche region may become so large that the avalanche current minimum becomes comparable with the thermally generated current. In this event, the phase lag between avalanche current and voltage is reduced from its optimum value of 90°. If this reduction is by an angle $\delta$, then the degradation in power and efficiency is by a factor of approximately cos $\delta$.

Appreciable ionisation may take place in the drift region, leading to the creation of holes which drift back into the avalanche region where they are further multiplied, thus giving rise to a further increase in $\delta$.

The modulation of depletion width may be such that at the RF voltage minimum an appreciable length of the relatively lossy drift-region material becomes undepleted and therefore adds to series resistance. The effect becomes even more serious when the signal level is sufficiently large that the edge of the depletion region moves at a velocity approaching the carrier saturated velocity.

Taking these point together it has been shown experimentally that for a uniformly-doped silicon diode the maximum value of $Va/V_B$ is around 0.3 and that to attain this figure the structure should be designed such that the depletion region just reaches the boundary between the drift layer and epitaxial substrate at the mean operating voltage, close to $V_B$. For gallium arsenide uniformly-doped structures this limit is higher at around 0.4, probably due to the more favourable material parameters - high low field mobility and high peak velocity in the transferred electron region.

## Current Density Limits

Two possible limits exist on current density, one electronic and one thermal:

Electronic the passage of the charge bunch through the drift region will perturb the electric field. If the pulse is sharp this will give a field step of magnitude $J_{DC}/\varepsilon f$, which must be kept below a certian fraction, say 0.5, of the peak breakdown field, if ionisation in the drift region is to be avoided. The peak DC current density $J_{max}$, is then given by:

$$J_{max} \simeq 0.5 \ \varepsilon \ Ep \ f \qquad (22)$$

Thermal since device efficiencies are typically around 10% it is clear that a large fraction of the input power goes towards heating up the junction. The maximum safe operating temperature, $T_{max}$, is usually determined by the reliability of the metal contacts and is set at around 200°C for device lifetimes in the range $10^4$ - $10^6$ hours. To minimise temperature rise, devices are fabricated with the p-n junction very close to a high conductivity heat sink material such as gold or copper. The situation is summarised by:

$$T_{max} = R_{th}.A.J_{DC} \cdot V_B(1 - \eta) \qquad (23)$$

where $R_{th}$ is defines as the thermal impedance of the heat sink. For a device of cross-sectional area A on a semi-infinite heat sink of thermal conductivity $\sigma$:

$$R_{th} \simeq 1/\sigma \sqrt{\pi A} \qquad (24)$$

Area Limit

As device area is increased in order to increase output power, device impedance falls linearly and the fraction of power dissipated in fixed circuit losses increases. Although there is no hard limit associated with this loss of efficiency it is usual to ascribe to a circuit a minimum matchable impedance, $R_{min}$ which may be equated to the modulus of device negative resistance to give the limiting area.

Frequency Dependence of Limits

In practice a correctly designed device should simultaneously be at the maximum value allowed for each parameter, that is the area should be the maximum one consistent with the impedance matching, the current the maximum allowed for either electronic or thermal limits and the circuit impedance such as to match the device at the maximum allowable voltage swing. Solving the above equations simultaneously we find that at low frequencies the device current is thermally limited and the frequency-dependence of power output is approximately as $f^{-1}$. At high frequencies the current becomes limited by the electronic space-charge limit and the power falls as $f^{-2}$. (More exact versions of the power/frequency scaling laws are given by Scharfetter (5) and are set as problem 4). The transition between the two regions is not well defined but is expected to occur at about 100 GHz in silicon and at about half this frequency in GaAs. When devices are operated pulsed in such a way that the temperature rise in each pulse is small, the thermal limit disappears even at low frequencies, and the device limit becomes the electronic limit at all frequencies.

## COMPARISON OF MATERIALS AND DEVICE STRUCTURES

The above discussion has centred on the simple uniformly-doped p-n junction in silicon. Here we review and contrast the essential features of IMPATT diodes in gallium arsenide and in other device structures. Some useful performance comparisons, particularly at high frequencies, are given by O'Hara and Grierson (6).

Silicon vs Gallium Arsenide (uniformly-doped)

The comparative efficiencies of uniformly-doped devices in the two materials can be estimated from equations (19) and (21). There are two main differences which affect the parameters in these equations. Firstly, in gallium arsenide the electron and hole ionisation rates are nearly equal which leads at moderate doping values, for devices around 10 GHz to a narrower avalanche width. (Schroeder and Haddad (7)). However, because of the $\alpha$ vs E relationship in gallium arsenide as devices are scaled towards 100 GHz, the advantage passes to silicon. The other experimental observation is that the maximum Va/Vb ratios are larger in GaAs, as mentioned above.

In summary, for silicon, the factors: $I_1(x)/I_o(x)$, $(1 - WA/W)$, $(1 - \cos\theta)/\theta$ and $Va/V_B$ have values of around: 0.8, 0.65, 0.72 and 0.3 respectively, giving a predicted efficiency of 11%, and in gallium arsenide: 0.9, 0.75, 0.72 and 0.4 giving an efficiency of around 19%. Best efficiencies approaching these values have been seen, but it is somewhat more normal for efficiencies to be up to 8% and 12% respectively. The device impedance per unit area also improves in proportion to the efficiency improvement, that is, about 1.5 times, but this advantage is balanced out by the fact that the electron saturated velocity is lower in gallium arsenide, which requires a narrower drift width and hence bigger depletion capacitance. The net result is a CW output power advantage roughly in proportion to the efficiency advantage. In practice, since silicon devices have traditionally been easier to fabricate and control, the use of gallium arsenide has been confined to high-efficiency structures where the performance advantage is much greater. On present experimental results silicon performs better than gallium arsenide above about 100 GHz, but the detailed reasons are not known and more work remains to be done.

Silicon $p^+n$ vs $n^+p$ Structures

Because of the inequality of ionisation rates in silicon, which we have conveniently ignored here, with $\alpha$ for electrons greater than $\alpha$ for holes, it can be shown (7) that the $n^+p$ structure has a narrower avalanche region than the $p^+n$, so that the complementary device in which holes rather than electrons are the drifting species could show an advantage. In practice, this is outweighed by the lower saturated velocity for holes, which raises the depletion capacitance, and the greater resistance per unit length for the undepleted p-type layer material. These structures have not been much developed.

Double Drift vs Single Drift Structures

In order to overcome the power vs frequency constraints in silicon, the double drift structure was proposed by Scharfetter et al (8) and is now in widespread use. This is a symmetrical p-n junction with doping and field profiles shown in Fig 5, and it can be considered a back-to-back combination of two single drift structures, with approximately twice the breakdown voltage and depletion width for a given frequency of operation. Carriers are generated in a central avalanche region and electrons drift through the n side while holes drift through the p side. The central avalanche region gives a smaller value of $W_A/W$ than in single drift devices (7) and leads to an efficiency improvement of 1.5 times, that is reproducible values of around 12%. The device has a higher impedance per unit area because of the doubling of the depletion capacitance. However this advantage cannot be used when the device is thermally limited (the proof is set as problem 5) and the power output advantage is approximately the square of the efficiency, or about 2 times.

Gallium Arsenide High Efficiency Devices

The most spectacular performers among IMPATT diodes are the gallium arsenide high efficiency structures sometimes also known as Read diodes because of their similarity to that originally proposed by Read (2). The two main variants are the 'Hi-Lo' and 'Lo-Hi-Lo' so called after the n-type doping profiles they use, which are shown in Figure 6. These are very much larger signal devices, operating in modes and utilising effects not encountered in other IMPATT devices, and requiring complex xomputer programs for accurate analysis. Two recent reviews, by Blakey et al (9) and Pribetich et al (10) cover these points. The two main features of these structures are: firstly, the very significant narrowing of the avalanche region, and secondly, the large ratios of $Va/V_B$ which they can sustain. Both these effects, the first by raising the avalanche resonance frequency close to the operating frequency, and the soncond by modulating the depletion region by very large ratios, make the devices rather non-linear and somewhat more prone to instability than their uniformly-doped counterparts. Nonetheless, they offer very significant improvements in efficiency: the world-record result is 35%, but a more repeatable value would be 20-25% and X and J bands, with CW power approaching 10 watts in low X-Band. This performance tails off towards 20 GHz and it is not yet known whether this represents a fundamental limit. Active research is being aimed at answering this question, and hybrid double-drift Read diodes are being considered as a means of extending the upper frequency range of this mode of operation.

Performance Summary

We conclude with a summary of state-of-the-art performance for IMPATT devices as a function of frequency; this is shown in terms of output power and efficiency in Fig 7 for all the main classes of devices discussed. These plots underline several points, some of which have already been made in the text.

At low frequencies, around 10 GHz, the high efficiency diodes out-perform the others for both power and efficiency.

Devices have been successfully operated at well above 100 GHz. At present there is no hard theoretical limit to the process; rather a progression of efficiency reducing mechanism - avalanche width broadening, tunnelling as observed by Misawa (11) and Elta et al (12), and not least the technological struggle to define minute areas and minimise series resistance.

Between these two boards the silicon double drift device currently holds the stage. It can be seen that the power advanatge over single drift structure is close to the factor of 2 observed above.

Finally, it might be thought that the IMPATT diode will give way to the FET over the next few years, since powers of at least a few watts are becoming available with this device. This could indeed become the case at the lowest frequencies, but it is becoming increasingly clear that the IMPATT diode is the main hope for medium power levels, ie watts to tens of watts, from solid state devices in the millimetre-wave region for some time to come. Much effort is now being applied to devise newer structures and combine large numbers of existing devices so that IMPATT diodes can allow the next generation of applications - notably solid state millimetre radars and radiometers - to be realised.

REFERENCES


1. Sze, S.M., and Gibbons, G., 1966, Appl Phys Lett, 8, 111.
2. Read, W.T. Jr., 1958, Bell Syst Tech J, 37, 401.
3. Gilden, M., and Hines, M.E., 1966, IEE Trans Electric Devices, ED-13, 169.
4. Tager, A.S., 1967, Soviet Phys Usp, 9, 892.
5. Scharfetter, D.L., 1971, IEEE Trans Electric Devices, ED-18, 536.
6. O'Hara, S., and Grierson, J.R., 1974, Solid State Electron, 17, 137.
7. Schroeder, W.E., and Haddad, G.I., 1971, Proc IEEE, 59, 1245.
8. Scharfetter, D.L., Evans, W.J., and Johnson, R.L., 1970, Proc IEEE, 58, 1131.
9. Blakey, P.A., Culshaw, B., and Gibbin, R.A., 1978, IEEE Trans Electron Devices, ED-25, 674.
10. Pribetich, J., Chive, M., Constant, E., and Farrayre, A., 1978, J Appl Phys, 5584.
11. Gewartoski, J.W., 1979, IEEE Trans Microwave Theory Tech, MTT-27, 434.
12. Misawa, T., 1972, Solid State Electron, 15, 457.
13. Elta, M.E., and Haddad, G.I., IEEE Trans Microwave Theory Tech, MTT-27, 442.


GENERAL REFERENCES AND REVIEWS


14. Carrol, J.E., 1970, 'Hot Electron Microwave Generators', Edward Arnold, London.
15. Gibbons, G., 1973, 'Avalanche-Diode Microwave Oscillators', Clarendon Press, Oxford.
16. Howes, M.J., and Morgan, D.V. (Eds), 1976, 'Microwave Devices - Device Circuit Interactions' John Wiley and Sons, London.
17. Sze, S.M., and Ryder, R.M., 1971, Proc IEEE, 59 1140.
18. 1979, 'Special Issue on Solid-State Microwave/ Millimetre-Wave Power Generation, Amplification and Control' IEEE Trans Microwave Theory Tech, MTT-27.




PROBLEMS

1. If the ionisation rate in silicon is given by $\alpha(E) = \alpha_o E^6$, where $\alpha_o = 2.7 \times 10^{-30}$ cm$^{-1}$, calculate the peak field at breakdown and hence the depletion width at breakdown and breakdown voltage for a p-n junction with n-type doping concentration of $10^{16}$/cc. Assume $q = 1.6 \times 10^{-19}$ coulombs and $\varepsilon = 1.1 \times 10^{-12}$ Farads/cm. [Hint: rearrange ionisation integral in equation 2 to read:

$$\frac{\varepsilon}{qN} \int_0^{Ep} \alpha(E)\, dE = 1]$$

2(a) What n-type doping is required for a silicon device, such that its optimum frequency is 20 GHz, given that $Ep = 5 \times 10^5$ volts/cm, $W_A/W = 0.25$ and $v_s = 8 \times 10^6$ cm/sec?

(b) For the same device, make a first order estimate of efficiency at frequencies of 15, 20 and 25 GHz, for a maximum voltage modulation ($Va/V_B$) of 0.25, by assuming $I_1(x)/I_o(x) = 1$. Discuss the effect of including the current value for this last factor.

(c) Estimate the avalanche resonance frequency for this device at a current density, $J_{DC}$, 500 amps/cm$^2$, assuming $\alpha(E)$ as in question 1.

3. By substituting for Va in equation 18, show that for fixed frequency and current, the device negative conductance falls with signal level proportionally to the function F in equation 16. Hence deduce that the small signal negative conductance $G_{SS}$ is given by

$$G_{SS} = (1 - \frac{W_A}{W}) \cdot I_{DC} \cdot \frac{k}{\omega W} (\frac{1-\cos\theta}{\theta})$$

For the device in problem 2, operating at 20 GHz, and embedded in a circuit of conductance 1.76 millimho, show that oscillation will commence for a device current of 10 mA.

If the current is increased to 20 mA, what is the voltage swing across the device and the power output? Repeat the calculation for 30 mA.

4. Derive an expression for the maximum allowable area for a device

a) which is simultaneously on the impedance (area) limit and the thermal limit and

b) simultaneously on the impedance limit and electronic (space-charge) limit.

Hence, by assuming that $V_B \propto f^{-0.81}$, derive the exponent n in the expression $P_D \propto f^n$ for each case, assuming device efficiency is constant. What is the effect on n of a circuit loss which increases as $f^{0.5}$?

5. Show that if a double-drift IMPATT diode has an efficiency of 1.5 times that of a single-drift device, the available output power is around a factor of two greater when the device operates with the same ratio $Va/V_B$ and is thermally limited.

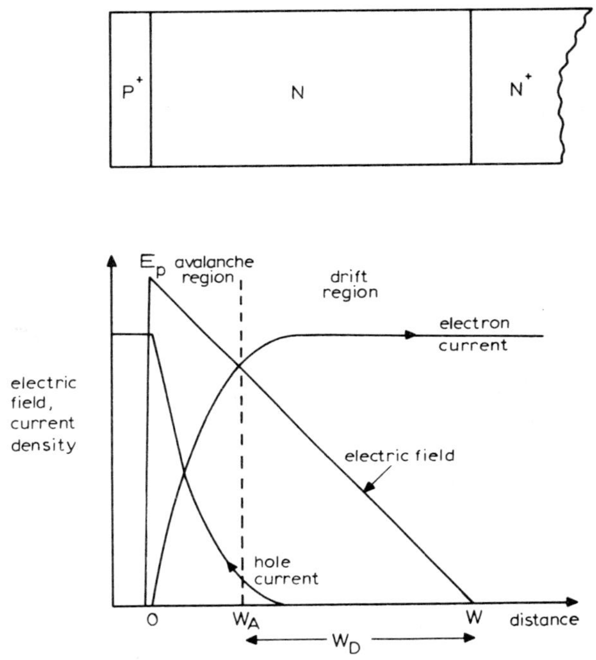


Figure 1. Electric field profile and electron and hole current densities at breakdown for a single-sided $p^+$-n-$n^+$ junction.

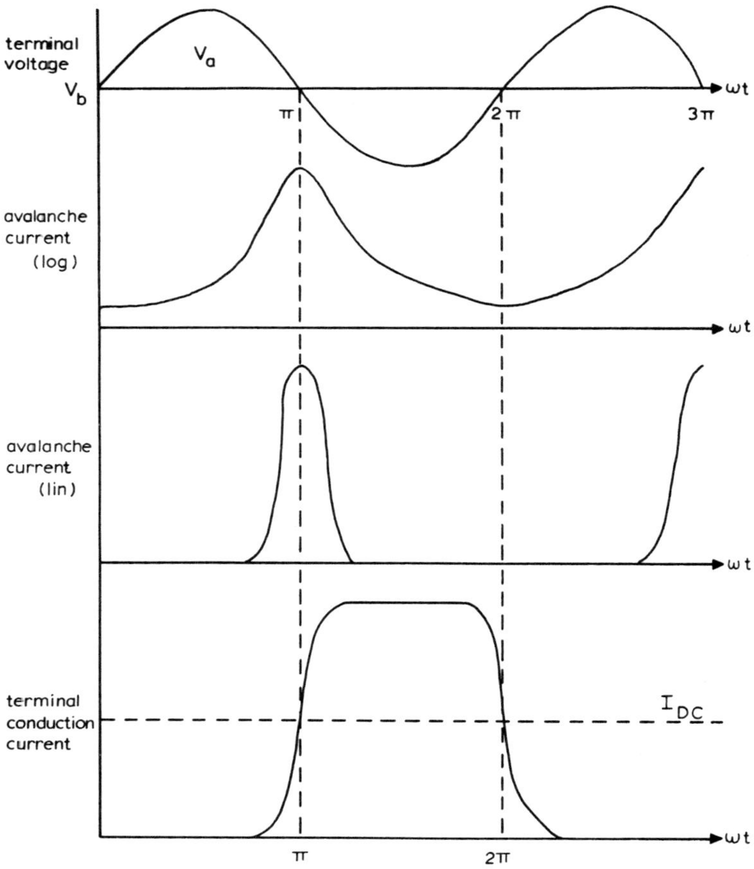


Figure 2. Current and voltage waveforms in an IMPATT diode.

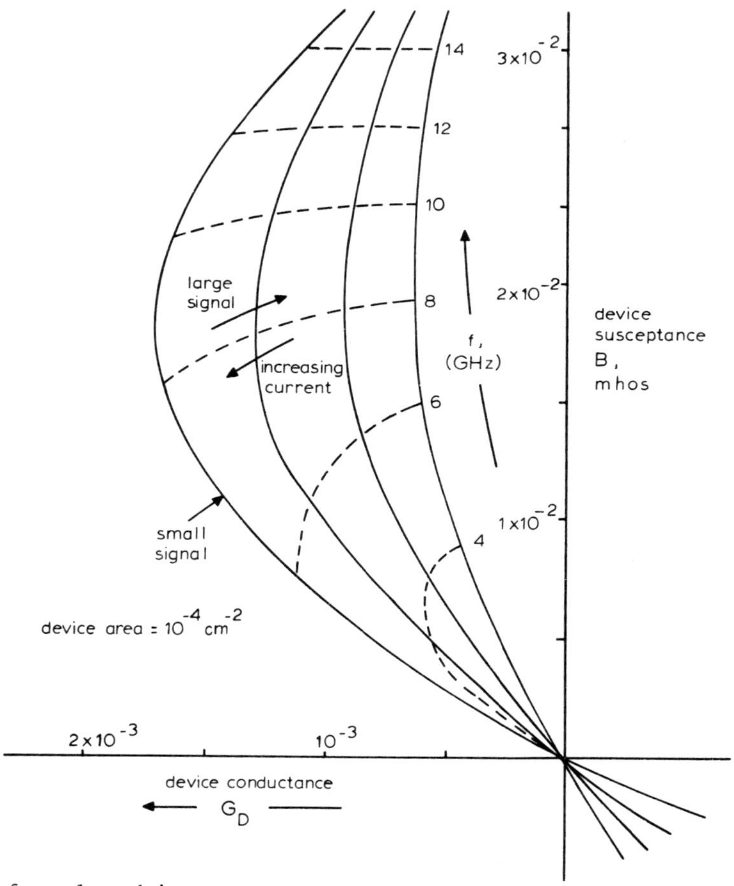


Figure 3. Schematic plot of complex admittance vs frequency at small and large signal levels for a typical X-Band IMPATT diode.

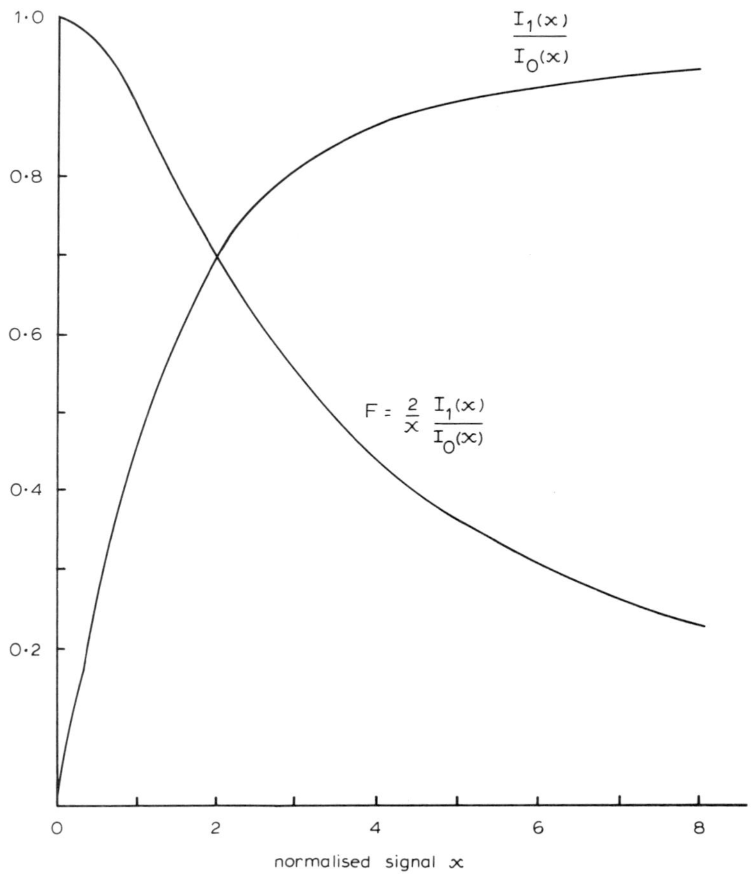


Figure 4. The factor F and the ratio $I_1(x)/I_o(x)$ vs x, the normalised signal level.

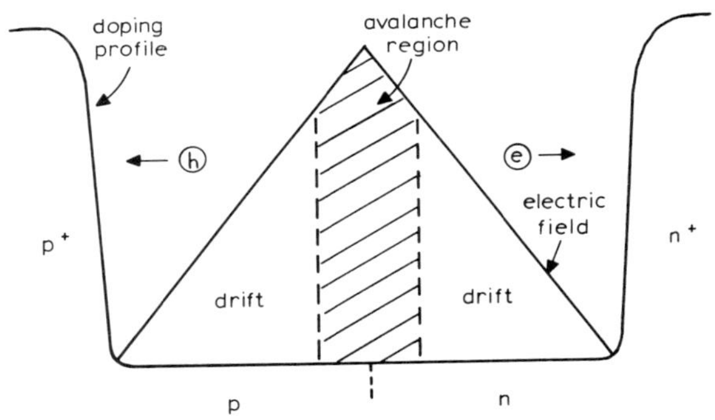


Figure 5. Doping and electric field profiles for a double-drift IMPATT diode.

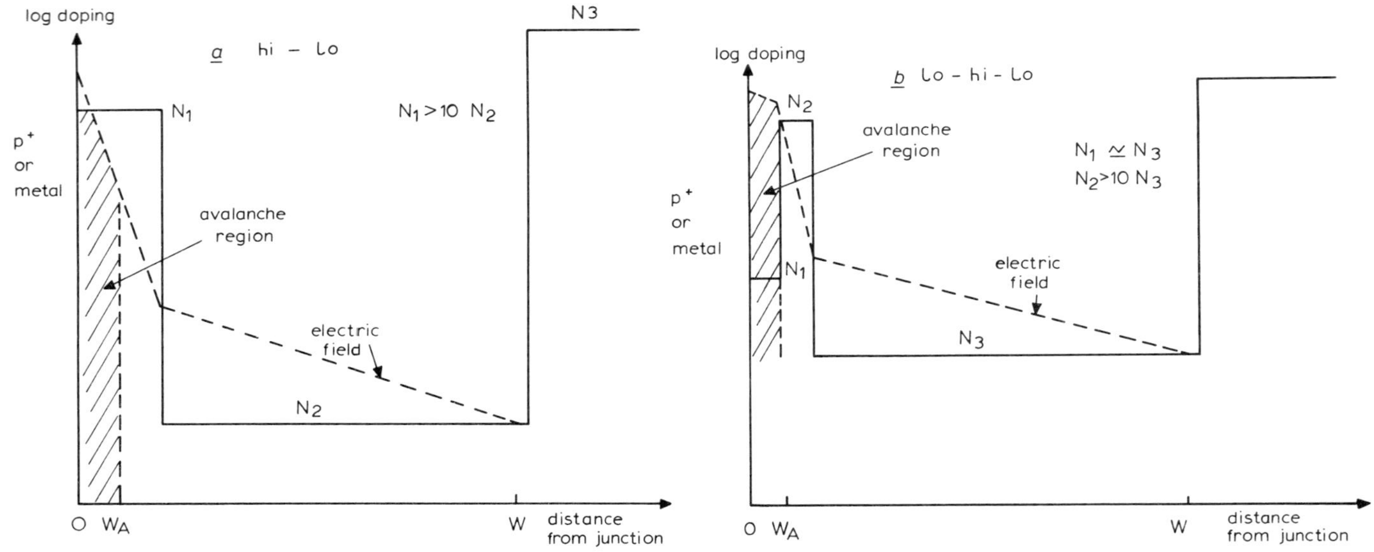


Figure 6. Doping and electric field profiles for Hi-Lo and Lo-Hi-Lo high efficiency IMPATT diodes.

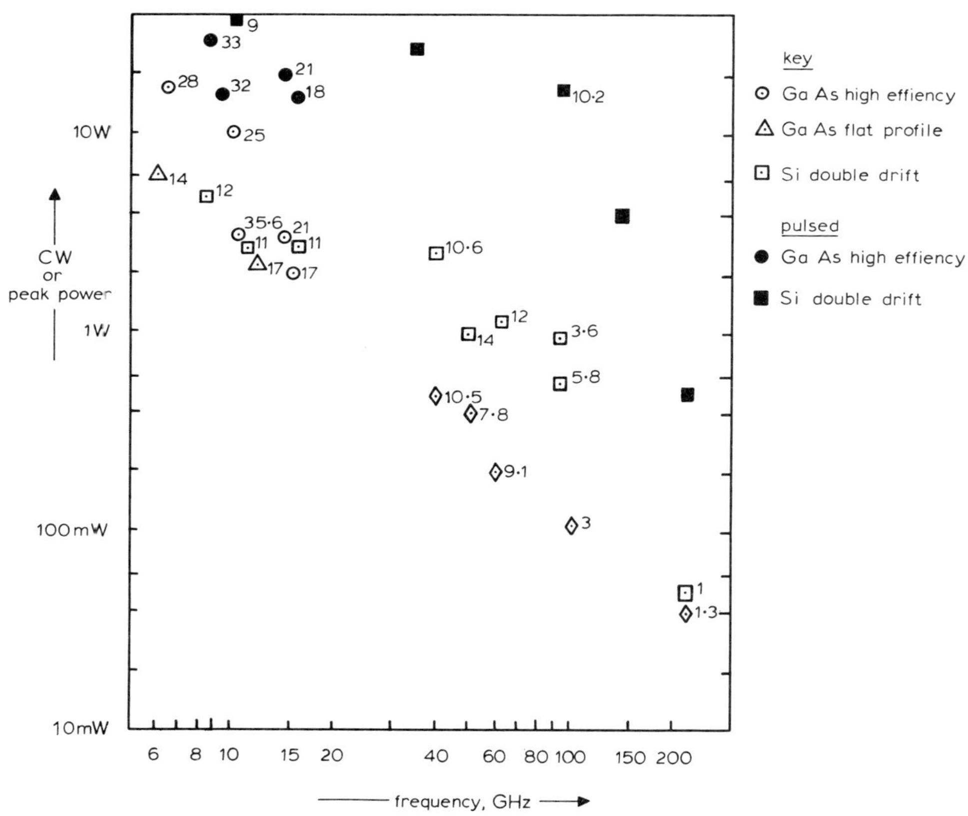


Figure 7. Summary of state-of-the-art peak and CW powers from IMPATT diodes (Figures by data points denote efficiency).

# CHAPTER 4

# Active microwave devices BJTs and FETs

Dr. J.S. Lamming

# ACTIVE MICROWAVE DEVICES - BJTs and FETs - Chapter 4

J.S. Lamming

GEC Hirst Research Centre, Wembley, Middlesex.

## INTRODUCTION

Nearly all microwave bipolar junction transistors (BJTs) are epitaxial planar silicon n-p-n types. To enable them to operate at microwave frequencies their dimensions must be very small. The very close control imposed by the requirements of such fine dimensions is one of the reasons for using silicon as the semiconductor material. This is because thermally grown silicon dioxide is eminently suitable to mask the diffusion of base and emitter dopants into the silicon and because it may be defined with 1 μm emitter diffusion slot widths by photolithography. Other aspects of silicon technology, such as epitaxial growth, ion implantation and diffusion are also under good control.

Microwave field effect transistors (FETs), on the other hand, are generally Schottky barrier gate gallium arsenide devices. Once again their dimensions must be very small. Gallium arsenide is preferred to silicon for these unipolar devices because its electron mobility is greater than that of silicon. A high electron mobility is important in the FET because, unlike the BJT, only one type of charge carrier (the electron) is responsible for conduction. The hole mobility in FET material is, therefore, unimportant. Other semiconducting compounds, like indium phosphide for example, could also be suitable materials for FET manufacture. Another advantage of these compound semiconductors is that they often have a semi-insulating form, which makes a suitable substrate on or in which a semiconducting layer can be provided, by epitaxial growth or ion implantation, to support the active FET behaviour. Diffusion is generally avoided in compound semiconductors, because of its complicated behaviour and masking-off difficulties, and for this reason the Schottky gate FET (MESFET) construction is used. Recently some metal/oxide (MOSFET) microwave gallium arsenide FETs have been made.

## MICROWAVE BIPOLAR TRANSISTORS

The equivalent circuit of a microwave silicon n-p-n epitaxial planar transistor is shown superimposed on a diagram of the transistor structure in figure 1. Under operating bias conditions current will be flowing in the branches of this circuit network and charge will be stored in the capacitors. If a microwave signal is fed into this transistor structure, via the emitter or the base, the circuit equilibrium will be disturbed and the speed with which the incoming microwave signal can be reproduced at the output terminals of the transistor will depend upon the ability of the transistor's equivalent circuit to respond quickly enough to the incoming microwave signal. The so-called frequency response of the transistor will depend therefore on the sum of the time constants associated with the various parts of the equivalent circuit.

### Signal Delay Time

Emitter delay time. Consider a microwave signal (a fluctuating voltage superimposed on the d.c. bias at the emitter terminal,e) which produces a charge fluctuation at the emitter junction. Equilibrium conditions at the emitter will be disturbed and the emitter transition region capacitance, $C_e$, will be either charged or discharged through the emitter resistance, $r_e$, and the emitter series resistance, $r_{es}$, giving rise to the first equivalent circuit RC time constant.

$$\tau_{eb} = (r_e + r_{es})C_e \qquad (1)$$

where the emitter resistance

$$r_e = \frac{kT}{qI_e} \qquad (2)$$

k is Boltzmann's constant, T is the junction temperature in $^{o}K$ and q is the electronic charge. $\tau_{eb}$ may be reduced by increasing the emitter current, $I_e$. $r_{es}$ may also include any emitter contact resistance, or current-sharing resistors in the case of a power transistor.

The microwave voltage fluctuation at the emitter will cause a corresponding flux of electrons to be injected into the base from the emitter and a much smaller flux of holes to be injected into the emitter from the base. The change of stored charge caused by the hole density fluctuation in the emitter gives rise to a time delay

$$\tau_e = \frac{x_e^2}{2D_{pe}\beta_o} \qquad (3)$$

where $x_e$ is the emitter junction depth, $D_{pe}$ is the diffusion coefficient of holes in the emitter and $\beta_o$ is the d.c. common emitter current gain. Equation (3) is obviously similar to equation (4), except that $\tau_e$ is very much reduced by $\beta_o$ which relates to the ratio of electrons to holes injected by the emitter junction. $\tau_e$ may be reduced by reducing $x_e$. The d.c. current gain, $\beta_o$, is proportional to the emitter efficiency, $\gamma$. When the current level reaches values such that the concentration of majority carriers in the base increases from the equilibrium value, the emitter efficiency drops. The decrease in emitter efficiency takes place when the integral of the injected minority charge per unit area in the base becomes comparable to the majority carrier concentration per unit area. Since the injection efficiency for electrons into the base decreases, more of the emitter current will consist of holes injected into the emitter from the base and hole storage in the emitter will be increased. This will cause $\tau_e$ to increase. It is assumed that $D_{pe}$ will not alter much at high injection levels because the injected hole concentration in the emitter will still be much less than the majority electron concentration.

Base delay time. The time $\tau_b$, for the electrons injected by the emitter to cross the base is given by

$$\tau_b = \frac{x_b^2}{nD_{nb}} \tag{4}$$

The base width, $x_b$, must be as small as possible, because $\tau_b$ depends on $x_b^2$. Before the advent of planar transistor technology this used to be the limiting factor to the achievement of high frequency operation. Nowadays, with improved diffusion technology, base widths of only 0.1 to 0.2 μm are possible, but care must be taken to avoid the emitter-dip effect (see later). Very narrow, and reproducible, base regions for microwave transistors can also be produced by ion implantation. The diffusion coefficient of electrons in the base, $D_{nb}$, should be made as large as possible by ensuring that the base region is not excessively doped. In this respect it is an advantage to have the base region underneath the emitter less heavily doped than the base region between the emitter and the base contact; this may be realised with the isoplanar method of construction (1). However, for very small base widths, with light base doping, injection modulation (2) of the base charge will cause $r_e$ to increaseabove the value given by equation (2) resulting in an increase of $\tau_{eb}$. The value of n in equation (4) depends upon the dopant distribution in the base region. For a uniformly doped base n=2.43, but if the doping profile in the base is graded, as is the case for a diffused base transistor, then

$$n = \frac{m^2}{m-1+\exp\{-m\}} \tag{5}$$

where $m \simeq \ln \frac{N_{BE}}{N_{BC}}$

providing that the value of m is not too small. $N_{BE}$ and $N_{BC}$ are the base impurity concentrations near the emitter and collector, respectively. For silicon diffused base n-p-n microwave transistors n in equation (5) is typically 6. However, in a graded (i.e. diffused) base transistor the value of $D_{nb}$ will be reduced as the concentration of impurities in the base increases towards the emitter, because of increased impurity scattering of the electrons, so that the effective value of n in equation (4) will be reduced to about 3, if an average value of $D_{nb}$ is assumed.

When the injected minority carrier concentration in the base approaches that of the ionised base impurities then a deterioration of the built-in field of a graded base transistor will occur, leading to a reduction of n. Under these conditions $D_{nb}$, which is a suitably averaged diffusivity of minority carriers in the base, will be effectively doubled owing to the fact that the electric field gradient in the base will be determined by the injected carriers rather than the fixed ionised impurities. The critical current for the transition of the base transit time from the low to the high injection value is a complicated function of the built-in field. However, for an exponentially graded base impurity profile the overall effect of the transition from low to high injection level conditions should be that the $nD_{nb}$ product remains essentially the same. Thus the same values of n and $D_{nb}$ can be used both at moderate and high injection levels in the equation for $\tau_b$.

The injected electrons in transit in the base constitute a charge stored in this region. It is sometimes assumed that the injected electron density at the collector side of the base is zero because, naively, the collector is assumed to be a perfect 'sink' for electrons. However, in order to maintain continuity of current through the device the charge at the collector edge of the base cannot approach zero, otherwise the electron velocity would have to approach infinity. Since, in fact, the electron velocity cannot exceed a maximum velocity, $v_s$, then the corresponding electron density required to maintain current continuity constitutes additional stored charge in the base. This stored charge fluctuates with the microwave signal causing a delay time

$$\tau_b' = \frac{Q}{I} = \frac{x_b}{v_s} \tag{6}$$

Both $\tau_b$ and $\tau_b'$ can be reduced by keeping the base width, $x_b$, small.

Collector delay time. When the injected electrons have crossed the base they enter the depletion region of the reverse-biased collector junction and are swept across it by the strong electric field which exists there. The electric field in the collector depletion layer is appreciably greater than the value of about $10^4$ V cm$^{-1}$, at which the electrons reach their scattering limited velocity, $v_s$, of $8 \times 10^6$ cm s$^{-1}$ in silicon. The electrons are accelerated to this velocity so quickly that they may be assumed to travel through the entire collector depletion region, of width $x_c$, at the velocity $v_s$. The transit time through this region is therefore

$$t_c = \frac{x_c}{v_s} \tag{7}$$

The a.c. voltage across the depletion layer excites a displacement current within it and this causes a phase shift in the total current flowing across the depletion layer such that the signal delay time corresponding to the carrier transit time is given by $t_c/2$. The collector depletion layer delay time is therefore

$$\tau_c = \frac{x_c}{2v_s} \tag{8}$$

The collector depletion region width should be minimised to keep $\tau_c$ small and this may be accomplished, for a given collector voltage, by lowering the resistivity of the collector epitaxial layer. However, microwave transistors are generally designed so that the collector space-charge layer fully depletes the epitaxial collector region and under these conditions $\tau_c$ is given by

$$\tau_c = \frac{w_{epi}}{2v_s} \tag{9}$$

where $w_{epi}$ is the width of the epitaxial layer.

The depletion region capacitance of the collector active region (underneath the emitter) will be charged or discharged by the microwave signal fluctuation and this will constitute another RC time delay,

$$\tau_{bc} = (r_e+r_{es}+r_{cs})C_{ci} \tag{10}$$

where $r_{cs}$ is the spreading resistance of the collector body material. $r_{cs}$ is made small by the use of $nn^+$ epitaxial silicon for the collector.

Total signal delay time. The sum of all the time constants discussed so far constitutes the total signal delay time from emitter to collector, $\tau_{ec}$.

$$\tau_{ec} = \tau_e+\tau_{eb}+\tau_b+\tau_b'+\tau_{bc}+\tau_c \tag{11}$$

The reciprocal of the total signal delay time may be expressed as a characteristic frequency, $\omega_T$.

$$\omega_T = \frac{1}{\tau_{ec}} \tag{12}$$

The frequency $\omega_T = 2\pi f_T$ is a most important parameter in a microwave transistor since it dominates both the gain and the noise figure. $f_T$ is the frequency at which the magnitude of the grounded emitter current gain, $h_{fe}$, is equal to unity, it is known as the current gain bandwidth frequency.

Base Resistance

The base resistance arises because of the small base current which flows out of the base region contact. In a planar transistor this current flow is parallel to the emitter and collector junction planes. Because of the resistivity of the base region the base current flow will cause a transverse voltage drop to be developed in the base region. Both the current and the voltage will be distributed in the base depending upon the dimensions of the structure and the specific resistivity in different parts of the base region. Because of the transverse voltage gradient in the base, those parts of the emitter junction furthest away from the base contact will be operating at a lower bias than the parts nearest to the base contact. To overcome this emitter self-biasing effect the emitter is made in the form of a plurality of long narrow strips. The base resistance is made up of three components

$$r_b = r_{bi}+r_{bo}+r_{bcon} \tag{13}$$

where $r_{bi}$ is the resistance underneath the emitter, $r_{bo}$ is the resistance between the emitter edge and the base contact edge and $r_{bcon}$ is the resistance from the silicon to the base contact metal. All three components must be small for maximum frequency response and low noise.

Collector Capacitance

The collector capacitance consists almost entirely of the collector transition region capacitance. This capacitance is often considered, in an equivalent circuit model, to be distributed into elements forming an RC transmission line with the three associated components of the base resistance.

Figure of Merit

A figure of merit, K, may be written

$$K = f_{max} = \left(\frac{f_T}{8\pi r_b C_c}\right)^{\frac{1}{2}} \tag{14}$$

or

$$f_{max} = \frac{1}{4\pi(r_b C_c \tau_{ec})^{\frac{1}{2}}} \tag{15}$$

Equation (15) shows that to increase $f_{max}$, the maximum frequency of oscillation of a transistor, the base resistance, $r_b$, the collector capacitance, $C_c$, and the signal delay time, $\tau_{ec}$, must all be made small.

High Current Effects

In the figure of merit expression for a transistor (equation (14)) the $r_bC_c$ product should be independent of area, therefore it should be possible to scale up the transistor area to handle any value of current, provided the device has a heat sink which will dissipate the heat generated within the active part of the transistor. If the current density in the transistor is not increased then $f_T$ should ideally be independent of area. The ability to increase the emitter strip length (or the strip multiplicity, which is equivalent) at will, without deteriorating the frequency response, is the fundamental premise on which the design of high frequency high power transistors is based. Obviously, however, the scaling up cannot go on indefinitely since technological limits exist, both in the production of fault-free epitaxial silicon and in the control of uniform processing over a relatively large area. These limitations are of a statistical nature and the relation between the density of these defects and the area of the transistor will determine the percentage yield of satisfactory transistors. As the power and hence the area of the transistor is increased the yield due to these effects must diminish.

The most important change of $f_T$ at high current levels is the reduction brought about by the modification of the emitter and collector depletion layer boundaries by the charge in transit through them. This effect produces changes in $C_e$, $C_c$, $x_b$, $x_c$ and, possibly, $r_{cs}$. The greatest affect on $f_T$ is brought about by an increase of the effective base width, $x_b$. This is often referred to as the Kirk effect (3).

In the Kirk effect, for the case of an n-p-n transistor at high current levels, the electrons in transit in the epitaxial part of the collector depletion layer will reduce the overall charge density, $\rho$, according to

$$\rho = q(N_{DC}-n) \tag{16}$$

where $N_{DC}$ is the donor impurity concentration in the collector epitaxial layer (assumed constant) and n is the electron density. The electrons move across the space-charge region primarily by drift, with velocity v(x), and give rise to a current density

$$J_c = qv(x)n(x) \tag{17}$$

From equations (16) and (17) Poisson's equation for the electric field, E, may be written in the form

$$\frac{dE}{dx} = -\frac{q}{\varepsilon\varepsilon_o}\left[N_{DC} - \frac{J_c}{qv(x)}\right] \qquad (18)$$

This equation may be integrated, subject to the boundary condition that $E = E(O)$ at $x = 0$ and $v(x)$ has the constant value $v_s$, to give

$$E(x) = E(O) - \frac{q}{\varepsilon\varepsilon_o}\left[N_{DC} - \frac{J_c}{qv_s}\right]x \qquad (19)$$

where $\varepsilon$ = dielectric constant of Si and $\varepsilon_o$ = permittivity of free space.

The electric field distribution in the epitaxial region of the collector is linear, with a slope depending upon the current density, $J_c$. Figure 2 shows the field distribution as a function of current density for a fixed value of collector voltage, $V_{CB}$. Curves (a) to (e) represent increasing values of $J_c$. Curve (a) is for low to moderate current levels. Curve (b) represents the situation when, with increasing current density, the collector edge of the depletion layer has just reached the substrate. Further increases in the current cause a reduction of the field at $x = 0$ until it reaches a value approximately equal to zero (curve d). The critical current density required to create the conditions represented by curve (d) is given by

$$J_o = qv_s\left[\frac{2\varepsilon\varepsilon_o V_{CB}}{qw_{epi}^2} + N_{DC}\right] \qquad (20)$$

On the assumption that the current flow is strictly one-dimensional then, because the area cannot be increased, an increase of current beyond that depicted by curve (d) of figure 2 can only be accomplished by increasing the current density. To achieve this large numbers of holes enter the collector in order to eliminate the space-charge due to electrons in this region. These extra charges must be taken into account in the integration of Poisson's equation and increase the gradient of the field distribution, this leads to a decrease in the length over which electrons travel by drift in the collector. Part of the collector region is thus electrically equivalent to the base region, since the carriers in it move by diffusion instead of drift. The reduction in the length of the electric field region is shown schematically by curve (e) of figure 2. That part of the epitaxial layer between the base-collector p-n boundary and the intersection of curve (e) with the distance axis of figure 2 is called the current induced base region.

## Electrical Characterisation

The maximum frequency of oscillation, $f_{max}$, has been defined in equation (14). More strictly, for d.c. common base current gain $\alpha_o \neq 1$,

$$f_{max} = \left(\frac{\alpha_o f_T}{8\pi r_b C_c}\right)^{\frac{1}{2}} \qquad (21)$$

The maximum unilateral gain, U, is given by

$$U \simeq \left(\frac{f_{max}}{f}\right)^2 \qquad (22)$$

It is independent of package reactances and common lead feedback elements and is, therefore, a unique measure of the intrinsic device performance.

A more practical measure of power gain in a transistor amplifier is the maximum available gain, MAG.

$$MAG \simeq \left(\frac{f_{MAG}}{f}\right)^2 \qquad (23)$$

where

$$f_{MAG} = \left(\frac{\alpha_o f_T}{8\pi(r_b+r_e+r_{es}+\omega_T L_e)C_c}\right)^{\frac{1}{2}} \qquad (24)$$

and $L_e$ is the emitter lead wire inductance. Equation (24) includes the feedback elements $r_e$, $r_{es}$ and $L_e$ in the common emitter branch of the equivalent circuit.

The transducer cut-off frequency is given by

$$f_{trans} = \frac{\alpha_o f_T}{\alpha_o+\omega_T C_c Z_o}\left(\frac{2Z_o}{Z_{in}+Z_o}\right)^{\frac{1}{2}} \qquad (25)$$

where $Z_{in}$ is the input impedance of the transistor and $Z_o$ is the positive real line impedance of a lossless transmission line system connected to the input and output ports of the transistor. The transducer gain is given by

$$G_T = |S_{21}|^2 \qquad (26)$$

where $S_{21}$ is the forward insertion gain with the output terminated in a matched load ($Z_L = Z_o$).

Equations (22), (23) and (26) show that the gain falls at the rate of 6 dB/octave with frequency (figure 3). In practice departures from the idealised 6 dB/octave slope occur, which may be attributed to second order frequency terms neglected in the calculation of U, MAG and $G_T$.

## Small Signal, Low Noise, BJTs

The noise figure, F, is defined as the ratio of the signal-to-noise ratio at the input to the signal-to-noise ratio at the output. The noise of a transistor is therefore relative to the noise in the input signal generator, so that the noise figure of the transistor can never be less than unity. The two main sources of noise in microwave bipolar transistors are thermal noise and shot noise. Thermal noise is caused by thermal agitation of the current carriers in the bulk material of the transistor, giving them a random motion. Thermal noise consists of the summation of short random current pulses and is uniform over the whole frequency band. Shot noise is due to the fact that the transistor is no longer in thermal equilibrium under bias conditions and additional noise arises from the flow of electron and hole currents within the device. Shot noise is therefore current dependent. Fluctuations in the generation-recombination process of the charge carriers also contribute to the shot noise.

The noise figure is generally assumed to be given by Nielsen's equation

$$F = 1+\frac{r_b}{R_g}+\frac{r_e}{2R_g}+\frac{(r_b+r_e+R_g)^2}{2\alpha_o r_e R_g}\left[\left(\frac{f}{\kappa f_T}\right)^2+1-\alpha_o\right] \qquad (27)$$

where $R_g$ is the generator resistance and $f_\alpha = \kappa f_T$. Equation (27) shows the importance of having a low base resistance, $r_b$, and a high base cut-off frequency, $f_\alpha$, in a microwave low-noise transistor.

Equation (27) may be differentiated with respect to $R_g$ to obtain the optimum value of generator resistance,

$$R_g(\text{optimum}) = \left[ (r_b+r_e)^2 + \frac{\alpha_o r_e(2r_b+r_e)}{\left(\frac{f}{\kappa f_T}\right)^2 +1-\alpha_o} \right]^{\frac{1}{2}} \tag{28}$$

Equation (28) may then be substituted in equation (27) to give the minimum noise figure, $F_{min}$. This results in a rather complicated expression and a simplified formula, which gives $F_{min}$ to a good approximation, is

$$F_{min} \simeq 1 + u + (2u + u^2)^{\frac{1}{2}} \tag{29}$$

where

$$u = \left[ 1 - \alpha_o + \left(\frac{f}{f_\alpha}\right)^2 \right] \frac{r_b q I_e}{\alpha_o kT} \tag{30}$$

To give the minimum noise figure the optimum value of $I_e$ is given by

$$I_e\ (\text{optimum}) = \frac{2\pi\alpha_o kT f_T C_e}{q} \tag{31}$$

The conditions for minimum noise figure are not necessarily the same as the conditions for maximum power gain and it is often necessary to compromise between the two.

In equations (27) and (30) $f_\alpha$ is strictly the cut-off frequency due solely to the base transit time (equation (4)),since in the derivation of these expressions the affect the emitter time constant $\tau_{eb}$ (equation (1)) was not included. Hawkins has shown (4) that if $\tau_{eb}$ is taken into account, as it should be for most microwave transistors, then the noise figure increases less rapidly with frequency than predicted by Nielsen's equation. Conversely, if Nielsen's expression is fitted to experimental results then artificially high values for $f_\alpha$ must be assumed. In Hawkins' expression for the noise figure the term in the square brackets of equation (27) is replaced by

$$\left[\left(1+ \frac{f^2}{f_\alpha^2}\right)\left(1+ \frac{f^2}{f_e'^2}\right) - \alpha_o\right]$$

where $f_e'$ is the cut-off frequency of the input circuit of the transistor.

The noise figure versus frequency curve for a silicon bipolar transistor is compared in figure 4 to that of a gallium arsenide FET, relative to the state-of-the-art in 1979.

## Power BJTs

The important performance parameters of a microwave power transistor are power output, power gain and efficiency. The power output is determined by the current and voltage handling capability of the transistor. The current handling capability is determined by the emitter periphery and epitaxial layer resistivity (because of current crowding to the edges of the emitter and the Kirk(3)effect, respectively and the voltage handling capability is determined by the breakdown voltage, which is limited by the resistivity of the epitaxial layer and the curvature at the edge of the base-collector junction. The power gain is determined by the ratio of $f_{max}$ to the operating frequency, so that for high power gain a high $f_{max}$ is required. The efficiency is the ratio of the power output to the power input and determines the heat dissipation within the device.

The operating current and, therefore, the power output of a transistor is dependent on the emitter periphery, because most of the current is injected near the edge of the emitter sites as a result of the transverse base current biasing effect. To optimise the frequency performance it is necessary to design the transistor to have the maximum emitter periphery, with the minimum emitter area, contained within the minimum base area. The emitter and base areas are kept small to keep $C_e$ and $C_c$ small. Typical values of current and power handling capability, in terms of emitter periphery are 0.4A $cm^{-1}$ and 4W $cm^{-1}$, respectively. These empirical values depend on the geometrical design of the transistor and the operating frequency. The three basic geometrical designs of power transistors are called interdigitated, overlay and mesh.

Power transistors are liable to hot-spot formation owing to thermal runaway in localised areas of the transistor structure, because the negative temperature coefficient of the emitter-base junction voltage leads to increased current injection at these sites. To promote more even current sharing ballast resistors are inserted in the metal track to each emitter,since the voltage drop in the resistor will offset the decreased junction voltage in the runaway emitter site. Collector ballasting also helps to prevent runaway under mismatch conditions.

A power transistor must be properly biased so that it does not operate in a current-voltage regime where 'second breakdown' of the collector junction is liable to occur. The current density in the metallisation pattern must also be limited to values less than $2 \times 10^5$ A $cm^{-2}$, so that electromigration of metal ions in the conductor track does not occur.

Any power device of uniform power density will have a higher temperature towards its centre. A good power transistor design will therefore be one in which the active region is divided into smaller heat sources separated by some distance. The active part of the device must be in good thermal contact with a heat sink.

To achieve maximum bandwidth, efficiency and power gain, microwave power transistors are being made with an electrical network inside the package, next to the transistor chip, to transform the device input and output impedance to match a 50 Ω microwave transmission line. These internal matching networks are designed to include the inductance of the bond wires to the transistor chip and the parasitic capacitances of the package. The bond wire length must be accurately controlled and a different length for each cell of a multi-cell chip may be required. A disadvantage of the internal matching network is that it is only matched over a certain fixed frequency band, whereas the normally packaged transistor chip can be externally matched over different frequency bands.

The power-frequency product of a microwave transistor should be given approximately by the relationship

$$P_{max} R_{in} f_{max}^2 \lesssim 1 \text{ kW } \Omega \text{ (GHz)}^2 \qquad (32)$$

where the real part of the input impedance, $R_{in} = r_b + \omega_T L_e$. Practical transistors have reached this theoretical limit, e.g. 600 W (peak) at 1 GHz. The power output of silicon bipolar transistors is compared with that of GaAs FETs in figure 5.

## Practical Limitations on Frequency Response

Pattern dimensions. The stored charge within the device has a capacitative effect and the amount of charge (or the capacity) is directly proportional to the active area of the device. The area must be kept to a minimum, therefore the dimensions in the plane of the surface of the silicon chip must be controlled accurately. It is important that the emitter width, the emitter-to-base contact spacing and the collector area, i.e. the transistor 'pattern', should be defined with the utmost precision. The figure of merit for a high frequency transistor may be defined in terms of the minimum width attainable in this pattern. Thus the figure of merit, K, for an interdigitated transistor pattern, may be expressed in terms of the emitter strip width, s.

$$K \simeq \frac{40}{s} \text{ GHz} \qquad (33)$$

where K, the $(\text{power gain})^{\frac{1}{2}} \times (\text{bandwidth})$ product, is equivalent to $f_{max}$, the maximum frequency of oscillation of the transistor (equation (14)), and s is in micrometres. The minimum dimensions attainable in the transistor pattern using photolithography can approach 1 μm, therefore transistors with $f_{max}$ approaching 40 GHz are possible. Smaller pattern dimensions are available with electron-lithography, but it would be difficult to maintain their very fine resolution during subsequent transistor processing. It should be possible, ultimately, to extend the useful electrical performance of microwave bipolar transistors into X-band using, perhaps, a combination of these techniques.

Emitter dip effect. An important phenomenon is the enhanced diffusion of the base impurity underneath the emitter. This has been referred to as the emitter-dip effect, the base push-out effect, or base run-on. It is important because it can limit the minimum base width attainable.

The emitter-dip effect is most pronounced in silicon n-p-n transistors with phosphorus doped emitters. The enhancement of the base impurity (boron) diffusion underneath the emitter is thought to occur because the tetrahedral covalent radii of the phosphorus atoms are significantly less than the radii of the silicon host crystal atoms. When a critical amount of misfit occurs between dopant and host crystal atoms, a dislocation network is created in the doped layer. If the Burger's vectors of these dislocations lie in the surface plane any subsequent propagation of the dislocations into the interior of the crystal during further heat treatment must occur by a climb mechanism, which generates vacancies. The vacancy concentration in the vicinity of the emitter will be supersaturated and the diffusion coefficient of the boron atoms in this region will be enhanced, since the boron atoms move by jumping into a vacancy.

The emitter-dip effect may be eliminated by changing the emitter dopant to arsenic. This is because the covalent radius of arsenic is close to that of silicon, therefore no misfit dislocation networks form during arsenic diffusion and excess vacancies are not created. In fact, owing to the formation of vacancy-arsenic complexes, vacancies are removed from solution in the silicon crystal lattice. Vacancy undersaturation occurs in the emitter and the emitter then acts as a sink for vacancies which diffuse in from the surrounding silicon, reducing the vacancy concentration there. The boron diffusivity in the vicinity of the emitter will be reduced and the base region underneath the emitter will be retarded. If it can be controlled accurately a retarded base structure should be better for a microwave transistor than one with the emitter-dip effect, because the intrinsic base width can be reduced to a minimum to decrease $\tau_{ec}$ and the extrinsic base can be made wider to reduce the base resistance. This should give a better high frequency figure of merit (equation (15)). A retarded base structure can also be produced by ion implantation, or by a combination of diffusion and ion implantation (as in the isoplanar transistor).

## MICROWAVE FIELD-EFFECT TRANSISTORS

GaAs metal semiconductor field effect transistors (MESFETs) have higher gain, higher power amplification efficiency and lower noise figure than Si BJTs above 4 GHz. Moreover, they have potential for further development in that other structures, like junction gate (JFETs) and insulated gate (IGFETs), may be suitable for microwave applications and other materials, like InP, $In_xGa_{1-x}As$ or $InAs_xP_{1-x}$, may be used for their fabrication. Monolithic integration of circuits on semi-insulating substrates will give device isolation with low parasitic capacitances, low-loss interconnections, and high packing density, at microwave frequencies.

The improvement in device performance obtained with GaAs, or other selected III-V compounds, is due to their higher electron mobility compared with Si. The result is that parasitic resistances are smaller, the transconductance is larger and the electron transit time through the high field region is shorter. These effects give the GaAs MESFET its very good high frequency and low noise performance.

### Principle of Operation

A diagram of a GaAs MESFET, with superimposed equivalent circuit, is shown in figure 6. On the GaAs semi-insulating substrate is first grown a GaAs buffer layer, generally by chemical vapour deposition (CVD). This layer has a high resistivity (approx. $10^7$ ohm cm) and contains very few mobile electrons. Consequently it acts effectively as an extension of the semi-insulating substrate; it serves the useful purpose of screening the n-type active layer, subsequently grown epitaxially upon it, from any deleterious electrical properties of the substrate. The n-type GaAs active layer has typically a doping level of about 1 or 2 × $10^{17}$ donors per $cm^3$ and a thickness of about 0.3 micron. The Schottky barrier gate often lies at the bottom of a trench etched into the active layer and the source and drain ohmic contacts may be on another more heavily doped (approx.

$10^{18}$ $cm^{-3}$) epitaxial layer. The last two features are not shown in figure 6.

In operation the source is grounded and the drain is positively biased. Ignore, for the present, the gate electrode and assume that the electrons carrying the current are confined to the n-type active layer, i.e. no current flows in the buffer layer. In practice this ideal situation will not be achieved and some electron current will flow in the buffer layer. However, in the ideal model under consideration electrons will flow from the source to the drain and for small drain voltages the device will behave like a linear resistor. For larger voltages the electron drift velocity does not increase at the same rate as the electric field, E, and the current-voltage characteritic will fall below that of a simple resistor, until at a critical field $E_C$ (which is about 3 kV $cm^{-1}$ for GaAs) the electrons reach a maximum velocity (about $2 \times 10^7$ cm $s^{-1}$ for GaAs) and the drain current starts to suturate.

Consider now the effect of the gate electrode. Even at zero bias on the gate with respect to the source a region of the GaAs adjacent to the Schottky barrier will be depleted of mobile electrons and this will restrict the cross section of the n-type layer available for current flow. The n-type layer between the source and the drain is called the channel. The current $I_{DS}$ flowing in the channel is given by

$$I_{DS} = wqn(x)v(x)d(x) \qquad (34)$$

where w is the gate width (see figure 6), q is the charge of an electron, n the density of conduction electrons, v their drift velocity, d the thickness of the conductive layer and x the coordinate in the direction of the electron drift. The voltage in the channel increases from zero at the source to $V_{DS}$ at the drain. Hence the Schottky barrier gate becomes more reverse-biased at its drain end and its depletion region is correspondingly widened. This decreases the channel cross section still further and this is compensated by an increase of the electric field and electron velocity, v, to maintain a constant current through the channel. As $V_{DS}$ increases the electrons reach their saturated velocity $v_s$ at a channel thickness of $d_o$ under the gate, where the electric field reaches the critical value, $E_C$. A further increase of $V_{DS}$ will increase the gate depletion width towards the drain and the point $x_1$, where the electron velocity reaches $v_s$, will move slightly towards the source. This causes the voltage at $x_1$ to decrease, since the field remains at $E_C$ but the distance between the source and $x_1$ has diminished. Consequently, the channel thickness at $x_1$ increases and more current is injected into the velccity limited region. The $I_{DS}$ versus $V_{DS}$ curve therefore has a small positive slope beyond the current saturation knee; this is particularly pronounced in microwave MESFETs which have short gate lengths.

In the velocity limited region of the channel the voltage still increases towards the drain, the depletion region widens, and, since the electron velocity is saturated, the narrowing of the channel is compensated by an increase of n(x) to maintain $I_{DS}$ constant, according to equatin (34). An electron accumulation layer therefore builds up between $x_1$ and a point $x_2$, where, owing to channel widening again beyond the gate edge towards the drain, the channel thickness has returned to its value at $x_1$. Beyond $x_2$ a small region of positive space charge (a region partially depleted of electrons) builds up to maintain constant $I_{DS}$. A dipole layer is thus formed in the channel and extends beyond the drain edge of the gate.

In GaAs, and other III-V compound semiconductors with a similar velocity-field characteristic, the build up of the dipole layer is aided in the electron accumulation region by the reduction in electron velocity due to their transfer from the central to satellite energy valleys in the conduction band and in the electron depletion region by the reverse process.

In microwave FETs with very short gate lengths the situation is made more complicated by the fact that the electrons do not reach equilibrium transport conditions in the high field region of the channel. The electrons can not be transferred quickly enough into the satellite valleys; they remain in the central valley, retaining their high mobility, and are accelerated by the electric field to a velocity in excess of $v_s$. This "velocity overshoot" shortens the electron transit time through the high field region, thus improving the transit time limited frequency, $f_T$.

The application of a negative voltage to the gate will make the gate depletion region wider, thereby increasing the resistance of the channel and consequently reducing $I_{DS}$. As $V_{DS}$ is increased the critical field $E_C$ is reached at a lower drain current than the $V_{GS} = 0$ case, owing to the larger channel resistance. For a further increase of $V_{DS}$ the current remains saturated.

In operation, the result is that voltages applied to the gate of the FET produce corresponding changes in the drain current, which, although at first ohmic, saturates at a certain value of drain voltage. The family of $I_{DS}$ versus $V_{DS}$ curves, for different values of $V_{GS}$, is shown in figure 7. The transconductance, $g_m$, is the rate of change of $I_{DS}$ with $V_{GS}$. A microwave signal applied to the gate (input) will produce useful corresponding changes in the drain current (output), provided that the resistive elements in the input part of the equivalent circuit (figure 6) are not very high compared with the reactance of the input capacitor $C_{sg}$; otherwise the input signal becomes severely attenuated.

## Electrical Characteristics

Because of the electron drift velocity saturation with electric field, the drift velocity may be approximated by

$$v_d = \frac{\mu_o E}{1 + \frac{\mu_o E}{v_s}} \qquad (35)$$

where $\mu_o$ is the low field electron mobility.

The gate voltage required to pinch-off the channel completely, when $V_{DS} = V_{GS} = 0$, is given by

$$V_p = \frac{qnd^2}{2\varepsilon\varepsilon_o} \qquad (36)$$

where $\varepsilon$ = dielectric constant of GaAs and $\varepsilon_o$ = permittivity of free space.

In Shockley's original theory (5) of the FET, which did not consider velocity saturation, the drain saturation current with $V_{GS} = 0$ is given by

$$I_{DSS} = \frac{V_p wqn\mu_o d}{3L} \qquad (37)$$

L is the gate length. In the velocity saturated theory, however,

$$I_{DSS}(sat) = \frac{3I_{DSS}}{2z} = \frac{wqnv_s d}{2} \qquad (38)$$

where

$$z = \frac{\mu_o V_p}{v_s L} \qquad (39)$$

is the ratio of the low-field velocity, extrapolated linearly to the field $V_p/L$, and the saturated drift velocity. The factor of 2 in the denominator of (38) allows for the wedge shape of the channel under the gate depletion layer.

In the non-saturated velocity theory the transconductance, with $V_{GS} = 0$, is given by

$$g_{mo} = \frac{wqn\mu_o d}{L} \qquad (40)$$

In the velocity saturated theory

$$g_{mo}(sat) = \frac{wqnv_s d}{2V_p} \qquad (41)$$

The drain conductance is given approximately by

$$g_d \simeq \frac{I_d}{E_o L} \qquad (42)$$

where $E_o$ is the field at the beginning of the restricted channel region.

The cut-off frequency of the lumped component R-C network of the channel and gate is

$$\omega_T = 2\pi f_T = \frac{g_m}{C_{sg}+C_{dg}} \qquad (43)$$

The maximum frequency of oscillation is

$$f_{max} = \frac{f_T}{2}\left[\frac{R_i+R_s+R_g}{R_{ds}} + \omega_T C_{dg} R_g\right]^{\frac{1}{2}} \qquad (44)$$

The maximum available gain is

$$MAG \simeq \left(\frac{f_T}{f}\right)^2 \left[\frac{4}{R_{ds}}\left(R_i+R_s+R_g+\frac{\omega_T L_s}{2}\right) + 2\omega_T C_{dg}(R_i+R_s+2R_g+\omega_T L_s)\right]^{-1} \qquad (45)$$

In equations (37), (40), (42) and (43) the values of $I_{DSS}$, $g_m$ and $g_d$ are reduced as the drain and gate voltages are increased, as well as being reduced by the velocity saturation conditions which depend on the value of z. For small gate and drain voltages and a large value of z, $f_{max}$ becomes similar to the cut-off frequency, $f_\tau$, determined by the limiting value of the electron transit time through the channel, $\tau = L/v_s$.

$$f_\tau = \frac{1}{2\pi\tau} = \frac{v_s}{2\pi L} \qquad (46)$$

Figure of Merit

It has been found empirically, for GaAs FETs with gate length of about one micrometre, that

$$f_{max} \simeq \frac{40}{L}\ (\text{GHz}) \qquad (47)$$

where L is in micrometres. $f_{max}$ is a figure of merit for the microwave FET and is comparable with the figure of merit for the microwave BJT (equation (33)); both depending on the minimum dimension available in the transistor "pattern" (either s or L). In the FET $f_{max}$ is directly related to the saturated drift velocity, $v_s$, and it is for this reason that GaAs is preferred to Si, since a best fit value for $v_s$ (GaAs) = $1.4 \times 10^7$ cm s$^{-1}$, whereas $v_s$(Si) = $8 \times 10^6$ cm s$^{-1}$.

Low Noise FETs

Intrinsic noise sources

Two sources of noise in the intrinsic FET are thermal noise in the channel and induced noise at the gate. The latter is a consequence of the former, since every disturbance of the channel potential induces a disturbance of the gate voltage. Because these two noise currents have the same origin they have a correlation factor, and all three depend upon the bias conditions. Under velocity saturation conditions the noise temperature of the electrons in the channel will increase, thereby increasing the two noise currents and their correlation coefficient.

In GaAs an additional source of noise is generated by intervalley scattering of electrons from the central to the satellite valleys in the conduction band.

Extrinsic noise sources. Three noise sources extrinsic to the active part of the transistor are associated with the gate metallisation resistance, $R_g$, the source resistance, $R_s$, and the gate bonding pad resistance. These resistances have the normal thermal noise current.

The Miller capacitance, $C_{dg}$, and the source bonding wire inductance have a feedback effect which reduces the noise figure slightly, but, as in the BJT, lossy parasitic elements associated with the header will increase the noise figure. Header parasitics have a relatively stronger influence in the MESFET because its intrinsic noise figure is so low.

Minimum noise figure An equation relating the minimum noise figure to material and geometrical parameters of the FET has been given by Hewitt et al (6).

$$F_{min} = 1+KfL^{5/6}\left(\frac{n}{d}\right)^{1/6}\left[\frac{3.3w^2\rho}{hL} + \frac{1.8L_{SG}}{nd_1} + \left(\frac{0.18R_c}{nd_2}\right)^{\frac{1}{2}}\right]^{\frac{1}{2}} \qquad (48)$$

where L = gate length (μm),

$L_{SG}$ = source to gate spacing (μm),

w = gate width (mm),

n = electron concentration in the channel ($10^{16}$ cm$^{-3}$)

d = channel thickness beneath gate (μm),

$d_1$ = channel thickness source to gate (μm),

$d_2$ = channel thickness beneath source (μm),

h = gate metallisation thickness (μm),

ρ = gate metallisation resistivity ($10^{-6}$ Ωcm),

$R_c$ = specific contact resistance ($10^{-6}$ Ωcm),

K = noise coefficient (0.033 to 0.036) for good epitaxial GaAs.

Another expression, relating $F_{min}$ to electrically measured parameters, has been given by Fukui (7).

$$F_{min} \simeq 10 \log [1+0.27fL\sqrt{g_m(R_g+R_s)}] \quad (49)$$

where $F_{min}$ is in dB, f in GHz, L in microns and the intrinsic value of $g_m$ is related to the measured value $g_m'$ by the relation

$$g_m \simeq \frac{g_m'}{1 - g_m'R_s} \quad (50)$$

Equations (48) and (49) illustrate the importance of having a low source and gate resistance in order to obtain a low noise figure.

## Power FETs

Power amplification with good gain, linearity and efficiency is possible with MESFETs. In a power FET the gate width is increased to give a larger drain current and measures are taken to increase the drain to source breakdown voltage to enable the device to operate at high $V_{DS}$, since the maximum output power is given approximately by

$$P_{max} \simeq \frac{I_{max}}{4} (V_{DS}-V_k) \quad (51)$$

where $V_k$ is the voltage at the knee of the $I_{DS}$ - $V_{DS}$ curves of figure 7.

Power FETs are not so liable to hot-spot formation as power BJTs, because $I_{max}$ decreases with an increase in temperature. Nevertheless, power FETs must be provided with a good heat sink and must be designed to avoid electromigration in their metallisation. The power output performance of GaAs FETs is shown in figure 5.

## ACKNOWLEDGEMENT

Acknowledgement is given to the GEC Hirst Research Centre for permission to publish.

## REFERENCES

1. Archer, J.A.: 1974, Solid State Electronics, 17, 387-393.

2. Beale, J.R.A. and Slatter, J.A.G.: 1976, Solid State Electronics, 19, 549-556.

3. Kirk, Jnr. C.T.: March 1962, IRE Transactions Electron Devices, ED9, 164-174.

4. Hawkins, R.J.: 1977, Solid State Electronics, 20, 191-196.

5. Shockley, W.: 1952, Proc. IRE, 40, 1365-1376.

6. Hewitt, B.S. et al, June 1976, Electronics Letters, 12, 309-310.

7. Fukui, H.: March 1979, Bell System Technical Journal, 58, 771-797.

## PROBLEMS

1. Calculate the total base time delay for a silicon n-p-n transistor with a uniformly doped base width of 0.2 μm. Assume electron diffusion coefficient = 8 $cm^2 s^{-1}$ and scattering limited velocity = 8 × $10^6$ cm $s^{-1}$.

2. A silicon n-p-n microwave transistor has $f_{max}$ = 10 GHz and $f_T$ = 8 GHz, calculate $f_{MAG}$. This transistor has a collector capacitance of 0.1 pF and an emitter lead inductance of 0.3 nH. It is operated at an emitter current of 1 mA and is designed so that $\alpha_o \simeq 1$ and $r_{es} \ll r_e$. (Boltzmann's constant k = 1.38 × $10^{-23}$ joules/$K^o$, junction temperature, T = 300°K, electronic charge, q = 1.6 × $10^{-19}$ coloumbs).

3. Calculate the minimum noise figure at 3.2 GHz for the transistor of question 2, if κ = 1.2.

4. What would be the effect on $f_T$ of increased emitter current density, assuming that the frequency response was limited by either

   (a) emitter resistance, or

   (b) base width.

5. Why do recent bipolar power transistor designs use multiple, isolated base areas?

A GaAs microwave FET has the following parameters:-

L = 1 μm, $L_{sg}$ = 1 μm, w = 0.3 mm,
n = $10^{17}$ $cm^{-3}$, d = 0.2 μm, $d_1$ = 0.4 μm,
$d_2$ = 0.4 μm, h = 0.7 μm, ρ = 3 × $10^{-6}$ Ωcm,
$R_c$ = 2 × $10^{-5}$ Ωcm, K = 0.035, $C_{sg}$ = 0.2 pF,
$C_{dg}$ = 0.02 pF, $C_{ds}$ = 0.08 pF, $R_{ds}$ = 400 Ω,
$R_i$ = 4Ω, $R_g$ = 2Ω, $R_d$ = 6Ω, $R_s$ = 3Ω, $g_m'$ = 20 mS,
$L_s$ = 0.1 nH, $v_s$ = 1.4 × $10^7$ cm $s^{-1}$,
$\mu_o$ = 4500 $cm^2V^{-1}s^{-1}$, q = 1.6 × $10^{-19}$ coulombs,
ε = 12.5, $\varepsilon_o$ = 8.85 × $10^{-14}$F $cm^{-1}$.

6. Calculate the drain saturation current for this FET from equations (37) and (38). Why is one answer much closer to the measured value of 70 mA than the other?

7. Calculate $f_T$ using $g_m$ from equations (41) and (50). Compare with the measured value of 15 GHz.

8. Calculate $f_{max}$ and compare it with the transit time frequency, $f_T$, and the figure of merit. Should the gate length be made smaller?

9. What is the maximum available gain of this FET at 7.5 GHz?

10. Compare $F_{min}$ at 10 GHz calculated from equations (48) and (49) with the measured value of 2.8 dB.

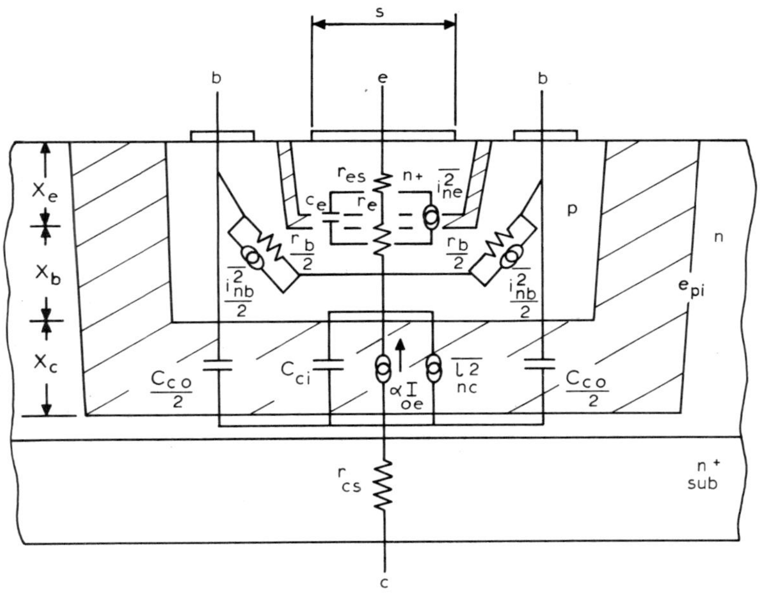


Figure 1

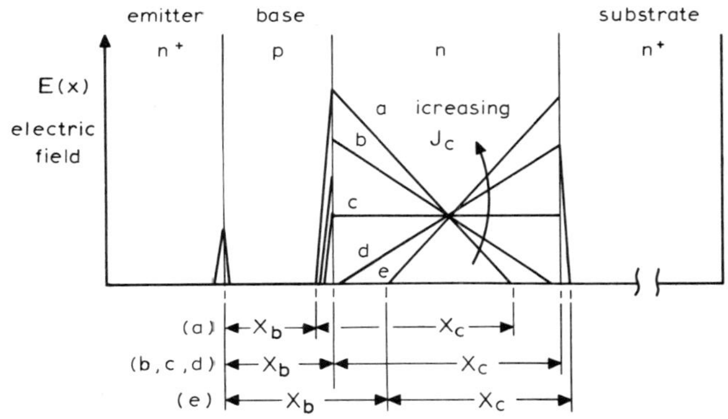


Figure 2

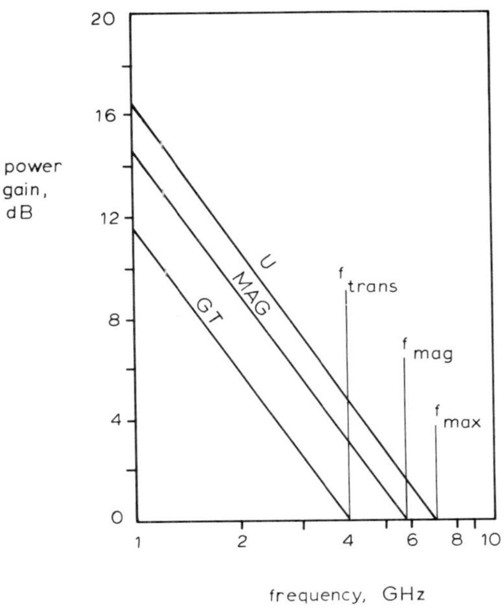


Figure 3

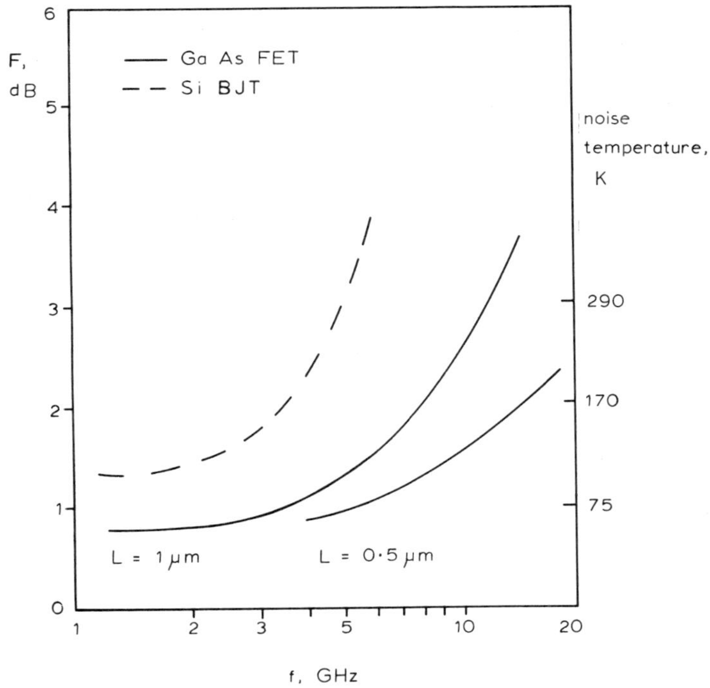


Figure 4

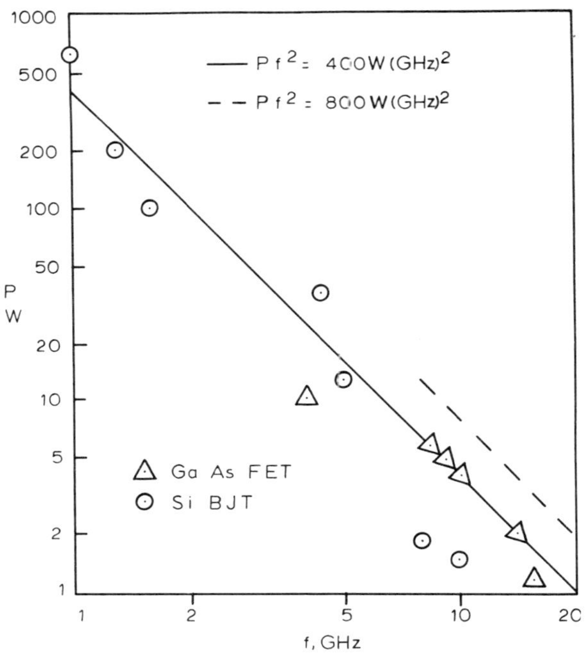


Figure 5

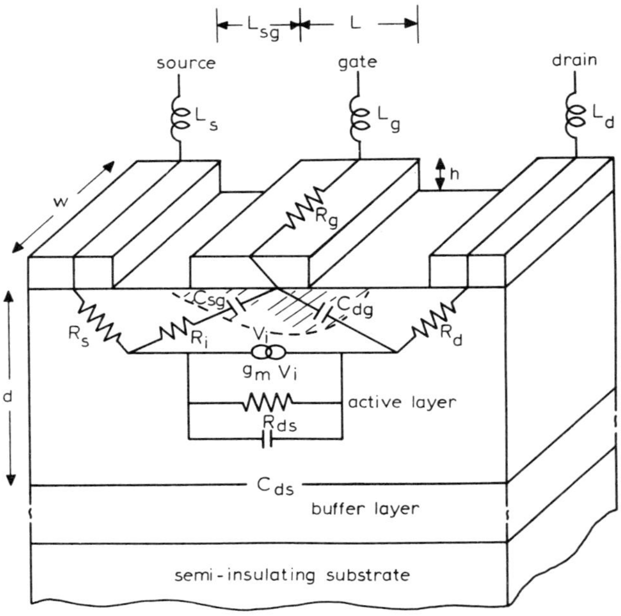


Figure 6

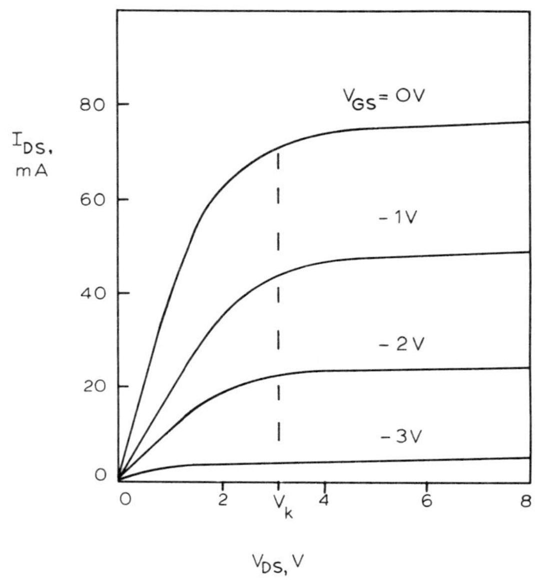


Figure 7

# CHAPTER 5

# Microwave circuits

Prof. J.O. Scanlan   Prof. J.D. Rhodes

# MICROWAVE CIRCUITS 1

J.O. Scanlan

UNIVERSITY COLLEGE DUBLIN

## INTRODUCTION

At microwave frequencies the well-tried techniques of lumped element circuit analysis and synthesis are often inappropriate and we must seek to develop methods more suited to such frequencies. At lower frequencies one may assume that all circuits are "lumped." This implies that the concept of distance does not enter into consideration and all circuit elements are at the same point in space, at least to an acceptably good degree of approximation. It is evident that this concept fails at microwave frequencies since one can for example readily conceive that to place two capacitors at the same point on a transmission line will yield different results compared to the case where they are placed at different points on the line. There is, of course, the question of when the "lumped" approximation ceases to be acceptable and this is mainly a matter of when the physical distances become significant compared to the wavelength at the frequency of interest. However, it is clear that once distance becomes significant some new approach to analysis and synthesis is needed.

The alternative viewpoint is to consider the solution of the field equations for any proposed structure and then to determine the circuit behaviour from such solutions. Although this is possible, in principle, it is not an attractive method in practice for a variety of reasons. Firstly, it is necessary to prescribe the precise physical form of the structure in advance in order to obtain the solution, whereas in practice many different structures may be capable of producing the same circuit behaviour. Secondly this approach makes it difficult to classify components on the basis of their circuit behaviour and finally, synthesis is virtually impossible using this approach.

A further illustration of the problems encountered is to consider the wide variety of (passive) microwave components which may be used in realizing a circuit. They include: lengths of waveguide propagating various modes; lengths of transmission line; stubs of various kinds; iris and post discontinuities; striplines; microstrips, etc. To be able to analyse and synthesise circuits containing all or some of these in various combinations is indeed a formidable task.

In the light of all the above considerations it is important to have a clear understanding of what can and cannot reasonably be accomplished in the present state of development. In the case of familiar lumped circuits the state of knowledge is such that any arbitrary connection of linear resistors, inductors and capacitors, driven in any arbitrary fashion can be completely analysed with great generality and efficiency. From a synthesis viewpoint any allowed circuit function can be realized in a variety of ways which permit the designer to choose optimum values and topologies to meet various requirements. This is the target to be aimed at in the case of microwave circuits. The first difficulty is that whereas in the lumped element case there are only three different elements, R,L,C, the number of possible elements in the microwave case is extremely large if all the possibilities previously mentioned are to be incorporated. The second difficulty is that of characterisation of the circuit elements. In the lumped case each element is described by a simple relationship between the current and voltage at its terminals whereas such a description of the various possible microwave elements would be far from simple, and indeed might not exist.

In view of the above discussion it is evident that we are unlikely to be able to formulate a circuit theory which will embrace all possible microwave elements and at the same time provide the full generality of analysis and synthesis methods which are available for the lumped element case. However, by having regard to the range of situations likely to be encountered in practice, and to the tasks we may reasonably be asked to perform it is possible to develop a circuit theory which is sufficiently powerful and flexible to yield most of the necessary results.

## FREQUENCY BEHAVIOUR

In view of the great variety of microwave circuit components mentioned in the introduction it is necessary to attempt some form of classification. One possibility is to classify in terms of the particular medium in which the component is to be realised. However, from the circuit viewpoint this is not very desirable since the same circuit behaviour might be achievable in a number of different media and we are primarily interested in the circuit properties. The classification to be adopted may also be influenced by the frequency range over which operation is of interest. For example if operation is effectively "single frequency" i.e. one is not interested in the performance as a function of frequency but only in the achievement of a result at one frequency, then the familiar simple Smith Chart methods are usually sufficient and the difficulties mentioned previously generally do not arise. We shall not be concerned with that situation in this lecture. If the range of frequencies of interest is sufficiently small then many useful approximations can be made, and these will be alluded to in later sections.

However, it is not prudent to assume such approximations in laying the foundation for a generally useful circuit theory.

In approaching the problem of defining a circuit theory the first task is to define the set of circuit elements to which the theory is to apply. Each element is defined by relationships between the currents and voltages and its terminals. For the purposes of this lecture we are only concerned with elements which are linear, passive and time-invariant and the most convenient descriptions relate to steady-state sinusoidal excitation and response or to the Laplace transform (complex frequency) description. Thus, for example, in the lumped case we have

| | | | |
|---|---|---|---|
| Resistor | $V = RI$ | | |
| Inductor | $V = j\omega LI$ | $V = pLI$ | (a) |
| Capacitor | $I = j\omega CV$ | $I = pCV$ | |

As is well known these are sufficient to completely characterise any circuit.

In the microwave case we shall confine this lecture to elements which are either one-ports or two ports. This covers a very large class and avoids difficulties associated with multi-mode structures which complicate matters considerably. We shall also, initially, confine discussion to structures which propagate a single TEM mode, although we shall see later how to extend the ideas to structures propagating other modes. Thus the initial class of circuit elements is the class composed of various lengths of TEM mode transmission line having various characteristic impedances. The first task is, therefore, to obtain a description of the circuit element in terms of the currents and voltages at its terminals. This is accomplished, in principle, by solving Maxwell's equations for the element and then converting this into a circuit description. However, in the case of a TEM line, shown schematically in Fig.1., it is well known that

$$\begin{pmatrix} V_1 \\ I_1 \end{pmatrix} = \begin{pmatrix} \cos\beta l & jZ_o\sin\beta l \\ jY_o\sin\beta l & \cos\beta l \end{pmatrix} \begin{pmatrix} V_2 \\ I_2 \end{pmatrix} \qquad \text{(b)}$$

Where sinusoidal excitation is assumed, l is the length of the line, β is the propagation constant and the line is assumed to be lossless. $Z_o$ is the characteristic impedance Now

$$\beta = 2\pi/\lambda$$

where λ is the wavelength and

$$f\lambda = c$$

where c is the velocity of propagation on the line.

Thus

$$\beta = \omega/c$$

and

$$\beta l = \omega l/c = \tau\omega$$

where $\tau = l/c$ is the one-way delay of the line.

Equation (b) can then be written as

$$\begin{pmatrix} V_1 \\ I_1 \end{pmatrix} = \begin{pmatrix} \cos\tau\omega & jZ_o\sin\tau\omega \\ jY_o\sin\tau\omega & \cos\tau\omega \end{pmatrix} \begin{pmatrix} V_2 \\ I_2 \end{pmatrix} \qquad \text{(c)}$$

to show explicitly the frequency dependence.

Rewriting equation (c) as

$$\begin{pmatrix} V_1 \\ I_1 \end{pmatrix} = \begin{pmatrix} \cosh(j\omega\tau) & Z_o\sinh(j\omega\tau) \\ Y_o\sinh(j\omega\tau) & \cosh(j\omega\tau \end{pmatrix} \begin{pmatrix} V_2 \\ I_2 \end{pmatrix}$$

and replacing $j\omega$ by p, we obtain the complex frequency description

$$\begin{pmatrix} V_1 \\ I_1 \end{pmatrix} = \begin{pmatrix} \text{Cosh}p\tau & Z_o\sinh p\tau \\ Y_o\sinh p\tau & \cosh p\tau \end{pmatrix} \begin{pmatrix} V_2 \\ I_2 \end{pmatrix} \qquad \text{(d)}$$

The length of line is then regarded as a two-port network having an ABCD matrix as given by equations (c) or (d), and shown in Fig.2.

Short-circuited and open-circuited stubs are formed by terminating one end of the line appropriately to give:

Short-Circuited Stub

$$V = jZ_o\tan\tau\omega I \qquad V = Z_o\tanh p\tau I \qquad \text{(e)}$$

Open-Circuited Stub

$$I = jY_o\tan\tau\omega I \qquad I = Y_o\tanh p\tau I \qquad \text{(f)}$$

which are illustrated in Fig.3.

We can, therefore, consider a circuit to be composed of arbitrary interconnections of such stubs and lengths of line and we can analyse its behaviour, as a function of frequency, by any of the usual methods of circuit analysis.

It is important to mention here that the circuit elements defined are idealizations of their eventual realizations. Firstly, they have been assumed to be lossless and to propagate only the TEM mode. In practice some loss is always present, and at sufficiently high frequencies other modes may exist. Secondly, when such elements are joined together to form a circuit, discontinuities will be created leading to excitation of higher order modes (generally non-propagating) which result in local energy storage at the junctions. Such effects may need to be taken into account in determining the behaviour of the circuit as realized, compared to the behaviour calculated by considering the connection of the ideal circuit elements. However, in formulating a circuit theory we shall always assume ideal circuit elements and then treat the various effects mentioned above as perturbations on the ideal behaviour.

TRANSMISSION LINE CIRCUITS - ANALYSIS

We consider arbitrary interconnections of stubs and lengths of transmission line described by equations (d), (e) and (f) together with lumped resistors. Analysis in such situations is relatively straightforward using the well-developed methods of conventional circuit analysis. The stubs are impedances given by

$$Z = Z_o \tanh \tau p$$

and

$$Z = Z_o / \tanh \tau p$$

for the short-circuited and open-circuited stubs respectively and the length of transmission line is a two-port whose ABCD matrix is

$$\frac{1}{\{1 - (\tanh \tau p)^2\}^{\frac{1}{2}}} \begin{pmatrix} 1 & Z_o \tanh \tau p \\ Y_o \tanh \tau p & 1 \end{pmatrix}$$

from which any other two-port description such as Z or Y matrix may easily be derived.

Now, although the frequency variable, P, is the same for every element and the characteristic impedance, $Z_o$, plays the same role as the various L and C in lumped element circuits the parameter $\tau$ may, in general, be different for each transmission line in the circuit. If each element is realized in the same medium then $\tau$ is directly proportional to the length of the line. This does not present too much difficulty in analysis although it has a crucial effect in synthesis as we shall see later. A simple example will serve to illustrate the methods of analysis.

Consider the circuit shown in Fig.4, and let us compute the ratio of the power in the load resistor $R_2$ to the available power from the source. The circuit consists of a cascade of a shunt short-circuited stub of length $l_1$ and characteristic impedance $Z_1$ followed by a length of transmission line described by $Z_2$, $l_2$ and a series connected open-circuited stub. described by $Z_3$, $l_3$. We can thus compute the ABCD matrix of the network (excluding the terminations) by multiplying the ABCD matrices of the various elements. To provide a convenient notation we put

$$\lambda = \tanh \tau p$$

and for lines with different lengths $l_i$ and corresponding one-way delays $\tau_i$ we put

$$\lambda = \tanh \tau_i p$$

Thus for the circuit under analysis the overall ABCD matrix is given by

$$\gamma_2 \begin{pmatrix} 1 & 0 \\ Y_1/\lambda_1 & 1 \end{pmatrix} \begin{pmatrix} 1 & Z_2\lambda_2 \\ Y_2\lambda_2 & 1 \end{pmatrix} \begin{pmatrix} 1 & Z_3/\lambda_3 \\ 0 & 1 \end{pmatrix}$$

where

$$\gamma_2 = 1/(1 - \lambda_2^2)^{\frac{1}{2}}$$

This gives

$$A = \gamma_2$$

$$B = \gamma_2(Z_2\lambda_2 + Z_3/\lambda_3)$$

$$C = \gamma_2(Y_2\lambda_2 + Y_1/\lambda_1)$$

$$D = \gamma_2(1 + Y_1Z_2\lambda_2/\lambda_1 + Y_2Z_3\lambda_2/\lambda_3 + Y_1Z_3/\lambda_1\lambda_3)$$

and the required power ratio is

$$\frac{4R_1R_2}{|AR_2 + DR_1|^2 + |B + CR_1R_2|^2}$$

with $p = j\omega$

We can thus readily calculate the response of the given circuit as a function of frequency. In applications where the circuit behaviour over a range of frequencies is of interest the lengths of the various transmission lines are usually related in some way and are most often found to be equal. If we consider the previous example with equal lengths then the various $\lambda$ are identical and we can write

$$A = \lambda$$

$$B = \lambda(Z_2\lambda + Z_3/\lambda)$$

$$C = \lambda(Y_2\lambda + Y_1/\lambda)$$

$$D = \lambda(1 + Y_1Z_2 + Y_2Z_3 + Y_1Z_3/\lambda^2)$$

and if we put

$$\lambda = j\Omega$$

when $p = j\omega$, then the power ratio is

$$\frac{4R_1R_2(1 + \Omega^2)}{\{R_1 + R_2 + R_1Y_1Z_2 + R_1Y_2Z_3 - R_1Y_1Z_3/\Omega^2\}^2 + \{(Z_2 + R_1R_2Y_2)\Omega - (Z_3 + R_1R_2Y_1)/\Omega\}^2}$$

We note that the frequency variable, $\Omega$, is now periodic ($\tan \tau\omega$), so that the circuit response is also periodic and, for example, the point $\Omega = 0$ corresponds to $\omega = n\pi/\tau$ so that in the above example there will be no power transmitted to the load at each of these frequencies. We can also see that the detailed behaviour of the circuit as a function of frequency is readily calculated from the given expression.

In general, therefore, the analysis of circuits consisting of lengths of transmission line may readily be accomplished by the usual methods of circuit analysis and indeed it is often a simple matter to modify existing computer programs to provide a capability for such transmission line circuits.

In the vast majority of problems of microwave circuit analysis the structure to be analysed is in cascade form or may be reduced to that form by various circuit transformations. In such cases the simplest method of analysis is to compute the ABCD matrix of each circuit element and then multiply these matrices to obtain the overall ABCD matrix. Any desired overall circuit property can then be calculated from this in an efficient manner. This method also has the advantage that if one wishes to consider the effect of variation of some particular element only that particular component of the ABCD matrix need by changed.

Suppose for example we consider the circuit of Fig.5(a) and wish to calculate its input impedance. The overall ABCD matrix is

$$\begin{pmatrix} A_1 & B_1 \\ C_1 & D_1 \end{pmatrix} \begin{pmatrix} 1 & 0 \\ Y_1\lambda_1 & 1 \end{pmatrix} \begin{pmatrix} A_2 & B_2 \\ C_2 & D_2 \end{pmatrix}$$

and the input impedance is

$$\frac{AR + B}{CR + D}$$

If we now wished to consider the effect of replacing the shunt opencircuited stub by a series shortcircuited stub at the same position, as shown in Fig.5(b), it is not necessary to analyse the complete network again. The ABCD matrices of the unchanged parts of the network remain the same and the overall ABCD matrix is

$$\begin{pmatrix} A_1 & B_1 \\ C_1 & D_1 \end{pmatrix} \begin{pmatrix} 1 & Z_2\lambda_2 \\ 0 & 1 \end{pmatrix} \begin{pmatrix} A_2 & B_2 \\ C_2 & D_2 \end{pmatrix}$$

from which the input impedance can again be computed, with minimum labour.

Another feature of this approach to analysis is the ability to incorporate mixtures of lumped elements and transmission lines. If in the example of Fig.4 the final series opencircuited stub is replaced by a lumped shunt capacitor to give the circuit of Fig.6 and we wish to calculate again the ratio of the power in the load to the available power from the source then we proceed, as before to calculate the overall ABCD matrix as

$$\gamma_2 \begin{pmatrix} 1 & 0 \\ Y_1/\lambda_1 & 1 \end{pmatrix} \begin{pmatrix} 1 & Z_2\lambda_2 \\ Y_2\lambda_2 & 1 \end{pmatrix} \begin{pmatrix} 1 & 0 \\ C_o p & 1 \end{pmatrix}$$

to give

$$A = \gamma_2(1 + Z_2C_op\lambda_2)$$
$$B = \gamma_2Z_2\lambda_2$$
$$C = \gamma_2(C_op + Y_2\lambda_2 + Y_1/\lambda_1 + C_oY_1Z_2p\lambda_2/\lambda_1)$$
$$D = \gamma_2(1 + Y_1Z_2\lambda_2/\lambda_1)$$

from which the power ratio can be calculated using the same formula as before.

In summary, therefore, the method of analysis most frequently needed is to compute the ABCD matrix of the circuit in question (excluding its terminations) and then to use the various standard formulae expressed in terms of the overall ABCD matrix to calculate the required performance parameter. This is a method of great flexibility and is easily automated on a programmable calculator or computer.

## TRANSMISSION LINE CIRCUITS - SYNTHESIS

Very powerful methods of circuit synthesis now exist for lumped element circuits and we shall now consider the extent to which they may be used in the case of transmission line circuits. The classical lumped element methods allow specification of some performance parameter of the circuit; its translation into a form which is guaranteed to be realizable; and the generation of one or more circuits which meet the specified performance criteria. In the course of developing such methods very deep insight has been gained into general aspects of circuit behaviour which have proved to be of great significance in a number of respects. A particularly striking example is that fruitless searches for non-existant solutions may often be avoided.

The key feature of lumped element circuits which has enabled this remarkable degree of progress to take place is that the number of circuit elements is only three (together with the ideal transformer) and only one frequency variable, p, is involved. If we are to adapt these results to the microwave case in any simple manner then we again must have at least a single frequency variable. Thus the various $\lambda_i$ encountered in our previous examples must be replaced by a single variable $\lambda$. In the case where all elements are to be realised in the same medium this means that all line lengths must be the same. Such a network is called a commensurate transmission line network. Only one frequency variable, $\lambda$, will appear. The number of circuit elements is, however, larger than in the lumped element case. We have:

| | |
|---|---|
| Short-circuited stub | $Z = Z_o\lambda$ |
| Open-circuited stub | $Y = Y_o\lambda$ |
| Resistor | $Z = R$ |

and the length of transmission line (or **"unit element"**) with ABCD matrix

$$\frac{1}{(1 - \lambda^2)^{\frac{1}{2}}} \begin{pmatrix} 1 & Z_o\lambda \\ Y_o\lambda & 1 \end{pmatrix}$$

We then ask whether a synthesis procedure similar to the lumped element case can be evolved for this set of circuit elements. The first deservation is that if we compare the two types of stub with inductors and capacitors where the inductor has $Z = Lp$ and the capacitor has $Y = Cp$ then if, in some LC circuit we replace every L by a s/c stub having $Z_o = L$ and every C by an o/c stub having $Y_o = C$, the effect will merely be to replace p by λ (or ω by tanτω). For example the circuit of Fig.7 (a) has the ratio of the power in the load to the available power from the source (the transducer power gain) given by

$$G = \frac{1}{1 + \omega^6}$$

Now if we carry out the substitution just mentioned to obtain the circuit of Fig.7(b) then the new transducer power gain is

$$\hat{G} = \frac{1}{1 + \Omega^6}$$

where $\Omega = \tan\tau\omega$

The responses of the two circuits are shown in Fig.8 as a function of ω. This shows clearly the periodic nature of the response of all commensurate transmission line networks. Note, however, that if $\hat{G}$ is plotted against Ω, the two are the same.

The above result would suggest that to solve the synthesis problem for microwave networks one might start with the solution for an appropriate lumped network and then carry out the substitution given above. However, a major problem is that in lumped element networks all connections are at the same point in space while it is clearly impossible, for example, in Fig.7(b) to physically connect the various stubs at a single point in space. Various methods are known for avoiding this problem while preserving the same performance but it is found that the resulting circuit is generally not very efficient and so the approach, although conceptually simple is rarely used in this context. In some microwave applications, particularly in narrowband waveguide circuits, the synthesis may be based on a particular lumped element realization which is then transformed into a form suitable for the required circuit and application. In general, the main deficiency of the approach is that it makes no use of the unit element as a two-port. It is, therefore, better to attempt to construct a synthesis based on non-restricted interconnections of stubs and unit elements. The full development of such a synthesis cannot be undertaken within the confines of this lecture but an outline of the main considerations will be given.

Since the basic elements consist of one-ports whose impedance (or admittance) is proportional to λ and two ports whose ABCD matrix has entries proportional to λ (apart from the factor $(1-\lambda^2)^{\frac{1}{2}}$) it is clear that any circuit function of an arbitrary interconnection of such elements will be a rational of λ (apart from the factor $(1-\lambda^2)^{\frac{1}{2}}$) since analysis will consist only of rational operations. Thus the class of realizable functions is similar to that encountered in lumped element networks with p replaced by λ. Again, since the left-half λ plane corresponds to the left half p-plane the stability conditions are also similar and so we can state that the conditions for realizability of an impedance function for a commensurate microwave network are:

(1) $Z(\lambda)$ a rational polynomial function

(2) $Z(\lambda)$ real for λ real

(3) $\mathrm{Re}Z(\lambda) > 0$ for $\mathrm{Re}\lambda > 0$

Now, indeed, as has been alluded to $Z(\lambda)$ can then be realized by any of the classical methods of lumped element synthesis but this does not make any use of the unit element. The key to the use of the unit element lies in Richard's Theorem which states that if $Z(\lambda)$ obeys the previous conditions then

$$Z_1(\lambda) = Z(1)\,\frac{Z(\lambda) - \lambda Z(1)}{Z(1) - \lambda Z(\lambda)}$$

also obeys the conditions and is not of higher degree than $Z(\lambda)$. The situation depicted in Fig.9 obeys the above relationship so that we can remove a unit element from any impedance $Z(\lambda)$ to leave a remainder $Z_1(\lambda)$ which is still realizable. It is clear that $Z_1(\lambda)$ has a factor $(1-\lambda)$ in numerator and denominator which cancels to ensure that the degree of $Z_1(\lambda)$ is not higher than that of $Z(\lambda)$. If, in addition, $Z_1(\lambda)$ has a factor $(1+\lambda)$ then the degree of $Z_1(\lambda)$ is lower. This occurs if $Z(\lambda) + Z(-\lambda)$ has a factor $(1-\lambda^2)$. Now if $Z(\lambda)$ is of degree n and if $Z(\lambda) + Z(-\lambda)$ has a factor $(1-\lambda^2)^n$ then repeated application of the above procedure will produce a remainder which is a constant so that $Z(\lambda)$ may be realized as shown in Fig.10. This is a cascade of unit elements and is practically very useful.

The required synthesis may, in general, require both stubs and unit elements in the cascade and consists in alternately removing the required stubs and unit elements.

The choice of the impedance to be realized may depend on many factors e.g. providing given values at specified frequencies; tuning some resonance over a wide bandwidth; matching one impedance to another; providing a desired filtering function, etc. Once the choice is made a suitable function $Z(\lambda)$ is constructed and then realized as outlined above.

## CONCLUSIONS

This lecture has given an outline of the main methods of analysis and sunthesis of microwave circuits which operate over a band of frequencies. The circuit approach has been emphasised and wherever possible related to the well-known ideas of lumped element circuits.

## EXAMPLES

a1. Calculate the input impedance of the circuit shown in Fig.11 where the lengths of the lines are equal and

(a) $Z_1 = 2 \quad Z_2 = 3$

(b) $Z_1 = Z_2 = 2$

(c) $Z_1 = 2 \quad Z_2 = 1$

Repeat for the case where $l_2 = l_1$

Comment on the results in each case.

a2. For the circuit of Fig.11 calculate the input impedance when (a) a short-circuited shunt stub is inserted between the cascaded transmission lines, with all lengths equal and the stub characteristic impedance, $Z_3 = 1$. (b) a series open-circuited stub of similar characteristic impedance is inserted.

a3. The RC combination to the right of Fig.12 represents the input impedance of a semiconductor device. It is desired to resonate this at a frequency $f_o$ and to transfer maximum power from the generator into the device. Consider the choice of the impedances $Z_1$ and $Z_2$ and the lengths $\ell_1$ and $\ell_2$ in order to accomplish this over as wide a bandwidth as possible.

a4. The circuit of Fig.7 has a transducer power gain as given in the text. Find another circuit where the stubs are separated by commensurate unit elements having the same transducer power gain.

a5. Show that the transducer power gain of a cascade of unit elements terminated in equal resistors is given by

$$G = 1/1 + P_n^2(\sin\omega\tau)$$

where $P_n$ is an $n^{th}$ degree polynomial.

Figure 1

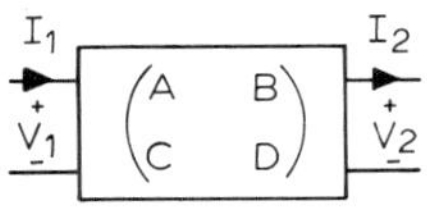


Figure 2

Figure 3

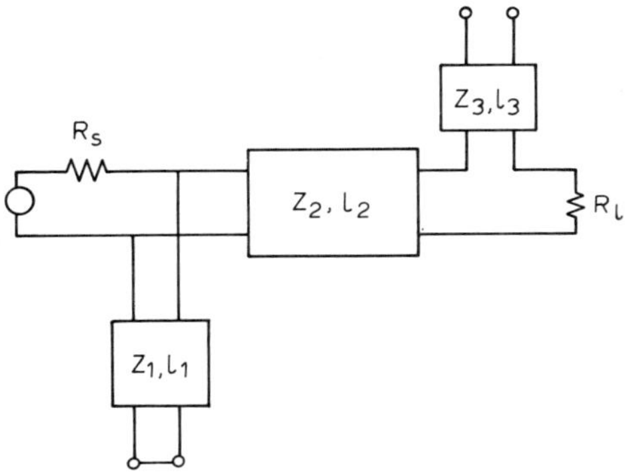


Figure 4

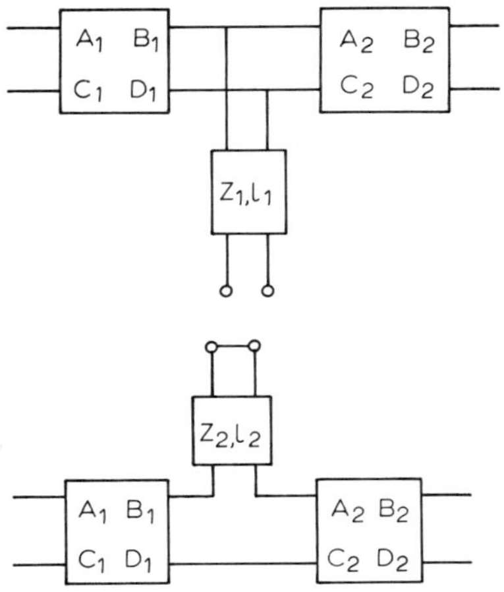


Figure 5

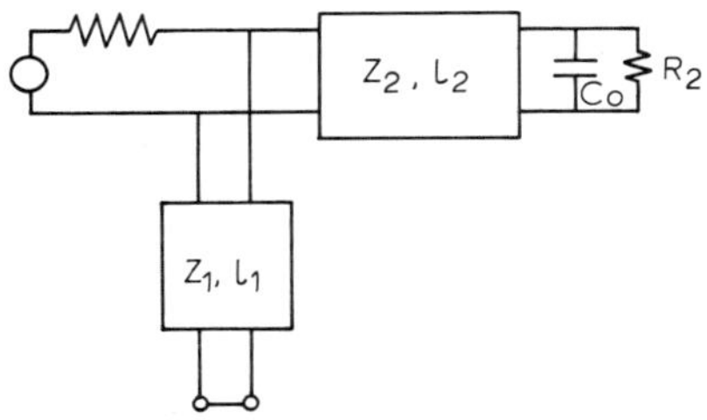


Figure 6

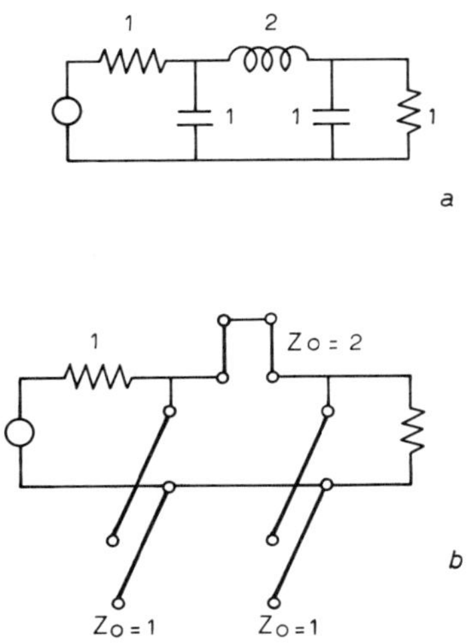


Figure 7

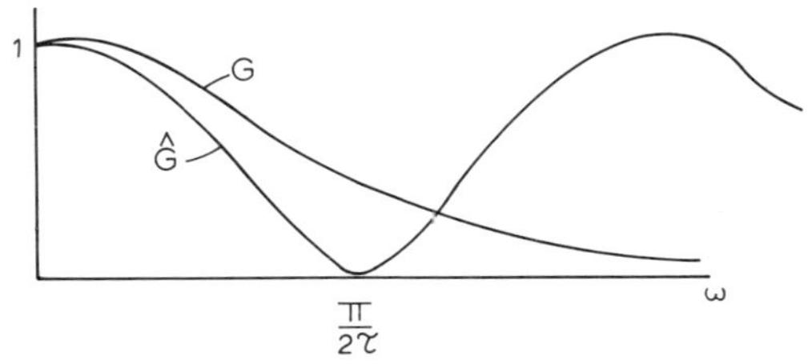


Figure 8

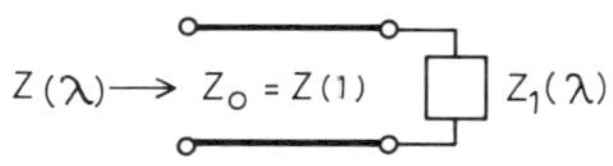


Figure 9

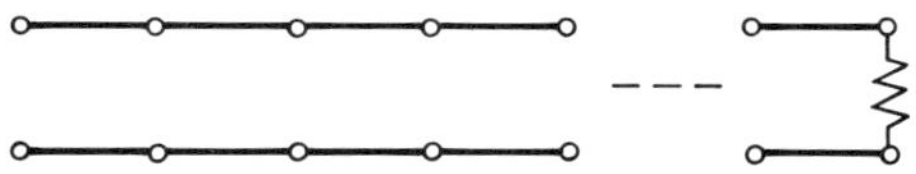

Figure 10

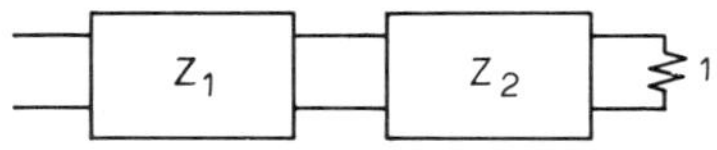


Figure 11

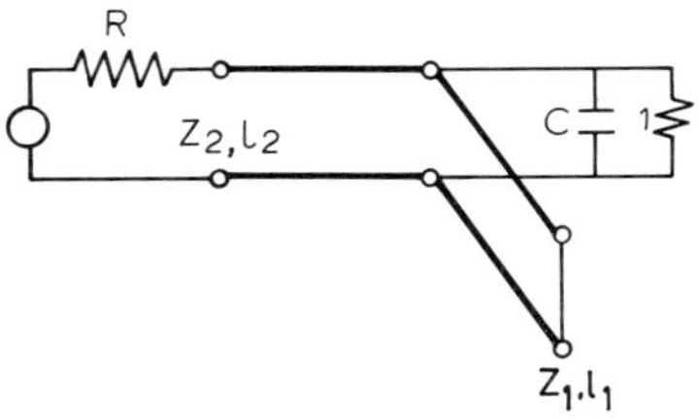


Figure 12

# MICROWAVE CIRCUIT REALIZATION

J D Rhodes

Dept of Electrical and Electronic Engineering, Leeds University

## INTRODUCTION

The realization of microwave circuit elements is not an exact science and many factors must be considered before the optimum solution can be found. One of the main factors is to have the ability to design a microwave circuit to perform a particular task rather than to be able to analyse a particular microwave circuit. There is a vast amount of information on particular discontinuities and junctions in waveguide coaxial and printed circuit microwave devices. This information is normally obtained from a numerical solution to Maxwell's equations and results in equivalent circuits containing several elements. From an analysis viewpoint this may be acceptable, but from a design viewpoint these equivalent circuits are generally of little value, due to their complexity. First , there are too many interdependent elements, and second for certain applications the approximations used in the numerical solution to the equations have unacceptably high errors.

Other points of importance are that if one imposes restrictions to simplify the theoretical aspects of the design then one frequently implies an unnecessary complex physical structure. Also, one would like to have designs which require the minimum amount of electrical tuning and are capable of being produced in quantity at a relatively low cost.

The basic elements which occur in most of the theoretical design work on microwave circuits are short and open circuited stubs, lengths of transmission line or unit elements, and possibly lumped capacitors and inductors. The distributed elements normally have to be the same electrical length in order to formulate a design procedure. If one attempts to realize these circuits directly whether in waveguide, coax or printed circuit form, then one has to attempt to take account of the discontinuity effects between every element and in certain cases the fact that two elements have to be connected at the same physical point. In some cases, this type of design approach can succeed but one must have the ability to analyse the real complex structure to see if the final realization will meet the design specification.

In addition to the basic distributed elements one may also introduce idealized circuit elements. The most important are the transformer, the impedance inverter and the equivalent reference planes for a particular discontinuity. The philosophy in this approach is to investigate any type of discontinuity which must be present in the final realization and approximate in such a manner that over broadbands only these three types of component (including reference planes) are required in the equivalent circuit. This approach is adopted here.

Initially, certain types of waveguide components will be considered followed by a particular important class of Suspended Substrate Stripline devices. These results will be applied to the design of broadband waveguide filters, and broadband Suspended Substrate Stripline lowpass and highpass filters.

## WAVEGUIDE STRUCTURE

### 1. Waveguide Discontinuities

In analysing waveguide structures one normally chooses a particular rectangular or circular or ridge guide, and the appropriate mode of propagation. The mode is defined by $\lambda_g$ the guide wavelength and normalizing all elements with respect to terminating impedances equivalent to an infinite uniform guide we may define a section of guide by a transfer matrix at real frequencies of the form

$$\begin{bmatrix} \cos\theta & j\sin\theta \\ j\sin\theta & \cos\theta \end{bmatrix} \qquad \text{.................... (1)}$$

where

$$\theta = \frac{\pi\lambda_{go}}{\lambda_g} \qquad \text{.............................. (2)}$$

and $\lambda_{go}$ is proportional to the length of the piece of uniform guide. In this particular case we have chosen $\lambda_{go}$ to be the guide wavelength when the section is one half of a wavelength long.

To form networks using this uniform waveguide one can connect pieces of waveguide in series or parallel, or introduce discontinuity into the guide. In the case of T junctions one may always obtain an equivalent circuit in terms of references planes and ideal transformers applicable over fairly broad bandwidths. For discontinuities, series effects are extremely difficult to introduce and hence we shall concentrate upon shunt discontinuities.

For a shunt inductive discontinuity we may simply insert any number of posts across a cross-section in contact with both the upper and lower sides of the waveguide for a dominant $TE_{01}$ mode in rectangular waveguide or TE11 mode in circular waveguide. The rectangular waveguide case is illustrated in Fig 1. Other configurations are irises formed by holes in plates across the guide. If the reference planes are chosen correctly, then this section is a shunt inductor in the guide wavelength described by the transfer matrix

$$\begin{bmatrix} 1 & 0 \\ \frac{-jB\lambda_g}{\lambda_{go}} & 0 \end{bmatrix} \qquad \text{.........................(3)}$$

where B is the inductive susceptance at $\lambda_g = \lambda_{go}$. The reference planes are not along the centre of the posts but are only shifted slightly to one side and are within the diameter of the posts as shown in Fig13 Thus the number of posts and their diameters determine the parameter B and the location of the reference planes. For any design, we shall be given B and then we have to determine the post structure and the reference plane locations. This problem will be considered later.

Other parts of the waveguide structure will consist

of lengths of transmission line and it is very difficult to design with a mixed lumped and distributed structure. Thus, we modify the inductive iris section by symmetrically embedding it in a uniform piece of waveguide of electrical length

$$\psi = \frac{\psi_o \lambda_{go}}{\lambda_g} \qquad \ldots\ldots\ldots\ldots\ldots\ldots\ldots\ldots(4)$$

as illustrated in Fig 14

The overall transfer matrix for this new section is

$$\begin{bmatrix} \cos\frac{\psi}{2} & j\sin\frac{\psi}{2} \\ j\sin\frac{\psi}{2} & \cos\frac{\psi}{2} \end{bmatrix} \begin{bmatrix} 1 & 0 \\ -jB\frac{\lambda_g}{\lambda_{go}} & 1 \end{bmatrix} \begin{bmatrix} \cos\frac{\psi}{2} & j\sin\frac{\psi}{2} \\ j\sin\frac{\psi}{2} & \frac{\psi}{2} \end{bmatrix}$$

$$= \begin{bmatrix} \cos\psi + \frac{B}{2}\frac{\lambda_g}{\lambda_{go}}\sin\psi & j\left(\sin + \frac{B\lambda_g}{\lambda_{go}}\frac{\sin^2\psi}{2}\right) \\ j\left(\sin\psi - \frac{B\lambda_g}{\lambda_{go}}\frac{\cos^2\psi}{2}\right) & \cos\psi + \frac{B}{2}\frac{\lambda_g}{\lambda_{go}}\sin\psi \end{bmatrix} \quad ..(5)$$

We now approximate this to an ideal impedance inverter in cascade with a 1:-1 transformer around $\lambda_g = \lambda_{go}$ defined by a transfer matrix.

$$\begin{bmatrix} 0 & -j\frac{\lambda_{go}}{K\lambda_g} \\ -jK\frac{\lambda_g}{\lambda_{go}} & 0 \end{bmatrix} \qquad \ldots\ldots\ldots\ldots\ldots\ldots(6)$$

where K is its characteristic admittance at $\lambda_g = \lambda_{go}$.

Comparing the 'A parameter entries of eqns 5 and 6 at $\lambda_g = \lambda_{go}$ yields

$$\cos\psi_o + \frac{B}{2}\sin\psi_o = 0$$

i.e. $$\psi_o = -\tan^{-1}\frac{2}{B} \qquad \ldots\ldots\ldots\ldots\ldots\ldots(7)$$

and from the difference between the 'B' and 'C' parameters we have

$$B = K - \frac{1}{K} \qquad \ldots\ldots\ldots\ldots\ldots\ldots\ldots\ldots(8)$$

The first point to note is that B is positive and therefore K>1. Normally K>>1. From eqn (7) the electrical length of the associated line is negative and hence not realizable in isolation. However, we shall be connecting this element to the remaining parts of a network using lengths of transmission lines. Thus the general statement is that a shunt inductive iris may be represented by an impedance inverter where the reference planes are defined by eqn (7) and when viewed from one side, are effectively at the opposite side of the iris.

The question remains as to how good this approximation is over a broadband since at the moment it is only valid at $\lambda_g = \lambda_{go}$. To investigate this we look at the A parameter in eqn (5) and express it in the form

$$F(x) = \cos(\psi_o(1-x)) + \frac{B\sin(\psi_o(1-x))}{2(1-x)} \qquad \ldots\ldots\ldots(9)$$

where $x = 1 - \frac{\lambda_{go}}{\lambda_g}$

Expanding F(x) as a power series around x = 0 yields

$$\cos\psi_o = \psi_o x\sin\psi_o - \cot\psi_o\left[ + \sin\psi_o + x(\sin\psi_o - \psi_o\cos\psi_o)\right] + \varepsilon(x^2) = x\left(\frac{\psi_o}{\sin\psi_o} - \cos\psi_o\right) + \varepsilon(x^2) \quad \ldots..(10)$$

If K>>1 then B>>1 and $\psi_o$ is small. In eqn 10 for $\psi_o$ small the coefficient of the x term is of the order of $\psi_o^2$ and hence is very small. Thus we may say that the 'A' parameter of the transfer matrix (5) will closely approximate to zero over a very broad bandwidth.

To check whether or not the frequency dependence in eqn (6) is correct, we may subtract the difference between the 'B' and 'C' parameters in (5) and (6) to give

$$H(x) = \frac{K}{1-x} - \frac{(1-x)}{K} - \frac{B}{1-x}$$

$$= x\left(\frac{2}{K}\right) + \varepsilon(x^2) \qquad \ldots\ldots\ldots\ldots\ldots\ldots\ldots(11)$$

and for K>>1 the x term is small.

Thus we have shown that the impedance inverter is a broadband representation of a shunt inductive iris symmetrically embedded in a negative length of uniform transmission line. Thus in any design technique the only parameter which is required is $K_1$ the characteristic admittance of the inverter given in eqn (6) from which B and the reference planes may be determined. The physical dimensions may be obtained experimentally as will be shown later.

For a shunt capacitive iris the transfer matrix is,

$$\begin{bmatrix} 1 & 0 \\ jB\frac{\lambda_{go}}{\lambda_g} & 1 \end{bmatrix} \qquad \ldots\ldots\ldots\ldots\ldots\ldots\ldots(12)$$

and introducing lengths of line at either side produces an overall transfer matrix similar to (5) where the 'A' parameter is:-

$$A = \cos - \frac{B}{2}\frac{\lambda_{go}}{\lambda_g}\sin\psi \qquad \ldots\ldots\ldots\ldots\ldots\ldots(13)$$

and equating to an impedance inverter without the 1:-1 transformer gives

$$B = K - \frac{1}{K} \qquad \ldots\ldots\ldots\ldots\ldots\ldots\ldots\ldots(14)$$

and $$\psi_o = \tan^{-1}\frac{2}{B} \qquad \ldots\ldots\ldots\ldots\ldots\ldots\ldots(15)$$

This appears to be very similar to the shunt inductive case apart from the fact that the length of line is positive. However, expanding the 'A' parameter gives

$$F(x) = \cos(\psi_o(1-x)) - \frac{B}{2}(1-x)\sin(\psi_o(1-x))$$

$$= \cos\psi_o + \psi_o\sin\psi_o - \cot\psi_o(\sin\psi_o - x(\sin\psi_o + \psi_o\cos\psi_o) + \varepsilon(x^2)$$

$$= x\left(\frac{\psi_o}{\sin\psi_o} + \cos\psi_o\right) + \varepsilon(x^2) \qquad \ldots\ldots\ldots\ldots\ldots(16)$$

Thus for K>>1, B>>1 and $\psi_o$ is small and the coefficient of x in 16 is approximately 2. Thus, the 'A' parameter quickly moves away from zero when $\lambda_g$ deviates from $\lambda_{go}$. Hence, the inverter approximation is only a narrowband approximation and is of little value in a general broadband design.

For the case of series discontinuities, the impedance inverter analogy is good for a capacitive discontinuity but is a narrow bandwidth approximation for series inductive discontinuities. This is used in TEM mode devices.

One of the main applications of shunt inductive irises is in the design of waveguide filters and we shall now turn to this example.

<u>Design of Waveguide Bandpass filters of Arbitrary Bandwidth</u>

The simplest type of waveguide bandpass filter is one consisting of a uniform waveguide with shunt inductive irises synchronously spaced along the guide to produce direct coupled cavities. A three cavity filter is shown in Fig 15. The inductive irises are taken with a negative length of guide to approximate to an ideal impedance inverter over a broadbandwidth. Since the connecting elements are simply lengths of line, the entire network may be viewed as a cascade of lines separated by impedance inverters. Apart from the $\lambda_{go}/\lambda_g$ frequency dependence the inverters only change the impedance level in the network and hence the entire network is simply a cascade of transmission lines. In order to produce an analytical design procedure, the lengths of line must be assumed to be equal and hence the network is a cascade of unit elements as shown in Fig 16 valid around the finite frequency given by $\lambda_g = \lambda_{go}$.

For a cascade of u.e.'s an optimum equiripple bandpass response occurs around $\theta=\pi$ when

$$|S_{12}|^2 = \frac{1}{1+\varepsilon^2 T_n^{\,2}(\alpha\sin\theta)} \qquad \text{...............(17)}$$

where

$$\theta = \pi\frac{\lambda_{go}}{\lambda_g} \qquad \text{............................(18)}$$

and

$$T_n(x) = \cosh(n\cosh^{-1}x) \qquad \text{..............(19)}$$

is the Chebyshev polynomial of degree n.

To take account of the original frequency dependence of the impedance inverter, we have to modify the response to:-

$$|S_{12}|^2 = \frac{1}{1+\varepsilon^2 T_n^{\,2}\left(\alpha\frac{\lambda_g}{\lambda_{go}}\sin\left(\pi\frac{\lambda_{go}}{\lambda_g}\right)\right)} \qquad \text{......(20)}$$

To design the filter one is normally given the ripple level $\varepsilon$, the degree n, and the two passband edge frequencies yielding $\lambda_{g1}$ and $\lambda_{g2}$. From $\lambda_{g1}$, and $\lambda_{g2}$ one can determine the parameters $\alpha$ and $\lambda_{go}$ as follows from the eqns.

$$\alpha\frac{\lambda_{g1}}{\lambda_{go}}\sin\left(\pi\frac{\lambda_{go}}{\lambda_{g1}}\right) = 1$$

$$\text{..................(21)}$$

$$\alpha\frac{\lambda_{g2}}{\lambda_{go}}\sin\left(\pi\frac{\lambda_{go}}{\lambda_{g2}}\right) = -1$$

Adding the two eqns gives

$$\lambda_{g1}\sin\left(\pi\frac{\lambda_{go}}{\lambda_{g1}}\right) + \lambda_{g2}\sin\left(\pi\frac{\lambda_{go}}{\lambda_{g2}}\right) = 0 \qquad \text{.........(22)}$$

If $\lambda_{g1} \simeq \lambda_{g2} \simeq \lambda_{go}$ then we may approximate as

$$\lambda_{g1}\pi\left(1-\frac{\lambda_{go}}{\lambda_{g1}}\right) + \lambda_{g2}\pi\left(1-\frac{\lambda_{go}}{\lambda_{g2}}\right) = 0$$

giving

$$\lambda_{go} = \frac{\lambda_{g1}+\lambda_{g2}}{2} \qquad \text{........................(23)}$$

However, for broad bandwidths this will not be sufficiently accurate and a better solution is obtained by applying the Newton-Raphson technique, as follows. Let

$$F(\lambda_{go}) = \lambda_{g1}\sin\left(\frac{\lambda_{go}}{\lambda_{g1}}\right) + \lambda_{g2}\sin\left(\pi\frac{\lambda_{go}}{\lambda_{g2}}\right) \qquad \text{......(24)}$$

then differentiating w.r.t. $\lambda_{go}$ gives

$$F'(\lambda_{go}) = \pi\left[\cos\left(\pi\frac{\lambda_{go}}{\lambda_{g1}}\right) + \cos\left(\pi\frac{\lambda_{go}}{\lambda_{g2}}\right)\right] \qquad \text{.........(25)}$$

and the modified value of $\lambda_{go}$ will be

$$\lambda_{go} = \frac{\lambda_{g1}+\lambda_{g2}}{2} + \frac{1}{\pi}\left[\frac{\lambda_{g1}\cos\left(\frac{\pi}{2}\frac{\lambda_{g2}}{\lambda_{g1}}\right) + \lambda_{g2}\cos\left(\frac{\pi}{2}\frac{\lambda_{g1}}{\lambda_{g2}}\right)}{\sin\left(\frac{\pi}{2}\frac{\lambda_{g2}}{\lambda_{g1}}\right) + \sin\left(\frac{\pi}{2}\frac{\lambda_{g1}}{\lambda_{g2}}\right)}\right] \qquad \text{..............................(26)}$$

Then from eqns (21) one may determine $\alpha$ as

$$\alpha = \frac{\lambda_{g1}}{\lambda_{go}}\sin\left(\pi\frac{\lambda_{go}}{\lambda_{g1}}\right) \qquad \text{........................(27)}$$

For a cascade of u-e's with the transfer characteristic given by (17) one may synthesise the network to obtain the characteristic impedances of the unit elements. However, for up to moderate bandwidths one may use explicit formulas for the element values [1]. Modifying the network by introducing impedance inverters of characteristic impedances $K_{r,r+1}$ as shown in Fig 17, the explicit formulas for the impedance values are

$$Z_r = \frac{2\alpha\sin\left[\frac{(2r-1)\pi}{2n}\right]}{y} - \frac{1}{4y\alpha}\left[\frac{\left(y^2+\sin^2\left(\frac{r\pi}{n}\right)\right)}{\sin\left[\frac{(2r+1)\pi}{2n}\right]} + \frac{\left(y+\sin\left(\frac{(r-1)\pi}{n}\right)\right)}{\sin\left[\frac{(2r-3)\pi}{2n}\right]}\right]$$

$$r = 1 \rightarrow n \qquad \text{.........................(28)}$$

and

$$K'_{r,r+1} = \frac{\sqrt{y^2+\sin^2\left(\frac{r\pi}{n}\right)}}{y} \qquad r = 0 \rightarrow n \quad \text{.....(29)}$$

with $y = \sinh\left(\frac{1}{n}\sinh^{-1}\frac{1}{\varepsilon}\right)$

In the waveguide realisation, since the waveguide is uniform, we must scale the internal impedance level as shown in Fig 18 where

$$K_{r,\ r+1} = \frac{K'_{r,\ r+1}}{\sqrt{Z_r Z_r + 1}} \qquad r = 0 \rightarrow n \quad \ldots\ldots(30)$$

with

$$Z_0 = Z_{n+1} = 1$$

This enables the susceptance of the irises to be obtained and the negative length of guide into which they are embedded from which the distance between the irises can be calculated.

Summarising the desing eqns, we have:-

Normally, the waveguide size and the passband edge frequencies $f_1$ and $f_2$ with the allowable passband ripple level are given. The latter is frequently obtained from the minimum passband return loss defined by

$$L_R = 10\log(1 + \frac{1}{\varepsilon^2}) \quad \ldots\ldots\ldots\ldots\ldots\ldots\ldots(31)$$

For the particular waveguide $\lambda_{g1}$ and $\lambda_{g2}$ are calculated, and from eqns (26) and (27), $\lambda_{go}$ and $\alpha$ are obtained. From (20), the insertion loss is,

$$L = 10\log(1 + \varepsilon^2\, T_n^{\,2}(\alpha \frac{\lambda_g}{\lambda_{go}} \sin(\pi \frac{\lambda_{go}}{\lambda_g}))) \quad \ldots\ldots(32)$$

and hence the minimum value of n, the degree of the filter, is obtained from the most severe constraint on the stopband insertion loss level.

One may now calculate the $Z_r$ and $K'_{r,\ r+1}$ in eqns (28) and (29) which may then be used to calculate the susceptance of the irises

$$B_{r,\ r+1} = \frac{\sqrt{Z_r Z_r + 1}}{K'_{r,r+1}} - \frac{K'_{r,r+1}}{\sqrt{Z_r Z_r + 1}} \quad r = 0 \rightarrow n \ \ldots(32)$$

and the electrical distance between them

$$\psi_r = \pi - \tfrac{1}{2}\cot^{-1}(\frac{B_{r-1,r}}{2}) - \tfrac{1}{2}\cot^{-1}(\frac{B_{r,r+1}}{2})$$

$$r = 1 \rightarrow n \quad \ldots\ldots\ldots\ldots\ldots\ldots\ldots\ldots\ldots\ldots\ldots(33)$$

where the electrical distance $\pi$ corresponds to the physical distance $\frac{\lambda_{go}}{2}$

We still have the problem of converting these electrical parameters into physical sizes. For the shunt inductive irises, the simplest physical method of producing them is to use posts, preferably of the same diameter, in line across the waveguide. Typically, one may use 2, 3, 4, 5, or 6 posts, depending upon the magnitude of the susceptance and a symmetrical location across the guide will reduce the excitation of higher ordered modes at higher frequencies. The advantage of using equal diameter posts is that the entire filter may readily be drilled on a numerically controlled machine. Over a particular range of susceptances using a fixed number of po ts, the distance between the posts across the guide may be varied to realize any particular value of susceptance.

The experimental procedure for obtaining the graphical information relating electrical susceptance to physical dimensions may be obtained in several ways. One of the most accurate is to construct a single cavity filter using two identical irises and from the measured insertion loss response the electrical values may be obtained. In this manner, the irises can be constructed very accurately. Furthermore, the effective reference planes for the irises can be deduced.

The resonant frequency of the cavities have to be accurate and typically as accurate as 1 part in $10^4$ Since it is physically impossible to achieve this order of accuracy on the distance between the irises along the waveguide it is necessary to introduce capacitive tuning screws at the centre of the cavities on the broadwall. Thus the final waveguide filter will be as shown in Fig 19 for a cavity device.

## SUSPENDED SUBSTRATE STRIPLINE FILTERS

Printed circuit filters have the advantages of small volume and weight and inherent reproduceability. The suspended substrate stripline configuration is shown in Fig 20 where the dielectric is 0.005" RT/duroid, and meets most environmental conditions, and b>>0.005" By minimising the dielectric, the filter has good temperature properties and low loss performance.

The filters designed here are based on the nth degree generalised Chebyshev lowpass prototype (2) which has an optimum equiripple passband, with (n-1) transmission zeros at a single frequency, $\omega_o$, close to bandedge and a single zero at infinity (see Fig 21). Its selectivity is therefore close to the same degree optimum elliptic function filter, but is easier to realise physically. This is due to the fact that impedance variations in the prototype are of the order of 2:1 as compared with greater than 10:1 in a normal elliptic function design.

### Design of the lowpass filters

To transform the lumped lowpass prototype into a distributed lowpass prototype, Richards transformation must be used:

$$p \rightarrow t/\alpha_L \quad \ldots\ldots\ldots\ldots\ldots\ldots\ldots\ldots\ldots\ldots(34)$$

where p os the lumped complex frequency variable, $t = \tan(a_L p)$ and $a_L$ and $\alpha_L$ are constants. Under this transformation the admittance of each shunt resonant section of the distributed lowpass prototype is,

$$Y = \frac{C_r t/\alpha_L}{1 + \dfrac{t^2}{\alpha_L^2\,\omega_o^2}} \quad \ldots\ldots\ldots\ldots\ldots\ldots\ldots\ldots\ldots(35)$$

If we now make the condition:

$$\alpha_L = 1/\omega_o \quad \ldots\ldots\ldots\ldots\ldots\ldots\ldots\ldots\ldots\ldots\ldots(36)$$

then (35) can be written

$$Y = \frac{C_r \omega_o}{2} \tanh(2a_L p) \quad \ldots\ldots\ldots\ldots\ldots\ldots\ldots(37)$$

That is, each resonator of the distributed prototype can be realised exactly by a uniform impedance shunt open circuited stub, which is half of a wavelength long at the stopband centre frequency, $f_{cL}$.

Using (34) each series element of the distributed lowpass prototype is a short circuited stub which is a quarter of a wavelength long at $f_{cL}$. This frequency is typically 2.2 times the bandedge frequency, $f_{bL}$, of the filter so for at least half an octave stopband bandwidth, the series elements can be approximated by short lengths of high impedance transmission line. The final printed form of the lowpass filter is, therefore relatively simple and is shown in Fig 22.

Design of the physical dimensions

The design begins by using the tables given in Ref (2) to find suitable element values of the lowpass prototype which will meet the required specification. This determines $\omega_o$ and n; $f_{bL}$ will be specified. Note that for the micr8wave filter, the frequency closest to bandedge where the minimum stopband insertion loss is first attained is, using (34),

$$f_{mL} = \frac{\tan^{-1}(\alpha_L \omega_m)}{\tan^{-1} \alpha_L} \cdot f_{bL} \qquad \text{.....(38)}$$

where $\omega_m$ is the corresponding frequency for the lumped lowpass prototype.

The following choices are now made for the ground plane spacing and the width of the unit elements:

$$b = 0.07'' \qquad \text{.....(39)}$$

$$w_L = 0.025'' \qquad \text{.....(40)}$$

Choice means the effect of the duroid is negligible whilst maintaining reasonable dimensions for the resonators. Since the critical resonant sections are now effectively in an air medium, the filter will be relatively insensitive to temperature. Given this choice, (40) ensures negligible electromagnetic coupling between adjacent resonators.

The widths of the resonators needed to give the characteristic admittances in (37) are found using Getzinger's graphs (3):

$$w_{rL} = \frac{(b-t)}{4}\ 7.534 \left[\frac{C_r\omega_o}{2} - \frac{4C'_f}{\varepsilon_o}\right] \qquad \text{.....(41)}$$

Using (34), the stopband centre frequency is,

$$f_{cL} = \frac{(\pi/2)}{\tan^{-1}\alpha_L} \cdot f_{bL} \qquad \text{.....(42)}$$

and the lengths of the resonators are given by

$$\ell_{rL} = u/(2f_{cL}) \qquad (r = 2, 4, \ldots, (n-1)) \qquad \text{..(43)}$$

where u is the velocity of light in free space.

Each unit element is designed by making the series impedance of its pi-network approximately equal to the impedance of the corresponding series stub of the csitributed prototype, at the bandedge frequency:

$$j\, Z_L\, \ell_{rL} \frac{2\pi f_{bL}}{u} = \frac{j\, L_r}{\alpha_L} \tan(a_L 2\pi f_{bL}) \qquad \text{...... (44)}$$

where $Z_L$ is the characteristic impedance of the u.e., normalised to 1Ω terminations, and $\ell_{rL}$ is the length of the u.e.

$$\text{or } \ell_{rL} = \frac{L_r u}{Z_L 2\, f_{bL}} \qquad (r = 1, 3, \ldots, n) \qquad \text{....(45)}$$

Using Getzinger's graphs and (40),

$$Z_L = \frac{7.534}{4W_L/(b-t) + 2C'_f + 2C'_{fo}} \qquad \text{..... (46)}$$

By accounting for the discontinuities in the structure, the physical lengths, $\ell'_{rL}$ (r=1, 2, ..., n), can be found.

Fine tuning of the resonators is achieved through capacitive tuning screws in the main housing, enabling a low passband VSWR to be achieved near bandedge.

Result on Lowpass Filters

A ninth degree 8 GHz lowpass filter has been designed to have a minimum passband return loss of 20 dB and a minimum stopband insertion loss of 40 dB. Experimental results, which agree well with a circuit analysis, are shown in Figs 23 and 24. Maximum passband insertion loss is 0.5 dB, and the substrate is only .45" x .64".

Recent work on the generalised Chebyshev prototype having 3 transmission zeros at infinity and (n-3) zeros at $\omega_o$ (see Fig 25) has now led to microwave lowpass filters having the theoretical minimum stopband insertion loss maintained up to 2½ times the bandedge frequency, for the same minimum passband return loss. There is also a slight improvement in selectivity. The printed form of this lowpass is shown in Fig 26, and is designed using the methods discussed previously. The only difference is that the end resonators are a quarter of a wavelength long at the stopband centre frequency, with each resonator producing a transmission zero at this frequency, and hence broadening the stopband.

Design of the broadband highpass filters

To form a distributed highpass prototype the following frequency transformation is applied to the lumped lowpass generalised Chebyshev prototype:

$$p \to \alpha_H/\tanh(a_H p) \qquad \text{..... (47)}$$

Following the same method as for the lowpass filters, (37) also gives the admittance of the highpass shunt resonant sections if $a_L$ is replaced by $a_H$; this is true, provided

$$\alpha_H = \omega_o \qquad \text{..... (48)}$$

Therefore, the highpass resonators can be designed in the same way as the lowpass resonators except that their lengths are, using (47),

$$\ell_{rH} = \frac{(\pi/2)}{\tan^{-1}\alpha_H} \cdot f_{bH} \ (r=2, 4, \qquad \text{..... (49)}$$

where $f_{bH}$ is the bandedge frequency of the highpass filter. The suspended substrate stripline realisation of the distributed highpass filter is shown in Fig 27. From (47) the series elements of the distributed highpass prototype are open circuited stubs a quarter of a wavelength long at the passband centre frequency, $f_{cH}$, which are approximated in this design by inhomogeneous open circuited digital sections. These sections produce the significant amount of series capacitance required for a good match at the bandedge frequency, but they also resonate near bandcentre to produce a good match in that region.

Correct design equations for the inhomogeneous sections of a nth degree filter are:

$$\left.\begin{aligned} \text{Im } Z'_{in} &= \text{Im } Z_{in} \\ Z'_H &= \frac{L_s\, \alpha_H}{j \tan(a_H \omega)} \end{aligned}\right\} \begin{aligned} &\text{at bandedge} \qquad \text{.....(50)} \\ &\text{frequency} \qquad \text{.....(51)} \end{aligned}$$

$$\left.\begin{aligned} \text{Im } Z'_{in} &= 0 \\ Z'_H &= 0 \end{aligned}\right\} \text{ at } f_c < f_{cH} \qquad \begin{aligned} &\text{.....(52)} \\ &\text{.....(53)} \end{aligned}$$

where fc is the solution of

$$\frac{j C_r \omega_o}{2} \tan (4\pi a_H f) + 2Y'_H(f) = o \quad \text{..........} \quad (54)$$

where s = m-1, r = m for m even and s = m, r = m-1 for m odd and m = (n-1)/2 is the number of resonators in the filter. Note that for m = 2, and m = 3, $2Y'_H(f)$ in (54) and for m = 2, s = 3.

In this design, identical inhomogeneous sections are used except for the end ones. This simplifies the design and can be done because the middle series inductances of the generalised Chebyshev lowpass prototype are almost equal. $Z'_H$ and $Y'_H$ are the series impedance and shunt admittance, respectively, of the pi-network of the identical middle inhomogeneous sections. $Z_{in}$ represents the series impedance produced by the end series open circuited stubs and the source or load resistance in the distributed highpass prototype and $Z'_{in}$ represents the corresponding impedance produced by the end inhomogeneous sections in the printed circuit highpass filter. Equations (50)-(54) ensure firstly, that the series impedances produced by the inhomogeneous sections are equal to the corresponding series impedances in the distributed highpass prototype, at bandedge. Secondly, these impedances will resonate when the total shunt admittances produced by the inhomogeneous sections and the shunt resonators become zero or almost zero. These conditions will ensure a good match over more than an octave.

The ABCD parameters of open circuited digital sections have been found by G I Zysman and A K Johnson (4) and are used here to find $Z'_{in}$, $Z'_H$ and $Y'_H$. These parameters are expressed in terms of the even and odd mode characteristic impedances and phase velocities and the lengths of the coupled lines. Equations (50)-(54) can then be solved by numerical techniques for the unknown widths, $w_{rH}$, and lengths, $\ell_{rH}$, (r=1 and 3) of the inhomogeneous sections.

## Results on Highpass Filters

A ninth degree highpass filter having a theoretical minimum passband return loss and stopband insertion loss of 20 dB and 40 dB, respectively, has been designed and experimental results are shown in Figs 28 and 29. Maximum passband loss was 0.6 dB and the substrate is only .58" square.

## Further results on both types of filters

Temperature tests have shown that the 3 dB frequency of both the lowpass and highpass filters shift by only ± 0.3% for a 100°C change in temperature. Many devices have been made to the same specification and as expected a high level of population tracking is achieved for both amplitude and phase. By increasing the degree of the filters to thirteen, a stopband insertion loss of 60 dB apart from within 5% of the 3 dB frequency is achieved for the same minimum passband return loss; the device has less than 1 dB passband loss and is still very compact.

## ACKNOWLEDGEMENT

The author would like to thank Dr J E Dean, who performed the work on the Suspended Substrate Stripline Structure as part of the requirements for a PhD degree at Leeds University, the Ministry of Defence, and Ferranti Ltd for their support on this work.

## REFERENCES

1. Rhodes, J. D., 1976, 'Theory of Electrical Filters'

2. Rhodes, J D., and Alseyab, S.A., January 1980, The Generalised Chebyshev Lowpass Prototype', International Journal of Circuit Theory and Applications.

3. Getzinger, W.J., January 1962, 'Coupled Rectangular Bars between Parallel Plates', IRE Trans. on Microwave Theory and Techniques, Vol. PGMTT-10, pp 65-72.

4. Zysman, G.I., and Johnson, A.K., 'Coupled Transmission Line Networks in an Inhomogeneous Dielectric Medium', IEEE Trans. on Microwave Theory and Techniques, Vol. MTT-17, No. 10, pp 753-759 October 1969.

## TUTORIAL QUESTIONS

b1. Design a broadband waveguide filter using shunt inductive irises to meet the following specification:-

| | |
|---|---|
| Waveguide WG16(WR90) | Internal Dimensions .4"x.9" |
| Passband | 9.0 → 10.0 GHz |
| Passband Return Loss | >20 dB |
| Stopbands | 0→8.5, 10.7→12.4 GHz |
| Stopband Attenuation | >20 dB |

b2. Design a lowpass Suspended Substrate Stripline filter with a passband from 0→8 GHz from the following lowpass prototype Generalised Chebyshev filter which has a passband cut-off frequency at w = 1, return loss of 20 dB, and a minimum stopband level of 40 dB.

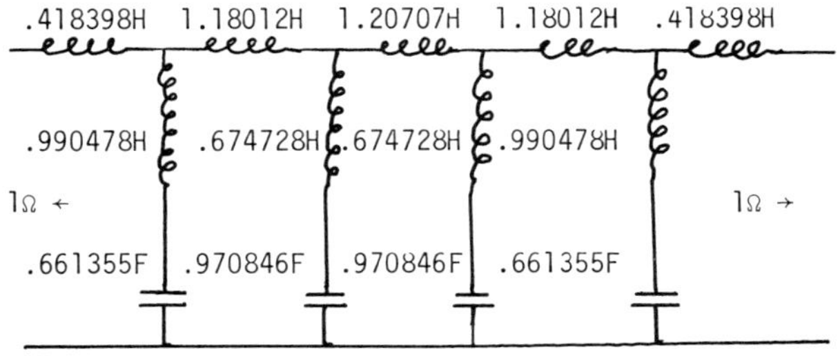


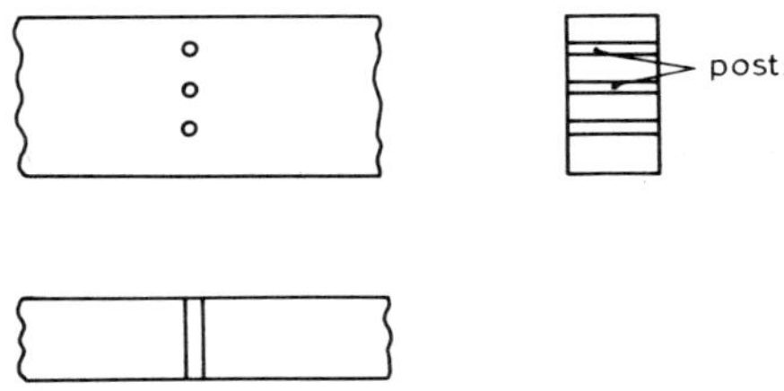


Figure 13 Waveguide shunt inductive iris

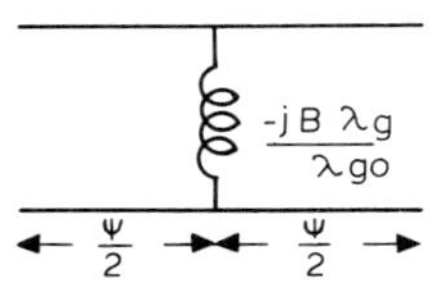


Figure 14 Shunt inductive iris in length of waveguide

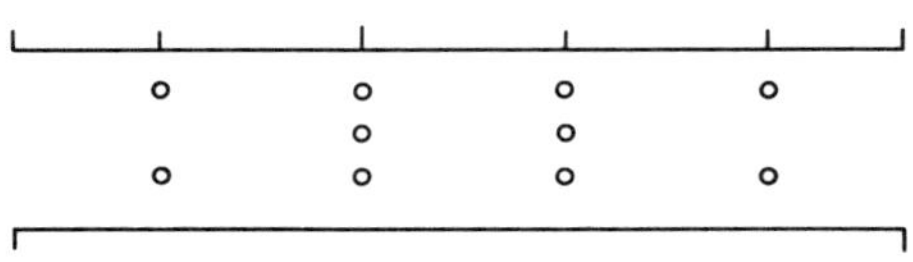

Figure 15 Three cavity waveguide filter

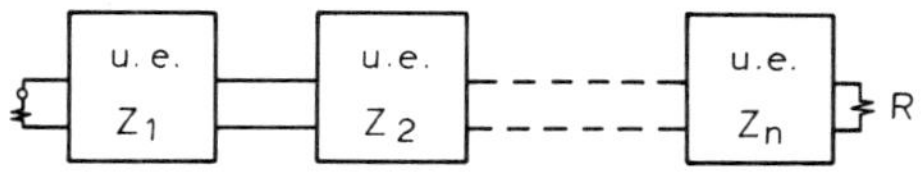


Figure 16 Cascade of unit elements

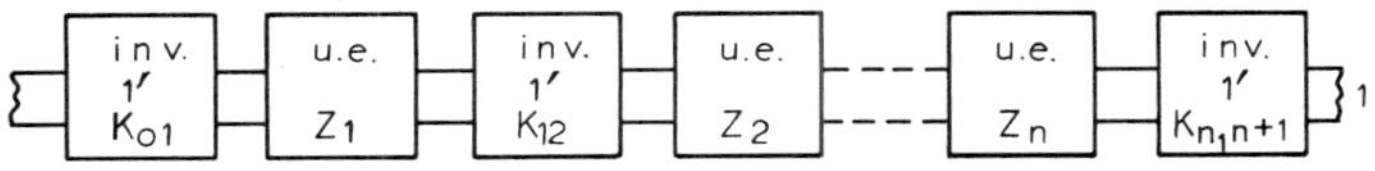


Figure 17 Cascade of unit elements and inverters

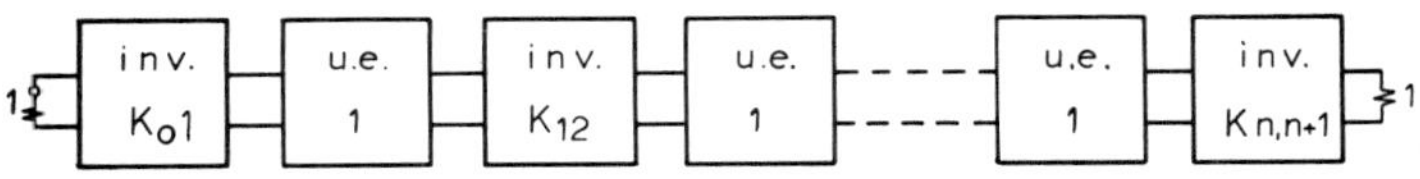


Figure 18 Final network after impedance scaling

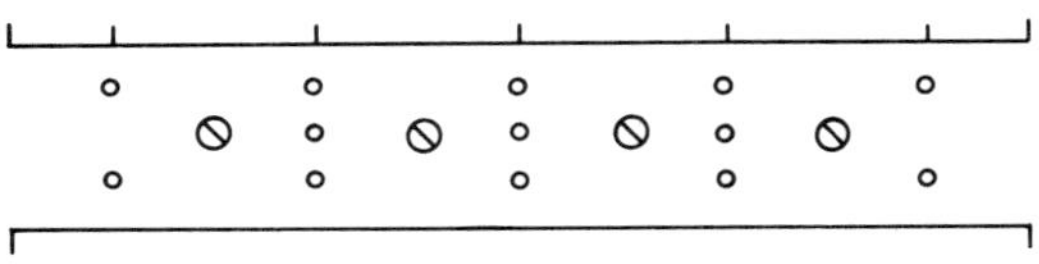

Figure 19 Four cavity filter with tuning screws

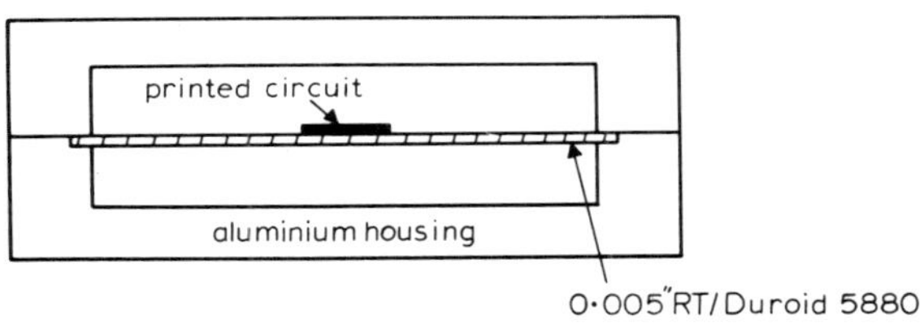


Figure 20 Basic suspended substrate stripline configuration

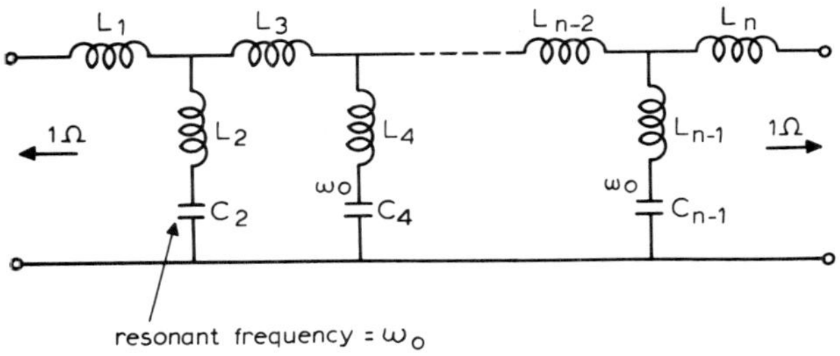


Figure 21 Generalised Chebyshev lumped lowpass prototype

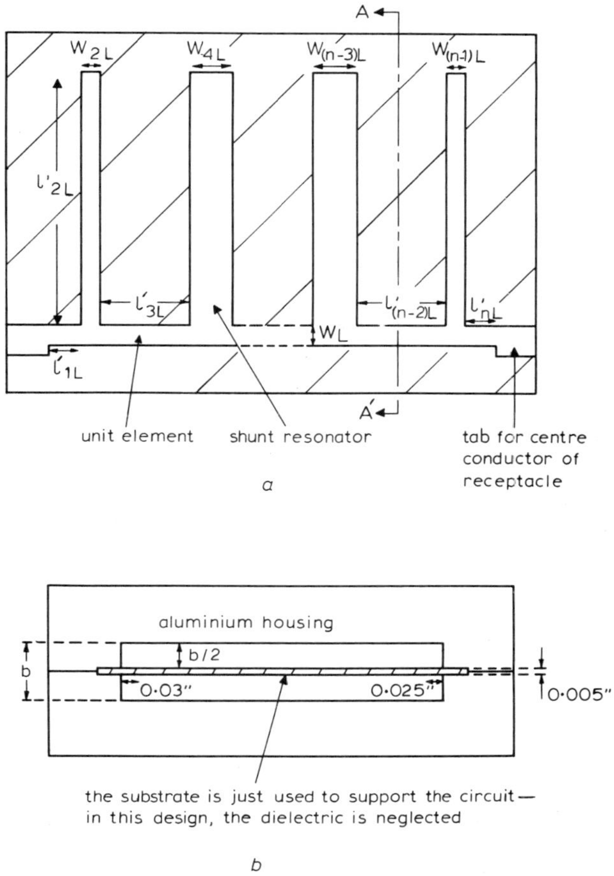


Figure 22 Suspended substrate stripline realisation of lowpass filter

(a) Printed circuit board
(b) Cross-section of filter box along AA'

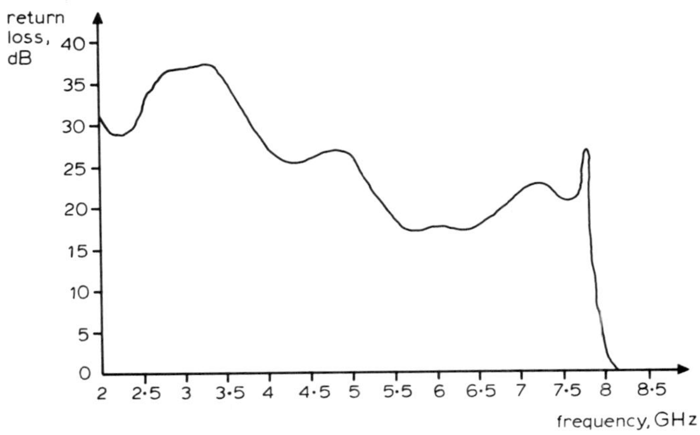


Figure 23 Measured return loss of 8GHz lowpass filter

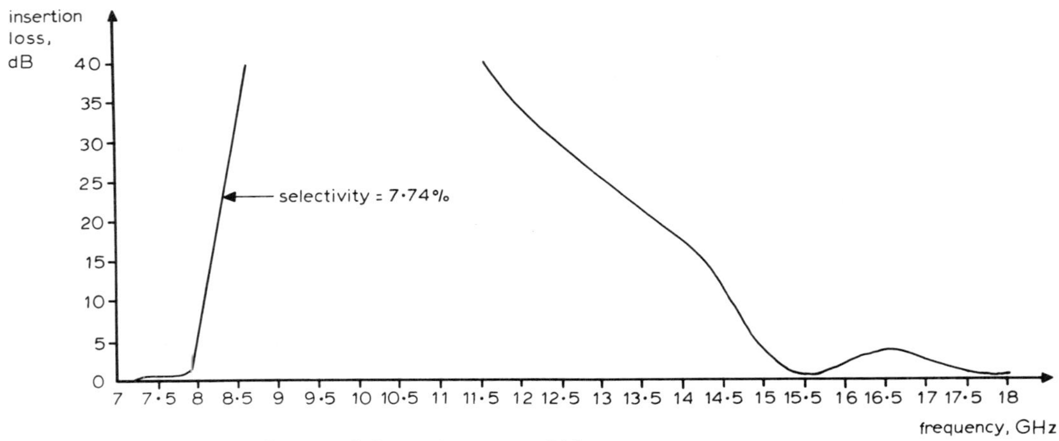


Figure 24 Measured insertion loss of 8GHz lowpass filter

Selectivity expresses 40 dB frequency in terms of the 3 dB frequency

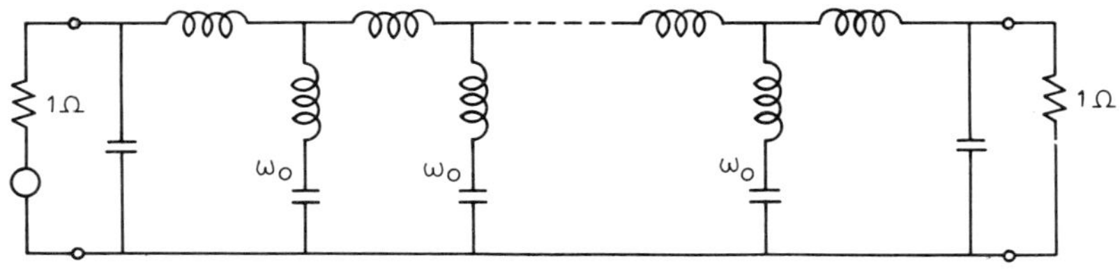


Figure 25 Generalised Chebyshev lumped lowpass prototype filter with n-3 transmission zeros of $W_o$ and 3 zeros at infinity

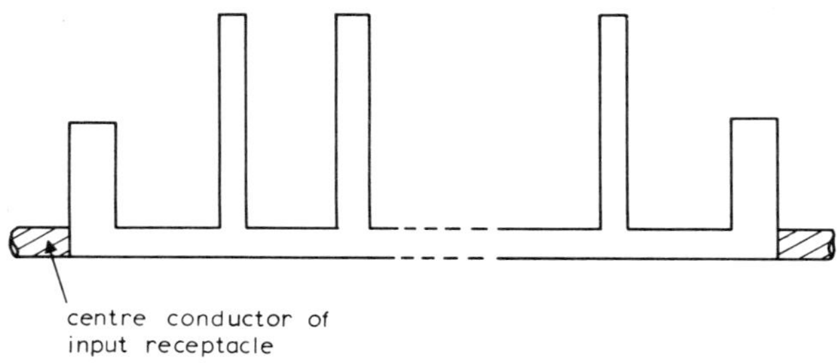


Figure 26 Printed circuit of the distributed lowpass filter based on the prototype of Figure 25

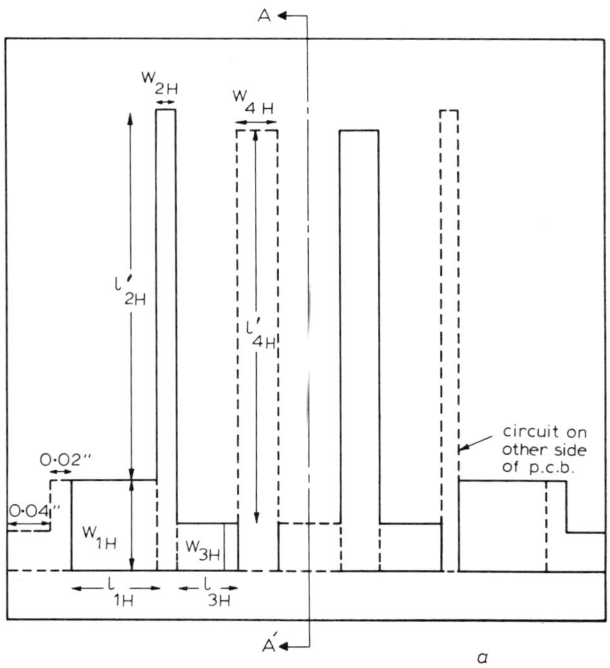


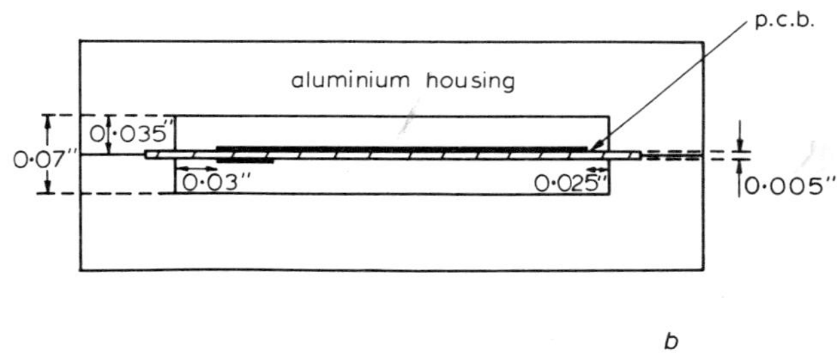


Figure 27 Suspended substrate stripline realisation of highpass filter

(a) Printed circuit board
(b) Cross-section of filter box along AA'

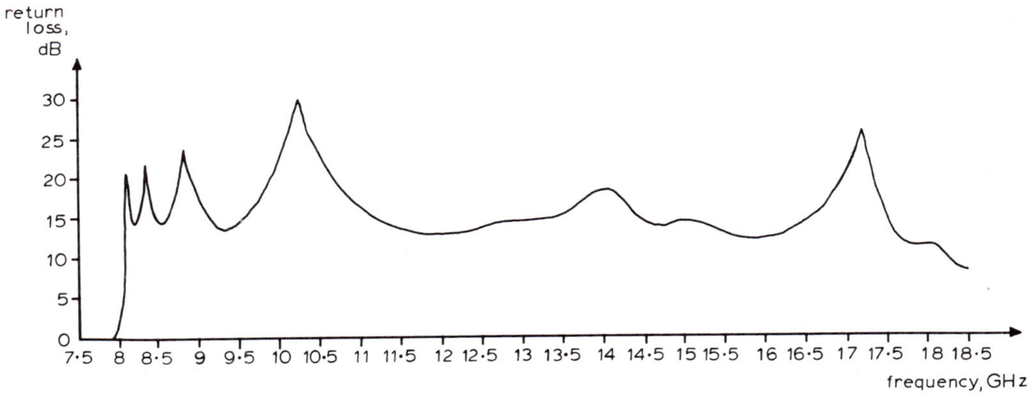


Figure 28 Measured return loss of 8 GHz highpass filter

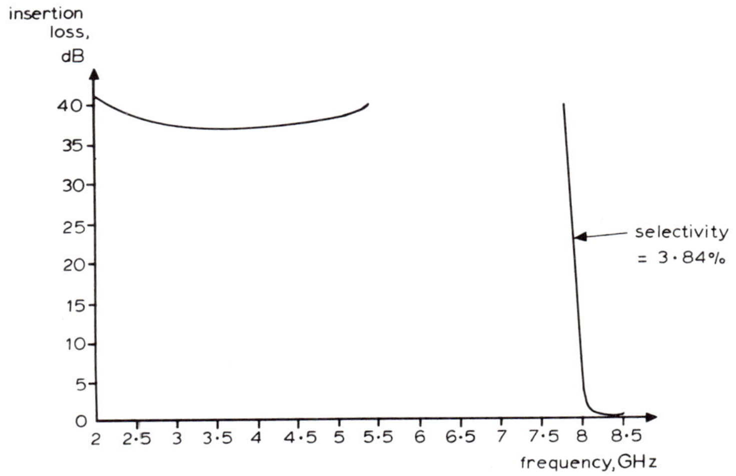


Figure 29 Measured insertion loss of 8GHz highpass filter

# CHAPTER 6

# Microwave circuit characterisation

R. D. Pollard

# CHAPTER 6 - MICROWAVE CIRCUIT CHARACTERISATION

Roger D Pollard

Department of Electrical and Electronic Engineering, The University of Leeds, Leeds LS2 9JT, U.K.

## INTRODUCTION

This chapter aims to provide an overview of some methods used in microwave measurement with emphasis on some more recent techniques made possible by advances in microelectronics. The essential feature of microwave measurements in general is the inability to directly read current and voltage so that, except for power measurements, some form of sampling is invariably required. Five independent types of microwave measurement can be identified:

1. Frequency - can be measured with greater accuracy than any other quantity.
2. Power - difficult to achieve high accuracy because of problems with good detectors and maintenance of impedance matched conditions. Spectrum analysis is also required if all the power in a signal is not at a single frequency.
3. Attenuation - simple transmission coefficient measurement usually achieved by substitution. Accurately characterised devices in this category (attenuators, couplers, etc.) are required for a wide range of other measurements.
4. Impedance - more difficult than at low frequencies because of the transmission line nature of microwave circuits. Usually measured in terms of VSWR or reflection coefficient at the end of a lossless, uniform transmission line of known characteristic impedance. Measurements of scattering parameters provide both impedance and attenuation information.
5. Noise - the ultimate performance of a system is determined by the noise presnt in the signal and that added during signal processing.

The material in this chapter will concentrate primarily on measurements on passive circuits and omit detailed examination of signal analysis techniques. Some introductory material on basic microwave theory is provided in view of its importance to later sections.

## BASIC MICROWAVE CONCEPTS

The lower frequency bound of the microwave spectrum is usually regarded as some point where low-frequency lumped-circuit techniques begin to give rise to problems; this is generally regarded as being around 1 GHz. The majority of the components employed have a size comparable with the wavelength at the frequency of operation and therefore the essential tool in the microwave regime is transmission line theory (1,2).

The usual techniques of analysis assume that a microwave system is comprised of sections of uniforms transmission line broken by "discontinuities". Voltage and current waves will be propagated along a transmission line but the fact that their values will be constantly changing and the difficulties in measuring their values leads to the adoption of other parameters to characterise the behaviour of transmission line circuits.

### Transmission Line Constants

A transmission line may usefully be considered in terms of the lumped-circuit equivalents of a short length. Each section of a two-wire line will possess self-inductance, capacitance, series resistance and shunt conductance with an equivalent circuit as shown in Figure 1. The transmission line is then regarded as being made of an infinite cascade of infinitely short sections of this type. These parameters allow the definition of two constants:

(a) The propagation constant

$$\gamma = \sqrt{(R + j\omega L)(G + j\omega C)} = \alpha + j\beta \qquad (1)$$

where $\alpha$ is the attenuation constant (nepers/unit length; 1 neper = 8.69 dB) and $\beta$ is the phase constant (radian/unit length).

In practice the attenuation in transmission lines is attributable to a number of physical factors including conductor (skin effect) loss, dielectric loss, hysteresis loss, mismatch loss and radiation.

(b) The characteristic impedance

$$Z_o = \sqrt{\frac{R + j\omega L}{G + j\omega C}} \text{ ohms} \qquad (2)$$

With R and G negligible, $Z_o = \sqrt{L/C}$
i.e. a real quantity independent of frequency.

### Waves and Discontinuities

A sinusoidal voltage applied to a transmission line will cause a wave to travel down the line and the voltage at any point is given by

$$E = Ve^{\gamma x} \qquad (3)$$

where V is the applied voltage $V'e^{j\omega t}$

There is, of course, an associated current wave. The line is terminated at some point with an impedance Z and if Z is not equal to the characteristic impedance of the line then it will be unable to absorb all the energy transmitted down the line and some part of that energy will be reflected. The situation is illustrated in Figure 2. The ratio of the incident and reflected voltages is the reflection coefficient

$$\Gamma = \frac{E_r}{E_i} \qquad (\rho = |\Gamma|) \qquad (4)$$

and is a vector quantity having magnitude and phase information. The magnitude of the reflection coefficient is determined by the degree of mismatch of the terminating impedance and its phase depends on the nature of the termination and the distance along the line from the point of measurement.

The effect of two waves travelling in opposite directions is to produce a standing wave pattern with alternate maxima and minima. In the situation described, the maximum and minimum values of the voltage on the line will be given by

$$E_{max} = |E_i| + |E_r| \quad (5)$$

and

$$E_{min} = |E_i| - |E_r| \quad (6)$$

and these are used to define the voltage standing wave ratio

$$VSWR = \sigma = \frac{E_{max}}{E_{min}} \quad (7)$$

Re-arranging, VSWR and reflection coefficient may be deduced in terms of each other, viz

$$\sigma = \frac{1 + \rho}{1 - \rho} \quad \text{and} \quad \rho = \frac{1 + \sigma}{1 - \sigma} \quad (8)$$

Note that a matched load has a reflection coefficient of zero and VSWR of 1 whilst a short or open circuit has a reflection coeffient magnitude of unity and infinite VSWR. A number of useful subsidiary expressions may be readily derived from these definitions.

(a) Power standing-wave ratio

$$PSWR = \sigma^2 \quad (9)$$

a consequence of constant characteristic impedance.

(b) It is often useful to express values in decibels where large values are encountered.

$$\begin{aligned} SWR\ (dB) &= 20 \log(\sigma) \\ Return\ loss\ (dB) &= -20 \log(\rho) \\ Mismatch\ loss\ (dB) &= -10 \log(1-\rho^2) \end{aligned}$$

(c) In terms of the terminating impedance

$$\Gamma = \frac{Z_\ell - Z_o}{Z_\ell + Z_o} \quad (10)$$

## The Smith Chart

The most common transmission line problem is that of finding the input impedance given the impedance at some point elsewhere on the line. This may be derived as

$$z_{in} = \frac{z_\ell + \tanh(\gamma l)}{1 + z_\ell \tanh(\gamma l)} \quad (11)$$

or, in a lossless system

$$z_{in} = \frac{z_\ell + j\tan(\theta)}{1 + jz_\ell \tan(\theta)} \quad (12)$$

where $\theta = 2\pi l/\lambda$ is the electrical length of the line and $z_\ell$ is the normalised impedance $Z_\ell/Z_o$.

With expressions of this complexity, there are obvious advantages in a graphical representation. The simplest form of impedance chart may be observed to be impractical because passive impedances can take any value between zero and infinity. The solution devised by Smith (3,4) was to plot impedance on a polar chart of reflection coefficient; all passive impedance values are then mapped within a circle of unit radius. The variation of reflection coefficient for movement along a lossless transmission line is simply a rotation of a vector of constant magnitude. The process is the mapping

$$\Gamma = \frac{z - 1}{z + 1} \quad (13)$$

and the familiar Smith Chart consists of sets of circles of constant resistance and reactance plotted in the reflection coefficient plane (Figure 3).

## Scattering Parameters

Two major problems are encountered in attempts to use the familiar sets of low frequency parameters (z,y,h) to characterise networks at microwave frequencies.

(a) The necessary measurements require pairs of terminals to be short- and/or open-circuited; a condition which may be difficult to realise.

(b) The voltages and currents on which the parameters depend vary along a transmission line.

The first step is the replacement of voltage and current by a new set of complex variables 'a' and 'b' which relate to the incident and reflected power waves and have magnitude that is constant with distance along a lossless line (5). At the ith port of a network, therefore

$$\text{incident wave} \quad a_i = \frac{V_i + I_i Z_o}{2\sqrt{Z_o}}$$

$$\text{reflected wave} \quad b_i = \frac{V_i - I_i Z_o}{2\sqrt{Z_o}} \quad (14)$$

Each wave is accompanied by power equal to the square of its magnitude and the net power flow is given by

$$P_i = |a_i|^2 - |b_i|^2 \quad (15)$$

The waves entering and leaving a two-port network are as indicated in Figure 4a. A convenient way of illustrating the signal paths is by means of a flow graph where each port is represented by two nodes and the possible paths to each other node which define the scattering parameters of the network (6) are indicated by lines as shown in Figure 4b. The equations describing the network are then readily constructed as

$$b_1 = a_1 s_{11} + a_2 s_{12} \quad (16)$$

$$b_2 = a_1 s_{21} + a_2 s_{22} \quad (17)$$

It may be seen that the four s-parameters can be measured by imposing certain, easily produced conditions on the network which does not conflict with requirements to maintain an impedance match at the ports. For example, terminating port 2 with its characteristic impedance ensures that no signal enters that port. Under these conditions

$$b_1 = a_1 s_{11} \quad \text{and} \quad b_2 = a_1 s_{21}$$

By reversing the network, the remaining two parameters are extracted. In summary

$$s_{11} = \left. \frac{b_1}{a_1} \right|_{a_2=0} \quad \text{input reflection coefficient} \tag{18}$$

$$s_{22} = \left. \frac{b_2}{a_2} \right|_{a_1=0} \quad \text{output reflection coefficient} \tag{19}$$

$$s_{21} = \left. \frac{b_2}{a_1} \right|_{a_2=0} \quad \text{forward transmission coefficient} \tag{20}$$

$$s_{12} = \left. \frac{b_1}{a_2} \right|_{a_1=0} \quad \text{reverse transmission coefficient} \tag{21}$$

The simple form of the parameters is noteworthy, they are either gains or reflection coefficients; the latter being readily plotted on a Smith Chart and converted to impedance. Scattering parameters are readily converted to other sets of parameters if required, the major benefit is the ability to make the measurements embedded in a transmission line environment. The use of flow graph representation provides a simple technique for carrying out network calculations in a physically meaningful way (7).

Measurement of Scattering Parameters

The full characterisation of a microwave network requires a set of parameters describing the complex reflection coefficient at each port and the forward and reverse transmission coefficients of each pair of ports. All of this data is normally required over a range of frequencies. In many instances considerable simplifications in the measurement procedures are possible where only limited information is required. Amongst the more common requirements are magnitude only (scalar) measurements - often the case for attenuators and filters; one-port measurements (input impedance) and single frequency measurements.

SCALAR (ATTENUATION AND GAIN) MEASUREMENTS

The measurement of attenuation as a simple ratio is almost always carried out by a substitution technique. The simplest is r.f. substitution as illustrated schematically in Figure 5. The technique employs a calibrated precision attenuator which is adjusted from an established level and directly read after it has replaced the unknown device. The major advantage of the technique is that the detector law need not be considered because the adjustments are made to maintain a constant output. The method does, however, require microwave components of considerably greater accuracy than the desired result.

A second method employs low frequency (d.c.) substitution (Figure 6) where the attenuator operates on the d.c. signal after detection and can therefore be more readily calibrated to a high degree of accuracy. The detector law also, however, requires to be established to a similar accuracy.

A major disadvantage of both methods described thus far is a limited dynamic range (approximately 50dB) inherent in the use of square-law detection. Employing a superheterodyne system (i.e. local oscillator and mixer) an i.f. substitution measurement can be implemented. This has considerably higher sensitivity, provides a large dynamic range and, using an accurately calibrated i.f. attenuator can provide high accuracy and linearity.

Swept Frequency Techniques. It is usually desired to make measurements over a range of frequencies using a swept-frequency source. The traditional technique employed an oscilloscope or pen recorder on which calibration curves were marked for a range of fixed settings of the standard attenuator with the frequency response of the device under test being derived by interpolation between the calibration lines. This type of measurement highlights the sources of error in these techniques. Whilst the main systematic errors can be reduced by techniques such as accurate calibration, drift compensation etc., the greatest source of error in a scalar measurement may, ironically, be due to vector errors. This type of error arises in a transmission line system where all the components are not accurately impedance matched and two or more reflections will beat together resulting in an error term which varies with frequency.

The availability of cheap digital storage has made it possible to routinely calibrate out many of the sources of error; a technique often described as storage normalisation. The response of the network without the device to be tested is measured, the results converted to digital form and stored. The device to be tested is then inserted and the stored calibration can be subtracted from the measured response and the corrected result displayed. The accuracy of the measurements depend largely on the detectors, attenuators and amplifiers (often logarithmic) associated with the storage and display unit.

Impedance Measurements

Slotted Line Techniques. In principle, the measurement of VSWR using a slotted line is straightforward - measure the ratio of the maximum to minimum voltage in the standing wave pattern using a moveable probe and detector inserted into a slot in a section of a (lossless) transmission line connected to the device being measured. In practice, for a VSWR in excess of about 10, the method leads to highly erronious results due to effects such as probe loading and the characteristics of the detector. Other sources of error in slotted line measurements include spurious signals from the source, mismatch of the slotted line section and losses.

A significant improvement in accuracy can be achieved with the double minimum technique. The electrical length between the points at which the probe output is double the minimum is measured and then (Figure 7)

$$\text{VSWR} = \frac{\lambda_L}{\pi \Delta x} \tag{22}$$

Alternatively a calibrated attenuator can be used between the signal source and slotted section to keep the detector output constant and the VSWR obtained from the difference in the attenuator settings.

To determine the phase of the reflection coefficient (and hence the impedance) it is necessary to convert the physical location of the minimum into electrical length (i.e. fractional part of a wavelength). The position can be readily referenced to the plane of the device under test by calibration with a short circuit. The information can be plotted directly onto the Smith Chart and thus determine the normalised impedance.

Swept Frequency Techniques. It is more usual for measurements to be needed at more than one frequency and swept frequency methods which permit a display of response as a function of frequency are preferable to laborious point-by-point measurements.

The swept slotted line technique employs a levelled sweep oscillator and movement of the probe in the slot over a least one-half wavelength to ensure that maximum and minimum are covered. The output from the detector displayed on an oscilloscope yields an envelope with bounds determined by the peaks of the detector output as the probe is moved. The amplitude of the envelope gives the VSWR as a function of frequency. This technique is still subject to the sources of error inherent in slotted line methods.

The most widely used technique for swept reflection coefficient measurements employs a reflectometer in which directional couplers separate the incident wave from the reflected wave. The arrangement is shown in Figure 8. The ratio of the outputs of the detectors of the incident and reflected signals will, by definition, yield the reflection coefficient directly. To achieve good accuracy, however, the specification of the reflectometer itself becomes very stringent. The directional couplers require high directivity and they and the associated detectors must track closely. The principle of a reflectometer measurement is illustrated in Figure 9. The ratiometer is unnecessary if the incident signal from the reflectometer can be used to accurately level the output from the sweep oscillator. The system is calibrated with a short circuit at the measurement port and a set of calibration lines can be plotted for a range of settings of a standard attenuator in the reflected signal arm. A calibration technique employing digital storage as described above can be used for these measurements. Usually the same equipment as used for attenuation measurements is employed with the addition of a dual directional coupler or a bridge to provide the required signal separation. The major sources of error in this type of measurement are coupler directivity, attenuator accuracy, coupler/detector tracking and mismatch.

## VECTOR MEASUREMENTS

### Network Analysis

In an increasing range of situations, vector information (i.e. magnitude and phase) is required over a wide range of frequencies. Not only may the complex impedance be required over a broad band but the complete set of parameters of a two-port network are often needed to be measured under varying external conditions as with, for example, the characterisation of a transistor. A network analyser should be capable of measuring both reflection and transmission parameters of a linear two-port network.

The principle of such measurements is shown in Figure 10. The key to the operation of a microwave network analyser is the receiver technique employed to detect the (conplex) ratio between the reference and test signals. To achieve a very broad frequency range, the receiver employs sampling and heterodyning to convert the input frequency to an intermediate frequency that can be readily processed to derive rectilinear or polar displays of magnitude and phase. An internal phase-lock loop is used to lock a harmonic of the local oscillator to the incoming signal and therefore allows considerable relaxation of the specifications demanded of the signal source (8).

A network analyser permits transmission and reflection measurements with equal ease, all that is required is a suitable transducer to split the signal appropriately to the reference and test inputs of the receiver. Calibration is usually carried out with a short circuit (reflection) or through connection (transmission) in conjunction with a line stretcher which allows the user to equalise the electrical lengths of the paths to the test and refernce ports and thus establish zero phase. The results are obtained directly in magnitude and phase; the use of a Smith Chart overlay on a polar display permits direct conversion to impedance.

The sources of error in a microwave network analyser are many and complex. Some of the more important are:

1. Tracking between test and reference channels as a function of frequency.
2. Accuracy of attenuators, amplifiers, etc.
3. Noise
4. Cross-talk between reference and test channels; can be minimised by suitable padding.
5. Mismatch error; occurs to some extent at each connection.
6. Directivity errors (in reflection measurements); can be cancelled at any single frequency using a sliding load (9).

The overall uncertainty in a network analyser measurement depends on frequency and the magnitude of the quantity being measured. The uncertainty in both magnitude and phase increases with increasing frequency, that in the phase is highest where the signal reaching the test channel is lowest whilst the uncertainty in magntude is least when the signal itself is small.

### Error Correction Techniques - Automatic Network Analyser

The principle behind the overwhelming majority of error correction techniques employed in conjunction with microwave network analysis is to assume that an 'ideal' measurement system is connected to the measurment port via a network into which all the errors in the system are concentrated (10,11). This is illustrated, for a one-port measurement in Figure 11 where the error network is characterised by 4 s-parameters as shown in the flow graph. The error parameters may be identified with the physical sources of error in the system.

$e_{11}$ Directivity
- loss of signal before reaching the test port due to presence of adaptors

and finite coupler directivity.

$e_{22}$ Source match
- reflections due to connections, mismatch of device under test, etc. Includes effects of adaptors and cables.

$e_{21}, e_{12}$ Frequency tracking
- variation of frequency response of couplers, cables, mixers, detectors, etc.

Analysing the flow graph of Figure 11, we obtain

$$\Gamma_m = e_{11} + \frac{e_{21}e_{12}\Gamma_L}{1 - \Gamma_L e_{22}} \qquad (23)$$

If the error parameters are known, the unknown reflection coefficient can be obtained from

$$\Gamma_L = \frac{\Gamma_m - e_{11}}{e_{22}(\Gamma_m - e_{11}) + e_{12}e_{21}} \qquad (24)$$

(note that $e_{21}$ and $e_{12}$ appear in all the equations as a product)
Since these equations involve three unknown error parameters, the system can, in principle, be calibrated given any three known devices. The calibration standards are, however, limited by the range of devices that can be accurately manufactured. A typical calibration procedure is:

1. Measure a matched load (reflection coefficient = 0)

$$\Gamma_{m1} = e_{11} \qquad (25)$$

It is not necessary to assume zero reflection coefficient (which is difficult to achieve in practice); a sliding load with low reflection is used measured at a number of positions.

2. Measure a short circuit

$$\Gamma_L = -1 \qquad (26)$$

$$\Gamma_{m2} = e_{11} - \frac{e_{21}e_{12}}{1 + e_{22}} \qquad (27)$$

3. Measure a shielded open circuit (small fringing capacitance) or offset short circuit (about quarter wavelength at the centre of each octave band).

$$\Gamma_L = e^{jx} = \Gamma_s \qquad (28)$$

$$\Gamma_{m3} = e_{11} + \frac{\Gamma_s e_{21}e_{12}}{1 - e_{22}\Gamma_s} \qquad (29)$$

Solving to extract the error parameters, these are stored at each frequency of interest and the measurement of an unknown is corrected using the stored data and equation (24).

The correction of errors for two-port measurements is the same in principle but includes error terms due to leakage between the two ports as well as the errors at each port. The various error models in use provide different compromises between mathematical complexity and accuracy; the best correction technique to adopt depends on the nature of the measurements being made.

With good, highly stable, microwave hardware, this measurement technique is readily automated and the necessary instrument control and complex arithmetic can be done by a very small computer as indicated in Figure 12. An incidental benefit is that the computer can readily convert the results to any desired form, e.g. VSWR, impedance, admittance, return loss, group delay, etc.

Some sources of error still remain, however; these include

1. Inaccuracy of the "standards"
2. Noise; reduced by averaging readings
3. Attenuator and amplifier gain errors
4. Connector repeatability
5. Detector non-linearities
6. Harmonic skip

The last of these errors is due to the network analyser internal phase-lock loop locking to a different harmonic during measurement to that used during calibration. This source of error can be entirely removed by providing a synthesised signal source which phase-locks the network analyser's internal oscillator. The accuracy obtainable after error correction is of the order of $\pm$0.1dB magnitude and $\pm$1 degree phase (11).

## Six-port Techniques

The conventional network analyser techniques are essentially "four-port" in nature (Figure 13) and thus four complex constants are required to characterise the intervening network (equation 23). If only reflection coefficients are required, only the ratio of any three constants to the fourth need be determined. The two detectors in this system are thus required to measure magnitude and phase - this necessitates superheterodyning and all the associated problems.

It can be readily appreciated that the direct measurement of phase can be eliminated by incorporating two additional detectors, making the intervening network a six-port, four of the ports being terminated by amplitude sensing detectors (Figure 14).

This does, however, lead to additional mathematical complexity since there are now four equations characterising the network (12,13).

$$P_3 = |Aa_2 + Bb_2|^2 \qquad (30)$$

$$P_4 = |Ca_2 + Db_2|^2 \qquad (31)$$

$$P_5 = |Ea_2 + Fb_2|^2 \qquad (32)$$

$$P_6 = |Ga_2 + Hb_2|^2 \qquad (33)$$

In principle, if A...H can be evaluated it is possible to solve for the reflection coefficient at port 2.

The key to the use of such an apparently complicated technique is the advent of very cheap computing and the microprocessor. A general and powerful theory (13) has been developed that enables the calibration of an arbitrary six-port junction to achieve results of standards laboratory accuracy (14). It is easiest to first consider the simple case where C=0, equation (31) becomes

$$P_4 = |D|^2|b_2|^2 \tag{34}$$

writing $q_3 = -B/A$, $q_5 = -F/E$, $q_6 = -H/G$,

$$P_3 = |A|^2|b_2|^2|\Gamma_\ell - q_3|^2 \tag{35}$$

$$P_5 = |E|^2|b_2|^2|\Gamma_\ell - q_5|^2 \tag{36}$$

$$P_6 = |G|^2|b_2|^2|\Gamma_\ell - q_6|^2 \tag{37}$$

Using equation (34) to eliminate $|b_2|^2$ gives equations such as:

$$|\Gamma_\ell - q_3|^2 = |D/A|^2.P_3/P_4 \tag{38}$$

Thus, having measured $P_3$, $P_4$ and obtained $q_3$ and $|D/A|$ from calibration, a circle in the complex $\Gamma_\ell$ plane with centre at $q_3$ and radius given by equation (38) defines the locus of possible values for $\Gamma_\ell$ (Figure 15). Similarly, equation (36) may be used to define another circle containing $\Gamma_\ell$ with centre $q_5$. $\Gamma_\ell$ is determined by the intersection of the two circles as shown in Figure 16 and if the unknown is passive, then the problem is solved by selecting the solution inside the unit circle. If the solution is extended by using equation (37) and $q_6$, the ambiguity is resolved. In practice the three circles do not intersect in a point because of measurement errors and a variety of techniques can be employed to assess the accuracy and/or improve the quality of the results.

The situation where C is non-zero is a bit complicated to illustrate. The locus of $\Gamma_\ell$ is still a circle but the position of its centre, as well as its radius, is a function of the measured power ratios. In practice it should be possible to design a network such that C = 0 to a good approximation, although this is by no means essential.

The design of a practical circuit for a six-port revolves around the selection of the q's. A circuit comprising hybrids and couplers (15) as shown in Figure 17 seems to be emerging as the most widely employed.

In order to provide a complete description of the six-port (i.e. the constants A...H), the calibration procedure requires the determination of 12 real parameters (11 if only impedance measurements are to be made). The essential principle of a calibration technique is to effect a reduction from the six-port to a four-port (5 parameters) and then carry out a standard four-port reflectometer calibration. In practice this involves collecting power meter data for a minimum of five positions of a sliding short, three of a sliding load, a reflection coefficient standard and a terminating power standard.

An extension to a dual six-port for the measurement of s-parameters has been demonstrated (16) and a variety of calibration techniques, including an extension of the TSD ("thru-short-delay") solution (17), have been used (18).

At the time of writing, six-port measurement techniques are time-consuming and require highly linear power sensors. Widespread use of the technique at frequencies below 18 GHz seems unlikely although there are applications where very high accuracy is required. It is probable, however, that six-port techniques represent the solution to providing an effective network analyser at frequencies well above 18 GHz where the use of conventional superheterodyne techniques is difficult.

## TIME DOMAIN REFLECTOMETRY

An ever-present difficulty with network analysis is the problem of locating and determining the properties of a device embedded in a network connected to the measurement port of an analyser ("de-embedding"). Time domain reflectometry (TDR) is a complementary technique to network analysis in the frequency domain. The system consists of a matched pulse generator with an extremely fast rise time (typically of the order of a picosecond), a tee with a high impedance probe and a suitable oscilloscope (19). The general arrangement is shown in Figure 18.

The physical separation between discontinuities is related to the time interval between successive echoes and the distance resolution of the system is then related to the rise time of the pulse generator. The great virtue of TDR is that the time response of the network to an impulse contains a complete representation of the properties of that circuit. The technique can therefore be used in situations involving more than one discontinuity and each mismatch producing a reflection can be analysed separately; this can be especially useful where the parameters of an active device chip are required derived from measurements on a packaged one. Once again the use of computational techniques permit a variety of information extraction procedures. In particular, the Fast Fourier Transform is widely employed for the extraction of frequency response functions from TDR measurements.

An interesting derived technique (20) consists of taking frequency domain measurements with an automatic network analyser and transforming into the time domain to obtain a simulated impulse response. The time domain response data appropriate to the required section of the network can be selected and its frequency response obtained by an inverse Fourier transform.

## REFERENCES


1. Kurokawa, K., 1969, "An Introduction to the Theory of Microwave Circuits", Academic Press.

2. Moreno, T., 1958, "Microwave Transmission Design Data", Dover Publications, U.S.A.

3. Smith, P.H., 1939, Electronics, 12, 29-31.

4. Smith, P.H., 1965, Microwave J., 8, No. 6.

5. Kurokawa, K., 1965, IEEE Trans., MTT-13, 194-202.

6. Hunton, J.K., 1960, Trans. IRE, MTT-8, 206-212.

7. Kuhn, N., 1963, Microwave J., 6, 59-66.

8. Anderson R.W. and Dennison, O.T., 1967,

Hewlett-Packard J., 18, No. 6.

9. Weinschel, B.O., Sorger, G.U., Raff, S.J. and Ebert, J.E., 1964, IEEE Trans., IM-13, No.12.

10. Hackborn, R.A., 1968, Microwave J., 11, 45-52.

11. Adam, S.F., 1968, IEEE Trans., IM-17, 308-313.

12. Engen, G.F., 1973, IEEE Trans., IM-22, 295-299.

13. Engen, G.F., 1977, IEEE Trans., MTT-25, 1075-1080.

14. Weidman, M.P., 1977, IEEE Trans., MTT-25, 1083-1085.

15. Engen, G.F., 1977, IEEE Trans., MTT-25, 1080-1083.

16. Hoer, C.A., 1977, IEEE Trans., MTT-25, 1070-1074.

17. Franzen, N.R. and Speciale, R.A., 1975, Proceedings of 5th European Microwave Conference, Hamburg, 69-73.

18. Engen, G.F., 1978, Microwave J., 21, p18.

19. Adam, S.F., 1969, "Microwave Theory and Applications", Prentice-Hall.

20. Hines, M.E. and Stinehelfer, H.E., 1974, IEEE Trans., MTT-22, 276-282.

PROBLEMS

1. Explain how a lossy line is represented on a Smith Chart. Using a Smith Chart, find the impedance of an unknown termination if a voltage standing wave ratio of 1.5 is measured using a slotted line with the minimum shifted by 0.2 wavelength towards the generator from its position with a short circuit at the end of the line. What would the impedance of the unknown termination be if the same measurements are made at the end of a cable with 2.5dB loss?

2. (a) How can the Smith Chart, as developed in Figure 3 for passive circuits, be extended to permit the plotting of impedances or admittances with negative real part (i.e. for use with active devices)? What is the disadvantage of using the chart to directly represent network analyser measurements of an active device?
(b) What is the effect of interchanging the TEST and REFERENCE inputs to a microwave network analyser? How does this relate to your answer to part (a)?

3. Extend the network analyser error model developed in equations (23)-(29) to account for errors in a transmission measurment (3 additional error terms). How can the one-port calibration procedure described in the text be extended to provide the additional corrections with the minimum number of additional standard devices?

4. With the assumption of ideal components in the circuit of Figure 17, determine the magnitude and phase of the q's [equations (34)-(38)] applicable to that configuration and show their positions on the reflection coefficient plane.

5. What is the maximum allowable rise time of a time domain reflectometry system if a resolution in position of 1 mm is to be achieved?

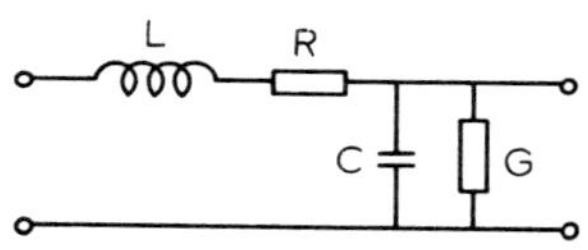


Figure 1 Equivalent lumped-circuit model of a short length of two-wire transmission line

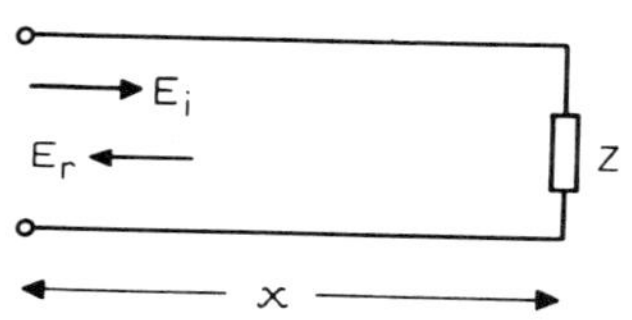


Figure 2 Incident and reflected signals on a transmission line terminated by an arbitrary impedance

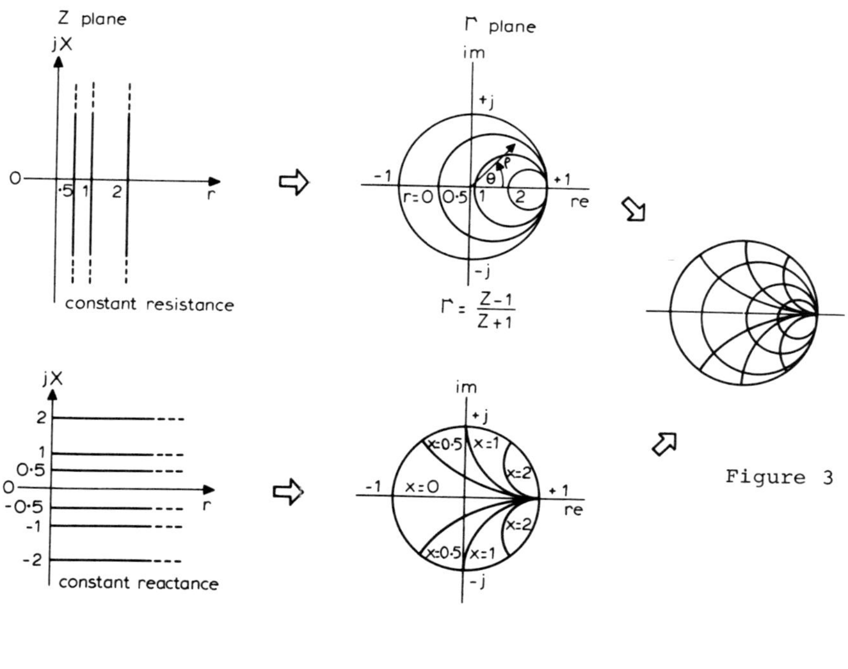


Figure 3 (a) Development of the Smith Chart by mapping from the impedance plane to the reflection coefficient plane

(b) Smith Chart with auxiliary scales

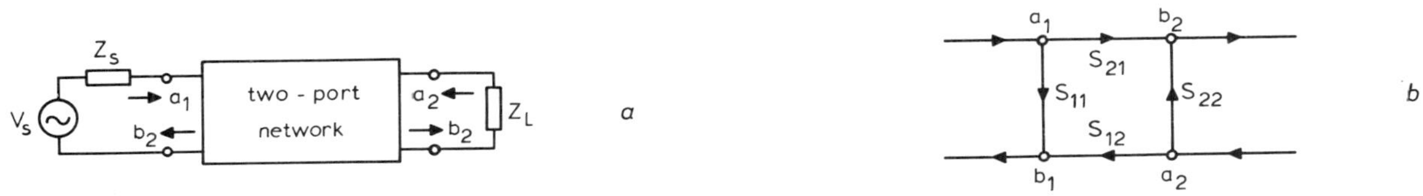


Figure 4 Two-port network and the flow graph representation of its scattering parameters

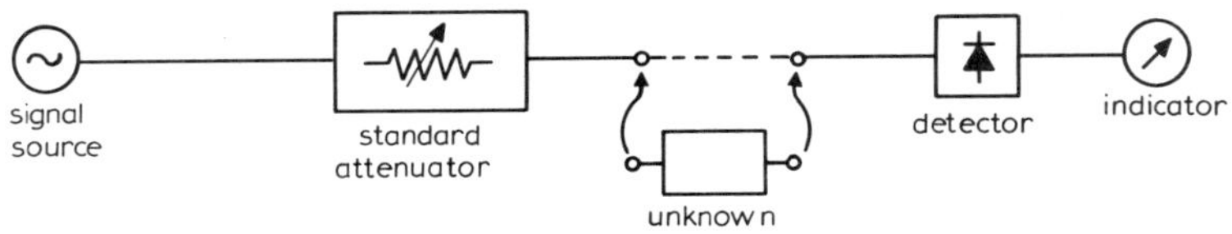


Figure 5 Attenuation measurement by r.f. substitution

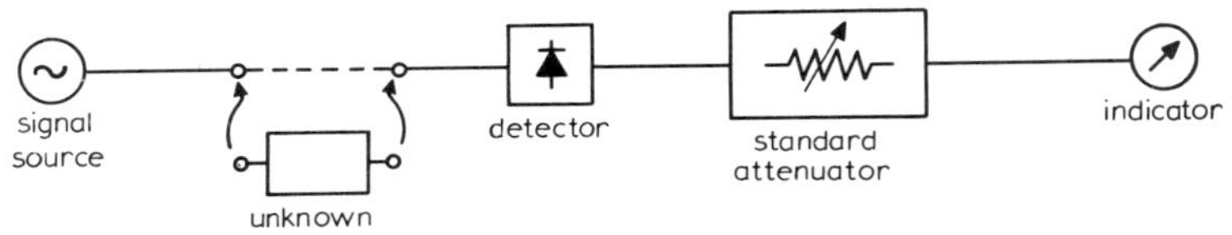


Figure 6 Attenuation measurement by d.c. substitution

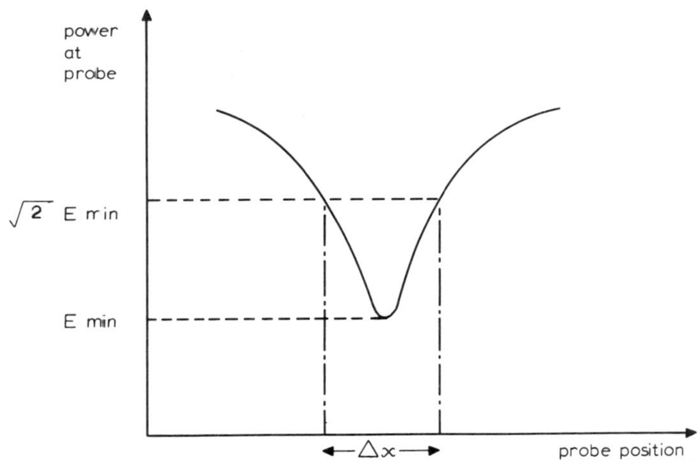


Figure 7 Double minimum technique for measurement of VSWR

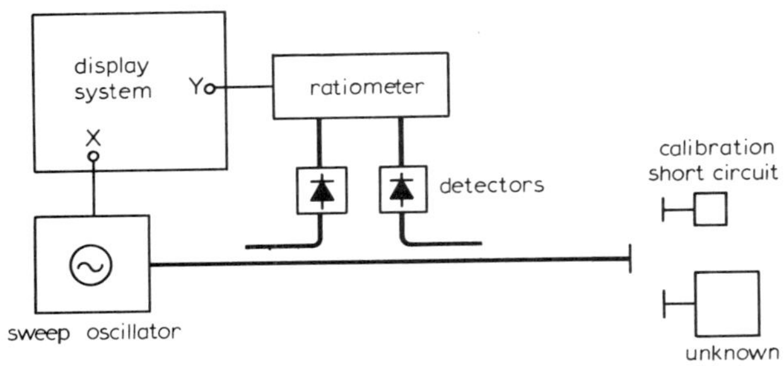


Figure 8 The reflectometer

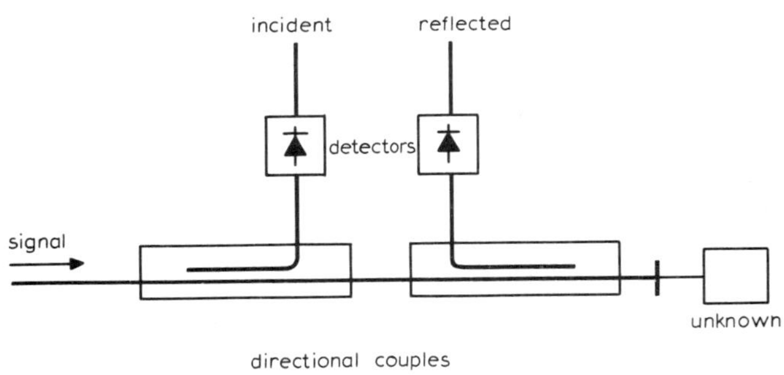


Figure 9 Basic reflectometer measurement set-up

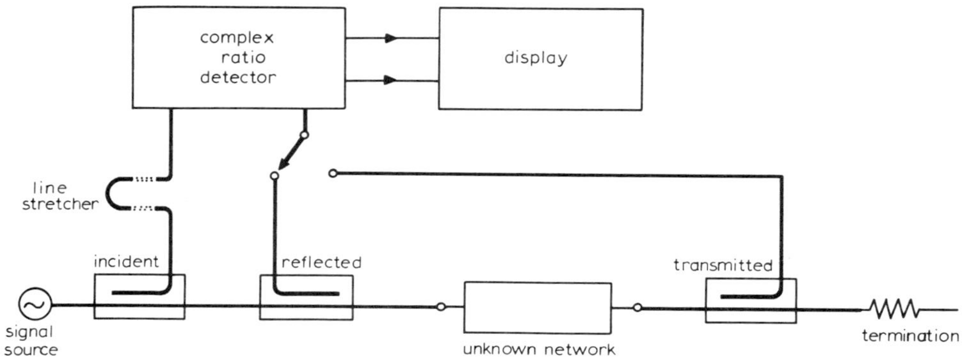


Figure 10 Basic arrangement of a microwave network analyser

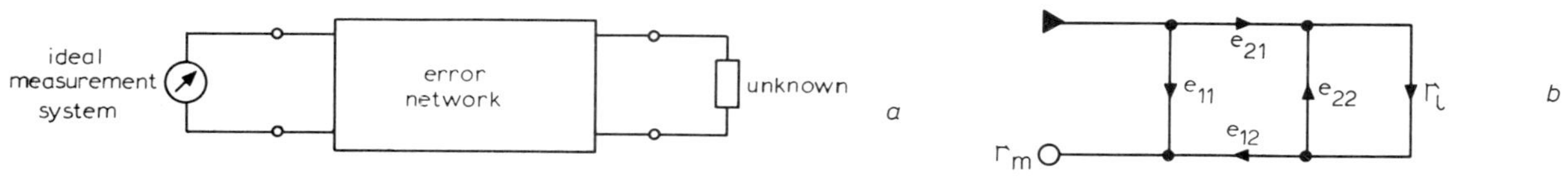


Figure 11 Error model for one-port reflection measurement

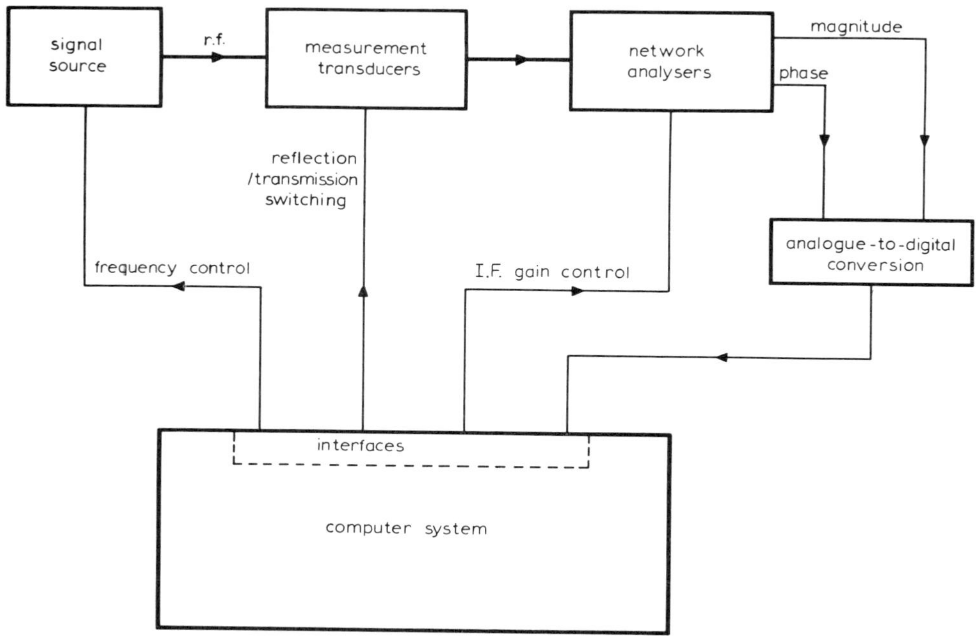


Figure 12 Schematic of computer controlled network analyser

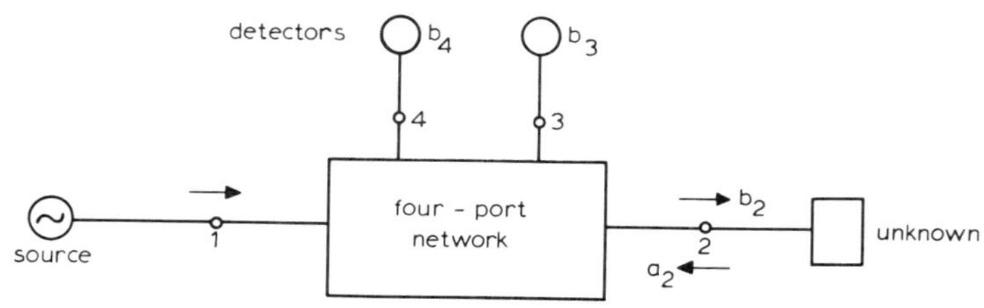


Figure 13 Determination of an unknown impedance using a four-port reflectometer

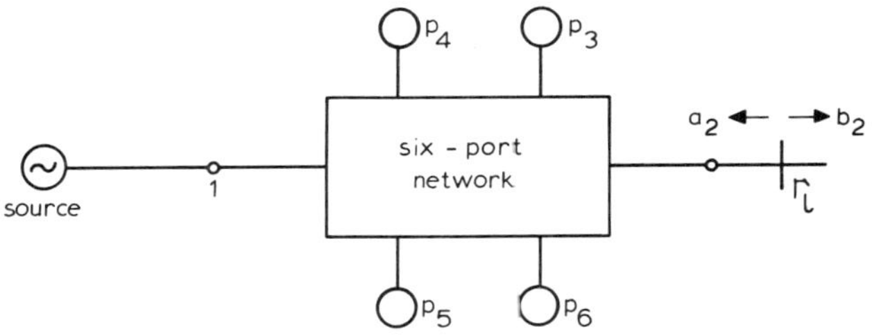


Figure 14 Arrangement of a six-port reflectometer for measuring complex impedance

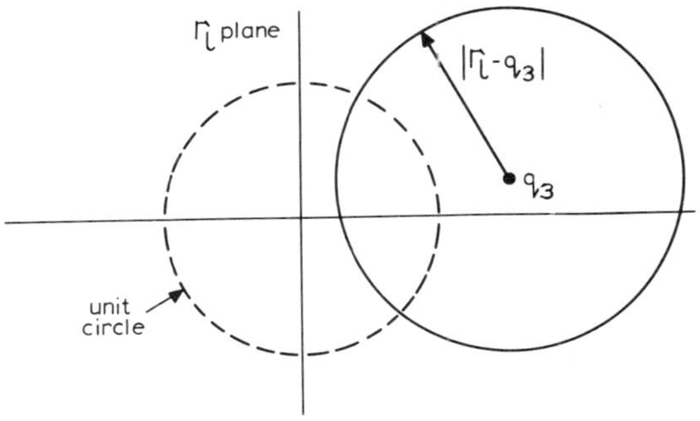


Figure 15 Locus of possible values for unknown determined by power measurements from two detectors on six-port reflectometer

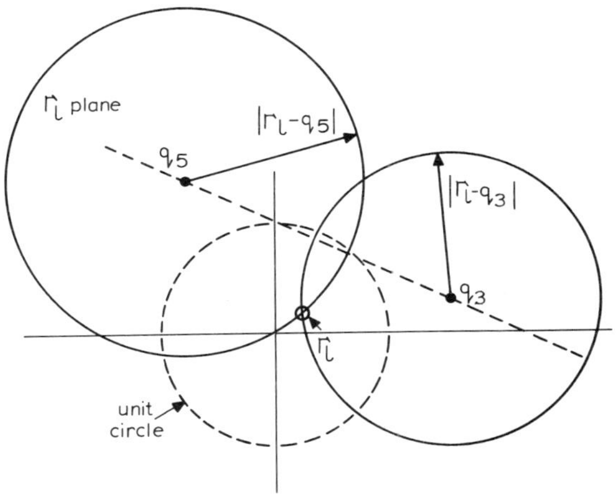


Figure 16 Determination of unknown passive impedance from the intersection of two circles

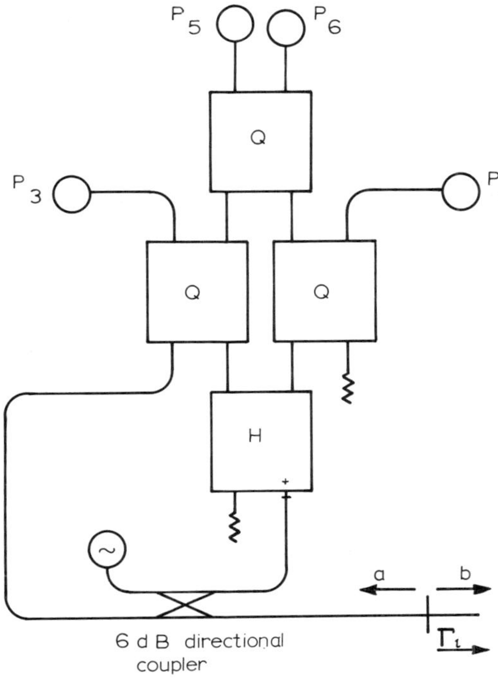


Figure 17 Six-port reflectometer circuit proposed by Engen (Reference 15) (Q indicates quatrature hybrid, H indicates 180° hybrid)

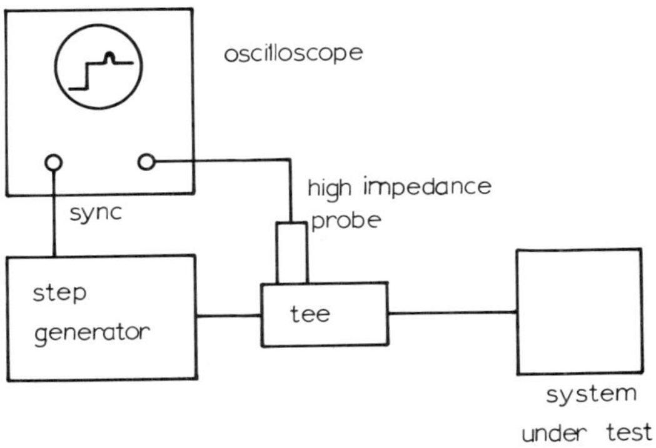


Figure 18 Basic arrangement for time domain reflectometry

# CHAPTER 7

## Device characterisation

Dr. J.A.C.Stewart

## 7.1 DEVICE CHARACTERISATION - INTRODUCTION

Many microwave components and systems use semiconductor devices as passive or active circuit elements. The design of such components requires a knowledge of the behaviour of the solid state device under varying d.c. bias and microwave operating conditions. Device characterisation is taken here to describe the process by which the microwave impedance of the solid state device is recovered from impedance measurements carried out at a suitable plane close to the device. We shall use active solid state microwave devices to illustrate the principles involved (IMPATT, GUNN, BARITT).

The active region of a microwave solid state device consists usually of a small "chip" of silicon, Ga As or other semiconductor material. This chip is placed into a microwave package, for ease of handling and also to prevent contamination of the chip by dust etc. The most common form of package is shown in Figure 1(a), and consists of a cylindrical pill, with metal end caps separated by a dielectric cylinder. The chip is mounted on a metal pedestal, to improve heat transfer from the active region. The connection from the chip to the top cap is provided by a thin gold wire or strap.

Figure 1(b) shows two ways in which the package can be inserted into a microwave circuit. Note that in each case provision must be made to apply d.c. bias to the package end caps and thus to the active device. The microwave circuit connecting the package to a convenient reference plane is called the mount. The design of an amplifier or oscillator using the active device involves matching the impedance of the packaged device to a known load in a controlled fashion. For this purpose, the input impedance presented by the active device must be known. Unfortunately the combination of package and mount modifies the active chip impedance considerably. Figures 2(a) and 2(b) show the active chip impedance, and input impedance at the reference plane, of a typical BARITT diode mounted in the coaxial circuit of Figure 1(b). The BARITT device chip can be represented in its active region by a capacitance and negative resistance in series. The input impedance measured at the reference plane shows two principal resonances; a series resonance and a parallel resonance, which indicate the dramatic transformation due to package and mount. A knowledge of this transformation is therefore required before the chip impedance can be inferred from microwave impedance measurement at the reference plane, or before the input impedance of a packaged chip of known impedance can be determined.

The first section of this chapter examines the form of lumped element equivalent circuits for both mount and package and the principles of the methods used to determine the circuit element values.

In linear amplifiers, the active device is operated in the small signal mode. This mode also provides useful starting information for oscillator design. Using the ideas introduced in Section 7.2, small-signal device characterisation is discussed in Section 7.3.

In Section 7.4 the special techniques used to determine device impedance under large signal conditions are described. The influence of large signal impedance on oscillator design will be examined.

## 7.2 PACKAGE AND MOUNT CHARACTERISATION

7.2.1. Mount and Package Transformation. Consider the configuration shown in Figure 1(b), of a packaged device chip mounted at the end of a coaxial line. It can be shown (Ramo, Whinnery and van Duzer (1)), that the impedance transformation from chip to measurement plane, at a fixed frequency, may be represented by a number of equivalent circuits two of which are illustrated in Figure 3. These circuits represent the transformation of microwave impedance between two ports, where a "port" implies a single mode of transmission at a specific reference plane. By carefully selecting the measurement plane, the effect of the length of transmission line in Fig. 3(a) can be avoided, leaving a simple circuit consisting of a transformer and shunt element.

These equivalent circuits suffer from two main disadvantages however: (a) they can be used only over a very narrow frequency band, (b) they are unique to a specific package-mount configuration and convey no information regarding the relative contributions of mount and package to the overall transformation.

By subdividing the volume between chip and measurement plane into small regions, each described by lumped equivalent circuit, an equivalent circuit may be derived which can be more readily identified with the physical configuration. Figure 4(a) illustrates the sub-regions chosen by Getsinger (2) in deriving an equivalent circuit for the coaxial mount. Note that the package transformation and mount transformation are considered separately.

It should be pointed out that the separation of package and mount equivalent circuits assumes that the radial line mode is fully developed at the package surface cylindrical plane. In practice this is unlikely to be so. The resulting discrepancy is accounted for by slight modifications to the measured package equivalent circuit parameters and results in a small variation in package parameters in different mounts.

7.2.2. Lumped Element Equivalent Circuits. The sub-regions discussed above comprise (a) a coaxial line segment, (b) a radial line segment (c) a transition region with distorted electric and magnetic fields. Consider these in turn.

(a) Coaxial line segment. The normal distributed transmission line equations which describe the transformation of impedance, result from integrating a lumped element description of an incremental length of line. The resulting line equations hold at a chosen frequency. For a lossless coaxial segment, provided a small volume is involved, the equivalent circuit shown in Fig. 5(a) may be used, where L and C are the coaxial inductance and capacitance per unit length.

(b) Radial line segment. Figure 5(b) represents a section of radial line, with the dominant mode propagating. Note that this is a TEM mode, with characteristic impedance a function of radial position. By using an incremental model for a length of radial line much less than one wavelength, the equivalent circuit shown in Figure 5(c) can be derived to represent the resulting impedance transformation. The circuit elements are the corresponding coaxial inductance and parallel plate capacitance.

(c) Transition region. From Figure 4(b), the electric field pattern in this region is distorted, in order to satisfy boundary conditions, while the magnetic field can approximately retain its coaxial form. The equivalent circuit inductance for this region can therefore be approximated by a coaxial inductance (as if the coaxial line centre conductor were extended to the end wall). The exact values of fringing capacitances $C_{f1}$ and $C_{f2}$ used to describe the electric fields in this region are open to discussion. Approximate equations are provided by Getsinger (3).

7.2.3 Mount Equivalent Circuit. The circuits just discussed are combined in the equivalent circuit of Figure 4(c) for the complete mount. $L_c$ and $L_T$ are coaxial inductances. $C_c$ (= $C_{c1}$ + $C_{c2}$) is the equivalent coaxial capacitance. For a standard $S_4$ type package in a 50 ohm 7 mm coaxial line, Getsinger's equations (2) for $C_{f1}$ and $C_{f2}$ give 0.1 pF and 0.043 pF respectively. Owens (3) obtains values 0.01 pF and 0.043 pF, while accurate finite difference method gave a total capacitance of 0.067 pF for the region. It would appear therefore that Owens' values agree better with the accurate computed values. $L_r$ and $C_r$ are the approximate radial line elements, as the radial line length is usually much less than one wavelength.

7.2.4 Package Equivalent Circuit. The package equivalent circuit may take different forms. In general the greater the frequency range over which the equivalent circuit is to be applied, the more elements are required. While it is not advisable to attempt to relate the equivalent circuit elements, which describe the distorted package fields, to physical package dimensions exactly, some justification for the common forms of the circuit can be obtained in this way.

Consider the package of Figure 1(a). In Figure 6(a), the form of package equivalent circuit used by Owens (3) is developed. Approximately, $L_2$ refers to the bonding wire, $L_1$ to the coaxial pedestal inductance, $C_2$ to the capacitance from pedestal to top cap and $C_1$ to the capacitance of the cylindrical dielectric spacer bounded by the top and bottom metal caps. Typical values, from Owens, are shown. From this circuit, for an S4 package (shown to be usable to 20 GHz), simpler versions can be derived.

It can be shown (Tutorial problem 3), that the impedance transformation due to a series L, shunt C combination remains approximately invariant when the order in which they appear in a circuit is reversed, provided $\omega^2 LC \ll 1$. Hence the position of $L_1$ and $C_1$ in Figure 6(a) can be interchanged, and $C_1$ combined with $C_2$ to produce the circuit of Figure 6(b). (Greiling and Haddad (4)). The effect of variation of bonding lead length and position from package to package is modelled by corresponding changes in $L_2$.

If the positional order of $L_2$ and $C_2$ are interchanged in the original circuit (Figure 6(a)), then the circuit of Figure 6(c) results, (Dean and Howes (5)), in which L represents the total package inductance.

Finally, if the order of $L_1$ and $C_2$ in Figure 6(b) are interchanged, the simple circuit of Figure 6(d) is obtained. This circuit is accurate over a relatively narrow frequency range, but can be used to identify the principal resonances in the device-package interaction.

7.2.5 Package Equivalent Circuit Evaluation. A wide variety of methods has been used to determine the appropriate element values for package circuits. To illustrate the principles behind the various methods, the evaluation of the elements of the simple package circuit of Figure 6(d) will be discussed.

Over a relatively narrow frequency range, the simple circuit of Figure 6(d) can represent either the complete impedance transformation from chip to coaxial measurement plane, or the package transformation alone. In the latter situation, input impedance measurements in the coaxial line, are transformed through the mount equivalent circuit, to give the input impedance at the cylindrival package surface for chosen chip reference impedances. Assume that an open-circuit and a short-circuit package are available. Then low frequency measurement of the capacitance of the open-circuit package will give C, while the microwave series resonant frequency of the short-circuit package will enable L to be found.

The short-circuit package is made by bonding the gold wire from the top cap directly on to the top of the pedestal, while in order to preserve the lead configuration, the open-circuit package bonding lead connects the top cap to a metal contact on a thin insulating layer deposited on top of the pedestal.

If an IMPATT diode, varactor diode, or BARITT diode are reverse biased, their depletion layer capacitance is a well defined function of bias voltage. This capacitance is invariant with frequency up to X-band frequencies. Hence accurate low frequency (1MHz) capacitance measurements on an open-circuit package, and then on a package containing the active device, enables the package capacitance, C, and device capacitance $C_j(V_o)$, to be found. The series resonant frequency provides an alternative method of finding L. Tutorial problem 4 illustrates a method by which L and C may be found without using special open-circuit and short-circuit packages.

As suggested briefly in the previous paragraph, it is not always possible to make special open-circuit and short-circuit packages. Hence methods which do not require such packages offer great advantages. By making use of the phenomenon of magnetoresistance, McBretney and Howes (6) have characterised a Gunn diode mount and package, without the use of special packages. Magnetoresistance manifests itself as a variation of low-field conductance in a semiconductor of suitable aspect ratio, on application of a transverse magnetic field. The low-field conductance is frequency independent, providing a reference chip impedance.

An alternative approach uses a computer minimisation routine to adjust the equivalent circuit parameters, to match measured and calculated input impedances at a suitable reference plane over a desired frequency range.

These methods, used to evaluate the small-signal impedance of active devices, will be described in detail in Section 7.3 of this chapter.

## 7.3 SMALL SIGNAL DEVICE IMPEDANCE

7.3.1 "Small Signal?". Active and passive microwave devices are used in amplifiers, oscillators, limiters, mixers, etc. To design the microwave circuits associated with the devices, a knowledge of the device input impedance as a function of frequency and bias conditions is usually necessary. Depending on the particular application the impedance of the packaged device may be required (a) at the input to the mount (b) at the input to the package or (c) at the chip itself. The methods, introduced in Section 7.2, for evolving the mount and package equivalent circuits, are used in the measurement of the device impedance.

Figure 8 shows the resistive part of the impedance of an IMPATT diode chip plotted as a function of applied r.f. current magnitude, for given d.c. bias currents. The magnitude of the chip resistance is seen to fall as the magnitude of the r.f. current increases. However below a certain value of r.f. current, the chip resistance becomes independent of r.f. drive level. In this region the device operates in the small-signal regime, in the sense that the r.f. current and r.f. voltage are linearly related.

$$V_{rf} = Z\ I_{rf}$$

In Figure 9, the measured small impedance characteristics of two microwave active devices (TE device and IMPATT) are presented. Note that both devices exhibit a band-pass negative resistance characteristic, with small signal resistance and reactance a function of microwave frequency and bias voltage or current. As explained previously this chip impedance will be considerably modified by package and mount effects.

As oscillators and amplifiers with large input signals employ solid-state devices operating under large-signal conditions, the use of small signal impedance to describe the device may be questioned. Some of the advantages and applications of small-signal impedance are listed below:

(a) Small-signal impedance is directly useful in linear amplifier design.

(b) S.S. Impedance is used in preliminary design of non-linear circuits (eg, oscillators).

(c) S.S. Impedance is independent of r.f. signal across the device, and hence of incident microwave power.

(d) S.S. Impedance requires low microwave power for its measurement, and hence can be found using a standard network analyser.

(e) S.S. Impedance is relatively easily computed for most solid-state devices.

(f) As a result of (e), theoretical predictions and small signal measurements can be compared, resulting in greater understanding of the operation of the solid-state device.

In 7.3.2 the consequences on the analysis of device impedance of making small-signal approximation will be explored. Three different examples of small-signal device impedance measurement are used to illustrate typical practical configurations. The section is completed by examining one application of small-signal impedance information.

7.3.2 Small-Signal Impedance Analysis. The meaning of "small-signal" is demonstrated by examining the type of approximation used to find small-signal impedances. Assume that a microwave device, an IMPATT diode, for example, has a d.c. bias current applied to it, with corresponding d.c. voltage. As shown in Figure 10 this d.c. voltage is related to the electric field profile across the diode. If a small sinusoidal perturbation is superimposed on the bias current, a variation in electric field profile and hence of diode voltage will result. For small-amplitude current change, the voltage change is also sinusoidal, the diode is linear, and the small signal impedance is found from the ratio of the sinusoidal changes in voltage and current at a specific frequency. It is not the intention here to derive a small-signal model for an IMPATT diode. However we shall examine one of the equations which is used in the model. This equation, derived from Maxwell's magnetic curl equation relates the total diode current to particle and displacement currents, as follows.

$$I_d = I_{particle} + I_{displacement}$$

Assuming that diffusion effects can be neglected, and that only one carrier (holes) exists,

$I_{particle} = qAvp$; $I_{displacement} = \varepsilon A\frac{dE}{dt}$ where $v$ = hole velocity, $p$ = hole density, $A$ = device area, $q$ = charge on hole, $\varepsilon$ = semiconductor material permittivity. The equation can therefore be written, for this simple situation:

$\frac{dE}{dt} = \frac{Id}{\varepsilon A} - \frac{q}{\varepsilon}\,vp$. This equation is solved together with the continuity equation for holes and Poissons equation with variables E (electric field) and p (hole density). Note that the equation is a non-linear equation, containing the product vp. (v is a function of E.) If an input diode drive current $I_d$ is assumed, the equations solved for E (and p) and hence diode voltage, $V_d$, the relation between $I_D$ and $V_d$ is non-linear and in general the large-signal diode impedance will result.

Let, however, each variable be separated into d.c. and a.c. parts, and assume that the magnitude of the a.c. quantities is much smaller than the d.c. quantities - ie, a small-signal approximation. For example $I_d = I_O + I_S$ where $I_S$ represents a small-magnitude sinusoidal input current. Then the equation becomes

$$\frac{d(E_O + E_S)}{dt} = \frac{I_O + I_S}{\varepsilon A} - \frac{q}{\varepsilon}\,(v_O + v_S)(p_O + p_S).$$

If the product of small a.c. terms is neglected, the small-signal equation results:

$\frac{dE_S}{dt} = \frac{I_S}{\varepsilon A} - \frac{q}{\varepsilon}\,(v_O p_S) - \frac{q}{\varepsilon}\,(p_O v_S)$. This is a linear equation in which the d.c. values of $v_O$ and $p_O$ are constant. A prior knowledge of $v_O$ and $p_O$ is assumed. Hence the relation between $I_S$ and $E_S$ (or $V_S$) is linear and the resulting small-signal impedance is independent of $I_S$.

7.3.3 Measurement - IMPATT diode in coaxial mount. The IMPATT diode chip, bonded in the package of Figure 1(a) is placed in the coaxial mount of Figure 1(b). Figure 11(a) represents a block diagram of a typical network analyser measurement system, while 11(b) illustrates the equivalent circuit of the package and IMPATT diode biased below reverse breakdown. Note that the reverse-bias depletion layer capacitance of the diode is invariant with frequency, and hence may be used as a reference chip impedance.

The procedure is to use the known chip reactance below avalanche breakdown, together with Getsinger's mount equivalent circuit, to find mount and package circuit values. The IMPATT diode is then biased into its operating range, and the device small-signal chip impedance found from impedance measurements at the input to the coaxial mount. One possible approach could be as follows:

(a) Low frequency capacitance measurement (1MHz) on an open-circuited package and a package containing the IMPATT chip biased below breakdown, gives C and $C_j$ (V).

(b) The diode when forward biased approximates a short-circuit. Hence measurement of the **parallel** resonance frequency of a forward biased packaged diode gives L2.

(c) With a chosen value of $C_j$(V), the resulting series resonant frequency will give L1. The values of L1, L2 and C can be checked by comparing calculated and measured resonance frequencies as $C_j$(V) is varied.

(d) The diode is biased above breakdown, and its chip impedance inferred from the mount input impedance. In this method, $R_S$ is assumed to be negligible when determining the package circuit. The losses in the package, mount and diode are found from the chip measurement just at breakdown.

By using a computer in conjunction with the network analyser, the method just described can be made much more flexible and accurate. Assume that the equivalent circuit remains the same, with $R_S$ representing package loss, bonding and contact losses. Assume also that the IMPATT reverse bias capacitance is known for a range of voltages. The mount input impedance is measured either (a) at a single frequency for a range of reverse bias voltages or (b) at a single reverse bias voltage for a range of frequencies. In either case, the computer is used to adjust the equivalent circuit values to fit the calculated mount input impedance to the measured impedance over the set of measurements.

Getsinger's values, and the method described above can be used to obtain initial values for the mount and package circuit. A minimisation routine is then used to adjust the circuit values to minimise:

$$F = \alpha \sum_{j=1}^{n} \{|\Gamma|_m - |\Gamma|_c\}^2 + \beta \sum_{j=1}^{n} (\theta_m - \theta_c)^2$$

where $\alpha$ and $\beta$ are scaling factors, $|\Gamma|$ is the input reflection coefficient magnitude and $\theta$ is the input reflection coefficient phase angle.

Figure 12 shows results for the small signal impedance of an H.P. IMPATT diode, measured using this approach. The package equivalent circuit is found over a frequency range 8 GHz to 12 GHz, for a diode reverse bias of 65 volts, just below breakdown.

7.3.4 <u>Measurement - Gunn diode in coaxial mount</u>. We have seen that for varactor diodes, Impatt and Baritt diodes, the reverse-biased depletion layer capacitance may be used to provide a reference chip reactance. For transferred electron devices, McBretney and Howes (6) have shown that the phenomenon of magnetoresistance can be used to provide reference chip conductances. Assume again that the Gunn diode is in the package and coaxial mount type of Figure 1(a) and Figure 1(b). If the diode is biased into its passive low electric field region (Figure 13(a)) and a transverse static magnetic field applied, then the d.c. conductance varies in the manner of Figure 13(b). McBretney and Howes have demonstrated that this conductance remains constant up to high microwave frequencies (20 GHz). Hence it can be used as a readily measured chip reference. Figure 12(c) illustrates a two-port network, which as mentioned in Section 7.2.1, can be used to describe the transformation from coaxial mount input port to chip. From the transfer matrix for this network, the admittance at the measurement plane, $Y_m$, and chip admittance, $Y_c$, are related by

$$Y_m = \frac{C + Y_c}{A + BY_c}$$

A, B and C are complex, and functions of frequency. At a specific frequency, three reference chip conductance values are needed to characterise the transformation. These are provided by varying the static magnetic field. Having determined A, B & C, the chip admittance can be evaluated from a subsequent small signal admittance measurement at the mount input.

This method has the prime advantage that the chip reference conductance values can be measured with the Gunn diode in its microwave circuit. In addition, no special packages are required.

It should be pointed out that the same two-port circuit can be used with varactors, IMPATT and BARITT diodes, with the depletion layer capacitance providing reference susceptances. The two-port network can be used to describe mount plus package, or just package, as required.

7.3.5 <u>Measurement - BARITT diode in microstrip</u>. The microstrip mount shown in Figure 14(a) is used frequently as the basis for microwave circuits and systems, to take advantage of the available small size, light weight and low cost. In this case the active device is a BARITT chip, placed in a mount comprising a 50 ohm microstrip line, and a launcher which transforms the planar microstrip geometry to coaxial geometry. Bias current for the diode is applied via a coaxial bias-Tee unit. For microstrip circuit design the package input impedance is often sufficient, and we will consider one method of measuring this impedance.

Following Wight et.al. (8) the impedance transformation due to the launcher can be represented by an L, C circuit. The mount equivalent circuit is shown in Fig. 14(b), where the transmission line section is assumed to be 50 ohms characteristic impedance, and initially to be lossless. The method proceeds as follows:

(a) measure impedance at the input to the mount, for the range of device bias current required above reachthrough.

(b) cut back the 50 ohm line, until $l = (2n+1)\frac{\lambda g}{4}$. If the open-circuit end capacitance of the line is neglected, short-circuit should be presented at the launcher, and hence from the measured input reactance, $L_L$ can be found. (See Figure 15(a) and 15(b)).

(c) cut back the line, until $l = n\frac{\lambda g}{2}$, when the line presents an open-circuit to the launcher. Knowing the value of $L_L$, $C_L$ can be determined (see Figure 15(a) and 15(b)).

(d) Knowing the mount equivalent circuit, the device input impedance at the package can be found from the results of step (a).

If necessary the method can be extended to determine chip impedances in microstrip, and to take microstrip line losses into account.

7.3.6 Application of Small-Signal Impedance. The use of small-signal device impedance for amplifier design will be discussed in other chapters of this course. One example illustrating the interaction of small-signal impedance measurement and computer simulation is provided by Decker and Dunn (9). Using the simple circuit of Figure 6(d) to represent both mount and package, they measured the precise small-signal chip admittance of a germanium n+ p p+ IMPATT diode over the frequency range 2-8 GHz. They accurately measured the diode physical parameters (epitaxial-layer doping and width, diode area) and used a small-signal computer model to predict the chip admittance for selected values of carrier velocity and ionisation rates. They found that the model predictions were very sensitive to the carrier parameters, and by adjusting the values to obtain best fit with the measured admittance, they produced new values, with an estimated error of ±5% for the velocity, and ±10% for the ionisation rates.

## 7.4 LARGE-SIGNAL DEVICE IMPEDANCE

7.4.1 Large-Signal Performance. Figure 8 shows how the negative resistance of an IMPATT diode varies with r.f. current magnitude. In the small-signal region of operation, the device resistance is independent of r.f. drive level. However in a solid-state oscillator, or amplifier with moderate or high input power, the active device operating point moves into the large-signal region, where the impedance becomes a non-linear function of r.f. signal level. For example, Figure 18 illustrates the large-signal admittance variation for a silicon IMPATT diode (Decker et.al. (10)).

The active device chip can be represented by either a series circuit, or a parallel circuit (Figure 19(a)). In each case the device parameters are a function of bias point ($I_o$, or $V_o$), and of r.f. drive level ($I_{rf}$ or $V_{rf}$). Consider the series equivalent circuit, and assume that a load impedance completes the series circuit. If the diode negative resistance variation with peak r.f. current is as shown in Figure 19(b), then on application of the diode d.c. bias, the diode negative resistance magnitude is greater than $R_L$ (osc.). The net resistance in the circuit is negative, and noise currents will grow exponentially, at a frequency determined by the reactive match of $X_L$ and $X_d$. As the peak r.f. current grows, the diode resistance magnitude falls, until when $|R_d| = R_L$ (osc.) the net circuit resistance is zero, and a steady-state oscillation results (Figure 19(c)). Note that stable oscillation requires that the device resistance magnitude falls as r.f. current increases. It is evident that for oscillations to grow,

$|R_{d_s}| > R_L(\text{osc.})$; $X_{d_s} = -X_L$ and for steady-state oscillation:

$$|R_d\ (I_o,\ I_{rf})| = R_L(\text{osc.});\ X_d(I_o,\ I_{rf}) = -X_L.$$

In general the device reactance (or susceptance) varies with $I_{rf}$, causing the frequency to change as the oscillation grows.

Assuming steady-state oscillation, the diode available power is given by

$P_d = \frac{1}{2} I_{rf}^2\ |R_d|$ where $I_{rf}$ = peak r.f. current.

From the condition for oscillation, $|R_d| = R_L$. Hence $P_d = \frac{1}{2} I_{rf}^2\ R_L$, and thus for any given value of $R_L$(osc.), the available power may be found. This represents the output load power, providing zero circuit losses and perfect matching can be achieved. In Figure 20, the available power of an IMPATT diode is plotted, for two values of diode series resistance ($R_S$). For zero series resistance, the maximum available power is approximately 800 mW at an r.f. current of 1.25 A and a load resistance of 1.0 ohm; for $R_S$ = 0.5 ohm, the maximum available power is 500 mW, at an r.f. current of 1 A and a load resistance of 1.5 ohm. It is evident therefore that for a load resistance less than the small-signal device negative resistance magnitude, $|R_{d_s}|$, oscillations can occur. The circuit performance may be derived from the device large-signal impedance.

If the load resistance is greater than $|R_{d_s}|$, then the solid-state device can perform as an amplifier. The circulator in Figure 21 acts to separate the incident and reflected power from an active device operating as a reflection amplifier. The power gain ($P_r/P_i$) is given by $|\Gamma|^2$, where $\Gamma$ is the reflection coefficient, i.e.,

$$g_a = \left|\frac{R_o + |R_d|}{R_o - |R_d|}\right|^2$$

$R_o$ is the load (or detector) impedance. By feeding a voltage proportional to sweep generator output frequency into the display unit, the gain-frequency response of the amplifier can be observed. In condition (a) (see Figure 21), the active device resistance is positive, and the device absorbs power. When the device active negative resistance just equals contact and substrate losses, $|R_d| = 0$, and $g_a = 1$ (b). Finally condition (c) represents the situation when $R_d$ is net negative, giving power gain. Figure 22 shows the variation of reflection amplifier gain with input power (and hence with diode r.f. drive level) for a single-tuned IMPATT diode (Scherer (11)). Note that, due to the large-signal variation of diode reactance, the amplifier centre frequency changes with input r.f. power level.

Having examined the effect of large-signal variation of device impedance on oscillator and amplifier performance, the methods used to measure large-signal diode impedance will now be discussed.

7.4.2 Direct Measurement of Large-Signal Impedance. In principle, the large-signal device impedance (or admittance) can be measured with a network analyser, using a similar procedure to that described in Section 7.3.3 for small-signal impedance. Several problems arise however. Consider Figure 23, which shows a typical layout for this measurement (Decker et.al. (10)). The coaxial mount of Figure 1(a) is made to fit directly on to the APC-7 connector of the network analyser reflection test unit measurement port. The transformation from diode chip to measurement plane (i.e., package and mount) is represented by the equivalent circuit shown in the insert of Figure 23. The values of this equivalent circuit can be determined using the methods of Sections 7.3.3 or 7.3.4. Assume that these values are known. The network analyser measures the reflection coefficient (magnitude and phase angle) at plane A-A. Knowing $\Gamma_A$ and the equivalent circuit values, $\Gamma_B$ and hence the diode impedance is found. The measurement is carried out at a single frequency, for a chosen value of diode d.c. bias current (or voltage), and for a range of input power and hence r.f. signal level. The diode r.f. current (for a series representation) is found from the input power, $P_i$, and $|\Gamma|_A$. If the package and mount circuit is lossless, then

$|\Gamma_A|^2 = |\Gamma_B|^2$ The output power from the amplifier consists of the input power plus power added due to the active diode negative resistance.

$P_r = P_i + P_{add}$, where $P_{add} = \frac{1}{2} I_{rf}^2 \, |R_d|$.

Hence $P_{add} = P_r - P_i = P_i\{|\Gamma_B|^2 - 1\} = P_i\{|\Gamma_A|^2 - 1\}$.

Knowing $|R_d|$, $P_i$ and $|\Gamma_A|$, $I_{rf}$ the peak r.f. diode current can be found.

The main problem is to ensure a large enough diode added power and hence r.f. current to enable the measurement to be useful. The large impedance discontinuity between the chip and the 50 ohm measurement line causes most of the input power to be reflected directly. Hence an input power of the order of several watts is needed. In Figure 23 this power level is achieved using a travelling wave tube amplifier. To make sure that the harmonic converter unit is not damaged, the coaxial pad attenuators preceding it must be increased from their normal 10 dB value.

In carrying out this procedure, care must be exercised to ensure that harmonic and other spurious responses from the device are negligible. This places an upper limit on the diode r.f. signal level.

Figure 24 illustrates the use of this method to determine the large-signal impedance of a BARITT diode. This low-power device requires only 200 mW input power to enable its large-signal impedance to be measured. The insert shows values for the chip-to-measurement plane transformation, evaluated using the computer-aided network analyser previously described. The measurement was carried out at a fixed frequency of 10.2 GHz, with a device bias current of 15 mA.

7.4.3 Amplifier Method. One main difficulty in direct measurement of large-signal impedance is to achieve adequate added power at the active device. The reflection amplifier of Figure 21 employs an impedance matching section to reduce the load resistance seen by the diode, thus increasing the reflection coefficient magnitude. As the added power is given by

$P_{add} = P_i\{|\Gamma|^2 - 1\}$, the r.f. signal at the diode is increased for a given input power $P_i$.

Assume that the reflection amplifier of Figure 21 is to be used in the diode large-signal impedance measurement. The small-signal diode impedance has already been found using, for example, the methods of Section 7.3.3. The matching section connecting the diode to the circulator port is lossless, and hence the power gain, $g = |\Gamma_B|^2 = |\Gamma_A|^2$.

At low input power levels, the amplifier gain is independent of input power, the diode is operating in its small-signal region, and the gain is given by

$$g_s = \left|\frac{R_L + |R_{d_s}|}{R_L - |R_{d_s}|}\right|^2$$

where $R_{d_s}$ represents the diode small signal resistance and $R_L$ the load resistance seen by the diode.

As the input power is increased, the gain reduces and the centre frequency changes, for a single tuned amplifier. Hence

$$g = \left|\frac{R_L + |R_d(I_{rf})|}{R_L - |R_d(I_{rf})|}\right|^2$$

where the diode r.f. current is given by $P_{add} = P_i \; g-1 = \frac{1}{2} I_{rf}^2 |R_d|$.

It can be shown (Tutorial question 15), that

$$\frac{R_d(I_{rf})}{R_{d_s}} = \frac{(\sqrt{g} - 1)}{(\sqrt{g} + 1)} \frac{(\sqrt{g_s} + 1)}{(\sqrt{g_s} - 1)}.$$

Hence by measuring the amplifier gain as a function of input power level, the large-signal diode resistance is found. As shown in Figure 22, the amplifier centre frequency changes with $P_i$. Knowing $X_{d_s}$, the variation of diode reactance with input power level can be found, by assuming a conjugate match at the chip.

An alternative approach uses the same amplifier matching section in conjunction with a network analyser (Figure 25). The transformation from chip through the lossless matching section is described by a two-port network equivalent circuit. As before, the network parameters are found using three standard chip impedances (which may be IMPATT capacitance, or TED conductance), at a single frequency. The reflection coefficient $\Gamma_1$, is found using the network analyser, as a function of input power, $P_i$. From a knowledge of $\Gamma_1$, B, D, C and $P_i$, $Z_d$ $(I_{rf})$ and $I_{rf}$ can be determined. (Figure 25)

7.4.4 Oscillator Method. In this method, the active device is allowed to generate steady state oscillations by matching it appropriately. Figure 26(a) demonstrates one implementation of the method, using a coaxial oscillator. The oscillator consists of an IMPATT diode mounted in a modified 50 ohm APC-7 coaxial connector, with a quarter-wave transformer matching section. The oscillator feeds a 50 ohm load, via a bias tee unit which allows d.c. current to be applied.

The conjugate matching relation required for steady-state oscillation, holds at any terminal plane within the oscillator circuit. Hence at the plane A-A, the chip impedance transformed by package and mount is the conjugate match of the input impedance to the right of A-A, i.e.,

$$|R_{in}\ (I_o,\ I_{rf})| = R_L; \quad X_{in}\ (I_o,\ I_{rf}) = -X_L$$

From a knowledge of the package and mount transformation, $R_d$ and $X_d$ may be found from $R_{in}$ and $X_{in}$.

Assuming that the oscillator is an amplifier with infinite gain, then the added power becomes the oscillator output power (provided the matching transformer is lossless), and

$P_o = P_{add} = \frac{1}{2} I_{rf}^2 |R_d|$, from which $I_{rf}$ can be found.

The procedure is as follows:

(a) Using the techniques previously described, the package and mount transformation from chip to plane A-A, is found.

(b) Device bias and matching transformer are adjusted to provide steady-state oscillation at the required frequency.

(c) The output power and frequency are measured.

(d) The diode holder is removed. The remaining circuit is compatible with the network analyser reflection test unit input APC-7 connector, enabling $R_L$ and $X_L$ to be measured by the network analyser.

(e) $R_d$, $X_d$ and $I_{rf}$ are calculated.

A modification to this method has been described by McBretney and Howes (6). The circuit connecting the active diode chip to the 50 ohm load, comprising package, mount, transformer, waveguide to coaxial transducer if the diode is mounted in waveguide, is represented at a single frequency by a two port network (Figure 26(b)). By using three chip reference impedances, the network parameters are found - for the matching section which has previously produced a known output power at a chosen measurement frequency. From the condition that for steady state oscillation the transformed chip impedance must match the load impedance, the diode impedance may be found. This method offers the advantage that the power meter and frequency meter load impedances are usually equal to the coaxial matched load impedance, making the measured diode chip large-signal impedance less sensitive to interchange of measuring equipment.

## 7.5 CONCLUSION

This chapter has described the principles behind the measurement of active device impedance at microwave frequencies. The initial section described the evolution and determination of package and mount equivalent circuits. The later sections demonstrate how these equivalent circuits are used in the measurement of device small-signal and large-signal impedance, and discuss briefly the meaning and use of these latter terms.

## REFERENCES


1. Ramo, S., Whinnery, J.R. and Van Duzer, J., 1965, "Fields and Waves in Communication Electronics", Wiley Toppan.

2. Getsinger, W.J., 1966, IEEE Trans. Microw. Th. and Tech., MTT-14, 58-69.

3. Owens, R.P., 1971, Elect. Lett., 7, 580-582.

4. Greiling, P.T., and Haddad, G.I., 1970, IEEE Trans. Microw. Th. and Tech., MTT-18, 842-852.

5. Dean, M., and Howes, M.J., 1973, IEEE Trans. Microw. Th. and Tech., MTT-21, 121-127.

6. McBretney, J., and Howes, M.J., 1979, IEEE Trans. Microw. Th. and Tech., MTT-27, 256-265.

7. Montgomery, C.G., Dicke, R.H. and Purcell, E.M., 1964, Principles of microwave circuits, Boston Technical Publishers Inc.

8. Wight, J.S., Jain, O.P., Chudobiak, W.J. and Makios, V., 1974, IEEE Trans. Microw. Th. and Tech., MTT-22, 48-52.

9. Decker, D.R. and Dunn, C.N., 1970, IEEE Trans. Electron. Devices, ED-17, 290-299.

10. Decker, D.R., Dunn, C.N. and Frank, R.L., 1970, IEEE Trans. Microw. Th. and Tech., MTT-18, 872-876.

11. Scherer, E.F., 1970, IEEE Trans. Microw. Th. and Tech., MTT-18, 922-931.


## TUTORIAL QUESTIONS

1. The transmission line equations derive from an incremental model for a short length of line. By considering an "L" section equivalent circuit for a short length dz, of lossless coaxial line show that

$$\frac{dv}{dz} = -jwLI; \quad \frac{dI}{dz} = -jwcV$$

where L and C are the inductance per unit length and capacitance per unit length respectively. Comment on the variation of characteristic impedance, $Z_o = \sqrt{\frac{L}{C}}$, with z.

2. The line equations describing the impedance transformation of an incremental length of radial line have been given in Reference (7) p. 256, by

$$\frac{dV}{dr} = -jw\sqrt{\varepsilon\mu}\, Z_o I; \quad \frac{dI}{dv} = -jw\sqrt{\varepsilon\mu}\, Y_o V$$

Compare these with the standard transmission line equations to find the equivalent line capacitance and inductance at a radial position, r. Hence show that the lumped equivalent circuit capacitance and inductance for an incremental length, dr, of radial line, are

$$C = \frac{2\pi\varepsilon r dr}{h} \; ; \quad L = \frac{\mu h}{2\pi} \ln \frac{\left(r + \frac{dr}{2}\right)}{\left(r - \frac{dr}{2}\right)}$$ where h is the height of the radial line.

3. Under certain conditions, the positional order of a series inductance and shunt capacitance in a ladder-type network can be interchanged, without materially affecting the network impedance transformation. Use the example of a single "L" section network to show that to allow order interchange, the condition $W^2LC << 1$ must be satisfied.

4. A varactor diode chip with negligible series resistance, and reverse bias capacitance given by the expression

$$C_j(V) = \frac{C_o}{\left(1 - \frac{V}{0.6}\right)^{\frac{1}{2}}}$$ is mounted in a microwave package with lumped element equivalent circuit similar to that of Figure 6(d).

A measurement at 1 MHz results in the package plus-varactor chip capacitance variation with bias shown in Figure 7. If the packaged varactor exhibits a series resonance in its measured microwave input impedance at 9.09 GHz, with a varactor reverse bias of 25 volts, determine the values of L and C in the package equivalent circuit.

5. The equivalent circuit of Figure 6(b) is to be used to describe the package shown in Figure 1(a). Assume that $L_1$ represents the coaxial inductance of the pedestal confined by the outer cylindrical surface of the package, and that the package dimensions are as given in Figure 1(a). The series resonance of an open-circuit package occurs at 26 GHz, while a short-circuit package (with the same lead geometry) exhibits a parallel resonance at 17.8 GHz. Find the value of the elements of the package equivalent circuit. What would be the effect on the equivalent circuit of increasing the number of bonding leads?

6. Small-signal device equations are derived by separating variables into d.c. terms, and a.c. terms which are assumed to be small in magnitude. If carrier velocity is given by

$$v = \mu E = \frac{\mu_o E}{\left\{1 + \left(\frac{\mu_o E}{vs}\right)^{\beta}\right\}^{\frac{1}{\beta}}}$$ , where $\mu_o$ = low-field mobility, vs = saturated velocity, $\mu$ = constant, separate the variables $\mu(E)$ and E into d.c. and a.c. components, and show that the small-signal velocity component is

$$v_{ac} = \mu_{dc}\left\{1 - \left(\frac{v_{dc}}{vs}\right)^{\beta}\right\}E_{ac}$$ - i.e. a linear relation.

$$v_{dc} = \mu_{dc} E_{dc} = \frac{\mu_o E_{dc}}{\left\{1 + \left(\frac{\mu_o E_{dc}}{vs}\right)^{\beta}\right\}^{\frac{1}{\beta}}}$$

7. When the computer is used to assist package equivalent circuit evaluation, as described in section 7.3.3, the minimisation operates on a function of $|\Gamma|$ and $\theta$, the magnitude and phase of the input reflection coefficient at the network analyser measurement plane. Derive the relationship between input impedance and input reflection coefficient magnitude and phase.

8. In McBretney and Howe's method for Gunn diode small-signal impedance measurement, the package transformation may be represented by the two-port network of Figure 13(c). Show that the input package admittances corresponding to O/C and S/C chips in lossless packages are $Y_{S/C} = \frac{1}{B}$, $Y_{O/C} = \frac{C}{A}$.

Use the results of Figure 16(a) to find the O/C package and S/C package resonant frequencies. Compare the O/C package resonant frequency with that measured by Owens (3), Figure 16(b), and comment on any discrepancy.

9. A BARITT diode is placed in the microstrip mount of Figure 14. The line impedance is 50 ohms and the wavelength in the microstrip is given by

$\lambda g = \frac{\lambda o}{\sqrt{\varepsilon eff}}$ , where the effective dielectric constant, $\varepsilon$eff is 1.92.

The measured mount input phase angle (for a lossless mount) as a function of "cut-back" line length is shown in Figure 15(c), at a frequency of 10.5 GHz. Find the values of the launcher equivalent circuit (Figure 14) (a) with $C_{O/C}$ = 0 (b) with $C_{O/C}$ = 0.03 pF.

10. Figure 17 shows an IMPATT diode placed in a conventional waveguide mount, together with an equivalent circuit for the mount. Discuss how the principles of the methods described in Sections 7.2 and 7.3 may be used to find the circuit element values, and hence the chip impedance from waveguide input impedance measurements.

11. Show, using the parallel oscillator equivalent circuit of Figure 19(a), that the active device available power may be written

$P = \frac{1}{2}|G_d|v_{rf}^2$ where $v_{rf}$ = peak r.f. diode voltage. Figure 18 shows the measured admittance of two IMPATT diode chips (A & B). Find the variation of available power for each diode as a function of r.f. voltage, and hence determine their maximum available powers. Comment on the results.

If diode A is to produce peak power at 6.5 GHz, find the required load admittance at the diode chip.

12. A reflection amplifier is designed using diode B of Figure 18 with a load conductance of 6 ms. Show that the centre frequency power gain can be expressed as

$$g = \left|\frac{G_L + |G_d|}{G_L - |G_d|}\right|^2$$ (assuming that the diode conductance is negative). Use the experimental results of Figure 18 to calculate the gain-input power curve and hence find the small-signal gain and 1 dB gain compression point for the amplifier. How does this compare with the corresponding values if diode A were used?

13. Figure 24 tabulates the results of a large-signal impedance measurement on a BARITT diode. Evaluate the diode chip added power, and hence determine the variation of chip impedance with peak r.f. current for the d.c. bias current of 25 mA (a) with $R_s$ = 0 (b) with $R_s$ = 1.4Ω.

14. A reflection amplifier has the large-signal power gain-frequency performance shown in Figure 22. The active device small-signal impedance ($P_i$ = 0.1 mW) is -4-j22 ohms. Assuming that the device impedance is constant with frequency over the amplifier pass-band, and that the load seen by the active device at the amplifier centre frequency is constant as the input power is varied, evaluate the large-signal device impedance as a function of device peak r.f. current.

15. Show that for a reflection amplifier, the ratio of large-signal active device resistance to small-signal device resistance is given by

$$\frac{R_d}{R_{d_s}} = \frac{(\sqrt{g} - 1)(\sqrt{g_s} + 1)}{(\sqrt{g} + 1)(\sqrt{g_s} - 1)}$$ where g and $g_s$ are the large-signal and small-signal amplifier gains respectively.

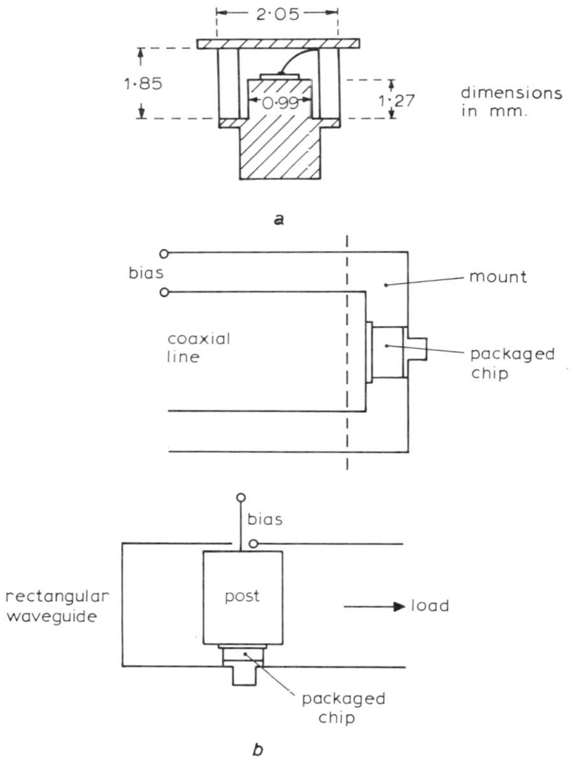


Figure 1 (a) Microwave package
(b) Typical microwave mounts

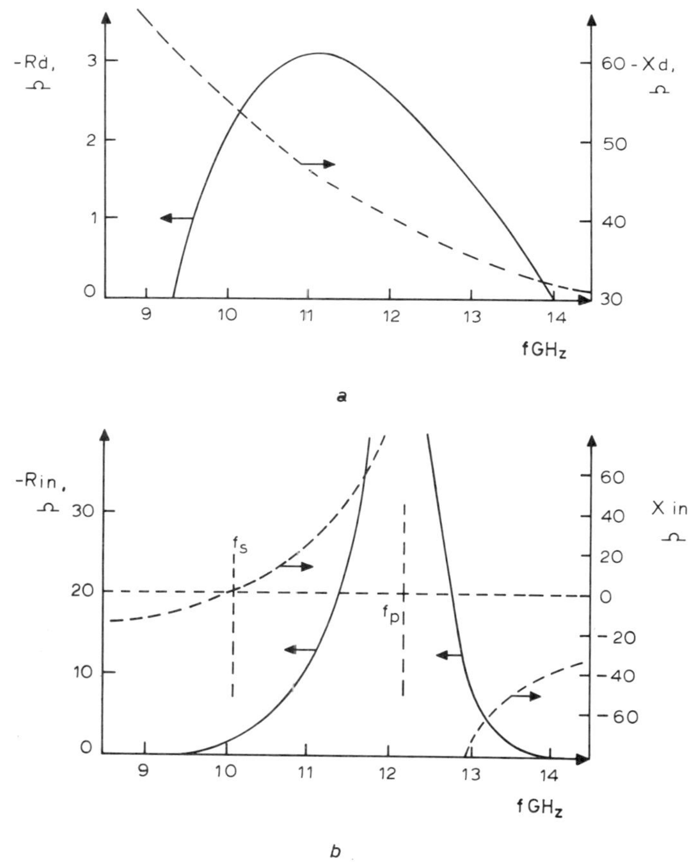


Figure 2 (a) BARITT diode chip small-signal impedance
(b) Corresponding coaxial mount input impedance

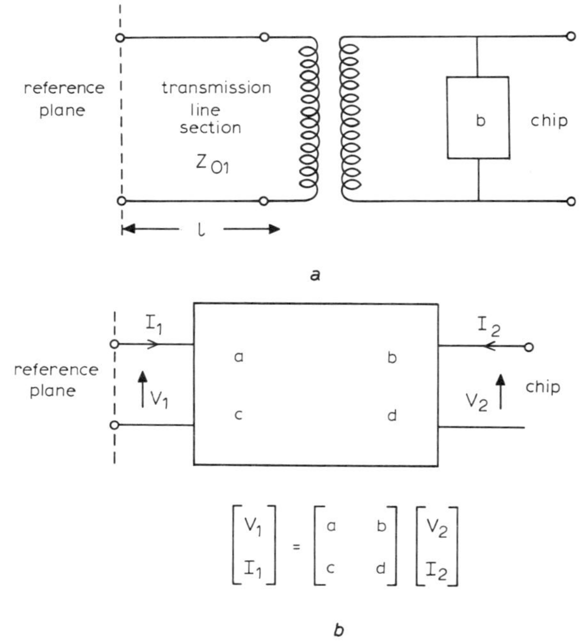


Figure 3 Chip-to-measurement plane equivalent circuits

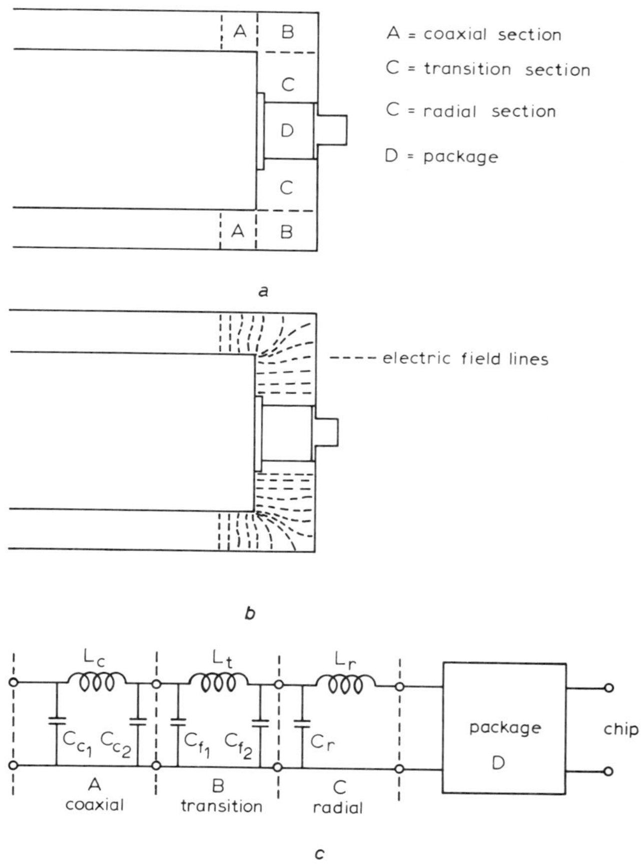


Figure 4

(a) Coaxial mount sub-regions
(b) Coaxial mount electric fields
(c) Coaxial mount equivalent circuit

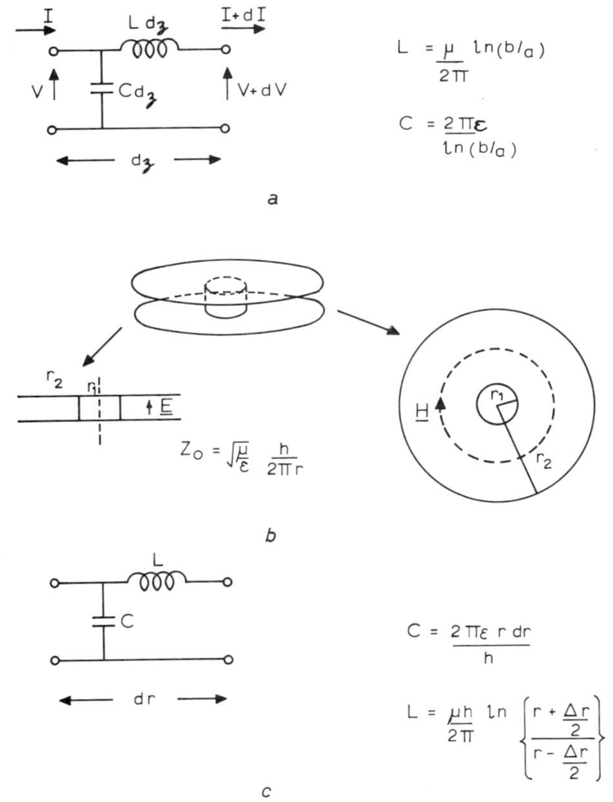


Figure 5 (a) Incremental coaxial line equivalent circuit
(b) Radial line section
(c) Incremental radial line equivalent circuit

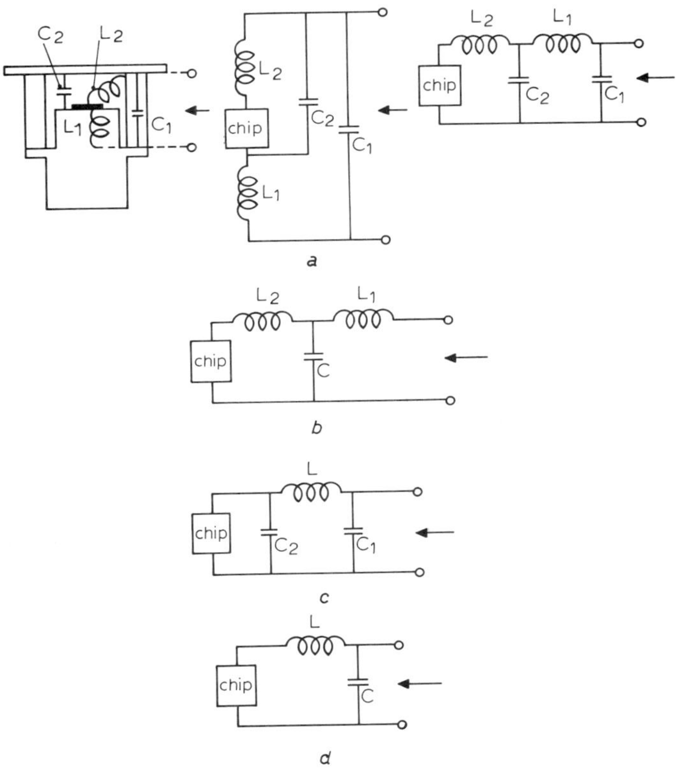


Figure 6 (a) Package equivalent circuit evolution
(b) T equivalent circuit
(c) π equivalent circuit
(d) L equivalent circuit

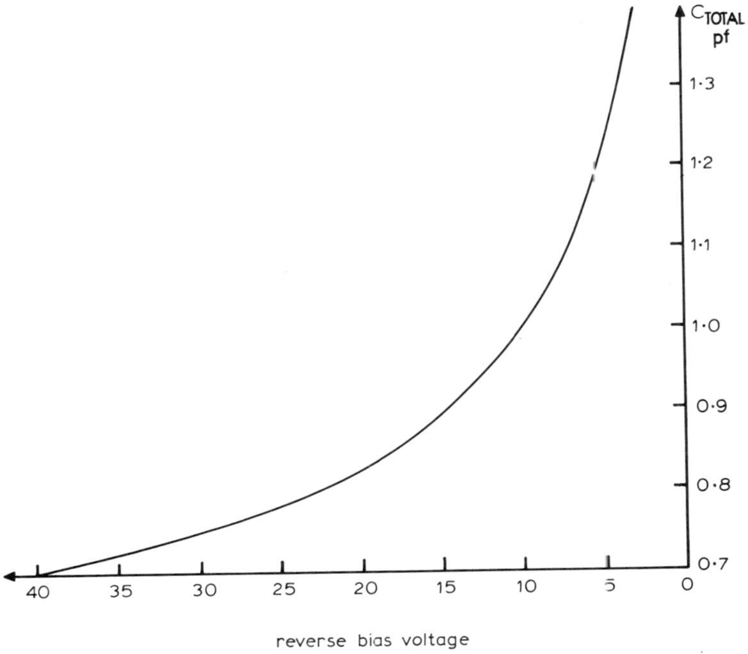


Figure 7 Packaged varactor diode capacitance versus reverse bias voltage

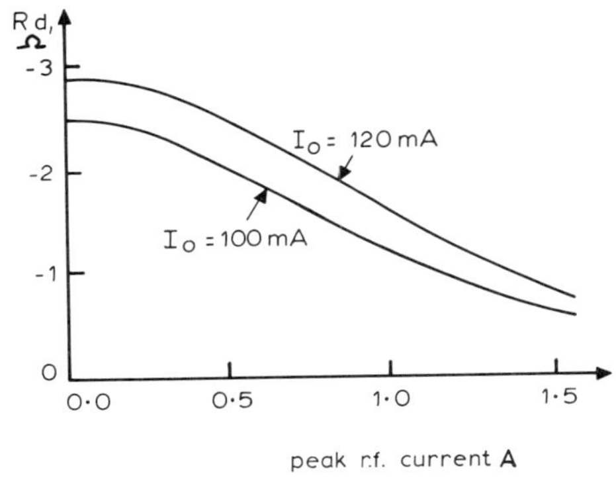


Figure 8 IMPATT diode chip resistance versus peak r.f. current

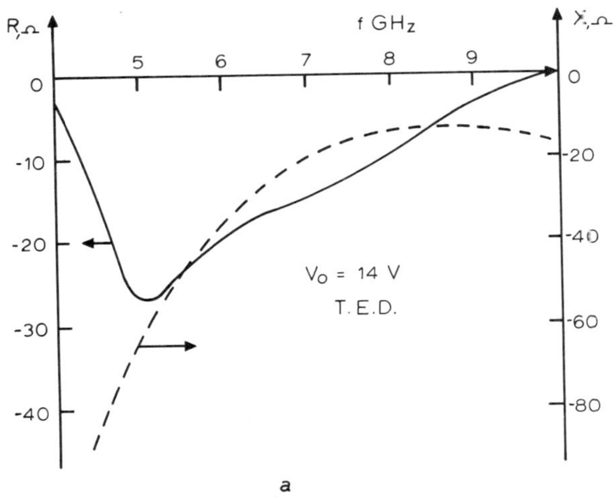


Figure 9 (a) Transferred electron device chip small-signal impedance

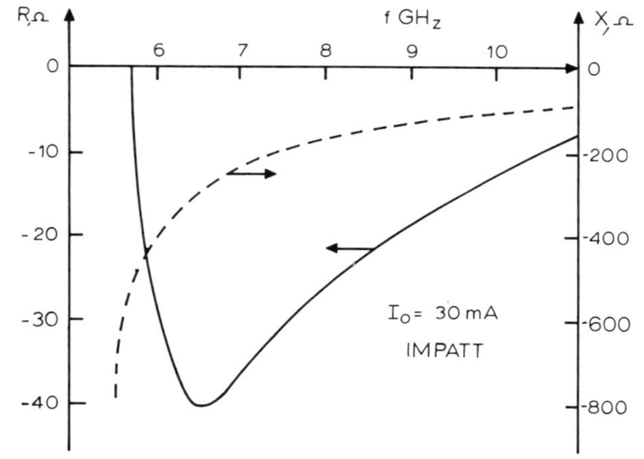


b

(b) IMPATT diode chip small-signal impedance

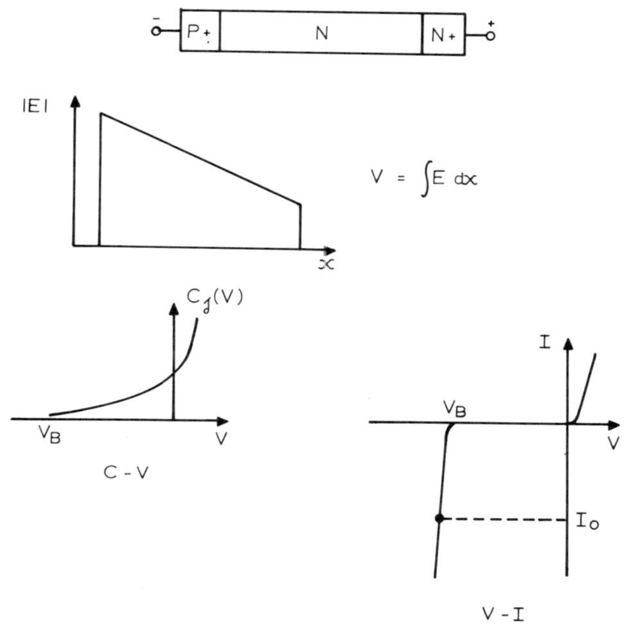


Figure 10 IMPATT diode parameters

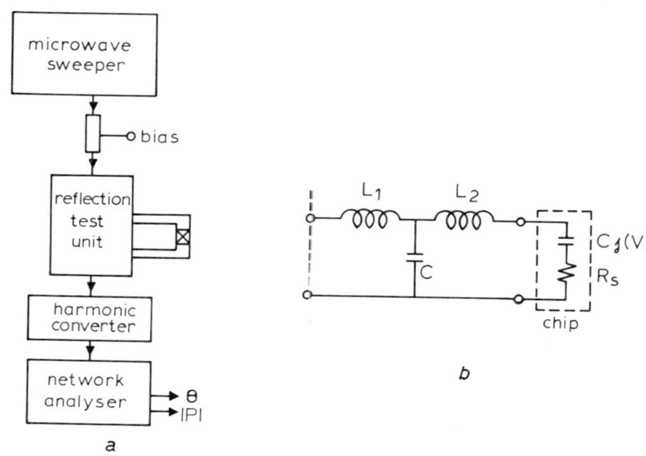


Figure 11 (a) Network analyser small-signal impedance measurement
(b) Package, mount and IMPATT chip equivalent circuit (below avalanche breakdown)

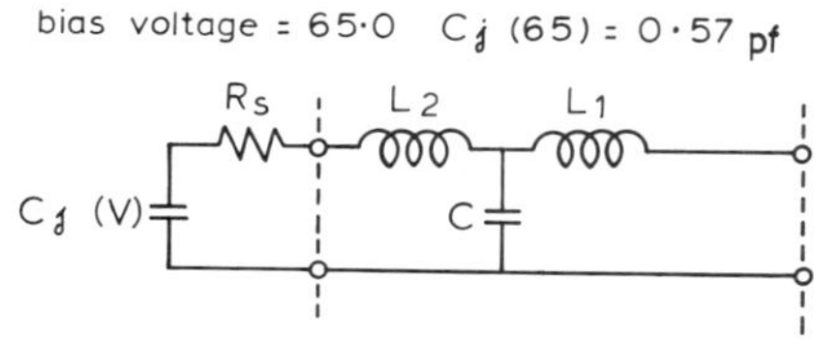


| | $L_1$ (nH) | $L_2$ (nH) | C (pF) | $R_s$ (Ω) |
|---|---|---|---|---|
| initial | 0·4 | 0·3 | 0·3 | 1·0 |
| final | 0·281 | 0·344 | 0·470 | 1·21 |

average error in |Γ| = 0·013

average error in θ = 0·72°

frequency range 8 GHz - 12 GHz

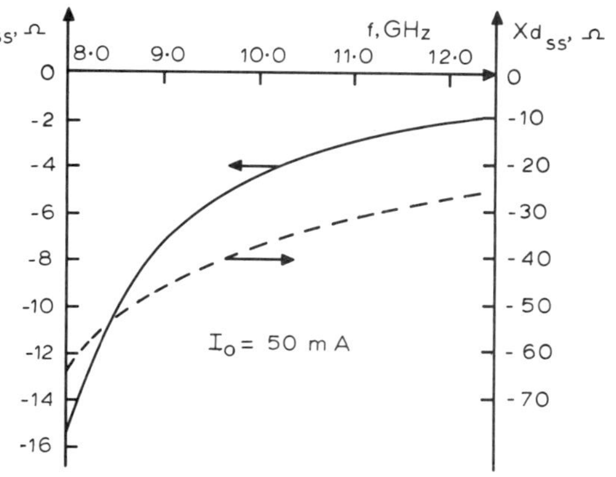


Figure 12 Results of computer-aided IMPATT diode chip small-signal impedance measurement

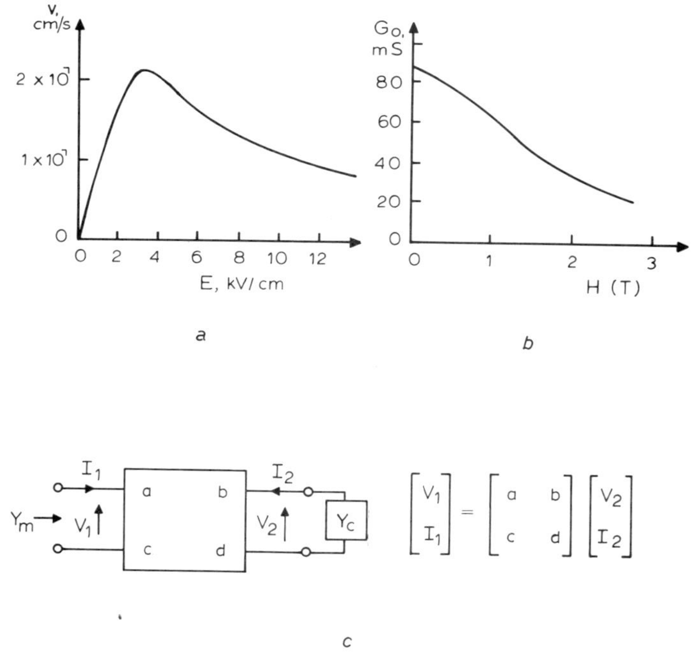


Figure 13 (a) Velocity - electric field characteristics of T.E.D.
(b) T.E.D. magnetic field-dependent conductance
(c) Mount and package two-port equivalent circuit

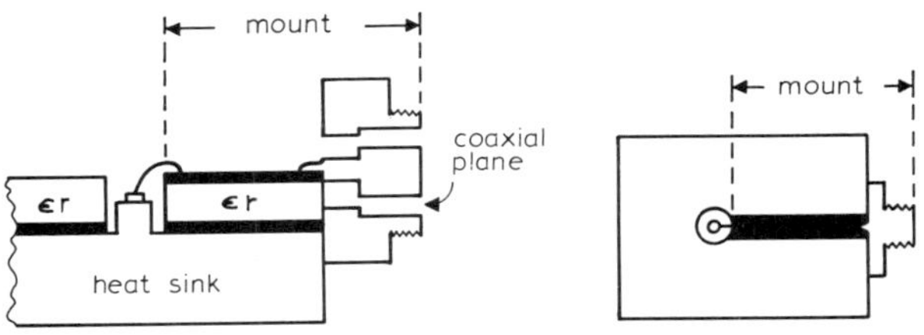


*a*

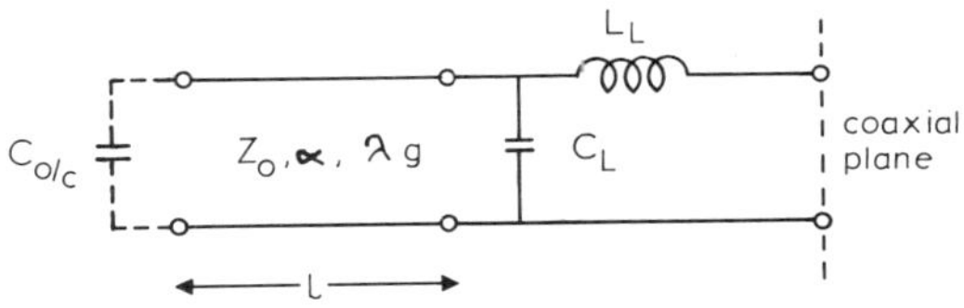


*b*

Figure 14 (a) Active device chip in microstrip mount
(b) Microstrip mount equivalent circuit

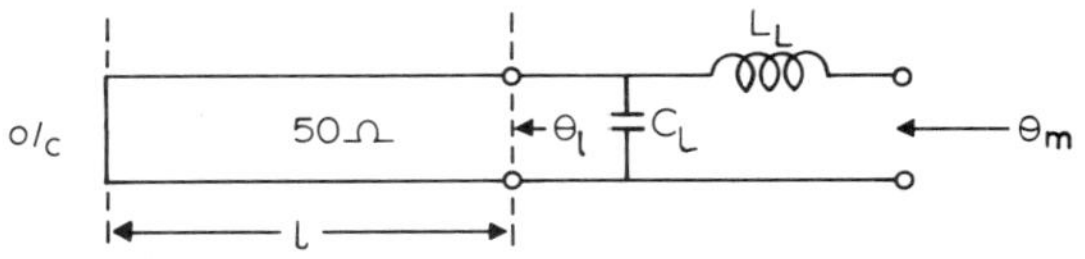


*a*

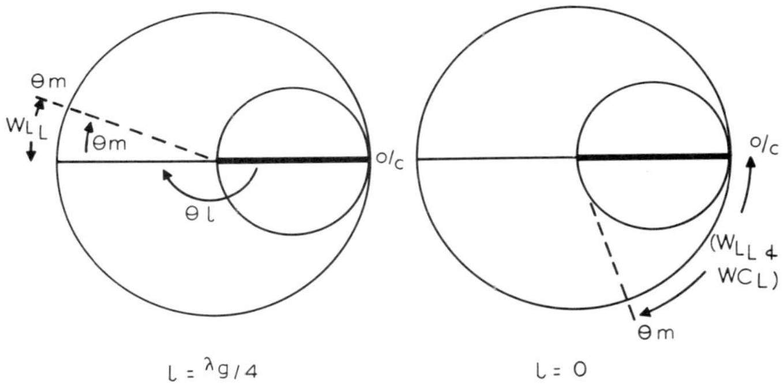


*b*

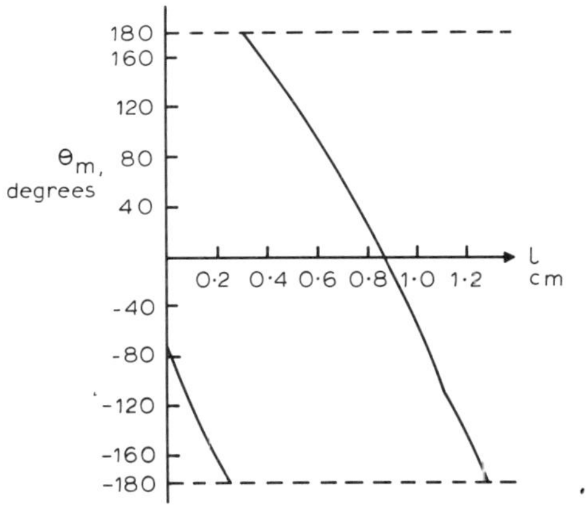


*c*

Figure 15 (a) Microstrip mount equivalent circuit values
(b) Smith Chart displays for 1 = 0, 1 = $\lambda_g/4$
(c) Measured input plane phase angle versus 1

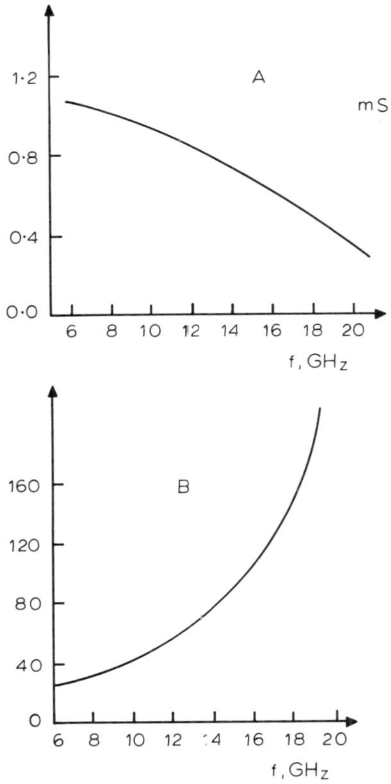


*a*

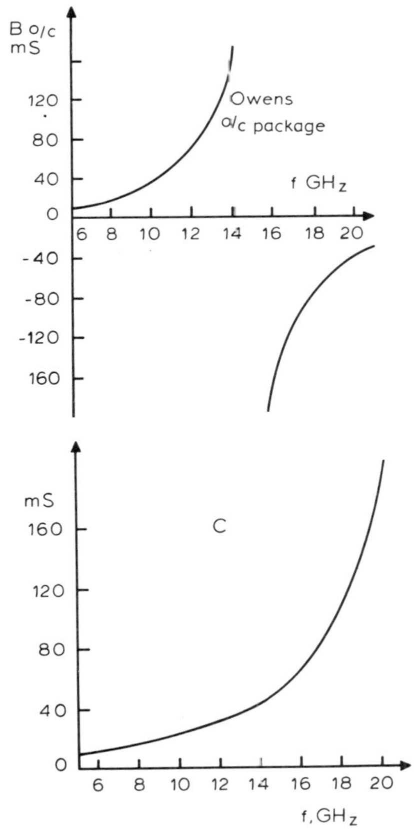


*b*

Figure 16 (a) Package two-port network parameters
(b) Owen's measured O/C package susceptance

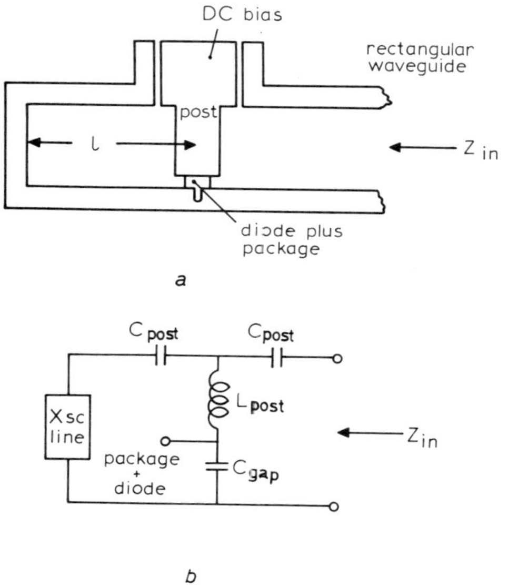


Figure 17 (a) IMPATT diode in waveguide mount
(b) Typical equivalent circuit for waveguide mount

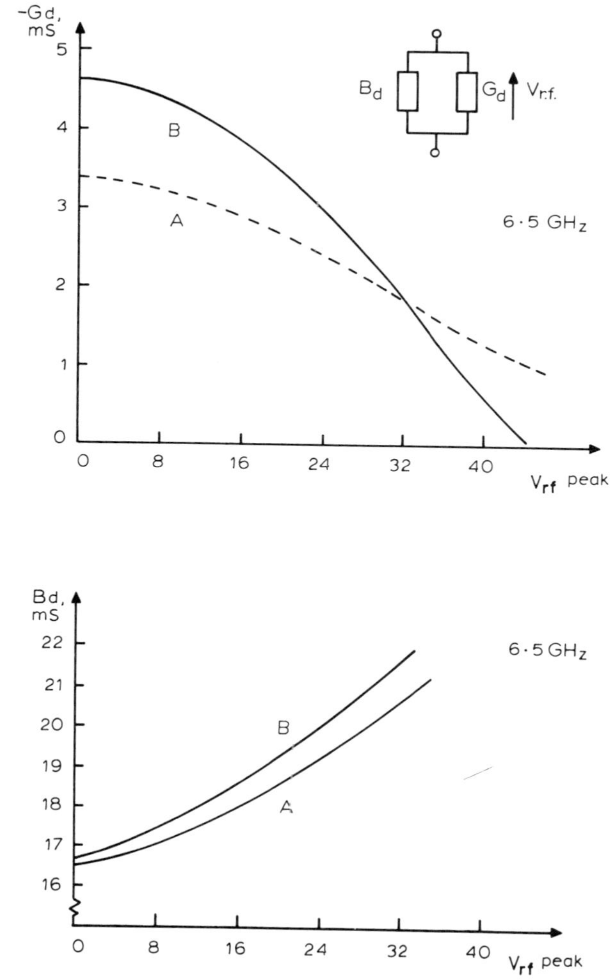


Figure 18 IMPATT diode chip large-signal admittance

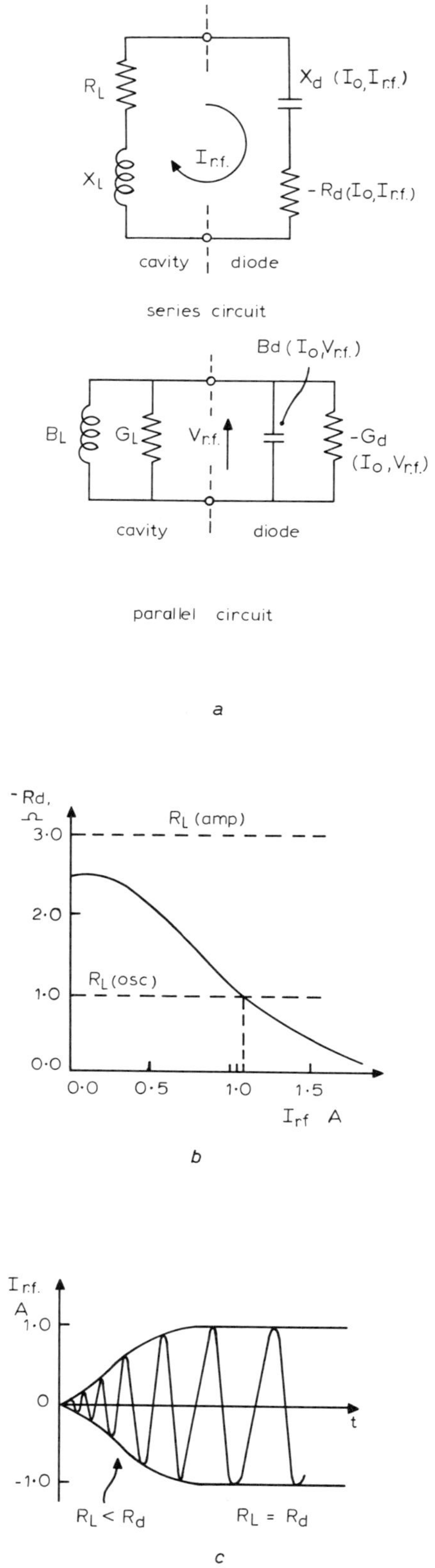


Figure 19 (a) Oscillator equivalent circuits
(b) Typical large-signal chip resistance versus r.f. current
(c) Growth of microwave oscillation

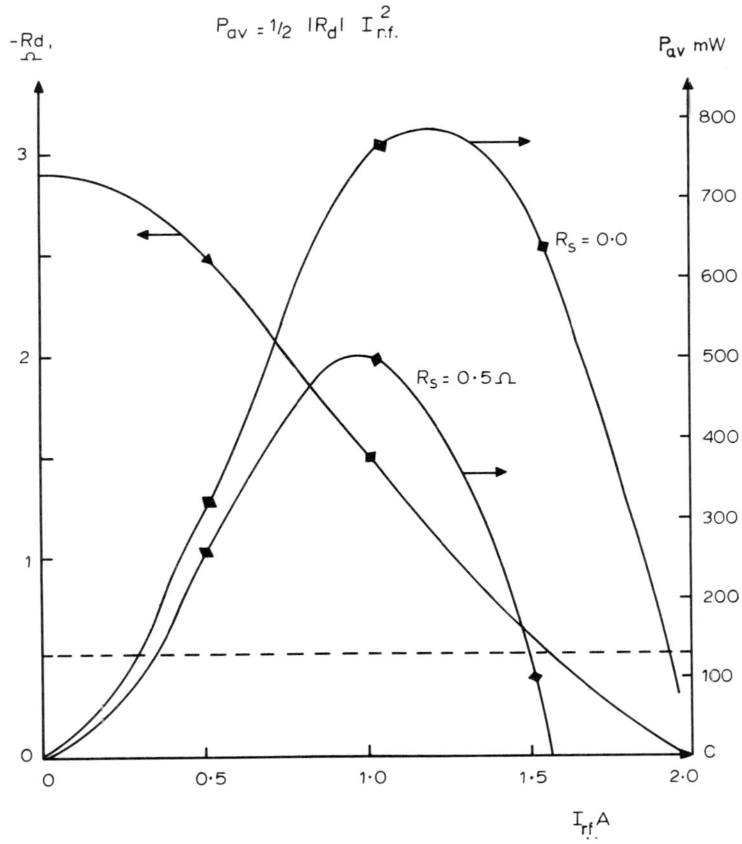


Figure 20 Computed available power for an IMPATT diode chip

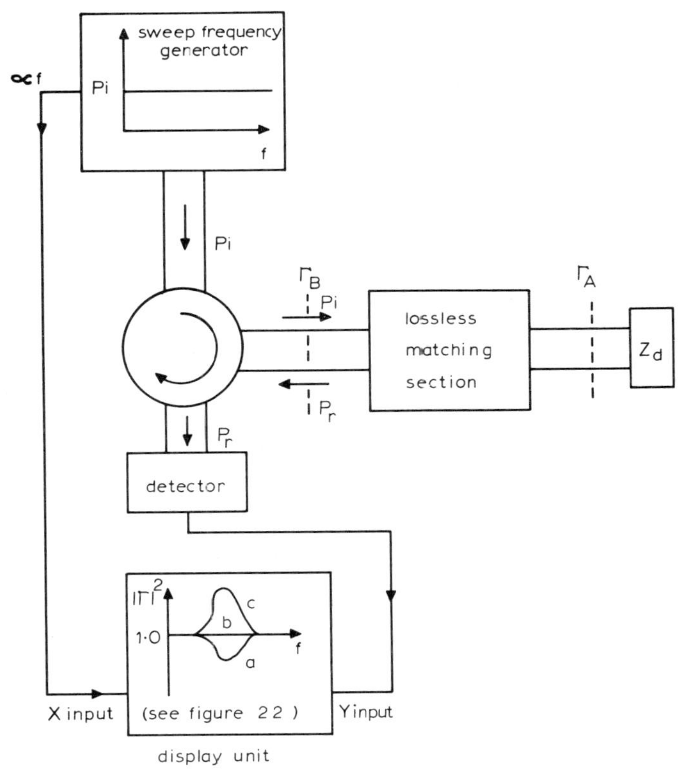


Figure 21 Microwave reflection amplifier

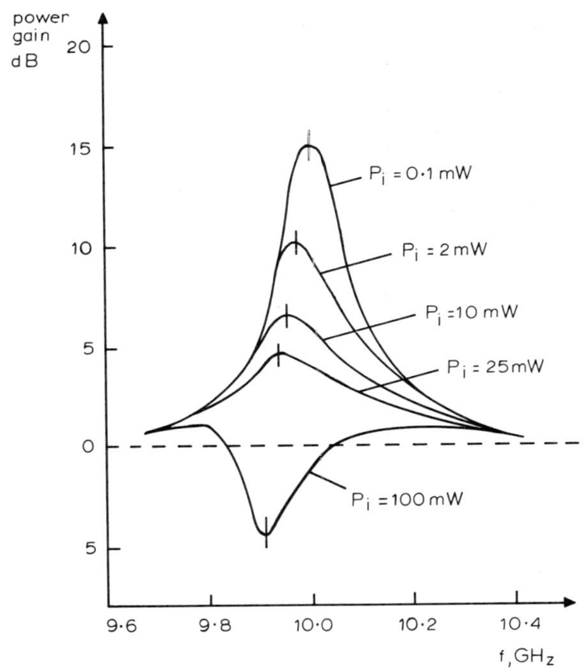


Figure 22 Measured large-signal performance of IMPATT reflection amplifier

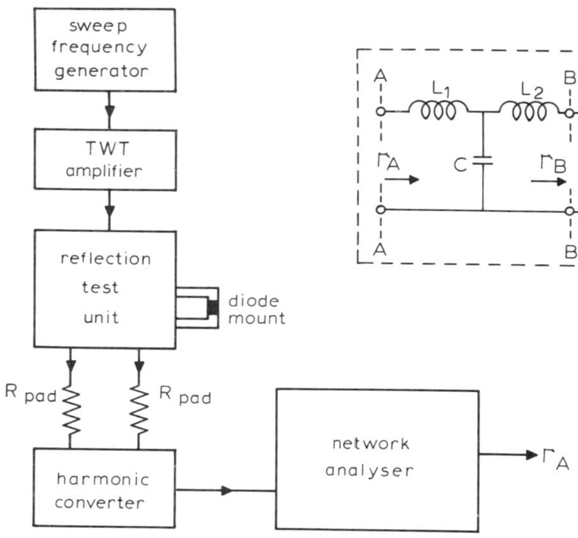


Figure 23 Network analyser layout for large-signal impedance measurement

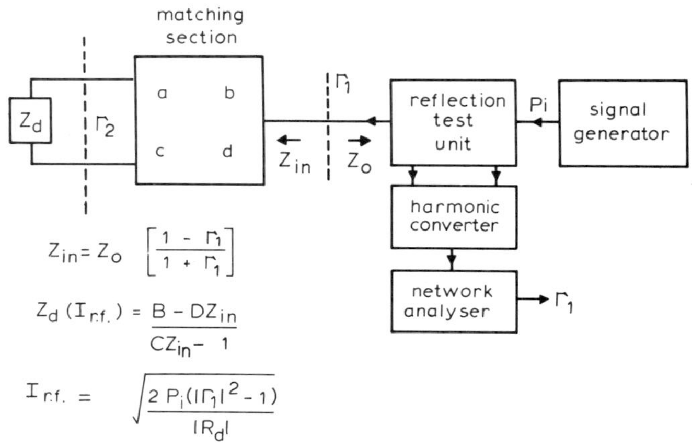


Figure 25 Modified amplifier method using network analyser

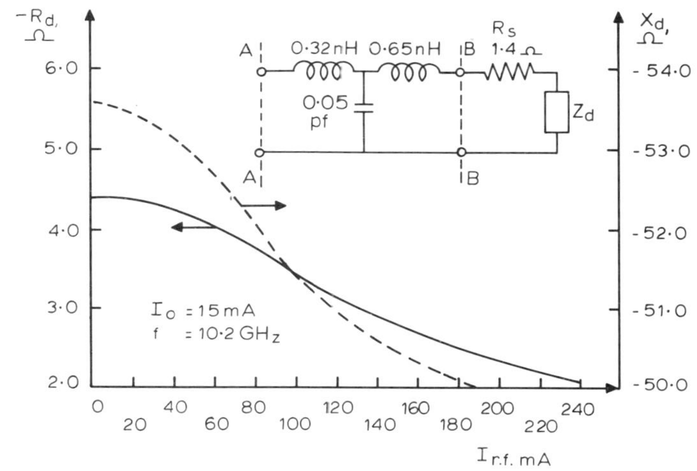


a

| Pi mW | $\vert\Gamma_A\vert$ | Rd Ω | Xd Ω |
|---|---|---|---|
| 2·5 | 1·06 | -3·05 | - 55·5 |
| 10 | 1·073 | -3·40 | -55·05 |
| 20 | 1·079 | -3·57 | -54·71 |
| 40 | 1·085 | -3·70 | -54·03 |
| 60 | 1·080 | -3·58 | -53·51 |
| 80 | 1·073 | -3·40 | -53·17 |
| 100 | 1·072 | -3·36 | -52·79 |
| 140 | 1·065 | -3·18 | -52·23 |
| 180 | 1·058 | -3·01 | -51·80 |

BARITT in coaxial mount
$I_o$ = 25 mA
f = 10·2 GHz

b

Figure 24 (a) Measured BARITT large-signal impedance at $I_o$ = 15 mA
(b) Tabulated results for $I_o$ = 25 mA

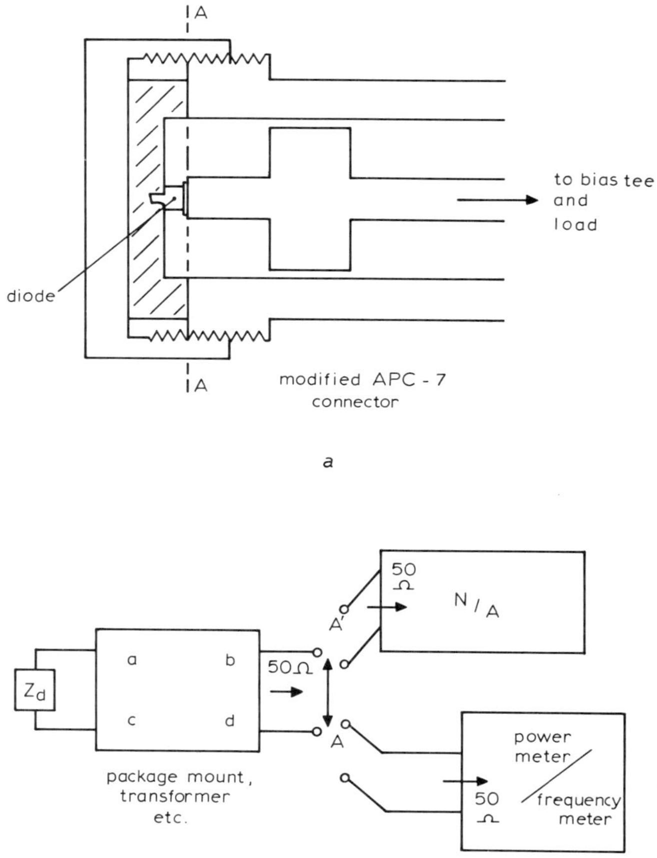


b

Figure 26 (a) Coaxial oscillator for large-signal impedance measurement
(b) Large-signal impedance measurement set up using two-port network

# CHAPTER 8

# Design of transistor amplifiers

Dr. J. Pengelly

# THE DESIGN OF MICROWAVE TRANSISTOR AMPLIFIERS

R.S. Pengelly

Plessey Research (Caswell) Ltd., Allen Clark Research Centre, Towcester, Northamptonshire, U.K.

## INTRODUCTION

The design of transistor amplifiers at microwave frequencies has recently taken up new impetus with the introduction in the early 1970's of the gallium arsenide Schottky-barrier gate field effect transistor. Modern GaAs FET devices are capable of producing noise figures as low as 3 dB at 18 GHz with almost 10 dB associated gain whilst at the lower frequencies noise figures of less than 0.5 dB are possible. Power FETs are now capable of producing 15 watts up to X-band in Class A operation with promise of considerably more power in the pulsed mode of operation.

The GaAs FET offers in its simplest form a device having somewhat more convenient impedances for matching than the bipolar device and since it is also capable of exhibiting much superior performance above approximately 4 GHz than the bipolar transistor this chapter deals exclusively with the theory and design of amplifiers using GaAs FETs. Reflection type amplifiers are not dealt with although some information on the common gate and source follower configurations is included.

## LOW NOISE/SMALL SIGNAL AMPLIFIERS

### S-PARAMETERS

The measurements which are usually made on a three terminal device are those where one of the terminals is effectively earthed. In the case of a microwave GaAs FET it is usual to measure the device in common source configuration, i.e. where the source electrodes are bonded to r.f. ground. This arrangement usually provides the design engineer with the most stable arrangement and low feedback capacitance between drain and gate. Alternative configurations such as common gate and common drain are dealt with later.

The gain parameters of a microwave transistor can be completely specified by a set of 2 port parameters, the so-called 'scattering (S) parameters'. Other descriptions of the transistor such as H or Y parameters require open- and short-circuited measurements that are difficult to establish accurately at microwave frequencies.

Because S-parameters are measured using travelling waves it is essential to specify the plane of reference of the measurements with respect to the device. Usually FET data sheets specify the reference planes at the ends of wire bond lengths furthest from the device in the case of bare chip transistors and at the package lead to substrate interface in the case of packaged devices.

The S-parameters of a GaAs FET whether chip or packaged have to be measured accurately over a large range of frequencies.

Since errors can occur in measurement due to equipment and also the test jigs used to hold the devices, all measurements made on FETs should be achieved using an automatic network analyser with post measurement processing to extract errors such as microstrip transmission line losses, coaxial to microstrip transition errors and dispersion with frequency.

Also a relatively large number of S-parameter measurements on a particular FET type need to be obtained so that a set of typical S-parameters can be generated for device data sheets. Since the maximum available gain of the device and the minimum noise figure are produced at different d.c. bias conditions at least two sets of S-parameters are needed. Also the S-parameter data for the device biased for maximum output power is often required. (This occurs at approximately 50% of the saturated drain current for a GaAs FET).

Table I shows typical S-parameters for a FET chip device (Plessey GAT6) over 2 to 18 GHz for a bare chip device whilst Table II also shows the same data over 4 to 16 GHz for a packaged GAT6. It may be seen that the package produces a considerable rotation in the phase of the S-parameters.

Bare chip devices should be used where the ultimate in noise figure or gain-bandwidth product is used. Packaged devices have the advantage of being pretested by manufacturers , being usually supplied in hermetically sealed enclosures and being more easily handled. Offset against these advantages are the disadvantages of narrower band applications and slightly inferior r.f. performance.

Chips or packaged devices should always be used with the grounding and mounting methods used by the manufacturer when measuring the S-parameters.

### STABILITY OF A 2 PORT

One important parameter in the design of microwave amplifiers is that of stability. An amplifier can be either unconditionally or conditionally stable. A circuit is unconditionally stable if its input and output resistances are positive for passive terminations. If the amplifier is conditionally stable then the input and output resistances may be negative.

The measured S-parameters of a transistor enable the maximum available gain, MAG, of the device to be determined.

The transducer gain, G, of a transistor is defined as:

$$G = \frac{|S_{21}|^2(1-|\Gamma_S|^2)(1-|\Gamma_L|^2)}{|(1-S_{11}\Gamma_S)(1-S_{22}\Gamma_L)-S_{12}S_{21}\Gamma_S\Gamma_L|^2} \qquad (1)$$

When the generator and load, $\Gamma_S$ and $\Gamma_L$ are image matched to the two FET ports the gain can be maximised to give:

$$G_{max} = \frac{|S_{21}|^2}{(1-|S_{11}|^2)(1-|S_{22}|^2)} \quad (2)$$

where $\Gamma_S = S_{11}^*$ and $\Gamma_L = S_{22}^*$, $S_{11}^*$ and $S_{22}^*$ being the complex conjugate reflection coefficients. In Eq.(2) $S_{12}$ is assumed to be zero, i.e. the device is assumed unilateral. The S-parameters also determine Rollett's stability factor, K (1).

$$K = \frac{1 + |S_{11}S_{22} - S_{12}S_{21}|^2 - |S_{11}|^2 - |S_{22}|^2}{2\,|S_{12}S_{21}|} \quad (3)$$

If K is larger than unity an optimum combination of $\Gamma_S$ and $\Gamma_L$ can simultaneously match the 2 FET ports to maximise the gain (2). If K is smaller than unity, the FET is only conditionally stable and $\Gamma_S$ and $\Gamma_L$ must be carefully chosen to operate the device in a stable region (2,3).

The I/P and O/P reflection coefficients with arbitrary source and load terminations are given by:-

$$\Gamma_{IN} = S_{11} + \frac{S_{21}S_{12}\Gamma_L}{1 - S_{22}\Gamma_L} \quad (4)$$

$$\Gamma_{OUT} = S_{22} + \frac{S_{21}S_{12}\Gamma_S}{1 - S_{11}\Gamma_S} \quad (5)$$

If we set equation (4) equal to one, a boundary is established beyond which the device is unstable

$$|\Gamma_{IN}| = \left|S_{11} + \left|\frac{S_{21}S_{12}\Gamma_L}{1 - S_{22}\Gamma_L}\right|\right| = 1$$

The solution to this equation is a circle whose radius $r_L$ is given by

$$r_L = \frac{S_{12}S_{21}}{|S_{22}|^2 - |\Delta|^2} \quad (6)$$

and whose centre $C_L$ is given by

$$C_L = \frac{(S_{22} - \Delta S_{11}^*)^*}{|S_{22}|^2 - |\Delta|^2} \quad (7)$$

where $\Delta = S_{11}S_{22} - S_{12}S_{21}$

The area either inside or outside the circle represents a stable operating condition. If $|S_{11}| < 1$ and $|\Gamma_{IN}| < 1$ then the shaded region in Figs. 1(a), 1(b) will enable stable gain to be realised. In Fig. 1(a) the stability circle encloses the origin of the Smith Chart and hence its interior represents the stable region. In Fig, 1(b) the origin is outside the circle and the exterior of the circle represents the stable region.

If $|S_{11}| > 1$, the origin (i.e. at $\Gamma_L = 0$) represents an unstable point. Fig.1(c) is the case where the stability circle encircles the origin and its exterior is the stable region. For the case where the circle does not encircle the origin, its interior represents the stable region as in Fig. 1(d).

For unconditional stability we must ensure that the magnitude of the vector $C_L$ the distance from the centre of the Smith Chart to the centre of the stability circle (Fig. 1(a)) minus the radius of the stability circle $r_L$ is greater than one. Table III lists the stability results for a packaged Plessey GAT6 from 6 to 14 GHz.

For the case where $K < 1$ the maximum stable gain or MSG of the device can be calculated as

$$MSG = \left|\frac{S_{21}}{S_{12}}\right| \quad (8)$$

In general, $MAG = \left|\frac{S_{21}}{S_{12}}\right|(K - \sqrt{K^2-1})$

Thus MAG = MSG when K = 1 and the MSG is the gain that can be obtained from the GaAs FET when the input and output reflection coefficients fall on the boundaries of the instability regions.

TRANSDUCER POWER GAIN

In the design of small signal amplifiers, a transducer power gain can be defined for the device terminated in arbitrary load and generator impedances, $Z_L$ and $Z_S$.

The transducer power gain is defined as

$$G = |S_{21}'|^2$$

where $|S_{21}'|^2$ is as defined in Eq.(1).

$$\Gamma_S = \frac{Z_S - Z_0}{Z_S + Z_0} \quad (9)$$

$$\Gamma_L = \frac{Z_L - Z_0}{Z_L + Z_0} \quad (10)$$

When $S_{12}$ is sufficiently small to be neglected (see Table I at the lower frequencies), the device is defined to have a unilateral transducer power gain

$$G_u = \frac{|S_{21}|^2(1-|\Gamma_S|^2)(1-|\Gamma_L|^2)}{|1-S_{11}\Gamma_S|^2\,|1-S_{22}\Gamma_L|^2} \quad (11)$$

The first term is related to the transistor used and remains invariant throughout the amplifier design. The other two terms, however, are not only related to the S-parameters $S_{11}$ and $S_{22}$ but also to the load and source reflection coefficients. These latter two quantities are used to control the design of the amplifier in terms of gain slope compensation etc.

The unilateral power gain can be divided into three independent gain blocks

$$G_u = G_S \cdot G_0 \cdot G_L$$

where

$$G_S = \frac{1 - |\Gamma_S|^2}{|1 - S_{11}\Gamma_S|^2} \quad (12a)$$

$$G_0 = |S_{21}|^2 \quad (12b)$$

and

$$G_L = \frac{1 - |\Gamma_L|^2}{|1 - S_{22}\Gamma_L|^2} \quad (12c)$$

The $G_0$ term exhibits a 6 dB/octave gain reduction characteristic with increasing frequency.

CIRCLES OF CONSTANT UNILATERAL GAIN

Considering Eq.12(a) for $\Gamma_S = S_{11}^*, G_S = G_{S\,max}$
for $\Gamma_S = 1 \quad , G_S = 0$
Thus for any arbitrary value of $G_S$ there is a value of $\Gamma_S$ which lies on a circle. The output gain term in Eq.12(c) is of a similar form and thus a similar set of circles can be generated. Such circles are called CONSTANT GAIN CIRCLES and can be generated from the FET S-parameters since

$$d_i = \frac{g_i |S_{ii}|}{1 - |S_{ii}|^2 (1 - g_i)} \qquad (13a)$$

$$R_i = \frac{\sqrt{1 - g_i}\,(1 - |S_{ii}|^2)}{1 - |S_{ii}|^2 (1 - g_i)} \qquad (13b)$$

where
$$g_i = G_i (1 - |S_{ii}|^2) = G_i / G_{i\,max}$$

where $G_i$ is the constant gain represented by the circle, $d_i$ is the distance from the centre of the Smith Chart to the centre of the constant gain circle along the vector $S_{11}^*$, $R_i$ is the radius of a particular circle (4).

Thus gain circles enable the design engineer to choose a matching topology for example, which will allow matching over a range of frequencies to produce constant gain or indeed gain of a specified slope with frequency. Fig. 2 is an example of the constant gain circles for a Plessey P103 packaged GAT5 at 10 GHz.

UNILATERAL FIGURE OF MERIT

Although the assumption that $|S_{12}| = 0$ is often made in initial amplifier design it is desirable to quantify the error involved in doing this.

The transducer power gain, $G$ , from Eq.(1) can be written as

$$G = G_u \cdot \frac{1}{|1 - X|^2}$$

where
$$X = \frac{\Gamma_S \Gamma_L S_{12} S_{21}}{(1 - \Gamma_S S_{11})(1 - \Gamma_L S_{22})} \qquad (14)$$

The ratio of the true gain to the unilateral gain is bounded by

$$\frac{1}{|1 + |X||^2} < \frac{G}{G_u} < \frac{1}{|1 - |X||^2} \qquad (15)$$

For complex image matching with $|S_{11}| < 1$ and $|S_{22}| < 1$

$$\frac{1}{(1 + X')^2} < \frac{G}{G_u} < \frac{1}{(1 - X')^2} \qquad (16)$$

where
$$X' = \frac{|S_{11}||S_{22}||S_{12}||S_{21}|}{(1 - |S_{11}|^2)(1 - |S_{22}|^2)}$$

A typical value for $X'$ at 10 GHz for the GAT5 in Fig. 2 is 0.1 giving

$$\frac{1}{(1 + X')^2} \equiv -0.8 \text{ dB}$$

$$\frac{1}{(1 - X')^2} \equiv +0.9 \text{ dB}$$

VARIATION OF GAIN WITH DRAIN CURRENT AND TEMPERATURE

The variation of maximum available gain of a GaAs FET with drain-source current $I_{DS}$ can be plotted at different frequencies. This enables the design engineer to calculate the gain that can be expected under certain bias conditions. Fig. 3 shows the gain v. drain current ratio $I_{DS}/I_{DSS}$ for a typical high frequency FET at 8 GHz. It may be seen that below a certain $I_{DS}/I_{DSS}$ ratio there is a sharp decrease in gain due to a rapid decrease in transconductance $g_m$ and consequently of $f_T$ as might also be seen by inspecting the drain current versus drain to source voltage characteristic of the device for gate voltages close to pinch off.

Fig. 4 shows the gain change with temperature for a packaged FET device over the temperature range -40 to +80°C.

Optimum Load Conditions for Output Power

In certain cases the FET is designed to operate in a linear (Class A) mode where the objectives are either lowest intermodulation distortion, largest added RF power or largest linear output power. In the last case, for example, the d.c. bias is graphically determined from the static drain-current versus drain-voltage characteristic (Fig. 5). The bias and load conductance line are chosen to maximise the product of the linear voltage and current swing. The limitations are determined by the maximum d.c. power dissipation, the drain-to-gate breakdown voltage and the positive gate bias ($V_{GS} \simeq 0.5$V) above which appreciable gate current flows. The optimum load conductance is typically much larger than the GaAs FETs output conductance. Consequently the GaAs FET is not matched to the load and does not deliver maximum gain. The output load conductance and susceptance determine the optimum reflection coefficient of the load, $\Gamma_{LP}$ . The generator reflection coefficient that provides a complex conjugate match to the input of the device is then

$$\Gamma_{SP} = \left[S_{11} + \frac{S_{12} S_{21} \Gamma_{LP}}{1 - S_{22} \Gamma_{LP}}\right]^* \qquad (17)$$

where the associated gain is then determined using Eq.(1).

EQUIVALENT CIRCUIT OF THE GaAs FET

It is convenient to use equivalent circuits of the GaAs FET for forming a clear view of microwave characteristics and noise characteristics of the device. The equivalent circuit can be used as the first step in amplifier design. Fig. 6(a) shows an equivalent circuit of the GaAs FET. This equivalent circuit is shown for the device operated in the saturated current region in common-source configuration. In the intrinsic FET model, the elements ($C_{dg} + C_{gs}$) represent the total gate-to-channel capacitance, $R_i$ and $R_{DS}$ show the effects of the channel resistance and $i_{ds}$ defines the voltage controlled current source. The transconductance $g_m$ relates $i_{ds}$ to the voltage across $C_{gs}$. The extrinsic (parasitic) elements are: $R_S$, the source resistance, $R_D$ the drain resistance, $R_g$ the gate metal resistance, and $C_{DS}$ the substrate capacitance. Typical element values for a 0.5$\mu$m gate length and 300$\mu$m gate width FET are shown in Table IV. Very often the input circuit of an

FET can be modelled as a series RC and the output circuit as a parallel RC for the case of a bare chip device. The equivalent circuit of a packaged transistor can also be found where the addition of package wire bond inductances and shunt parasitic capacitances on input and output together with an increased feedback capacitance are usually sufficient to model accurately the device reflection coefficients. (Fig. 6(b)). By adding the current source as an element giving gain to the transistor we have built up a simple equivalent circuit for the FET.

EXAMPLE OF NARROW BAND AMPLIFIER DESIGN

This example deals with the design of a simple narrow band amplifier to operate between a source and load of characteristic impedance of 50 ohms, which provides maximum gain at 2 GHz. The scattering parameters of the transistor used are given in Table I. The equivalent circuit of the transistor derived from the $S_{11}$ and $S_{22}$ parameters is given in Table IV. The transistor is assumed unilateral (i.e. $S_{12} = 0$) for the initial design.

In order to match the I/P circuit a series inductor is required to resonate with the 0.43 pF series capacitor together with a transformer to transform 10.75$\Omega$ to 50$\Omega$. The latter can be achieved using a simple single section quarter wavelength transformer if the amplifier is to be realised using transmission lines or a combination of series inductance and shunt capacitance if lumped element matching is to be used.

The scattering parameters of the transistor are

$$S_{11} = 0.95 \angle -38^\circ$$
$$S_{21} = 2.95 \angle 150^\circ$$
$$S_{12} = .02 \angle 75^\circ$$
$$S_{22} = .725 \angle -15^\circ$$

The stability factor K is calculated using equation (3) to be 0.33 i.e. the transistor is only conditionally stable. The device can be stabilised by the use of a 100 ohm shunt resistor at its output.

Input Matching Circuit

The source impedance for maximum gain is 10.75 + j 145.

First transform the 50$\Omega$ generator to 10.75$\Omega$ with a 23.2$\Omega$ quarter wave line

where $23.2 = \sqrt{50 \times 10.75}$

Then add a shunt open circuit line with a characteristic impedance of 25 ohm, $\lambda/4$ long. This essentially adds shunts capacitance to the matching structure lowering the imaginary part of the matching admittance. A high impedance series transmission line or inductance completes the input matching network. Fig. 7(a) shows the steps in the input matching sequence for the elements shown. These were optimised using CAD to give the optimum input VSWR and the values shown in Fig. 7(a).

Output Matching Network

The output matching is considerably simplified because of the use of the stabilising resistor which lowers the magnitude of the output reflection coefficient from .725 to 0.28. The output matching circuit consists simply of a series wire bond of 1.1 nH inductance followed by a single section transformer nominally 60 ohms impedance , $\lambda/4$ long.

Output source impedance needed is $0.725\angle 15^\circ$. 50 ohms is transformed to start with a unit element ( $\lambda/4$ long) of impedance 60 ohms. A series inductor of value 1.1 nH is then used followed by the shunt stabilising resistor. Fig. 7(b) shows the sequence of matching of the output of the device.

The design gives an input reflection coefficient of .01 and an output reflection coefficient of 0.1 with a gain of 18.9 dB and a stability factor of 1.2.

EXAMPLE OF A BROADBAND AMPLIFIER DESIGN

This example deals with the design of a broadband amplifier to operate over the frequency band 8 to 12 GHz. The amplifier is to be designed to give maximum gain with a bandpass characteristic having a gain ripple of less than 0.5 dB over the band of interest. The scattering parameters of the transistor are again given in Table 1. These parameters can be interpolated from the table in 500 MHz intervals.

The input and output matching networks are designed to give the best input and output VSWR's over the band. However, because of its bandwidth the VSWR's will only be optimum at certain frequencies and the main design philosophy is aimed at flat gain. To achieve this the device can be mismatched at the low frequency end of the band to reduce the contribution of $G_S$ and $G_L$ of Eq.(12) to the overall amplifier gain. The gain can also be reduced by using absorptive resistive loading preferably on the drain side of the FET.

Lumped Element Designs -
Design where stability is not considered

The equivalent circuit of the FET has been given in Table IV, and calculation of the stability factor at 10 GHz gives a value of 0.8. Again let us assume that the device is unilateral. A lumped element matching technique will be discussed. Fig. 8 shows the input constant gain circles at 8 and 12 GHz and also shows the sequence of input matching used at 8 GHz. The input impedance of the FET will in practice be determined by the output matching conditions when the device is conditionally stable as given by Eq. (4). The performance of the amplifier is shown in Fig. 9, together with a schematic of the lumped circuit. As may be appreciated from Fig. 8 it is essential in the design of broadband matching networks to ensure that the network is capable of synthesising the necessary changes in impedance such that a wanted gain slope compensation and gain ripple can be met. It is often necessary to optimise the initial design values on a computer to minimise gain ripples etc. Such an optimisation will also account for the non-unilateral behaviour of the transistor.

Design where stability is considered

A shunt resistor of value 500 ohms can be used to produce a stable gain condition with a K = 1.7 at 10 GHz. This lowers the amplifier gain by approximately 3 dB. Fig. 10 shows the output matching sequence again using lumped elements where in this example the gain slope compensation needed (approximately 3.8 dB per octave) is supplied by this matching network. Fig. 11 shows the complete matching networks together with the amplifier response. It is noticed, because the input network has been chosen to give 0 dB gain slope, that the input VSWR is much more constant than the output VSWR which is low at the high end of the band becoming progressively larger towards the low end, thus mismatching the output of the FET and lowering its gain.

It is possible to stabilise the previous example by using a balanced amplifier approach or using input and output isolators.

## Distributed Designs

It is possible to convert lumped element designs to a close distributed element equivalent by using the following equations which are based on the use of $\pi$ equivalent circuits for transmission lines

$$\omega L = Z \sin\theta \qquad (18)$$

and

$$\omega C = \frac{1}{Z_0} \tan \theta/2 \qquad (19)$$

It is also possible to design matching networks directly from transmission line theory as was done in the narrow-band amplifier design previously.

Considering the conversion of the circuit of Fig. 11 it is worth noting that a good approximation is that inductors can be considered as 120$\Omega$ characteristic impedance short circuit stubs or unit elements and capacitors can be considered as 25$\Omega$ characteristic impedance open circuit stubs. This approximation is particularly useful when implementing the resultant network in microstrip on alumina of 0.635 mm thickness since the two impedance levels correspond to the generally accepted extremes of line widths which are conveniently reproduced. Fig. 12 shows the result of converting inductors and capacitors to transmission lines. Note in Fig. 12 that a small amount of inductance has been left at the FET ports allowing wire-bonding to take place in the realisation. Fig. 12 also shows the resultant response of the circuit.

# DESIGNING AN AMPLIFIER FOR OPTIMUM NOISE FIGURE

## Introduction

In many applications such as for receiver preamplifiers the amplifier design objective is minimum noise figure rather than maximum gain. The main task is to synthesise the optimum source admittance for minimum transistor noise figure at the FET input plane.

The noise performance of GaAs FETs is superior to that of bipolar transistors above approximately 4 GHz. For example the Plessey GAT6 device exhibits chip gains of 10 dB with a noise figure of 2 dB at 12 GHz. A theoretical treatment of GaAs FET noise performance has been given by Hewitt et al ( 5 ) and the bias dependence of the FET device noise figure is predicted by the noise theory of Pucel et al ( 6 ). This theory predicts that the minimum noise figure is obtained at low drain currents of approximately 15% of the saturated drain current.

## Constant Noise Figure Circles

In many amplifier designs the overall noise figure of the amplifier is important since this defines the signal to noise ratio at the output of the amplifier.

If we consider an infinite cascade of transistors all operated at their minimum noise figure d.c. bias conditions, then the lowest cascaded noise figure (often referred to as the noise, measure, M+1) is given by

$$M+1 = (FG-1)/(G-1) \qquad (20)$$

where each stage has a noise figure F and associated gain, G.

In general, for an arbitrary source reflection coefficient $\Gamma_s$, the noise figure is determined by the equation

$$F = F_{min} + 4r_n \frac{|\Gamma_s - \Gamma_o|^2}{(1-|\Gamma_s|^2)|1+\Gamma_o|^2} \qquad (21)$$

where $\Gamma_o$ is the optimum reflection coefficient of the source which produces the minimum noise figure $F_{min}$ and $r_n$ is the equivalent noise resistance of the device. Eq. (21) has the form of a circle. To determine a family of noise figure circles the noise figure parameter $N_i$ is defined where

$$N_i = \frac{F_i - F_{min}}{4r_n} |1+\Gamma_o|^2 \qquad (22)$$

where $F_i$ is the value of the desired noise figure circle and $\Gamma_o$, $F_{min}$ and $r_n$ are as previously defined. The radius of the circle is given by

$$R_{Fi} = \frac{1}{1+N_i}\sqrt{N_i^2 + N_i(1-|\Gamma_o|^2)} \qquad (23)$$

and the centre by

$$C_{Fi} = \Gamma_o/(1+N_i) \qquad (24)$$

From these equations we see that $N_i = 0$ when $F_i = F_{min}$ and the centre of the $F_{min}$ circle with zero radius is located at $\Gamma_o$ on the Smith Chart. The equivalent impedance to $\Gamma_o$ is often referred to as $Z_{OPT}$, the optimum source impedance for minimum noise figure.

The value of $r_n$ can be found by measuring the noise figure when the source impedance is 50 ohms so that $\Gamma_s = 0$ and

$$r_n = (F_{\Gamma_s=0} - F_{min})\frac{|1+\Gamma_o|^2}{4|\Gamma_o|^2} \qquad (25)$$

Fig. 13 shows the noise circles at 8, 10 and 12 GHz for a GAT6 FET chip device. All these plots show $Z_{OPT}$ at a point where the gate wire bond is attached to the 50 ohm transmission line. Fig. 14 shows a plot of $R_N$ against frequency for the same device.

## Noise Modelling

It is useful to have the concept of a noise model when designing amplifiers which need to have optimum noise tracking over their frequency band. A device model can be derived that has an input impedance equivalent to the complex conjugate ($Z_{OPT}*$) of the optimum noise impedance, $Z_{OPT}$.

For a device terminated with a network that presents $Z_{OPT}$ to the device terminals, then the input impedance at the driving source terminal (50Ω) is not power matched, i.e. not 50Ω. However, if the device is replaced by the equivalent noise model ($Z_{OPT}*$), the input impedance is now 50Ω. This technique enables networks to be derived that provide optimum noise tracking. From the plots of $Z_{OPT}$ that have been measured or extracted from a data sheet, $Z_{OPT}*$ can be found and an equivalent RLC model derived that fits this $Z_{OPT}*$ data. This RLC model can be computer optimised.

## EXAMPLE OF BROADBAND AMPLIFIER DESIGNED FOR MINIMUM NOISE FIGURE

It is desired to design a low noise broadband 8-12 GHz amplifier for optimum noise performance over the band together with a flat gain response using the device whose S-parameters are given in Table V and whose noise data is given in Fig. 13.

### Input Matching

The equivalent noise input admittance can be approximated with a capacitance $C_{im}$ in series with a resistance $R_{im}$ in series with an inductor $L_{im}$ with good accuracy over the desired frequency range. The values $C_{im}$ = 0.23 pF, $R_{im}$ = 12Ω and $L_{im}$ = 0.41 nH are found. It is desired to produce a Tchebyshev impedance matching network to match the generator impedance $Z_o$ to the noise input impedance $Z_{OPT}$. Tchebyschev prototypes ( 7) can be selected to maximise the bandwidth and provide a good match at the band edges.

In order to establish the number of resonators required to synthesise $Z_{OPT}*$ with a specified accuracy between the lower and upper band edges a maximum noise figure increase from the optimum values has to be specified. Let us assume that 0.6 dB excess noise figure can be tolerated. For example, at 10 GHz by inspecting the constant noise figure circles it can be seen that a worst case VSWR of approximately 2:1 can be tolerated with respect to $Z_{OPT}*$. In synthesising $Z_{OPT}*$ with a Tchebyschev impedance matching network the bandwidth can be computed using the VSWR versus load decrement curves plotted in ( 7, fig. 4.09-3)

The network shown in Fig. 15 will provide wide-band input impedance matching.

The FET's equivalent noise input impedance is resonated with the series inductance $L_1$ and coupled to a network consisting of 2 quarter wavelength resonators. In this technique the resistance $R_{im}$ = 12Ω can be considered the load resistance and ($L_1$ + $L_{im}$) - $C_{im}$ the last resonator of a 3 stage bandpass filter coupling $R_{im}$ to the source resistance $Z_o$.

By choosing the resonator line impedances to be equal to $Z_o$ the inverter constants of Fig. 15 are given by ( 7 )

$$K_{12} = Z_o\left(\frac{\pi \omega R_{im}}{4\, g_1 g_2 Z_o \delta}\right)^{\frac{1}{2}} \quad (26)$$

$$J_{23} = \frac{1}{Z_o}\frac{\pi\omega}{4}\left(\frac{1}{g_2 g_3}\right)^{\frac{1}{2}} \quad (27)$$

and

$$K_{34} = Z_o\left(\frac{\pi\omega}{4 g_3 g_4}\right)^{\frac{1}{2}} \quad (28)$$

where $\omega$ is the relative bandwidth:

$$\omega = (f_u - f_L)/f_o \quad (29)$$

where $f_o$ is the centre frequency = $(f_u + f_L)/2$ and where $f_u$ and $f_L$ are the upper and lower amplifier band limits.

$\delta$ is the decrement of the load impedance at $f_o$ given by

$$\delta = (2\pi f_o R_{im} C_{im})/\omega \quad (30)$$

$g_1, g_2, g_3$ and $g_4$ are the normalized element values of the Tchebyschev impedance matching prototype for the decrement $\delta$ ( 7 ).

1) Series Inductance

$$L_1 = \frac{1}{(2\pi f_o)^2 C_{im}} - L_{im} \quad (31)$$

2) Impedance Inverters

$$X_{iK} = \frac{K_{iK}}{1-(K_{iK}/Z_o)^2} \quad (32)$$

$$\ell_{iK} = \frac{\lambda}{4\pi}\tan^{-1}\left(\frac{2X_{iK}}{Z_o}\right) \quad (33)$$

and

$$L_{12} \simeq K_{12}/2\pi f_o \quad (34)$$

3) Admittance Inverter

$$B_{23} = \frac{J_{23}}{1-(J_{23}Z_o)^2} \quad (35)$$

$$\ell_{23} = \frac{\lambda}{4\pi}\tan^{-1}(2B_{23}Z_o) \quad (36)$$

where $\lambda$ is the wavelength of the transmission line at $f_o$.

The synthesis for the X-band amplifier with

f = 8 to 12 GHz
$R_{im}$ = 12Ω
$C_{im}$ = 0.23 pF
$L_{im}$ = 0.41 nH (including 0.12 nH of wire bond)

is as follows.

$\omega$ = (12-8)/10 = 0.4
$\delta$ = (2 x10x$10^9$x12x0.23x$10^{-12}$)/0.4
∴ $\delta$ = 0.433

For a decrement $\delta$ , Fig. 4.09-3 of Ref. ( 7 ) indicates that an insertion loss maximum of 0.85 dB will be achieved with n = 2 with a corresponding Tchebyschev ripple of 0.4 dB. However, this will not fall within the criterion of 0.6 dB increase in noise figure. Thus with n = 3 we have an insertion loss maximum of 0.6 dB and a corresponding ripple of 0.2 dB.

For the decrement $\delta$ = 0.43 and n = 3 the normalized element values of the Tchebyschev impedance-matching prototype can be found from Fig. 4.09-7 of Ref. ( 7).

$g_1 = 2.3$, $g_2 = 0.68$, $g_3 = 1.5$ and $g_4 = 0.53$.

1) Series Inductance

$$L_1 = \left\{\frac{10^9}{(2\pi \times 10^{10})^2 \times 0.23 \times 10^{-12}}\right\} - (0.41 - 0.12)\ nH$$

$= \underline{0.81\ nH}$

2) Impedance Inverters

$$X_{12} = K_{12}/(1-(K_{12}/Z_0)^2)$$

where $K_{12} = Z_0\left(\frac{\pi \times 0.4 \times 12}{4 \times 2.3 \times 0.68 Z_0 \times 0.43}\right)^{\frac{1}{2}}$

$= 16.74$

$\therefore X_{12} = 18.85$,

the equivalent shunt inductor being $(18.85/2\pi \times 10^{10})$ nH, i.e. 0.3 nH.

$$\therefore l_{12} = \frac{\lambda}{4\pi}\tan^{-1}(2X_{12}/Z_0) = .05\lambda (\equiv 18.5^\circ)$$

and $L_{12} \simeq K_{12}/2\pi f_0 \quad = \underline{0.27\ nH}$

Thus the first inductor is (0.8-0.27)nH = 0.53 nH.

$$X_{34} = K_{34}/(1-(K_{34}/Z_0)^2)$$

where $K_{34} = Z_0\left(\frac{\pi \times 0.4}{4 \times 1.5 \times 0.53}\right)^{\frac{1}{2}} = 31.43$

$\therefore X_{34} = 52$,

the equivalent shunt inductor being $(52/2\pi \times 10^{10}) = \underline{0.83\ nH}$

and $l_{34} = \frac{\lambda}{4\pi}\tan^{-1}(2X_{34}/Z_0) = \underline{.09\lambda}(\equiv 32.4^\circ)$

3) Admittance Inverters

$$B_{23} = J_{23}/(1-J_{23}Z_0)^2$$

where $J_{23} = \frac{1}{Z_0}\frac{\pi \times 0.4}{4}\left(\frac{1}{0.68 \times 1.5}\right)^{\frac{1}{2}} = .0062$

$\therefore B_{23} = 13.1 \times 10^{-3}$, the equivalent series capacitor being $(13.1 \times 10^{-3}/2\pi \times 10^{10})$ = 0.21 pF

and $l_{23} = \frac{\lambda}{4\pi}\tan^{-1}(2 \times B_{23} \times Z_0) = .073\lambda (26.3^\circ)$

$$l_{12} + l_{23} = (.05 + .073)\lambda = 0.123\lambda$$

Thus, the resonator length is

$(0.25 - 0.123)\lambda \quad = \quad \underline{0.127\lambda}$

and $l_{23} + l_{34} = (.073 + .09)\lambda \quad = 0.163\lambda$, with the corresponding resonator length of:

$(0.25 - 0.163)\lambda \quad = \underline{0.087\lambda}$

Output Matching Network

In this example the transistor has been matched for minimum noise figure over the band so that the device will exhibit a gain slope of approximately 3 dB/octave. An ideal output network then has to couple all available power to the load at the upper band edge, $f_u$ and provide a frequency dependent attenuation that compensates the FET's gain slope for $f < f_u$.

A network suitable for this is shown in Fig. 16. The output impedance of the transistor $R_{om}$ (in this particular case equal to 250 ohms), in parallel with $C_{om}$ (in this case 0.17 pF) is resonated with the series inductance $L_o$ at the upper frequency, $f_u$. The transformer matches the resulting resistance to the load $Z_o$. At $f_u$, no power is dissipated in the resistor $R_1$ since it is connected in series with a short-circuited stub being $\lambda/4$ at $f_u$. Below $f_u$ the power transferred to the load decreases since the impedance of the resonated FET output changes rapidly and the shunt branch with the resistor becomes increasingly lossy. The resistance $R_M$ and the transmission line impedance $Z_M$ are chosen to give the best approximations to a 3 dB/octave slope.

The complete circuit is shown in Fig. 17.

The value of the resonating inductor $L_o$ is found at the upper band edge frequency of 12 GHz to be:

$$L_o = 1/(2\pi \times 12 \times 10^9)^2 \times 0.17 \times 10^{-12}$$

$= 1$ nH

The Q of the FET output is given by

$Q = 2\pi f_u R_{om} C_{om}$

$= 2\pi \times 12 \times 10^9 \times 250 \times 0.17 \times 10^{-12}$

$= 3.2$

Leichti has shown that for such a Q value (8)

$R_1/R_{om} = 0.21$ i.e. $R_1 = 52.5\Omega$

$Z_1/R_{om} = 0.38$ i.e. $Z_1 = 95\Omega$

and the transformer ratio $m^2 = (R_{om}/Z_o) \times 0.1$

$m = 0.707$

i.e. a quarter wavelength resonator of impedance 35Ω of length 0.21λ at the centre frequency of 10 GHz.

The accuracy of the design using the unilateral model has been checked by using computer analysis and optimization with the FET represented by its S-parameters and optimum noise reflection coefficients etc., using the COMPACT program. The resulting element values are shown in Fig. 17 in brackets. It may be appreciated that the initial element values are close to those finally adopted after optimization and this circuit gives a theoretical noise figure which is 0.6 dB higher than the optimum noise figure at 8 GHz. Worst case input and output VSWR's were 3.5:1 and the gain was 8.8 dB ± 0.25 dB.

It should be stressed that the above example is only one method that can be used to synthesise networks suitable for low noise amplifier realisation. The choice of a particular means of matching into a device is often dictated by the available technology and the device selected. For packaged devices, for example, it is desirable to add a short length of line to the packaged FET S-parameters to act as a soldering area. Such action moves the matching load impedance of the device and this should be assumed at the beginning of the design.

## COMPUTER-AIDED DESIGN PRACTICE

As we have seen in the last example the use of a computer optimisation routine is of great advantage in producing a final circuit design suitable for realisation. The computer should not be used as a replacement for good microwave engineering but an addition to the microwave engineer's range of aids to design.

### General Format of Microwave CAD Programs

There are a considerable number of microwave circuit analysis and optimization routines in use today ranging from specially written 'in-house' packages to wide-ranging 'bureaux' type programs (9).

All programs of the type mentioned have several items in common.

Circuit Elements. Circuit elements are input into the program as separate building blocks and are cascaded within the program using transmission matrices. Elements may be lumped or distributed and may be varied by the program in its optimisation mode. Feedback elements, series or parallel, nodal analysis etc. may also be features of the program. S-parameters and optimum noise data are usually input to the program as separate blocks of data or are stored in a separate data file and 'called-up' by the main working program.

The circuit is described using codes or abbreviations.

Frequency Input. There are usually two ways of entering frequency. Firstly as a series of separate frequencies or as a band with constant intervals.

Noise Figure Data. This is usually specified in terms of the optimum noise figure, the optimum reflection coefficient and either the noise figure at $\Gamma_S = 0$ (i.e. 50Ω) or the noise resistance $R_N$. There is also a weighting factor associated with noise figure as explained later.

One and Three Port S-parameters. It is often found possible to input single port data such as that used with a diode and also three port data such as that associated with a dual-gate FET. In the latter case the three port S-parameters of which there are nine elements are converted to 2 port parameters with a specified load on the third port.

Negative Components. Many programs allow the user to introduce negative elements, thus allowing the designer to 'strip-off' certain parts of a circuit such as transmission lines.

Optimization. In order to affect optimization the program needs to have an error function input to which it attempts to converge with the circuit description available. Usually the total error function, which is a sum of the sum of the squares of the individual error functions is weighted.

The total error function may be written as

$$E = \frac{1}{N}\sum_{F_1}^{F_2} W_1 \times |S_{11}|^2 + W_2 \times |S_{22}|^2 + W_3 \times |S_{21} - G|^2 + W_4 \times |F_{OPT} - F_{ACT}|^2 \quad (37)$$

where $S_{11}$, $S_{22}$ and $S_{21}$ are the scattering parameters of the resultant overall circuit; $F_1$ is the first and $F_2$ the last frequency of interest; N is the number of frequencies, $W_1$ to $W_4$ are the weighting factors; G is the desired amplifier gain and $F_{OPT}$ and $F_{ACT}$ are the optimum and actual noise figures.

Sensitivity Analysis and Mapping. Certain programs allow the user to analyse his circuit design in terms of its sensitivity to changes in component values. This can be done using a Monte Carlo analysis where specified components are altered in a random fashion within specified limits thus simulating, for example, production tolerances on transistor noise figures and associated gains. It is also possible to use some programs in a 'search' mode which will indicate the elements in a particular circuit which are the most sensitive or indicate whether the circuit is capable of performing the task asked of it.

A few rules worth bearing in mind when using computer aided analysis and optimization of microwave circuits are

1. Use conventional matching theory with a Smith Chart or synthesis techniques to get a starting circuit for optimization.
2. Avoid large numbers of frequencies and variables. This will result in excessive computer time and cost. It is often possible to select the most sensitive elements by a few trial analyses.
3. A stability analysis should always be achieved before an optimization run is performed. The addition of resistive loading or feedback will often produce a stable condition after inspection of the stability circles such that optimization will be successful.
4. Do not try and optimize a complicated circuit such as a multistage amplifier before analysing the individual circuits.
5. The selection of weighting factors is important and attention should be paid to the gain and noise figure parameters of the transistors used.

## NETWORK SYNTHESIS

The use of the Smith Chart and resulting calculations is somewhat time consuming. Passive network synthesis, has been used in the design of filters and can be applied to the design of interstage input and output matching networks for microwave amplifiers where the networks operate between unequal impedance levels and include parasitic elements associated with the active devices.

Synthesis routines (10 , 11) have been in use for many years and more recently have been extended to include the synthesis of networks of prescribed gain, bandwidth, ripple, slope and impedance transformation. Mellor (12) has produced a synthesis routine on which is based a commercial synthesis computer program, AMPSYN.

An example of the technique is given in the following:

Example. An interstage network for a 2 stage FET amplifier is to be designed to operate between a 135 ohm source and a 10.75 ohm load with a 6 dB/octave gain slope and a 0.4 dB ripple over the 6 to 12 GHz frequency range. A 300Ω shunt resistor is used at the output of the first FET to stabilise the device. The specification for good input and output match implies a frequency response that is flat over the passband of the amplifier.

The interstage matching networks must provide a positive-sloped gain with frequency (Fig. 18(e)) to compensate the transistor's rolloff (Fig. 18(c)) and give an overall flat gain.

The procedure to be described in this example is based on a method of synthesis of matching networks of arbitrarily specified passband slope with provision for inclusion of parasitic elements. The parasitic elements in this particular case are the output capacitance of a GAT6 namely 0.17 pF and the input capacitance namely 0.31 pF.

1) In general two resistors coupled by a lossless ladder network having all transmission zeros at zero and infinite frequency are called trapless filters. The insertion loss for such filters can be expressed as

$$IL = \frac{\text{power available from source resistance}}{\text{power delivered to load resistance}} = \frac{a_0 + a_2\omega^2 + a_4\omega^4 + \dots + a_{2N}\omega^{2N}}{\omega^{2J}} \qquad (38)$$

where N is the order of the network, i.e. the numbers of inductors and capacitors before impedance transformation and J is the number of high pass elements.

2) From this insertion loss function form

$$|\rho_1(\omega)|^2 = 1 - 1/IL \qquad (39)$$

3) Substitute $s = j\omega$ to obtain

$$|\rho_1(s)|^2 = \rho_1(s)\,\rho_1(-s) \qquad (40)$$

4) Obtain the poles and zeros of the resulting expression in the usual manner.

5) Form

$$\rho_1(s) = \pm 1 \frac{\prod_{i=1}^{N}(s - z_i)}{\prod_{i=1}^{N}(s - p_i)} \qquad (41)$$

where the poles of $\rho_1(s)$ are the poles of $|\rho_1(s)|^2$ that are in the left hand plane (LHP) and the zeros of $\rho_1(s)$ are chosen from the zeros of $|\rho_1(s)|^2$ that are in the LHP or the RHP (right hand plane).

6) Form

$$z_1(s) = R_s \cdot \frac{1 + \rho_1(s)}{1 - \rho_1(s)} \qquad (42)$$

where $R_s$ is the source resistance.

7) Determine the network element values by forming the appropriate expansion of $z_1(s)$ about zero and infinity.

In order to absorb the parallel capacitance at the FET drain and also the series capacitance at the second stage FET gate we require a network of the form shown in Fig. 19(a). This is a network of N = 4; J = 1 type. Insertion loss functions approximating 6 dB/octave gain slope are easily obtained from flat insertion-loss functions. The flat insertion-loss function is first normalised to an upper cut off frequency of 1 rad/sec and then divided by $\omega^2$. This operation results in a sloped insertion loss of exactly the same ripple and passband as the flat insertion loss case. The order of the IL functions remains unchanged (i.e. at 4) while the number of transmission zeros at d.c. (J) is increased by 1.

1) The insertion loss function is found using the methods of (12) and is given by:-

$$IL = \frac{0.8 - 5.8\omega^2 + 19.9\omega^4 - 23.9\omega^6 + 10.1\omega^8}{\omega^4}$$

2)

$$|\rho_1(\omega)|^2 = \frac{0.8 - 5.8\omega^2 + 18.9\omega^4 - 23.9\omega^6 + 10.1\omega^8}{0.8 - 5.8\omega^2 + 19.9\omega^4 - 23.9\omega^6 + 10.1\omega^8}$$

Substituting $\omega = s/j$ gives

3)

$$|\rho_1(s)|^2 = \frac{0.8 + 5.8s^2 + 18.9s^4 + 23.9s^6 + 10.1s^8}{0.8 + 5.8s^2 + 19.9s^4 + 23.9s^6 + 10.1s^8}$$

4) The Poles of $|\rho_1(s)|^2$ are

$S = \pm 0.19 \pm j0.48$
and $S = \pm 0.19 \pm j1.01$ } from the numerator

The zeros of $|\rho_1(s)|^2$ are given by

$S = \pm 0.18 \pm j0.51$
and $S = \pm 0 \pm j0.98$ } from the denominator

5) Thus

$$\rho_1 = \frac{(-1)(s+j.98)(s-j.98)(s-0.18+j.51)(s-0.18-j.51)}{(s+0.19+j.48)(s+0.19-j.48)(s+.19+j1.01)(s+j.19-j1.01)}$$

where the zeros of $\rho_1$ are chosen to be in the right hand plane.

Multiplying out we obtain

$$\rho_1 = \frac{-s^4 + 0.36s^3 - 1.2s^2 + 0.36s - 0.28}{s^4 + 0.74s^3 + 1.4s^2 + 0.5s + 0.28}$$

$$Y_1 = 1/z_1 = 1 - \rho_1(s)/(1 + \rho_1(s))$$

$$= \frac{2s^4 + 0.38s^3 + 2.7s^2 + 0.14s + 0.56}{1.1s^3 + 0.21s^2 + 0.86s}$$

$$Y_1 = 1.77s + \frac{1}{1.53s} + \frac{1}{2.42s + \frac{1}{0.538s} + 0.46}$$

The synthesised network is then as given in Fig. 19(b). The network is scaled to 135Ω ohm source and 12 GHz upper cut-off frequency to give Fig. 19(c).

The shunt-series connection of inductors is used to transform the found impedance level of 62 ohms to the wanted level of 10.75 ohms. The resulting TEE network is given in Fig. 19(d) where the new component values are given by

$L_3 = (1-n)L_2$

$L_4 = n^2(L_1+L_2) - nL_2$

$L_5 = nL_2$ where $n^2 = 10.75/62$

Thus, n = 0.416

$L_3 = 1.6$ nH

$L_4 = 0.084$ nH

$L_5 = 1.14$ nH

and the synthesis is completed by scaling the series capacitor from 0.0528 pF to 0.3 pF.

Now the output capacitance of the 1st FET can be absorbed into the 0.174 pF shunt capacitance whilst the 0.31 pF input capacitance of the 2nd FET is absorbed exactly by the series capacitance thus yielding the circuit of Fig. 19(e).

This method is obviously tedious especially where the IL function is to be calculated and the zeros and poles of the network identified. However such a synthesis technique is readily amenable to CAD programming.

This example is therefore completed by designing a complete 2 stage amplifier to operate over the 8 to 12 GHz band using the input network for the low-noise 1st stage as previously defined followed by the presently designed interstage network and completed with the design of an output stage network using the synthesis procedure described.

The GAT6 transistor chip on which the above example has been based has an approximate 4 dB/octave gain slope under maximum gain conditions. Thus a two stage network will have a 8 dB/octave gain slope. However, the input matching network has been designed for minimum noise figure and it may be shown by measurement that the available gain of the FET when matched for minimum noise figure exhibits a slope of approximately 3.2 dB/octave. Thus the output matching network has only to provide -(3.2 + 4-6)dB/octave gain slope, i.e. approximately 1.2 dB/octave slope where the 2nd stage is biased for maximum gain.

Fig. 20(a) shows the final complete circuit of the two stage amplifier together with its frequency response (Fig. 20(b)).

## THE USE OF SINGLE ENDED AND BALANCED AMPLIFIERS

The examples of amplifier design which have been given indicate that it is possible to design more than one stage of amplification to achieve gain flatness over a specified bandwidth. However for multistage amplifiers where gains of 30 dB or greater are required the use of such "single-ended" realisations can lead to difficulty in producing the required gain flatness etc. This is because of the mismatches occurring between gain stages causing gain and phase variations. In the single-ended case it is possible to significantly reduce such gain ripples by using interstage as well as input and output isolators. These devices can produce significant loss which will increase the overall noise figure of the unit. For integrated amplifiers such isolators can also be comparatively large unless they are integrated into the amplifier medium, e.g. the use of planar isolators with a microstrip amplifier.

An alternative approach which is often used in commercial amplifiers is the adoption of balanced amplifier techniques. Fig. 21 illustrates the concept. Two amplifier units of the same performance are arranged between the output and input ports of quadrature 3 dB couplers. The theory of balanced amplifiers has been given by Kurokawa (13) and it may be shown providing the two amplifiers are matched in their characteristics that the input VSWR and output VSWR are predominantly dependent on that of the couplers.

If the individual amplifiers in the balanced pair are not perfectly matched at certain frequencies then a signal in the $0^{\circ}$ arm of the coupler will be reflected from the corresponding amplifier and a signal in the $\pi/2$ arm of the coupler will be similarly reflected from its amplifier. On travelling through the coupler after reflection the signals will again be phased 0 or $\pi/2$ and the reflected power is 'dumped' into the isolated port (Fig. 21).

In a microstrip realisation for octave band-amplifiers one of the most popular 3 dB couplers is the so-called Lange coupler which achieves the coupling factor required by using interdigitated $\lambda/4$ fingers (14). Another coupler which is used for more narrow band applications is the 'branch-line' coupler (15).

The overall noise figure of the amplifier is the same as the equivalent single-ended amplifier with the addition of the loss of the input coupler. However, the amplifier is able to handle an increase of 3 dB in input r.f. power for the same compression characteristic and the intermodulation performance is increased by 9 dB assuming perfect phase matching between the 2 parallel and matched amplifier chains.

Fig. 22 shows the microstrip realisation of a balanced amplifier which is fabricated with bare chip FETs on an alumina substrate medium.

The balanced amplifier has the further advantage that if for some reason one of the FETs fails in a unit then the amplifier will still operate albeit with a reduced performance.

Balanced amplifiers enable the design engineer to produce a matched 'gain block' which is able to be cascaded with other matched 'blocks' such as other amplifiers, attenuators, detectors etc.

A major advantage of the balanced configuration is that of stability. As has already been demonstrated stable gain can be assured in a single-ended amplifier by using a stabilising resistor (or reactance in the source to ground connection). However, such action reduces the available gain of the transistor and does not allow the minimum noise measure to be achieved in a multistage amplifier. The use of balanced units ensures that the input and output of each FET stage 'sees' an impedance close to 50 ohms. Indeed the stability factor of such a balanced stage can be an order of magnitude higher than its single-ended equivalent, depending on the VSWR's and isolation of the quadrature couplers used.

The disadvantage of using the balanced approach is that the unit uses double the number of active devices in a single-ended unit, is larger and consumes more d.c. power. However the advantages to be gained from the balanced approach will usually outweigh the disadvantages.

## VARIATIONS IN AMPLIFIER PERFORMANCE

### Transistor Variations and Circuit Sensitivity

Transistor parameters which are given in manufacturers' data sheets are only typical sets of data collected from a large number of measurements. In order to allow the use of any transistor of a particular type and also to allow for circuit realisation tolerances it is often necessary to include some means of adjustment in the designed matching circuits.

For amplifiers realised in a planar transmission medium such as microstrip or triplate the use of adjustable open and short circuited series transmission line elements can often compensate for variations in the FET gate-source and drain-source capacitance due to process variations in channel doping density and variations in the input capacitance and gate resistance due to gate length variations. Fig. 23(a) and (b) show a typical spread of S-parameters and $Z_{OPT}$ data for a transistor together with a corresponding circuit which was designed to compensate for such variations without affecting the performance of the amplifier (15).

### Variations in Amplifier Performance with Temperature

Fig. 4 has shown the variation in the associated gain of a single gallium arsenide FET over the temperature range -40 to +80°C. Fig. 4 also shows the effect of temperature on the noise figure of a single transistor. For a multistage amplifier, therefore, substantial gain changes can occur with temperature as well as variations in the overall amplifier noise figure. Because of small variations in the S-parameters with temperature the gain ripple over the frequency band of interest may also alter its character. Thus for certain applications, particularly military, an amplifier must be constructed which has integrated compensation to maintain gain to within specified limits over a certain range of temperatures.

This can be achieved using three different techniques.

1. By varying the bias on single-gate FET's.
2. By using the AGC facility of a dual-gate FET by varying the bias on the second-gate
3. By using a PIN diode attenuator.

With all three methods a temperature sensing element is needed such as a thermistor or diode, a circuit enabling the gain change 'law' of the amplifier to be realised and a drive circuit.

## DESIGNING AN AMPLIFIER FOR A SPECIFIED LINEAR OUTPUT POWER

So far in this chapter we have described the design of maximum gain and low noise figure amplifiers. There are many cases where it is also desirable to be able to specify the r.f. power handling capabilities of the amplifier. This is usually achieved using the definition of 1 dB gain compression point. This is the point at which the output r.f. power minus the input r.f. power in dBm is 1 dB lower than the small signal linear gain of the device. For a GaAs FET the 1 dB gain compression point for a small signal device is that measured at a specified drain to source voltage (e.g. $V_{DS}$ = 5V) and a gate to source voltage which optimises the output power (usually such that $I_{DS}$ = 50% $I_{DSS}$).

Small signal FETs usually have 1 dB gain compression points of +10 to +13 dBm but certain small signal FETs will give much higher output powers, e.g. a Plessey GAT4/021 will give r.f. output powers of +20 dBm at 8 GHz.

If two signals at frequencies $f_1$ and $f_2$ are input to a FET device the third order intermodulation products occur at frequencies $2f_1-f_2$ and $2f_2-f_1$ where the products $2f_1+f_2$ and $2f_2+f_1$ are assumed to be out-of-band.

By observing the 3rd order intermodulation products (IP) on a spectrum analyser, for example, it is possible to plot the graph of 3rd order product power output versus input power. If this plot is compared with the fundamental output power plot a point of intersection occurs where the 3rd order IP amplitude equals the fundamental signal amplitude. This point is referred to as the 3rd order intercept point (Fig. 24).

An amplifier is designed for optimum output r.f. power handling using the same matching techniques already discussed. However, the S-parameters that are used in the design should be those measured with the transistor operating at $\frac{1}{2}I_{DSS}$. For amplifiers having output powers up to +13 dBm this technique of design is considered valid but for higher power amplifiers the small signal S-parameters become invalid since the transistor impedance levels change when subjected to larger r.f. powers. This is more fully covered in the following sections.

## POWER AMPLIFIERS

### Introduction

High power, high efficiency amplification of microwave power using GaAs FETs has been demonstrated over the past few years at frequencies in C, X and Ku band (17,18). Power FET devices are commercially available which enable linear Class A powers of several watts to be achieved up to X band. A power FET is essentially a multicell structure where the GaAs material and the fabrication techniques used are optimised for higher breakdown voltage than a small signal FET.

GaAs power FETs are very attractive solid state amplifiers. In comparison with silicon bipolar devices, Schottky barrier power GaAs FETs have higher maximum frequency of oscillation, low noise figures and large gains. Above 4 GHz they have higher efficiencies, do not exhibit secondary breakdown, are self-ballasting and have inherently higher input impedance. A typical power GaAs FET is shown in Fig. 25(a).

## D.C. CHARACTERISTICS

The device shown in Fig. 25(a) has a total gate width of 2400 m (compared to typically 300 m for a small signal device) comprising 4 unit cells each 600 m wide. Devices with well matched cell characteristics are required since such devices yield good power-added efficiencies (over 30%) and good thermal characteristics.

The maximum output power obtainable from a GaAs FET in class A operation with full gate modulation is determined from the characteristic of Fig. 25(b). The maximum output voltage and current swings give output power for a sinusoidal input of

$$P_{OUT} = I_F (V_B - V_S)/8 \qquad (43)$$

where $I_F$ is the value of $I_{DS}$ corresponding to a forward gate bias of 0.7V (for an aluminium to GaAs Schottky barrier), $V_B$ is the source-drain voltage at which breakdown occurs in the pinch-off region and $V_S$ is the saturation or 'knee' voltage of Fig. 25(b) usually between 1.5 and 2 volts. The optimum load resistance is

$$R_L = (V_B - V_S)/I_F \qquad (44)$$

From equations (43) and (44) the requirements for high output power may be seen. Large values of $I_F$ are obtained by placing in parallel a number of FET 'cells' to give a large total gate width. The number of gates put in parallel is limited by the minimum load resistance which can be presented to the device by equation (44).

The breakdown voltage $V_B$ is determined principally by the avalanche breakdown of the gate to drain diode. These are related by the equation

$$V_B = V_A - V_P \qquad (45)$$

where Vp is the pinch-off voltage.

## R.F. CHARACTERISTICS OF POWER GaAs FETs

### S-parameters

The small signal scattering parameters of power FETs are usually supplied on manufacturers' data sheets. This enables the design engineer to obtain a device bandwidth capability and also indicates the regions of instability of the device. The low frequency S-parameters of a power FET usually indicate, below 2 GHz, that the regions of instability are large particularly for the output of such a device so that it is necessary to use bias networks, for example, that present an impedance to the device close to 50 ohms. Table VI shows the small signal S-parameters for a 1 watt power FET from 2 to 12 GHz in 1000 MHz steps. The Table shows the maximum available gain of the device as well as the Rollett stability factor, K. It is apparent that the gain of a power FET is substantially lower than that of a small signal device, a result of the device being optimised for its power handling performance.

### Large Signal Parameters

The small signal S-parameters of a power FET, although useful design aids, are not so useful for accurate prediction of the large signal gain and power output of a device either at a spot frequency or over a bandwidth. The measurement of high power S-parameters is difficult. The measurement of the source and load impedances for the device operating at its 1 dB gain compression point are more easily measured and together with the small signal S-parameters give sufficiently accurate information for design. These impedances are measured by matching the device using for example, stub tuners at the r.f. power levels of interest and then separately measuring the tuner impedance using a conventional small signal network analyser. Fig. 26(a) shows the result of doing this for the large signal impedance at the output of a Plessey PGAT1000 FET. The output power of power GaAs FETs largely depends on the output being matched at the large r.f. signal levels. Fig. 26(b) shows the effect of increasing the r.f. power level into a 250 mW power GaAs FET from 10 dBm to 26 dBm. The latter input power is larger than would be used in practice but the trends shown in Fig. 26(b) may be used to modify the remaining small signal S-parameters. For example, $\angle S_{21}$ is virtually unchanged. However both the magnitude and angle of $S_{22}$ change significantly.

For operation over medium bandwidths (i.e. 20% or so) it is desirable to be able to calculate the variation in output power with frequency. This can be achieved by plotting the large signal gain circles or constant output power circles at discrete frequencies over the band of interest and superimposing on these the matching circuit impedance chosen for the amplifier.

The large signal output impedance is usually used in the design of the output matching network together with an $|S_{21}|$ which reflects the decrease in gain under large signal conditions. The small signal $S_{11}$ and $S_{12}$ can be used during design of the matching networks with final circuit adjustment being performed after the amplifier has been constructed.

## CIRCUIT TOPOLOGIES FOR MATCHING POWER FETs

To determine the suitable circuit topologies for microwave GaAs FET amplifiers in microstrip form, for example, consideration must be given to the bandwidth, output power and frequency of operation.

If high output power is required, a large gatewidth device will have to be used. This will result in low input and output impedance and the circuit designer will have to choose a circuit topology that can match to these low impedances within the physically realisable limit of the transmission line. One useful approach is the lumped element impedance matching techniques already discussed with reference to small signal amplifiers.

The impedance matching circuits that can be used for power GaAs FET amplifier design include the following:

### Quarterwave Impedance Transformers

For moderate bandwidth applications ( 10 to 20%) a single section of quarterwave impedance transformers may be used. A packaged device is most often used for power amplifiers (since to bond up a power FET is an intricate operation) and if the device impedance appears inductive, either a capacitive shunt stub or a series capacitor may be used for resonance prior to the necessary impedance transformation to the source or load.

### Matching Networks Based on Low Pass Filter Prototypes

For the broadband amplifier design, the lumped element equivalent circuit model of the FETs may be used to compute the input/output

Q's of the device. These Q values are then used to determine the complexity of the network (the 'n' value) and the prototype 'g' values (7).

Using the coupled-resonator matching technique Leighti (8) has designed a broadband amplifier. The broadband amplifiers with edge-coupled transmission line sections as described by Tserng (18) can be considered a modified version of this broadbanding technique, since the single-section edge-coupled line can be represented as a series impedance transformer with two open-circuited shunt stubs connected to each end of the transformer.

### Lumped-Element Impedance Matching Networks

Lumped-element impedance matching techniques can also be used for power FET amplifier design. For narrow to moderate bandwidths, a simple design procedure using a Smith Chart will be adequate. GaAs power FET amplifiers using lumped LC elements have been realized (19). For wideband applications the initial element values of the lumped element, low pass Tchebyschev impedance matching network can be obtained from the work of Matthaei (20). A low-pass to bandpass transformation can be made to obtain the correct element values for the desired frequency of operation.

## NARROW BAND POWER FET AMPLIFIER DESIGN

This example describes the design of a 9 to 10 GHz power amplifier. The amplifier is to provide 20 dB gain at 32 dBm output power with three cascaded FET amplifier stages.

The optimum large-signal circuit impedances for three FET power devices with various gate widths are shown in Fig. 27(a). The FETs have gate widths of 300$\mu$m, 1200$\mu$m and 2400$\mu$m to be used in the first, second and third stages respectively. The plot shows the effect of more cells lowering the impedance of the FET. The simple distributed matching networks shown in Fig. 27(b), are designed to give output powers of 120 mW, 500 mW and 1.5 watts with 9 dB, 7 dB and 5 dB of gain respectively. The 1 dB bandwidth was 1 GHz with power-added efficiencies of 23 to 30%. Based on the circuit topologies shown in Fig. 27 an integrated, 3 stage amplifier with 50 ohm interstage transmission lines was designed. The performance of the overall amplifier is 20 dB gain with +12 dBm and an output power of 1.6W is obtained at 9.5 GHz with a power added efficiency of 30%.

## BROADBAND POWER FET AMPLIFIER DESIGN

The use of distributed matching networks for broadband power amplifiers follows the same synthesis routines as for small signal amplifiers. Fig. 28(a) shows for example the use of a particular circuit topology for broadband matching of a 600$\mu$m gate width device over the frequency band 6-12 GHz giving 100 mW output power. Performance is shown in Fig. 28(b).

One of the most useful methods of broadband matching power FETs is to use 'internal' matching within a microwave package using wire bonds and chip capacitors. Such a circuit is shown in Fig. 29(a) with a set of performance curves in Fig. 29(b). It is seen that output powers of the order of 2 to 3 watts with 5 to 6 dB gain can be achieved with 1 dB bandwidths of approximately 2 GHz over the 7 to 10 GHz frequency range. Power-added efficiencies generally fall into the 20 to 30% range.

## DESIGNING AN AMPLIFIER FOR MAXIMUM SPURIOUS FREE DYNAMIC RANGE

The spurious free dynamic range of an amplifier is defined by the equation

Spurious Free Dynamic Range, D

$$= {}^{2}/_{3}\,(P_1 - P_0 - 10\log BW - NF)\,dB$$

where $P_1$ is the input intercept point (obtained by subtracting the amplifier gain from the output intercept point).

$P_0$ is the effective input noise power with no signal (-114 dBm/MHz).

BW is the system noise bandwidth in MHz (controlled by the amplifier bandwidth).

NF is the amplifier noise figure in dB.

Thus, for example, if an amplifier has a gain of 40 dB with an output intercept point of 30 dBm, a bandwidth of 500 MHz and a noise figure of 10 dB the spurious free dynamic range is

$$D = {}^{2}/_{3}\left\{(30-40)-(-114)-10\log 500-10\right\} dB$$

$$= \underline{44.7\ dB}$$

i.e. the range over which the amplifier will not have any 3rd order intermodulation products greater than the noise floor is 44.7 dB.

For any specified amplifier bandwidth the parameters available enabling an increase in the dynamic range of the amplifier are the overall noise figure and the output intercept point.

Thus, the use of power FETs with their high 3rd order intercept point and low-noise small signal preamplifier FETs will enable large dynamic ranges to be achieved. In order to accurately predict the distortion products of a power FET amplifier considerable attention has been paid to the use of power FET models for predicting gain-compression, harmonic and intermodulation distortion (21, 22). Non linear circuit-type device models have been derived which indicate that the third order intermodulation distortion on many GaAs power FETs does not follow a 3:1 slope. Indeed often the intermodulation products do not follow a monotonic relationship with the input power of the test tones (Fig. 30).

## POWER COMBINING TECHNIQUES

Often the design of power FET amplifiers is limited to the available r.f. power handling capability of the FETs available. Also, 1 dB gain compression point output powers are at single frequencies whilst for wideband operation power outputs are significantly below these levels. Wide gate devices have low input and output impedances which are difficult to match over wide bands. In addition gain is at a premium at the higher frequencies.

In search for a solution circuit designers

have been employing various types of hybrid power combiners to combine the output of several lower power amplifiers that incorporate more easily matched narrow gate width GaAs FETs. The main disadvantage with the use of hybrid power dividers is that of gain and power losses due to their insertion losses and that effect on overall power added efficiency.

Fig. 31 is an example of a 4 way power combiner, where the gain of the driver stages is 6 dB with an efficiency of 20% and the output devices have a gain of 5 dB with the power added efficiencies defined in Fig. 32. As may be seen if circuit losses and combiner losses are both equal to 0.5 dB and the output devices have power added efficiencies of 30% the overall efficiency of the circuit is only 15%.

Two simple planar combining techniques that combine the output powers of four GaAs FETs are shown in Fig. 32(a),(b). The first technique makes the use of two-way combiners of the Wilkinson type (23) while the second approach employs two hybrids similar to that described by Lange (14) and a simple 2 way splitter.

Fig. 32(a) shows the use of two-way combiners that have been connected to a branch like network by two cascaded transforming elements. In order to provide physical access to the single-ended amplifier submodules, additional line length can be inserted into the combiner and divider circuits. This is accomplished by means of the elements $Z_2$ and $Z_3$ as well as $Z_5$. In order to combine the output power of 2 balanced amplifier modules power combiners are used as shown in Fig. 32(b). In contrast to the four-way Wilkinson type combiner, the dual quadrature hybrid combiner has the advantage of the excellent matching conditions that are typical of balanced amplifiers. As a result the overall amplifier exhibits very low return loss and a smooth gain response (24).

## THERMAL CONSIDERATIONS IN POWER AMPLIFIER DESIGN

Unlike small signal FET devices attention must be paid to the heatsinking of power FET devices to avoid the channel temperatures of the devices exceeding the manufacturer's specified maximum operating point. For example, a 3W amplifier having 20 dB of gain using Plessey PGAT1000 devices over a 1 GHz bandwidth centred at 8 GHz would have an overall power added efficiency of approximately 15%, i.e. 6.7W of d.c. power is input to the FETs plus dissipation in other components such as regulators etc. 10W of heat could be generated in such an amplifier and this heat must be dissipated as efficiently as possible. Thus, it is necessary to heat sink the power FETs effectively. Power FET packages are designed to allow this to be achieved conveniently e.g. holes in the package base allowing the devices to be screwed into a baseplate made of a high thermal conductivity material such as copper. Power FET data sheets usually provide a derating curve if the amplifier is to be operated at temperatures higher than 25°C.

## THE USE OF FEEDBACK, COMMON-GATE AND SOURCE-FOLLOWER GaAs FET CONFIGURATIONS IN AMPLIFIER DESIGN

### Feedback Amplifiers

The use of negative feedback in bipolar transistor microwave amplifier stages is now a universally adopted design approach. The technique allows the designer to control the amplifier's performance and to design a configuration which allows the use of transistors with variable parameters.

Negative feedback can control the input and output impedances of a single-ended stage without resource to balanced techniques. Negative feedback produces an amplifier capable of multioctave bandwidths and gain variations held to a few tenths of a dB. A negative feedback amplifier circuit is shown in Fig. 33(a). For simplicity the DC bias components and coupling capacitors are excluded. Representing the FET by its low frequency model (Fig. 33(b)) the results of an analysis in terms of S-parameters is as follows (25):

$$S_{11} = S_{22} = 1 - \frac{g_m Z_0^2}{R_2(1+g_m R_1)} \bigg/ \left(1 + \frac{2Z_0}{R_2} + \frac{g_m Z_0^2}{R_2(1+g_m R_1)}\right) \quad (46)$$

$$S_{21} = 1 - \frac{2Z_0}{R_1} + \frac{2Z_0}{R_2} \bigg/ \left(1 + \frac{2Z_0}{R_2} + \frac{g_m Z_0^2}{R_2(1+g_m R_1)}\right) \quad (47)$$

and

$$S_{12} = \frac{2Z_0/R_2}{1 + 2Z_0/R_2 + g_m Z_0^2/R_2(1+g_m R_1)} \quad (48)$$

where $g_m$ is the transconductance of the FET and $Z_0$ is the characteristic impedance of the source and load. For the condition that $S_{11} = S_{22} = 0$, it may be shown from Eq.(46) that

$$1 + g_m R_1 = g_m Z_0^2 / R_2 \quad (49)$$

By substituting Eq.(49) into Eq.(47) and (48) expressions for $S_{21}$ and $S_{12}$ are found to be

$$S_{21} = \frac{Z_0 - R_2}{Z_0} \quad (50)$$

and

$$S_{12} = \frac{Z_0}{R_2 + Z_0} \quad (51)$$

and

$$R_1 = \frac{Z_0^2}{R_2} - \frac{1}{g_m} \text{ ohms} \quad (52)$$

Thus, unless the transconductance of the GaAs FET is large enough $S_{11} = S_{22}$ cannot equal zero with a positive value of $R_1$. The minimum required transconductance required occurs when $R_1 = 0$, and

$$S_{21} = |1 - g_m Z_0| \quad (53)$$

Thus, for example if a 10 dB amplifier gain is required with a characteristic impedance of 50 ohms a transistor with a $g_m$ of 83 mS is required with a feedback resistor of 207Ω.

Single-stage amplifiers using the described method of feedback have been fabricated and show extremely wide bandwidths from virtually d.c. to over 8 GHz. Input and output VSWR's can be maintained to less than 2:1 up to approximately 5 GHz, with a stability factor always greater than 1 (26).

Common-Gate and Source-Follower Configurations

A GaAs FET is most commonly used in common-source operation since the device is designed to give its maximum gain and minimum noise figure under this circuit condition.

However considerable advantage can be attained in terms of input and output reflection coefficients by using common-gate and common-drain (source-follower) configurations respectively. Table VII shows the S-parameters for the common gate and source follower connected GAT6 and may be compared directly with Table I. It may be seen that up to 6 GHz the input and output reflection coefficients of the respective configurations are low indicating that these configurations would be suitable as the input and output stages of an amplifier. It should also be noted that $|S_{11}|$ for the source follower configuration is very close to 1, whilst $|S_{12}|$ is large compared with a common source connection. It may also be shown that in the common gate configuration the optimum source admittance for low noise figure is close to 50Ω particularly at the low frequency end.

In order to provide flat gain over a wide bandwidth the circuit of Fig. 34(a) is of a typical type where a common gate FET is connected to a common source device. Fig. 34(b) shows the performance of the circuit of Fig. 34(a) where the devices used were of the GAT6 type (i.e. 300μm gate width and 0.5μm gate length. In order to flatten-out the gain it is necessary to include some simple matching between stages such as series inductance. Such circuits using impedance matching over broad bandwidths by active devices are particularly suitable where very small components are needed such as in the case of integrated circuits (27).

CONCLUSIONS

This chapter has dealt with the extensive subject of microwave amplifier design with particular reference to the use of gallium arsenide field effect transistors. The descriptions given are by no means complete and the student is referred to the references as a means to further reading.

REFERENCES

1. Rollett, J. "Stability and Power Gain invariants of linear two ports", IRE Trans. on Circuit Theory, Vol. CT-9, pp.29-32, March 1962.

2. Bodway, G. "Two port power flow analysis using generalised scattering parameters" Microwave Journal, Vol. 10, pp.61-69, May 1967.

3. Gledhill, C. and Abulela, M. "Scattering parameter approach to the design of narrow-band amplifiers employing conditionally stable active elements", IEEE Trans. Microwave Theory and Techniques, Vol. MTT-22, pp.43-48, January 1974.

4. - "S-parameter design" Hewlett-Packard Application Note 154, pp.25-34, April 1972.

5. Hewitt, B. et al, Electron Letters, Vol. 11, p.309, 1976.

6. Pucel, R., Haus, H. and Statz, H., "Signal and noise properties of gallium arsenide microwave field effect transistors" Advances in Electronics and Electron Physics, Vol. 38, pp.195-265.

7. Matthaei, G.L., Young, L. and Jones, E.M.T. "Microwave filters, impedance-matching networks and coupling structures" New York: McGraw-Hill, 1964.

8. Liechti, C.A. and Tillman, R.L., "Design and performance of microwave amplifiers with GaAs Schottky-gate field effect transistors" IEEE Transactions on microwave Theory and Techniques, Vol. MTT-22, No. 5, pp.510-517, May 1974.

9. Compact Version 4.8 - Compact Engineering Inc., 1088 Valley View Court, Los Altos, Calif. U.S.A.

10. Skwirzynski, I.F., "On synthesis of filters" IEEE Transcations on Circuit theory (Special Issue on Computer-Aided Circuit Design), Vol. CT-18, pp.152-163, Jan. 1971.

11. Saal, R. and Ulbrick, E., "On the design of filters by synthesis", IRE Trans. Circuit Theory, Vol. CT-5, pp.284-327, Dec. 1958.

12. Mellor, D.J. and Linvill, J.G., "Synthesis of interstage networks of prescribed gain versus frequency slopes" IEEE Transcations on Microwave Theory and Techniques, Vol. MTT-23, No. 12, December 1975, pp.1013-1020.

13. Kurokawa, K., "Design theory of balanced transistor amplifiers", Bell System Technical Journal, Vol. 44, No. 10, pp.1675-1798, October 1965.

14. Lange, J., "Integrated stripline quadrature hybrids", IEEE Trans. MTT Vol. MTT-17, December 1969, pp.1150-1151.

15. Reed, J. and Wheeler, G.J. "A method of analysis of symmetrical four port networks" IRE Transactions on Microwave Theory and Techniques, October 1956 pp.246-252.

16. Estabrook, P., Knowne, C.M. and Crescenzi, E.J., "A low noise single-ended GaAs Schottky FET amplifier for a 14 GHz satellite communication application" 1978 IEEE MTT-S International Microwave Symposium Digest, June 27-29, Ottawa, Canada, pp.129-131 (78CH1355-7).

17. Macksey, H.M., Adams, R.L. and Wisseman, W.R. "X-Band performance of GaAs power FETs" Electronic Letters, Vol. 12, pp.54-56, Jan. 22, 1976.

18. Tserng, H.Q., "Design and performance of microwave power GaAs FET amplifiers" Microwave Journal June 1979, pp.94-100.

19. Honjo, K. et al, "Broad-band internal matching of microwave power GaAs MESFETs" IEEE Transactions on Microwave Theory and Techniques, Vol. MTT-27, January 1979, p.3.

20. Matthaei, G.L., "Table of Tchebyschev impedance-transforming networks of low-pass filter form", Proc. IEEE, Vol. 52, p.939 (1964).

21. Tucker, R.S. and Rauscher, C. "Modelling the 3rd-order intermodulation-distortion properties of a GaAs FET" Electronic Letters, Vol. 13, No. 17, pp.508-510 (1977).

22. Willing, H.A., Rauscher, C. and de Santis, P. "A technique for predicting large signal performance of a GaAs MESFET" 1978 IEEE MTT-S International Microwave Symposium Digest, Ottawa, Canada, pp.132-134 (78CH1355-7).

23. Wilkinson, E.J., "An N-way hybrid power divider" IRE Trans. MTT, Vol. MTT-8, January 1960, pp.116-118.

24. Niclas, K.B., "Planar power combining for medium power GaAs FET amplifiers in X/Ku-bands" Microwave Journal, June 1979, pp.79-84.

25. Cooke, H.F., "Microwave FETs - A status report" IEEE International Solid State Circuits Conference Digest of Technical Papers, pp.116-117 (February 1978).

26. Ulrich, E., "Use negative feedback to slash wideband VSWR" Microwaves, October 1978, Vol. 17, No. 10, pp.66-70.

27. Abbott, D.A., Cockrill, J.R., Pengelly, R.S., Stubbs, M.G. and Turner, J.A. "Monolithic gallium arsenide circuits show great promise" MSN August 1979, Vol. 9, No. 8, pp.73-96.

## QUESTIONS

1. Using complex arithmetic derive an expression for the stability factor K and obtain values for this parameter at 2, 4, 8 and 16 GHz for the transistor characterised in Table 2.

2. Calculate the maximum unilateral and non-unilateral gains at 2 and 4 GHz for the transistor in Table 2. Verify that the difference in the two gains agrees with the unilateral figure of merit.

3. From the S-parameters of Table 2 derive an approximate equivalent circuit for the packaged FET at 4 GHz and from this produce simple matching networks giving low input and output VSWR's. A Smith Chart should be used.

4. Describe how you would go about designing a broadband amplifier to give flat gain over the band 4-8 GHz using the packaged FET of Table 2. Derive the constant gain circles for this transistor at 4, 6 and 8 GHz for both the input and the output and suggest suitable matching circuits using Table III as an aid.

5. Using the data of Table V and Figs. 13 and 14 describe the design of an amplifier having a bandwidth of 2 GHz centred at 8 GHz with a noise figure of less than 2.0 dB and a gain of 10 dB. Calculate the noise measure of an infinite cascade of amplifiers having such a noise figure.

6. If the optimum source reflection coefficient for minimum noise figure is $\Gamma_{OPT}$ and the transistor is assumed to be non-unilateral, derive an expression for the output reflection coefficient $S_{22}'$ for minimum noise figure. Use this expression to calculate the $S_{22}'$ at 16 GHz for the device of Table V assuming that $\Gamma_{OPT} = 0.6\angle{-175^{\circ}}$.

7. It is required to produce a broadband amplifier capable of delivering a minimum of 20 mW into 50 ohms over the frequency range 4 to 8 GHz. However the amplifier will be driving a cable having a 3 dB/octave increase in loss. Assuming that the transistors to be used have an MAG of 10 dB at 8 GHz and a slope of 4.5 dB/octave describe the design of the amplifier based on matching network synthesis. (The FET is capable of delivering 80 mW at 4 GHz into 50 ohms).

8. What method could be used in the design of question 7 to increase the power handling of the FETs by a factor of 4 without the need to change the transistor type? Comment on the effect of circuit losses on the technique.

9. Describe the main points to be given attention when designing a wideband amplifier having 60 dB gain, a 6 dB noise figure and an output 1 dB gain compression point of +16 dBm. Assuming the 3rd order intercept point to be 12 dBm higher than the 1 dB gain compression point calculate the spurious free dynamic range of the amplifier in a 10 MHz bandwidth.

10. From Table VI and Fig. 26(a) for a PGAT1000 power FET design a narrowband amplifier which will guarantee 1000 mW output power at the 1 dB gain compression point with 200 mW input power over the frequency band 7.9 to 8.4 GHz. Comment on the stability of the amplifier as the frequency is decreased.

11. Calculate the stability factors for the common gate and common drain FET configurations from the S-parameters given in Table VII. Calculate the constant gain circles for the common gate source and drain configurations at 2 GHz and estimate the overall gain of a common gate/source/drain cascade at 2 GHz as well as the overall stability factor.

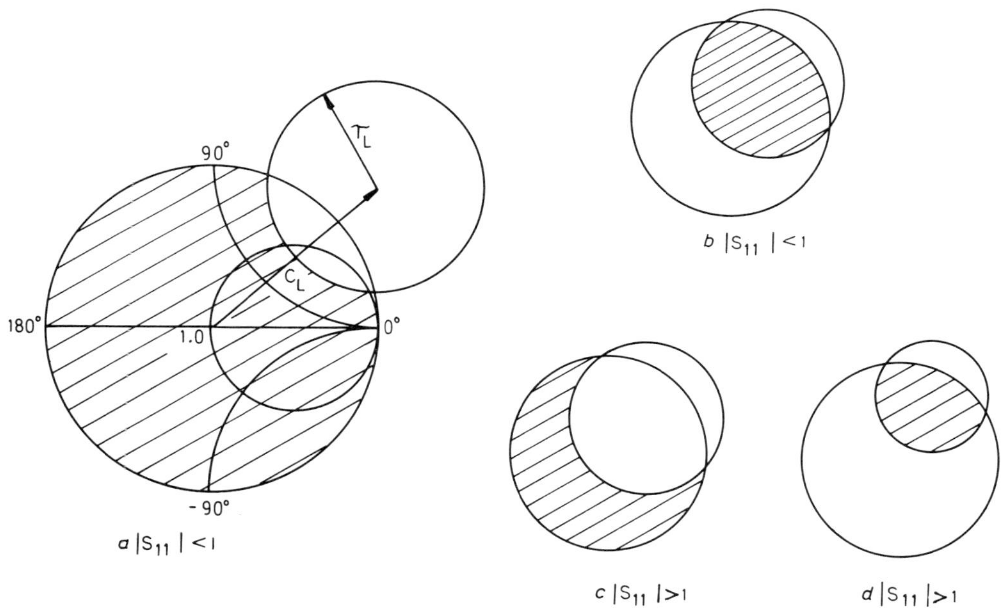


Figure 1 The four different regions of stability in the $\Gamma_L$ plane

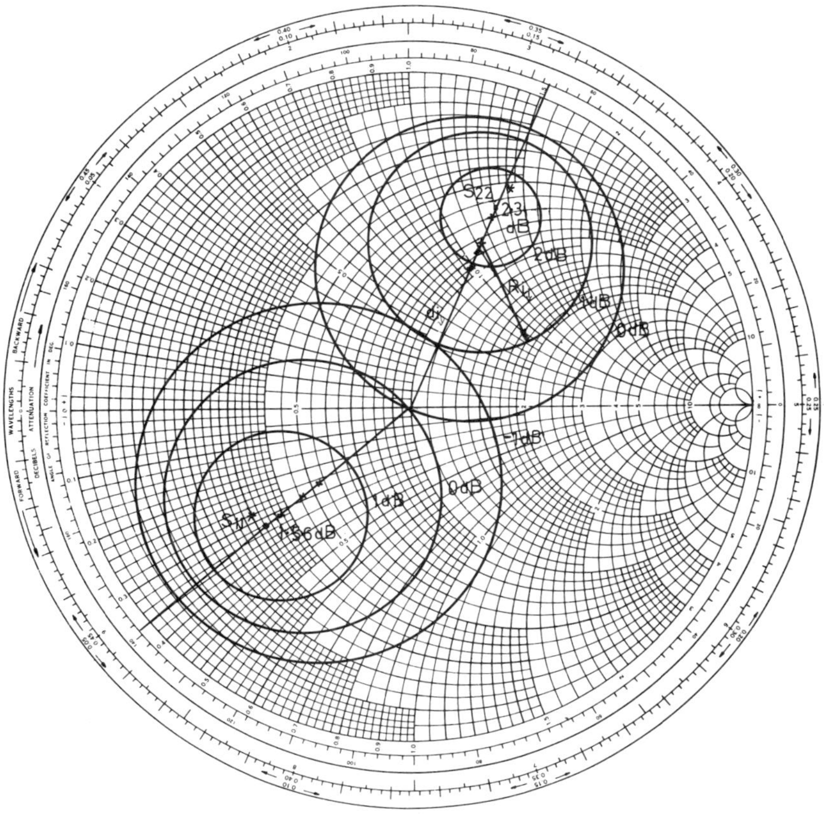


Figure 2 Constant gain circles for a P103 packaged GAT5 GaAs FET at 10 GHz

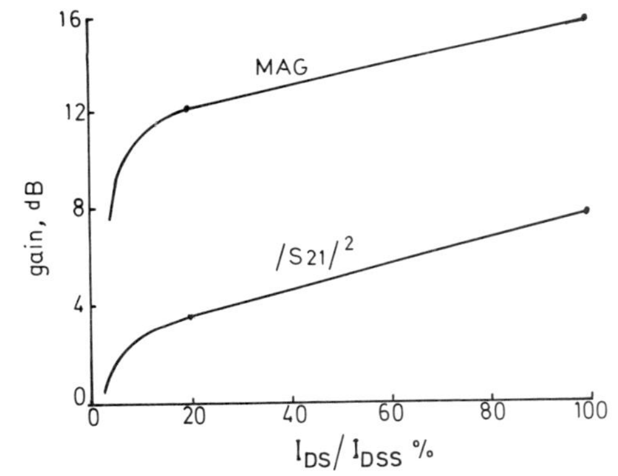


Figure 3 GaAs FET gain versus $I_{DS}/I_{DSS}$ ratio

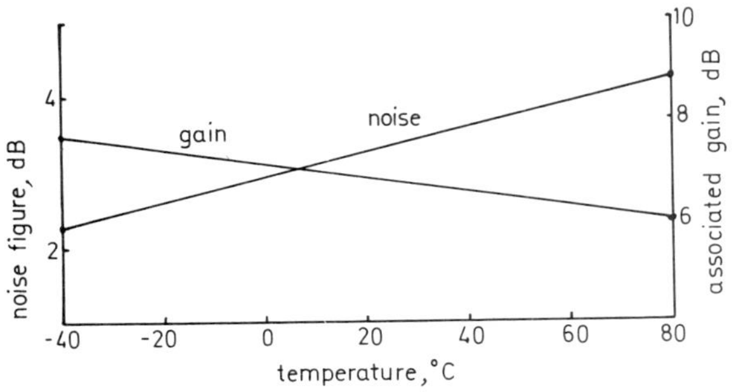


Figure 4 GaAs FET gain and noise figure versus temperature

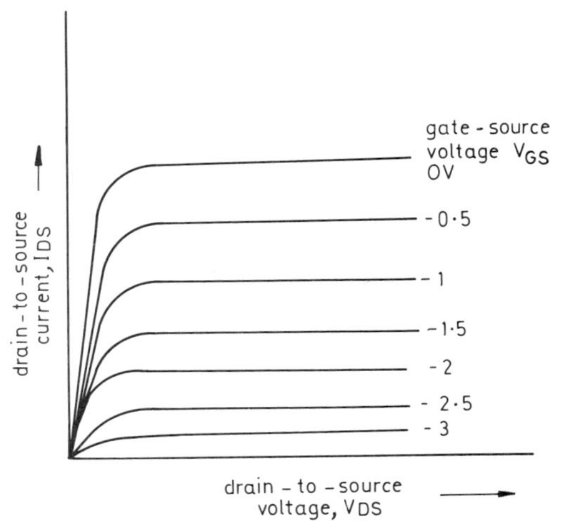


Figure 5 Static drain current versus drain voltage characteristic for a GaAs FET

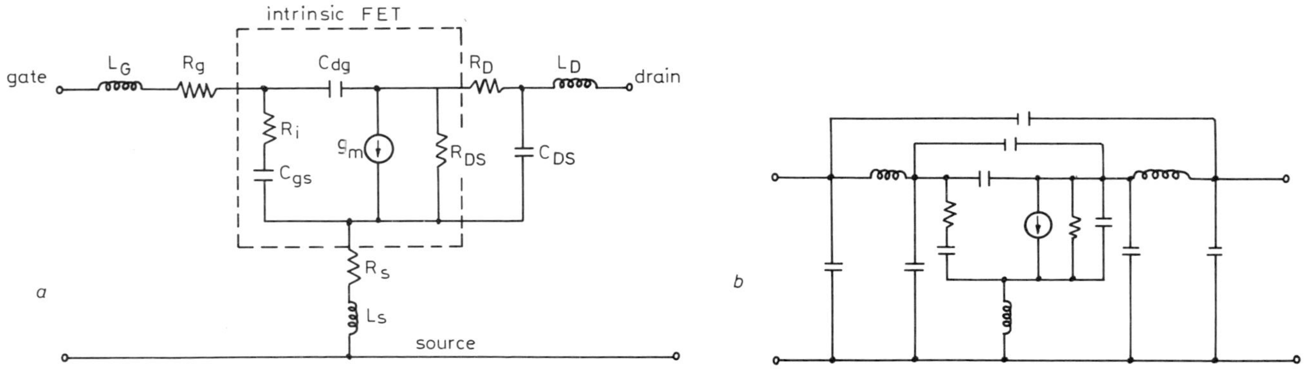


Figure 6 (a) Equivalent circuit of GaAs FET chip (b) Equivalent circuit of packaged GaAs FET

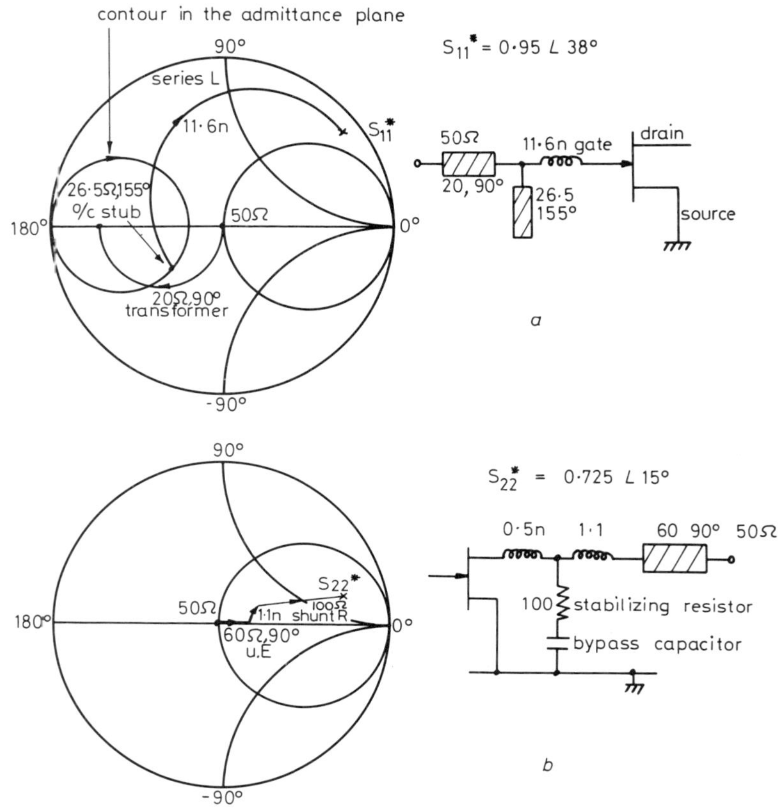


Figure 7 (a) Input matching network for narrow band small signal amplifier at 2 GHz
(b) Output matching network for narrow band small signal amplifier at 2 GHz

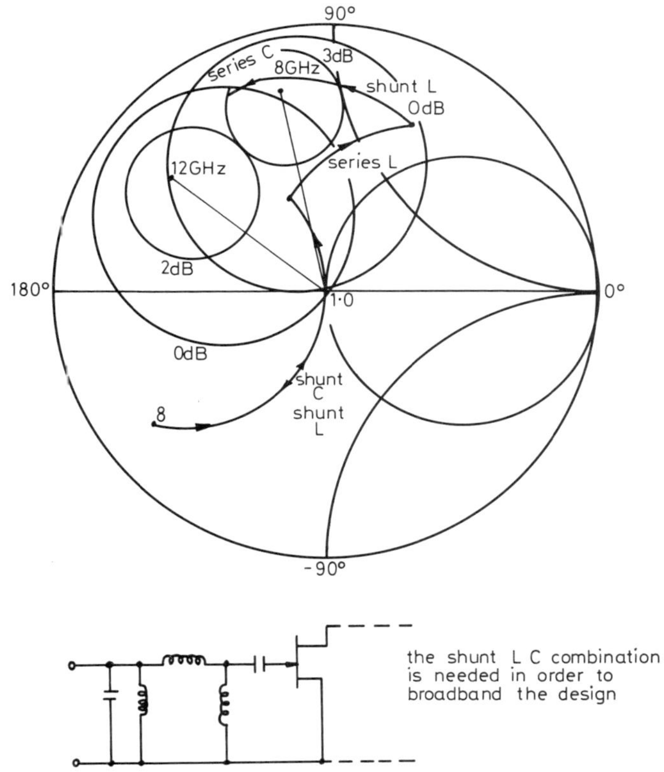


Figure 8 Input constant gain circles at 8 and 12 GHz

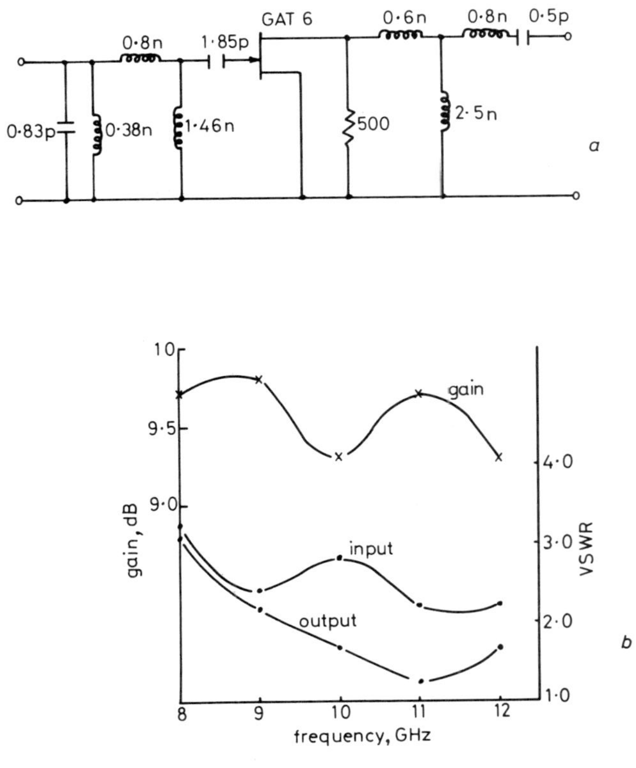


Figure 9 (a) Conditionally stable 8-12 GHz GaAs FET amplifier

(b) Response of 8-12 GHz amplifier having conditional stability

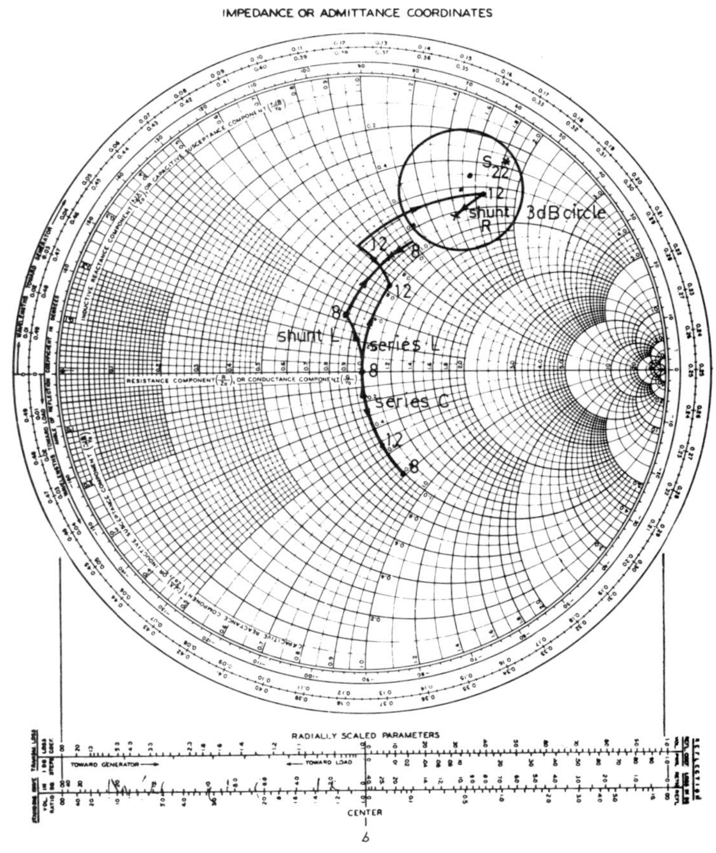


Figure 10 Output matching sequence for a stabilised 8 to 12 GHz amplifier

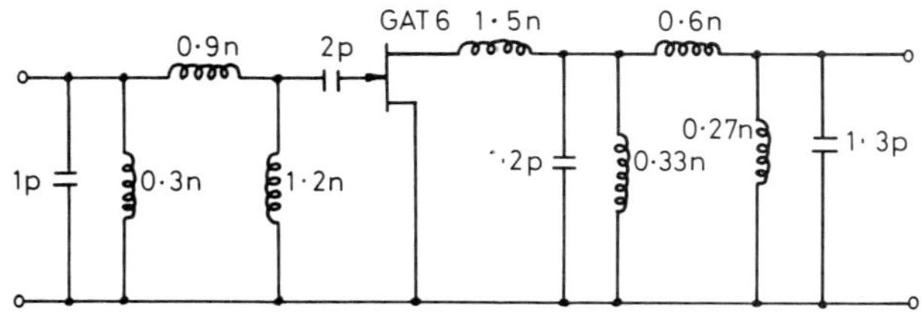


a

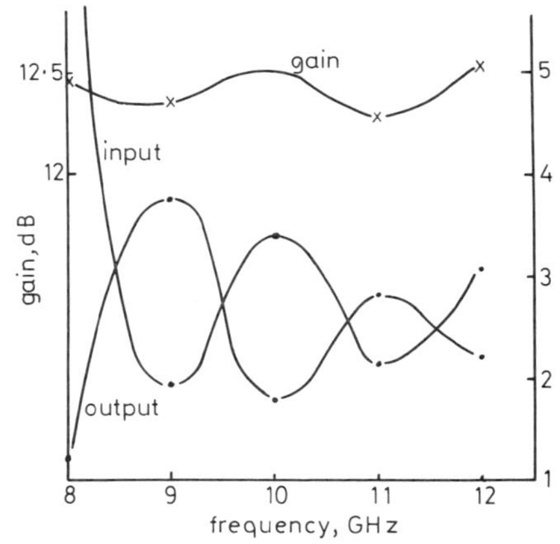


b

Figure 11 (a) Lumped element 8-12 GHz GaAs FET amplifier

(b) Response of lumped element 8-12 GHz GaAs FET amplifier

a

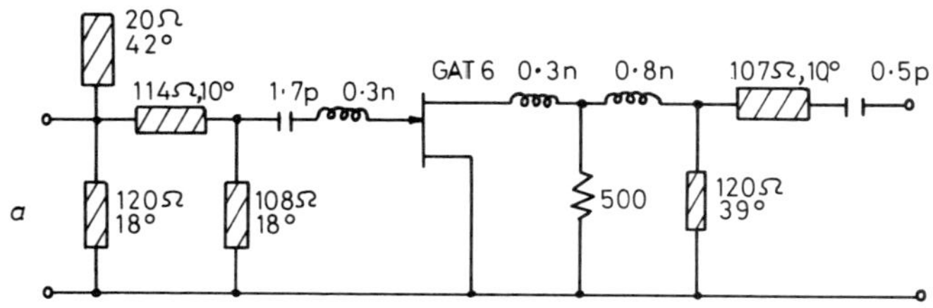


b

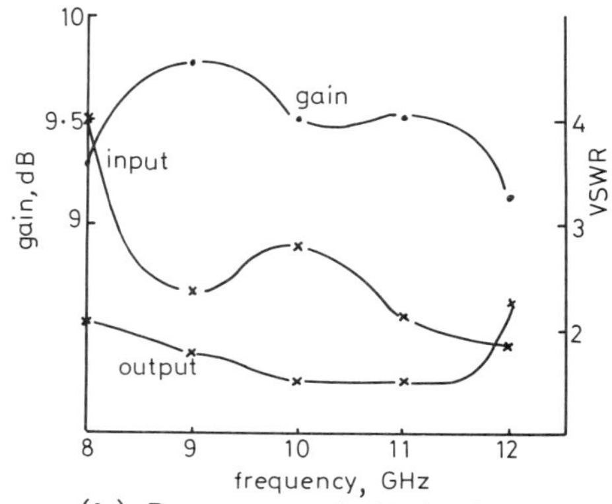


Figure 12 (a) Distributed element 8-12 GHz GaAs FET amplifier

(b) Response of distributed element 8-12 GHz GaAs FET amplifier

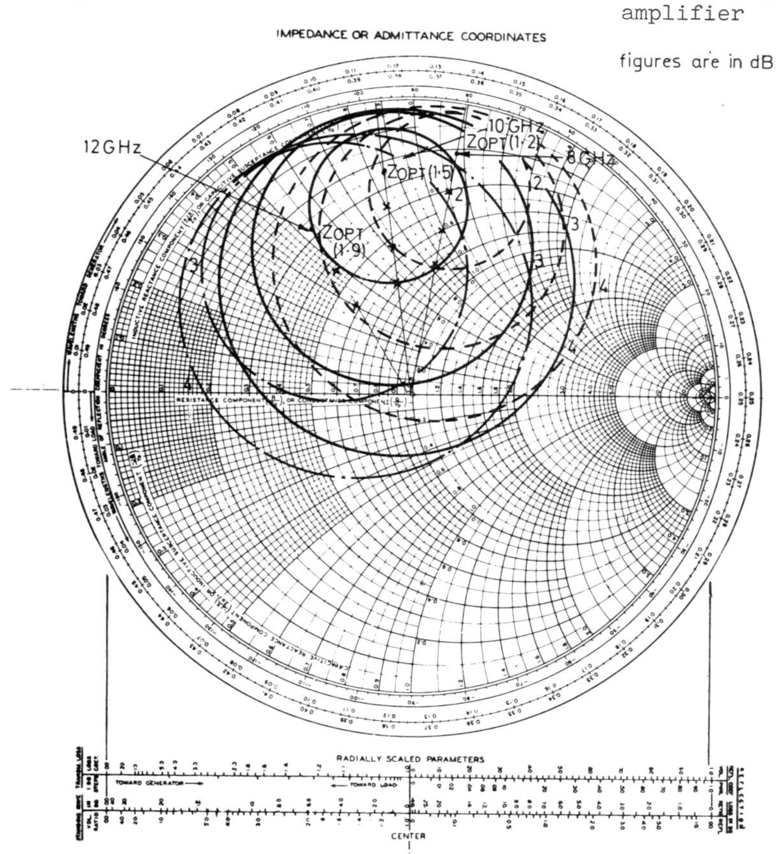


Figure 13 Constant noise figure circles for a GAT6 transistor chip

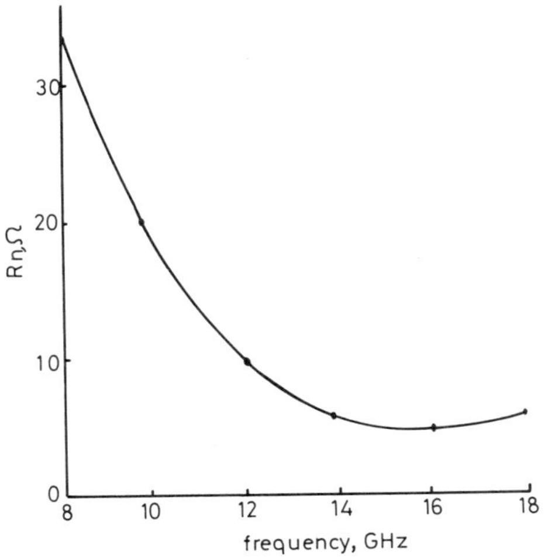


Figure 14 Plot of noise resistance versus frequency for a GAT6 transistor chip

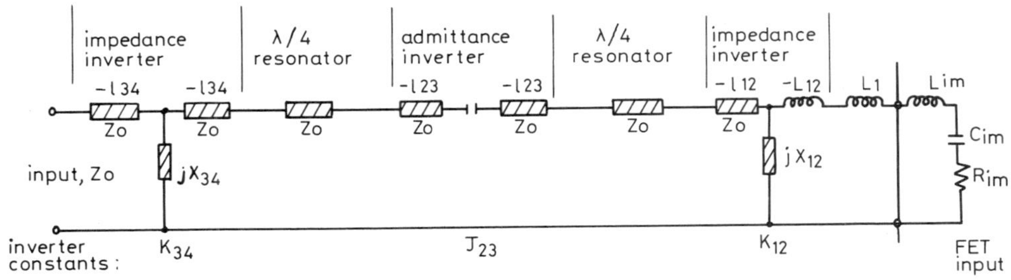


Figure 15 Wideband impedance matching network for minimum noise figure

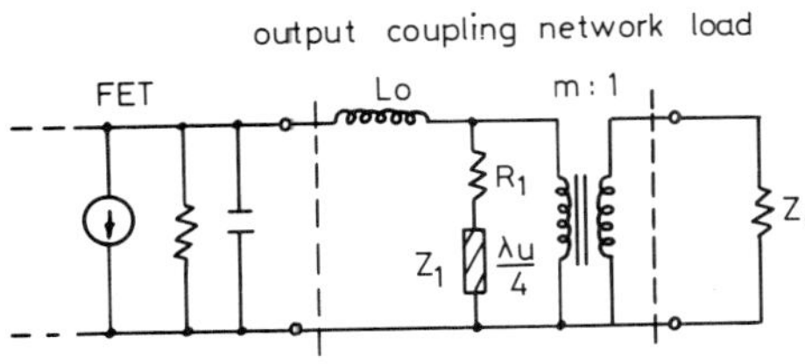


Figure 16 Wideband GaAs FET output matching circuit

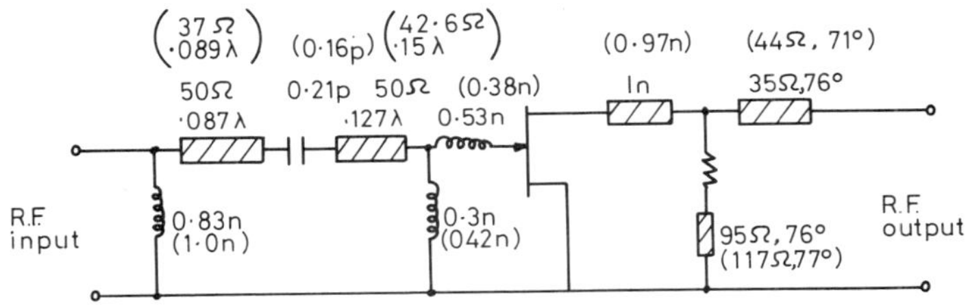


Figure 17 Single stage low noise X-band GaAs FET amplifier with absorptive output matching

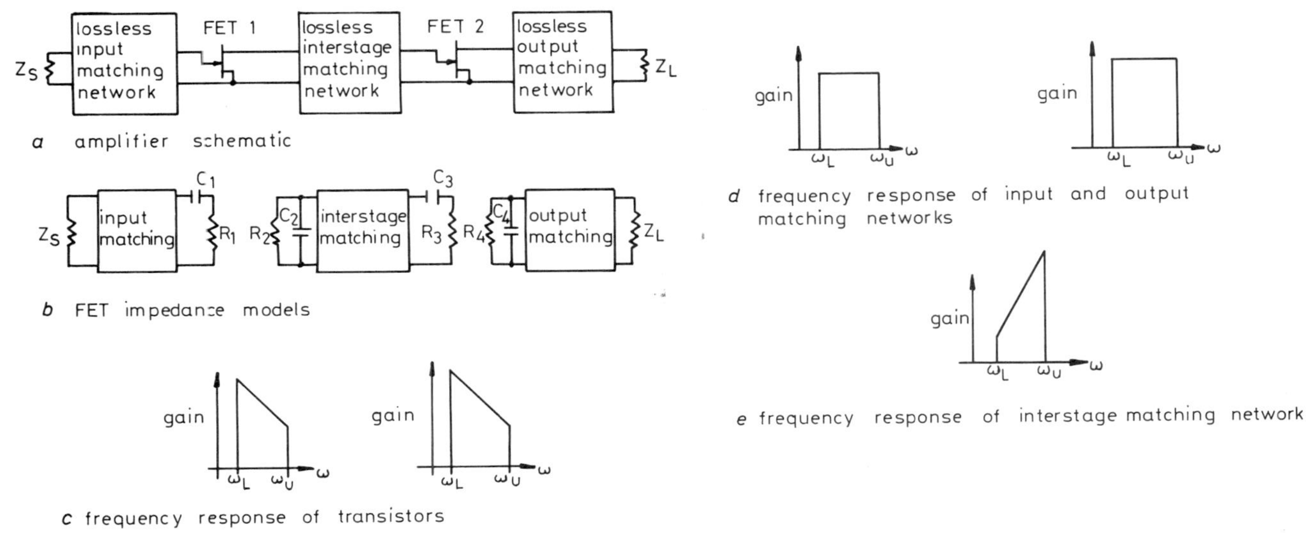


Figure 18 Typical amplifier design problem for network synthesis

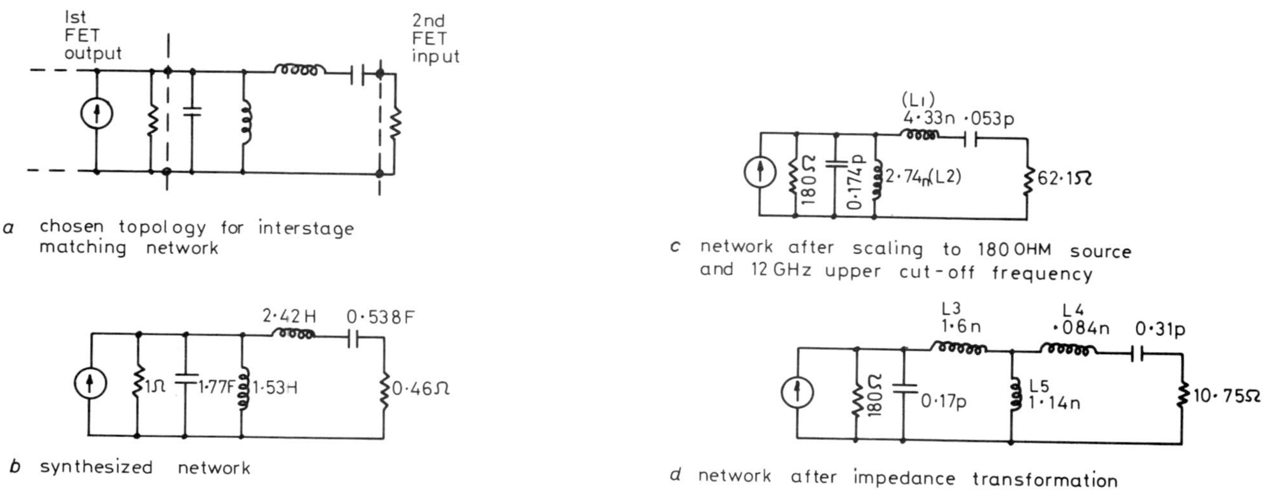


Figure 19 Synthesis routine for FET amplifier interstage matching

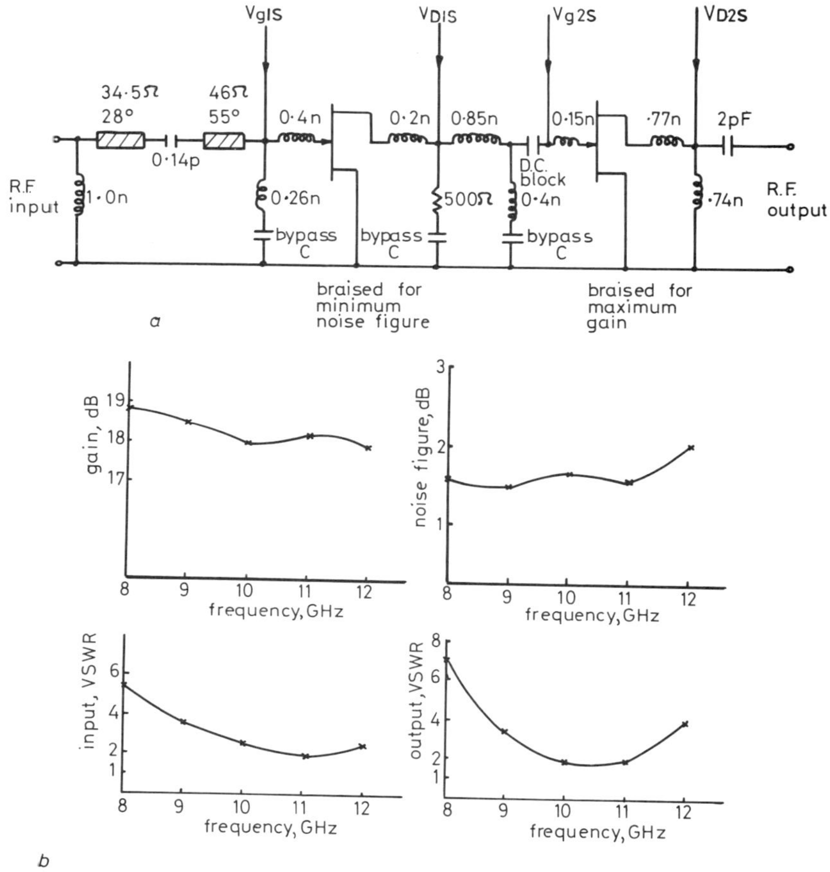


Figure 20 (a) 2 stage X-band wideband GaAs FET amplifier
(b) Response of 2 stage X-band GaAs FET amplifier

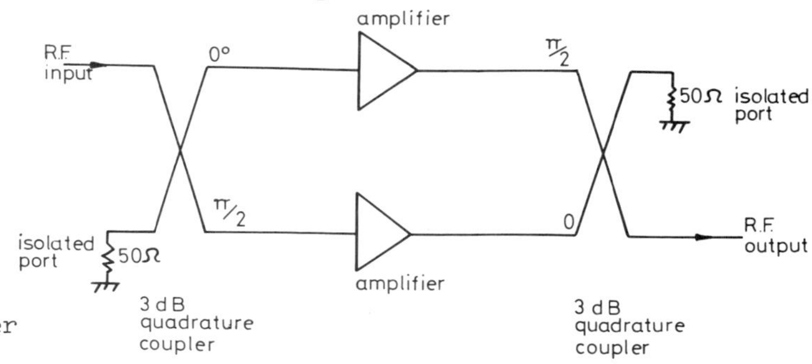


Figure 21 Balanced amplifier configuration

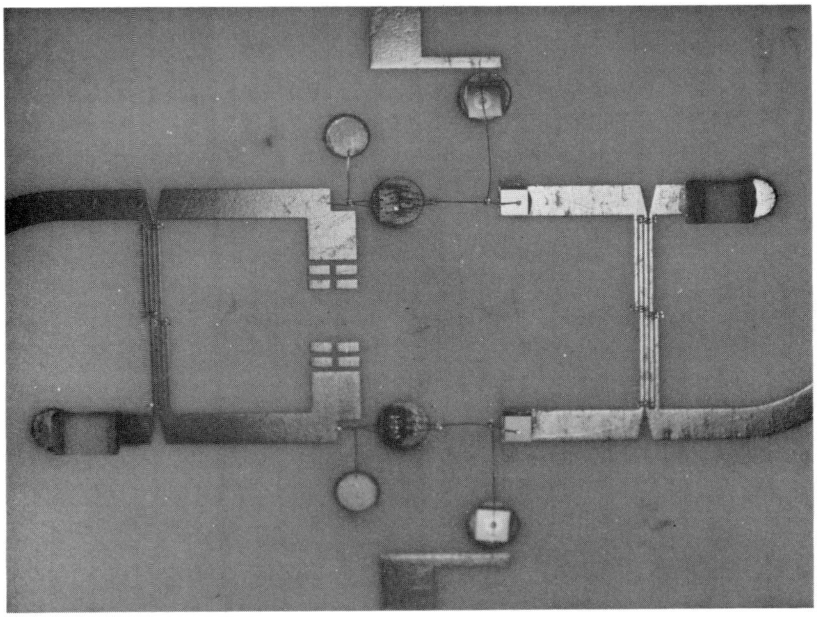

Figure 22 Example of microstrip wideband balanced amplifier (5 to 10 GHz)

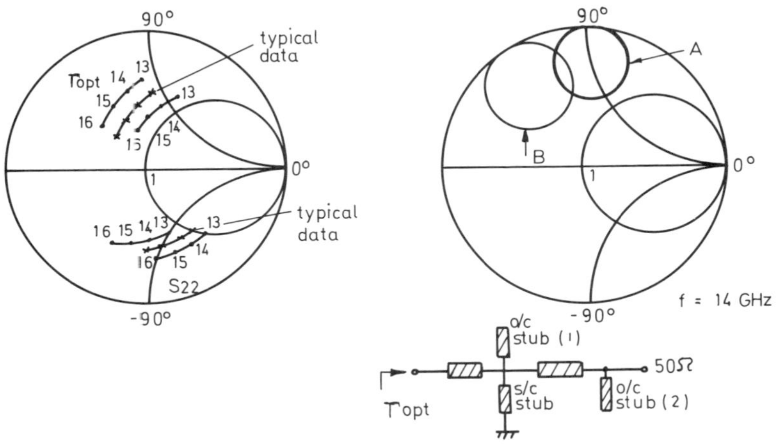


Figure 23 Typical method for adjustment of amplifier performance to account for transistor variations. Contour A is obtained by varying the length of element 2 (o/c stub); contour B is obtained by varying the length of element 1 (o/c stub). Intersection of these two contours defines the initial unaligned Γ opt value. Impedance coverage about that point is delineated by the contours emanating from that point. A similar method is used for S 22

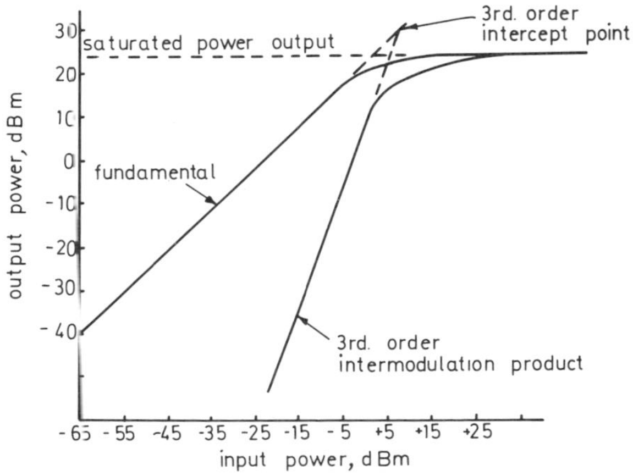


Figure 24 Intercept point response for a typical amplifier

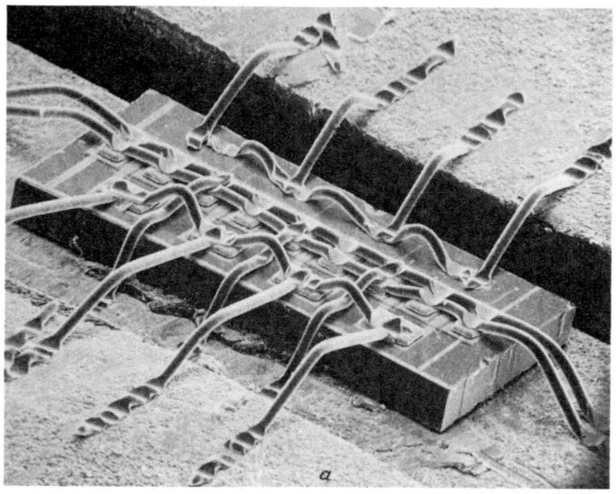

Figure 25 (a) Photograph of 4 cell 1W power GaAs FET

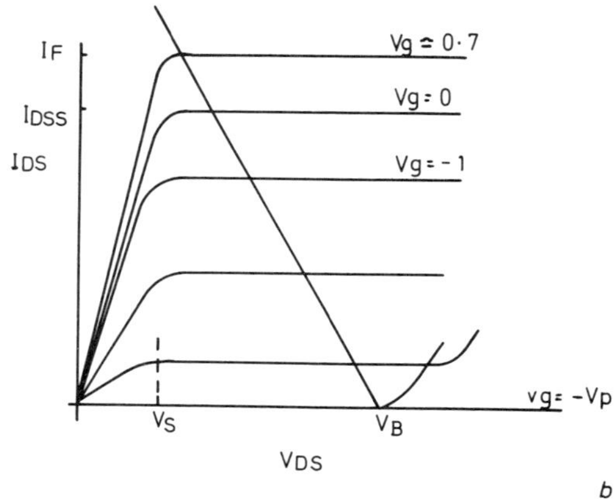


(b) I-V characteristics of a power FET showing the load line

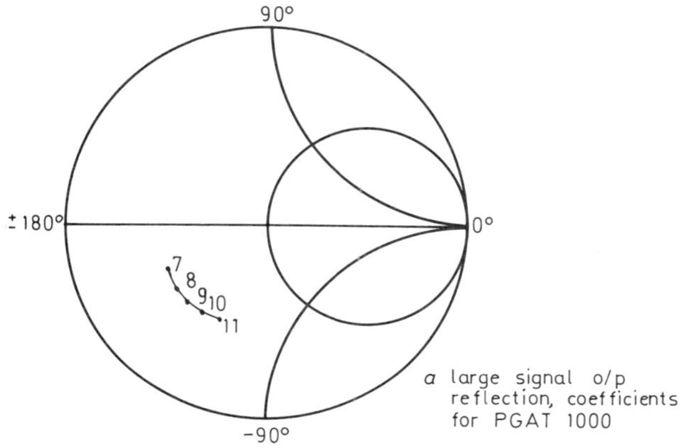


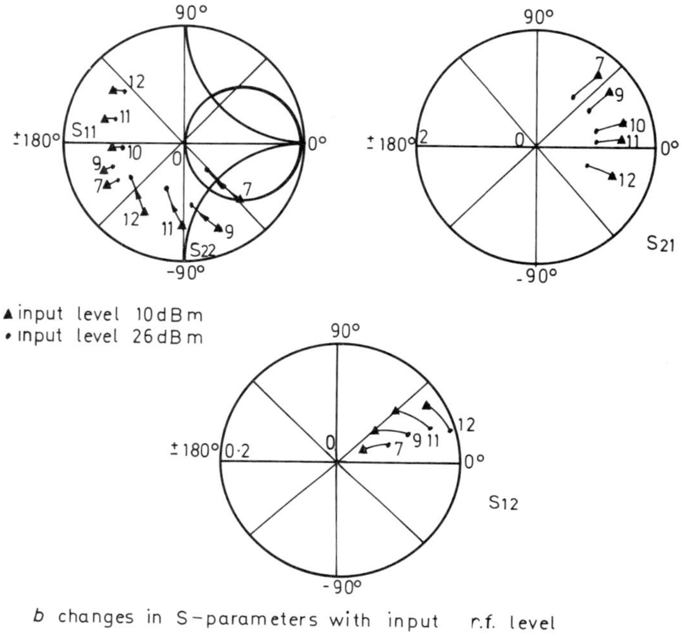


Figure 26 Large signal S-parameters of GaAs FET devices

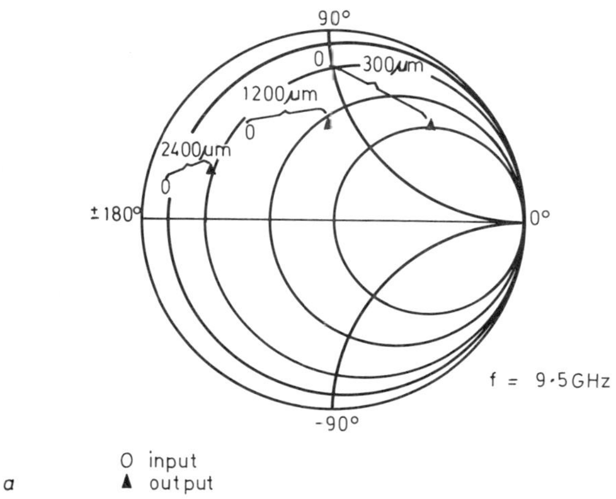


Figure 27 (a) Input and output circuit impedances for power GaAs FETs of varying gate widths

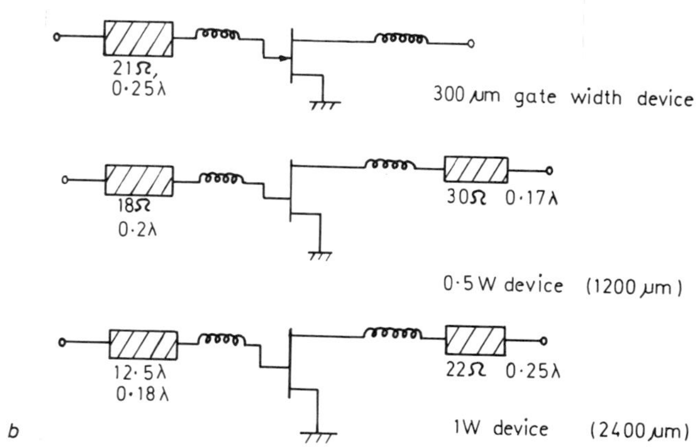


(b) Matching circuits for 100 mW, 0·5 W and 1 W FET

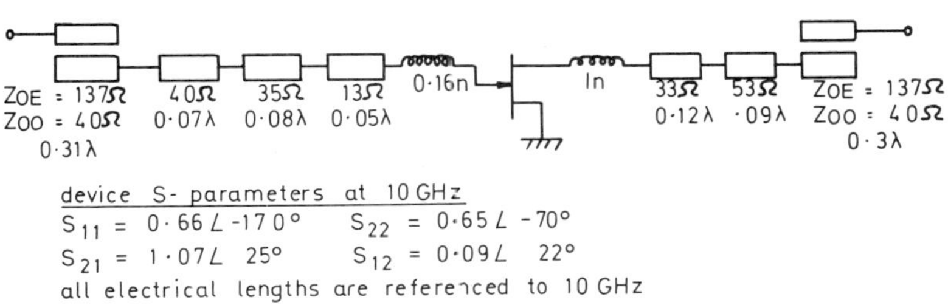


a

Figure 28 (a) Broadband power GaAs FET amplifier for 6-12 GHz operation

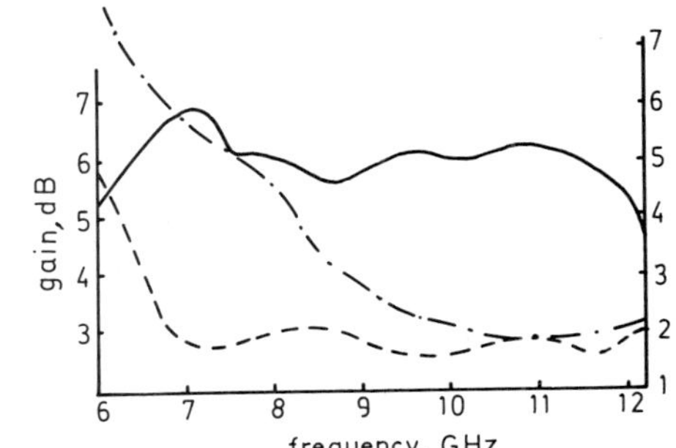


b

(b) Gain-frequency and VSWR characteristic of 6-12 GHz power FET amplifier

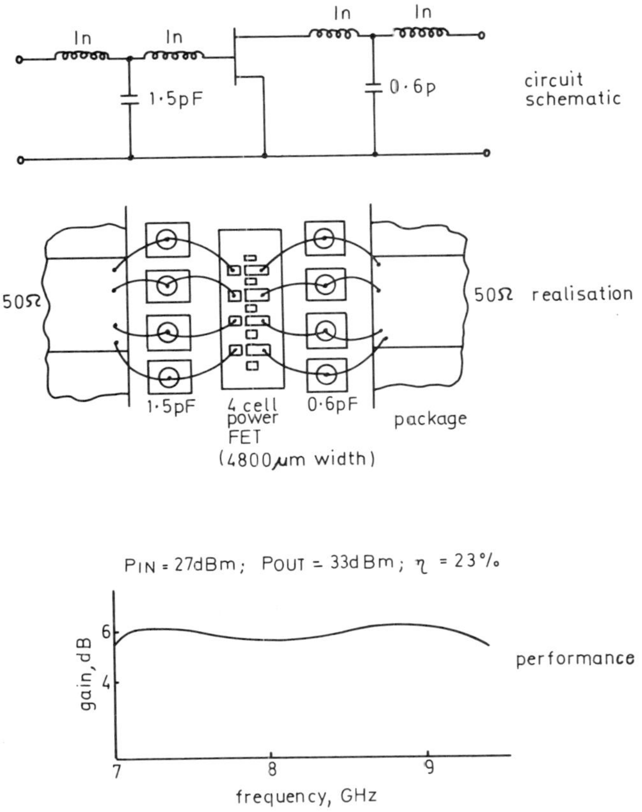


Figure 29 Broadband power FET with internal matching

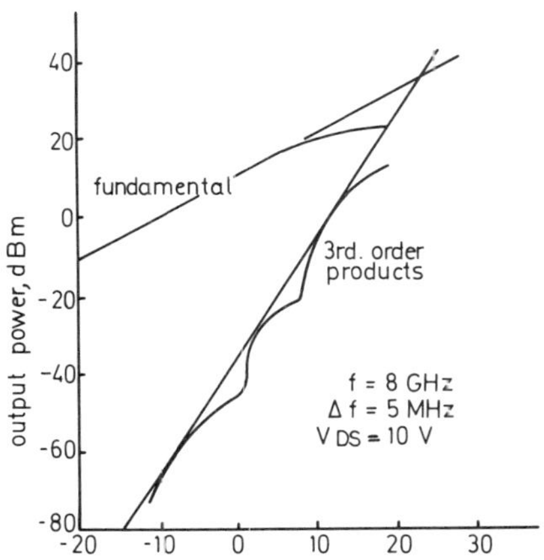


Figure 30 Measured values of output power level and IMD products for a power GaAs FET

efficiency of 4 combined amplifiers

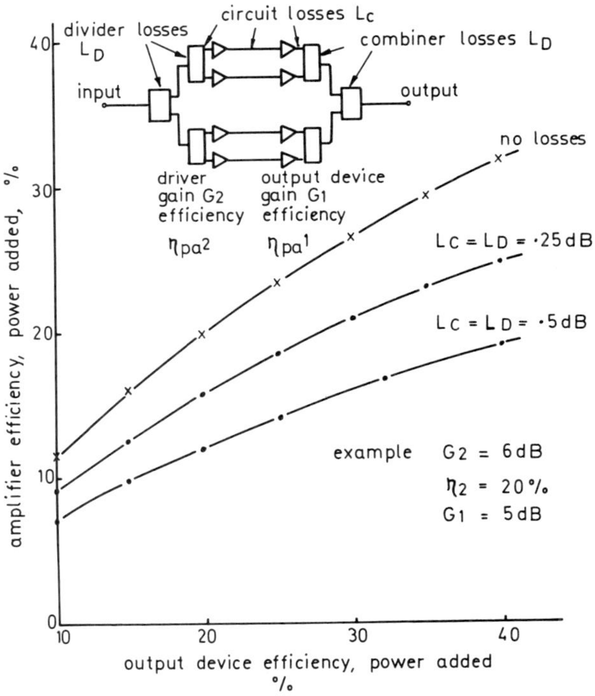


Figure 31 Efficiency of 4 combined amplifiers

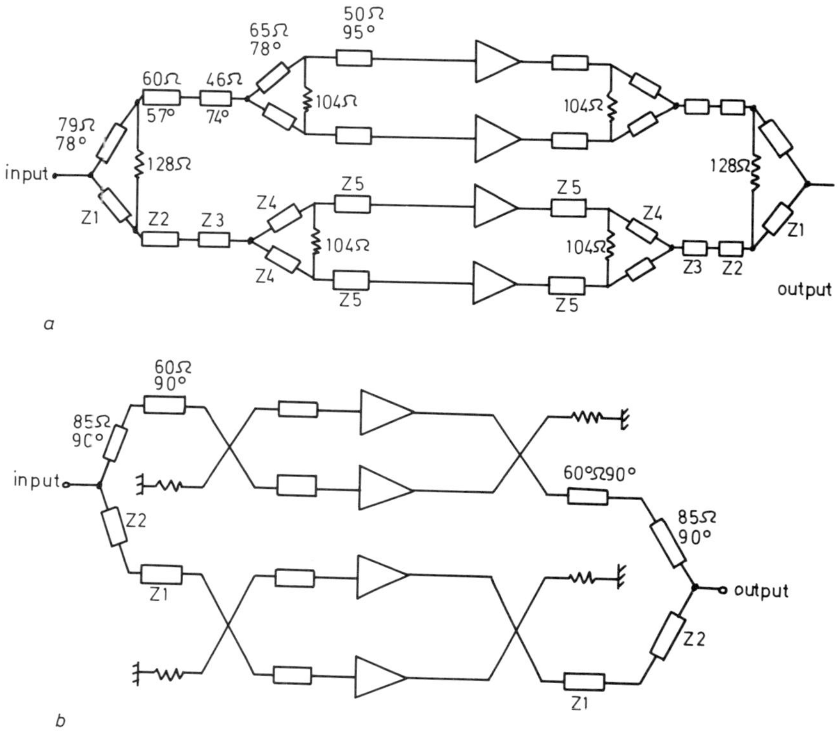


Figure 32 (a) Four way single ended amplifier
(b) Dual quadrature hybrid balanced amplifier

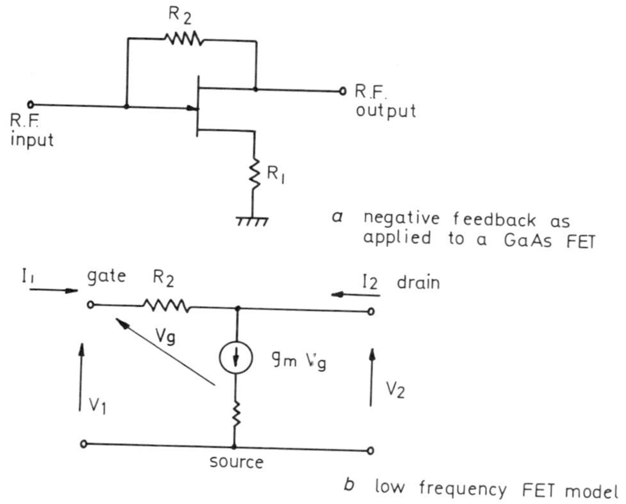


Figure 33 Negative feedback FET amplifier

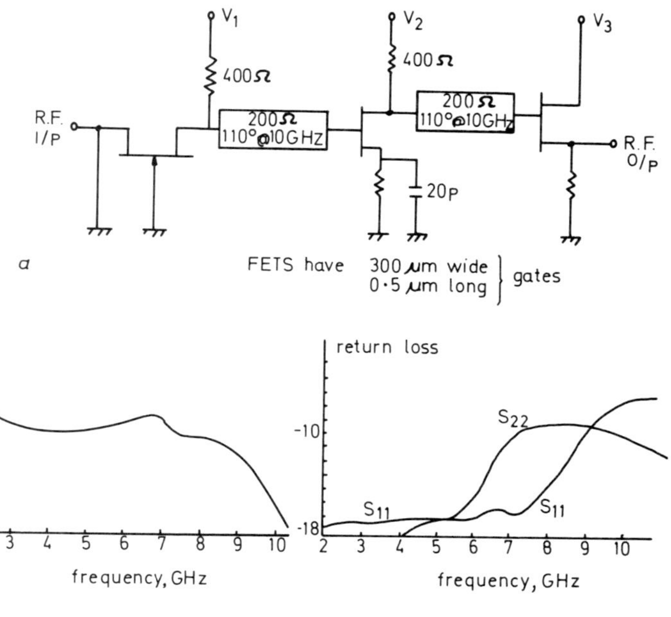


Figure 34 (a) Common gate/common source/ Source-follower amplifier
(b) Performance of circuit in (a)

TABLE I: TYPICAL S-PARAMETERS FOR GAT6 CHIP AT $I_{DSS}$, $V_{DS}$ = 5 VOLTS.

| Frequency GHz | $S_{11}$ | | $S_{21}$ | | $S_{12}$ | | $S_{22}$ | |
|---|---|---|---|---|---|---|---|---|
| | Mag | Phase | Mag | Phase | Mag | Phase | Mag | Phase |
| 2 | .95 | -38 | 2.95 | 150 | .02 | 75 | .725 | -15 |
| 3 | .92 | -48 | 2.77 | 140 | .02 | 74 | .725 | -17 |
| 4 | .89 | -60 | 2.6 | 130 | .02 | 78 | .725 | -21 |
| 5 | .85 | -72 | 2.41 | 120 | .02 | 83 | .725 | -26 |
| 6 | .825 | -84 | 2.25 | 110 | .025 | 90 | .725 | -31 |
| 7 | .8 | -96 | 2.09 | 100 | .03 | 97 | .73 | -36 |
| 8 | .775 | -107 | 1.96 | 90 | .035 | 104 | .735 | -42 |
| 9 | .75 | -118 | 1.84 | 81 | .04 | 110 | .74 | -47 |
| 10 | .73 | -128 | 1.73 | 73 | .045 | 114 | .75 | -52 |
| 11 | .71 | -136 | 1.64 | 64 | .053 | 115 | .755 | -58 |
| 12 | .695 | -145 | 1.55 | 56 | .06 | 115 | .765 | -62 |
| 13 | .68 | -153 | 1.47 | 47 | .065 | 113 | .77 | -68 |
| 14 | .665 | -161 | 1.4 | 40 | .073 | 110 | .78 | -72 |
| 15 | .655 | -168 | 1.33 | 33 | .08 | 106 | .79 | -77 |
| 16 | .64 | -175 | 1.26 | 27 | .09 | 100 | .8 | -81 |
| 17 | .63 | -180 | 1.2 | 22 | .1 | 93 | .805 | -86 |
| 18 | .62 | 174 | 1.14 | 16 | .11 | 86 | .815 | -90 |

TABLE II: TYPICAL S PARAMETERS FOR A P103 PACKAGED GAT6 AT $I_{DSS}$, $V_{DS}$ = 5 VOLTS

| Frequency GHz | $S_{11}$ | | $S_{21}$ | | $S_{12}$ | | $S_{22}$ | |
|---|---|---|---|---|---|---|---|---|
| | Mag | Phase | Mag | Phase | Mag | Phase | Mag | Phase |
| 4 | 0.865 | -78 | 2.37 | 112 | .04 | 43 | 0.74 | -35 |
| 5 | 0.84 | -94 | 2.24 | 99 | .04 | 37 | 0.72 | -44 |
| 6 | 0.81 | -112 | 2.14 | 85 | .05 | 37 | 0.71 | -53 |
| 7 | 0.78 | -113 | 2.04 | 70 | .05 | 34 | 0.692 | -62 |
| 8 | 0.76 | -150 | 1.94 | 55 | .05 | 37 | 0.68 | -70 |
| 9 | 0.74 | -164 | 1.85 | 44 | .06 | 47 | 0.666 | -76 |
| 10 | 0.724 | -178 | 1.77 | 30 | .06 | 46 | 0.654 | -84 |
| 11 | 0.71 | 166 | 1.68 | 20 | .07 | 49 | 0.644 | -95 |
| 12 | 0.695 | 150 | 1.6 | 6 | .07 | 44 | 0.634 | -100 |
| 13 | 0.685 | 134 | 1.53 | -4 | .08 | 43 | 0.626 | -114 |
| 14 | 0.672 | 121 | 1.47 | -15 | .09 | 37 | 0.618 | -125 |
| 15 | 0.664 | 110 | 1.4 | -25 | .11 | 38 | 0.615 | -136 |
| 16 | 0.66 | 100 | 1.34 | -34 | .125 | 34 | 0.61 | -149 |

TABLE III: STABILITY CIRCLES FOR A PACKAGED GAT6 HAVING THE S-PARAMETERS OF TABLE II

| Frequency GHz | INPUT PLANE | | | | OUTPUT PLANE | | | | K | MSG dB |
|---|---|---|---|---|---|---|---|---|---|---|
| | Location | | Radius | Stable Region | Location | | Radius | Stable Region | | |
| | MAG. | ANG. | | | MAG. | ANG. | | | | |
| 4 | 1.21 | 88 | 0.29 | Outside | 1.61 | 59 | 0.76 | Outside | .67 | 17.7 |
| 5 | 1.22 | 103 | 0.26 | Outside | 1.51 | 63 | 0.59 | Outside | .79 | 17.5 |
| 6 | 1.23 | 122 | 0.31 | Outside | 1.42 | 70 | 0.54 | Outside | .68 | 16.3 |
| 7 | 1.31 | 122 | 0.31 | Outside | 1.51 | 77 | 0.52 | Outside | .97 | 16.1 |
| 8 | 1.24 | 156 | 0.26 | Outside | 1.35 | 79 | 0.38 | Outside | .87 | 15.9 |
| 9 | 1.23 | 168 | 0.29 | Outside | 1.32 | 82 | 0.39 | Outside | .76 | 14.9 |
| 10 | 1.26 | -179 | 0.28 | Outside | 1.34 | 88 | 0.37 | Outside | .9 | 14.7 |
| 11 | 1.26 | -166 | 0.30 | Outside | 1.35 | 95 | 0.39 | Outside | .84 | 13.8 |
| 12 | 1.30 | -150 | 0.30 | Outside | 1.38 | 100 | 0.38 | Outside | 1 | 13.6 |
| 13 | 1.32 | -137 | 0.33 | Outside | 1.41 | 110 | 0.42 | Outside | .97 | 12.8 |
| 14 | 1.36 | -125 | 0.37 | Outside | 1.45 | 120 | 0.46 | Outside | .97 | 12.1 |
| 15 | 1.41 | -117 | 0.44 | Outside | 1.50 | 127 | 0.54 | Outside | .89 | 11.05 |
| 16 | 1.46 | -109 | 0.50 | Outside | 1.57 | 138 | 0.62 | Outside | .88 | 10.3 |

TABLE IV: TYPICAL EQUIVALENT CIRCUIT ELEMENT VALUES FOR A 0.5μm GATE LENGTH GaAs FET

| Intrinsic FET | | Parasitic Components | |
|---|---|---|---|
| $R_i$ | = 6.2 Ω | $R_g$ | = 0.8 Ω |
| $R_{DS}$ | = 400 Ω | $R_D$ | = 3 Ω |
| $C_{gs}$ | = 0.43 pF | $R_S$ | = 3 Ω |
| $C_{dg}$ | = 0.01 pF | $C_{DS}$ | = 0.16 pF |
| $g_m$ | = 30 m℧ | $L_G$ | = 0.2 nH |
| | | $L_S$ | = 0.05 nH |

TABLE V: TYPICAL S-PARAMETERS FOR A GAT6 CHIP AT $I_{DS}$ = 10 mA, $V_{DS}$ = 5 VOLTS (MINIMUM NOISE FIGURE CONDITIONS).

| Frequency GHz | $S_{11}$ | | $S_{21}$ | | $S_{12}$ | | $S_{22}$ | |
|---|---|---|---|---|---|---|---|---|
| | MAG | ANG | MAG | ANG | MAG | ANG | MAG | ANG |
| 4 | 0.93 | -43 | 1.66 | 134 | .039 | 64 | 0.79 | -24 |
| 5 | 0.9 | -54 | 1.59 | 125 | .042 | 62 | 0.79 | -29 |
| 6 | 0.88 | -64 | 1.52 | 115 | .045 | 61 | 0.78 | -34 |
| 7 | 0.85 | -75 | 1.46 | 106 | .048 | 60 | 0.78 | -39 |
| 8 | 0.83 | -85 | 1.4 | 97 | .05 | 60 | 0.77 | -44 |
| 9 | 0.81 | -93 | 1.34 | 90 | .052 | 61 | 0.77 | -49 |
| 10 | 0.79 | -101 | 1.29 | 80 | .055 | 63 | 0.77 | -54 |
| 11 | 0.77 | -110 | 1.24 | 72 | .057 | 65 | 0.77 | -60 |
| 12 | 0.75 | -116 | 1.19 | 64 | .059 | 67 | 0.77 | -65 |
| 13 | 0.74 | -123 | 1.14 | 55 | .061 | 69 | 0.77 | -70 |
| 14 | 0.72 | -130 | 1.1 | 47 | .063 | 71 | 0.77 | -75 |
| 15 | 0.7 | -137 | 1.06 | 38 | .066 | 72 | 0.77 | -81 |
| 16 | 0.68 | -144 | 1.03 | 31 | .068 | 71 | 0.77 | -85 |

TABLE VI: TYPICAL S-PARAMETERS FOR A PLESSEY PGAT1000 POWER FET (P105 PACKAGE)

| Frequency GHz | $S_{11}$ | | $S_{21}$ | | $S_{12}$ | | $S_{22}$ | |
|---|---|---|---|---|---|---|---|---|
| | MAG | ANG | MAG | ANG | MAG | ANG | MAG | ANG |
| 2 | 0.9 | -149 | 3.15 | 56.8 | .035 | 11 | 0.33 | -86 |
| 3 | 0.82 | -171 | 2.57 | 33.5 | .035 | -4 | 0.38 | -104 |
| 4 | 0.86 | 168 | 2.11 | 9 | .036 | -20.7 | 0.44 | -121 |
| 5 | 0.82 | 145 | 1.71 | -13.1 | .038 | -30.1 | 0.5 | -142 |
| 6 | 0.79 | 126 | 1.38 | -39 | .039 | -42.8 | 0.55 | -171 |
| 7 | 0.72 | 107 | 1.17 | -64 | .042 | -53 | 0.58 | 166 |
| 8 | 0.7 | 88 | 1.07 | -90 | .052 | -61.1 | 0.6 | 140 |
| 9 | 0.7 | 68 | 0.88 | -114 | .054 | -75 | 0.65 | 125 |
| 10 | 0.71 | 50 | 0.784 | -138 | .06 | -87.1 | 0.7 | 117 |
| 11 | 0.7 | 25 | 0.65 | -158 | .065 | -102 | 0.72 | 105 |
| 12 | 0.7 | 12 | 0.6 | -174 | .074 | -117 | 0.75 | 95 |

TABLE VII: COMMON GATE AND COMMON DRAIN CONNECTED GAT6 S-PARAMETERS

| Frequency GHz | $S_{11}$ | | $S_{21}$ | | $S_{12}$ | | $S_{22}$ | |
|---|---|---|---|---|---|---|---|---|
| | MAG | ANG | MAG | ANG | MAG | ANG | MAG | ANG |
| 2 | .27 | -175 | 1.26 | -9.9 | .116 | 23.6 | .91 | -7 |
| 3 | .26 | -176 | 1.25 | -13.9 | .124 | 27.8 | .92 | -8 |
| 4 | .26 | -175 | 1.25 | -18.2 | .139 | 32.1 | .94 | -10 |
| 5 | .27 | -173 | 1.25 | -22.8 | .159 | 35.6 | .96 | -13 |
| 6 | .28 | -173 | 1.25 | -27.9 | .178 | 36.3 | .98 | -16 |
| 7 | .29 | -171 | 1.27 | -33.8 | .196 | 36.5 | 1.02 | -20 |
| 8 | .3 | -172 | 1.29 | -40.2 | .218 | 36.8 | 1.07 | -24 |
| 9 | .33 | -173 | 1.32 | -47.1 | .232 | 36.4 | 1.13 | -28 |
| 10 | .36 | -175 | 1.35 | -54.4 | .244 | 36.1 | 1.2 | -33 |
| 11 | .36 | -178 | 1.41 | -63.3 | .263 | 36.3 | 1.29 | -39 |
| 12 | .38 | 178 | 1.49 | -73.8 | .251 | 37.5 | 1.38 | -42 |

| Frequency GHz | $S_{11}$ | | $S_{21}$ | | $S_{12}$ | | $S_{22}$ | |
|---|---|---|---|---|---|---|---|---|
| | MAG | ANG | MAG | ANG | MAG | ANG | MAG | ANG |
| 2 | .98 | -15 | 1.22 | -7.8 | .202 | 73.0 | .32 | 167 |
| 3 | .98 | -20 | 1.2 | -10.9 | .267 | 69.6 | .3 | 158 |
| 4 | .97 | -25 | 1.18 | -13.9 | .343 | 64.2 | .28 | 152 |
| 5 | .95 | -31 | 1.17 | -17.1 | .419 | 58.0 | .26 | 145 |
| 6 | .94 | -37 | 1.16 | -20.4 | .492 | 52.4 | .25 | 138 |
| 7 | .93 | -43 | 1.14 | -23.9 | .567 | 46.5 | .23 | 129 |
| 8 | .92 | -49 | 1.14 | -28.0 | .635 | 40.8 | .21 | 120 |
| 9 | .91 | -56 | 1.13 | -31.8 | .7 | 35.2 | .2 | 110 |
| 10 | .9 | -62 | 1.13 | -35.4 | .757 | 29.7 | .18 | 100 |
| 11 | .91 | -68 | 1.12 | -40.5 | .811 | 25.0 | .17 | 83 |
| 12 | .88 | -75 | 1.09 | -44.4 | .882 | 20.8 | .19 | 63 |

# CHAPTER 9

## Practical transistor amplifier

Dr. E. Huges/ Dr. S. Cripps

# PRACTICAL TRANSISTOR AMPLIFIERS (a) BJT AMPLIFIERS - Chapter 9

E. Hughes

GEC Hirst Research Centre, Wembley, Middlesex.

## INTRODUCTION

This section is concerned with the design and realisation of amplifiers using BJTs (bipolar junction transistors, which would normally be silicon npn types) from approximately 1 GHz up to 4 GHz. The gain of currently available BJTs tends to fall rapidly at a frequency of about 5 GHz, which is roughly the point at which FETs take over. However, recent work (1) on submicron emitter width devices, coupled with other technological improvements, have resulted in transistors capable of giving adequate gain up to 6 or 8 GHz, with the prospect of further improvements. Before going on to design methods, it is thus considered worth while to consider the merits of BJTs in microwave amplifiers.

## MERITS OF BJTs IN AMPLIFIERS

The question of adequacy of device performance in a given application is not the only criterion for choosing a particular type of transistor; matters such as cost, availability, ruggedness, reliability, etc. are often of equal importance. In appropriate situations, the BJT can be preferable to the FET, which is its only rival in the solid-state field, for the following reasons (considering now applications up to about 4 GHz);

(a) Lower cost.

(b) Well-proven reliability.

(c) Ruggedness; the BJT can withstand much larger overload pulses than currently-available FETs. This is a particularly important consideration in radar systems.

(d) BJT technology is mature and well-tested, and can allow optimisation of devices for a wide range of applications.

(e) BJTs have terminal impedances which are far easier for the circuit designer to deal with, particularly in wide-band applications.

(f) The BJT is usually unconditionally stable at the frequencies under discussion, whereas most FETs are unstable.

(g) The 'optimum noise' source impedance of the BJT in the range 1-4 GHz is of such a value that there is not a large trade-off between good noise performance and high gain.

(h) The bias currents of low-noise BJTs are much lower than those of FETs; this can be of importance where power consumption is a prime consideration.

It is thus evident that, although FETs are finding increasing use, the BJT still has plenty to offer in its own frequency range (i.e. up to 4 or 5 GHz).

The remainder of this part of the chapter will deal with the practical aspects of low-noise small-signal amplifier design and realisation, and will also give an introduction to power amplifier design.

## LOW-NOISE SMALL-SIGNAL AMPLIFIERS

### Basic Design Considerations

The previous chapter has covered the subject of s-parameters, which are used almost universally for the design of small-signal amplifiers; these parameters, in conjunction with the noise parameters defined below, completely characterise the device in terms of its circuit performance, and give all the information required to embark on a design to a given specification.

Amplifier specifications generally include a large number of items, the most important being the 'primary' characteristics (gain, gain flatness, terminal matches, and noise figure). Those which can be regarded as 'secondary' (distortion, output power, behaviour with temperature, etc) are less readily calculable and will not be considered in detail in this chapter.

In order to outline the main points of the design procedure, a hypothetical, albeit perfectly realistic, specification will be assumed for a single-stage amplifier, viz:

| | | |
|---|---|---|
| Frequency range | : | 3.7 to 4.2 GHz |
| Gain | : | 9 ± 1 dB over the band |
| Noise figure | : | 3 dB (max) |
| Output vswr | : | 1.5:1 (max) |

Selection of suitable device. This is a matter of consulting the data sheets on available transistors, which nowadays include full s-parameter data, noise information, and data on the maximum available gain (MAG) as a function of frequency. A device would be chosen on its ability to provide the required gain and noise figure at the highest frequency, making due allowance for the fact that there will be some circuit losses, which will degrade the noise figure and reduce the gain. For the hypothetical example, we have chosen a particular device for which we find the following information at 4 GHz;

$S_{11} = 0.55 \angle 170^{\circ}$ $\quad S_{12} = .05 \angle 25^{\circ}$

$S_{21} = 1.7 \angle 25^{\circ}$ $\quad S_{22} = 0.83 \angle -70^{\circ}$

MAG = 14.7 dB

Minimum noise figure, $F_{min} = 2.5$ dB

Source impedance, $\Gamma_o$, for $F_{min} = 0.48 \angle 168$

'Noise resistance', $R_n = 3.5$ ohms

We must now consider some fundamental BJT properties, together with some definitions of the noise quantities mentioned above.

Fundamental BJT properties.

(a) Figure 1 illustrates the variation of $F_{min}$ with operating current at 4 GHz. $F_{min}$ is defined as the minimum achievable noise figure, and is a function of the source impedance; there is a unique value ($\Gamma_o$) of the source reflection coefficient at which $F_{min}$ occurs (for a given frequency and current). The quantity $R_n$ has the dimensions of resistance, and is a measure of how rapidly the noise figure changes with variation of the source reflection coefficient, $\Gamma_s$. The general equation for the noise figure (F) as a function of $\Gamma_s$ is

$$F = F_{min} + 4R_n \frac{|\Gamma_s - \Gamma_o|^2}{(1-|\Gamma_s|^2)(|1+\Gamma_o|)^2} \qquad (1)$$

(b) Figure 1 also shows the 'associated gain' ($G_A$) obtained under the same measured conditions as used for $F_{min}$; $G_A$ is normally less than MAG, but is the appropriate figure to use for low-noise designs. It is seen that $G_A$ reaches a maximum and $F_{min}$ a minimum at about 4 mA operating current, so this would be the best choice of bias for the design. At this current, the transistor has $F_{min}$ = 2.5 dB and $G_A$ = 9.3 dB.

(c) It can be shown from equation (1) that F can be represented on the Smith chart by a system of coaxial circles surrounding $\Gamma_o$ (see Tutorial Question 1). If we choose, say, F = 2.6 dB, the locus of $\Gamma_s$ would lie on a circle, and any of the values of $\Gamma_s$ on this circle would yield F = 2.6 dB. Similarly, the circle for 2.7 dB would be of larger radius, and would enclose the 2.6 dB circle, and so on (these are the 'constant noise figure' circles). For BJTs, these circles are of quite large radius, and the noise figures are less sensitive to matching errors than are FETs in the same frequency range. If we construct the circle for F = 2.6 dB, we find that the magnitude of $\Gamma_s$ can change from .635 to .311 for only 0.1 dB increase in F over $F_{min}$. Thus, it is possible to choose a value of $\Gamma_s$ from this circle which gives a value of gain greater than $G_A$; in this case, we find we can choose $\Gamma_s$ to give a gain of 11 dB at F = 2.6 dB.

(d) The gain of the first stage is important as far as the total noise figure of a number of cascaded stages is concerned, since the second stage can give a significant contribution to the total; the noise figure ($F_{tot}$) of an input stage (F,G) followed by a stage having a noise figure $F_2$ is

$$F_{tot} = F + \frac{F_2 - 1}{G} \qquad (2)$$

Hence it can be an advantage to choose the conditions such that G is high, even though F may also be a little higher than $F_{min}$, in order to realise a low value of $F_{tot}$. This represents the trade-off between F and G, and it is useful to know that the minimum noise figure that can be obtained from an infinite cascade of identical stages ($F_c$) for a given device is

$$F_c = 1 + \frac{F_{min} - 1}{1 - 1/Ga} \qquad (3)$$

and M, defined as ($F_c$ - 1), is known as the 'noise measure' of the device (2), and is a useful figure of merit.

Calculation of the load impedance. Since one requirement of the design is for a good output match, the output circuit is designed to conjugately match the device's output impedance, and we proceed as follows:

(a) Calculate $S'_{22}$, the output reflection coefficient of the device with the chosen source reflection coefficient (assume this to be $\Gamma_o$ for the present purpose). The appropriate equation is

$$S'_{22} = S_{22} + \frac{S_{12}S_{21}\Gamma_o}{1-S_{11}\Gamma_o} \qquad (4)$$

(Note that the corresponding equation for $S'_{11}$ with a load termination of $\Gamma_L$ is obtained by interchanging the suffices 1 and 2 and putting $\Gamma_L$ in place of $\Gamma_o$).

(b) Devise a load reflection coefficient, $\Gamma_L$, equal to $S'_{22}{}^*$.

At this stage, some practical points regarding the circuit configurations in general would be useful.

Hints on circuit elements. The selection of a suitable circuit to provide a given reflection coefficient can take many forms. From the practical viewpoint, and assuming that the final circuit will be realised in distributed form on open microstrip, the following points are made, which are applicable to the input or output circuits.

(a) If possible, select a configuration including a shorted shunt stub fairly close to the device terminals, to provide a lead-in for the d.c. bias.

(b) Use the highest possible impedance for shorted stubs, since these are far less sensitive to the decoupling conditions at the 'earthy' end than are low-impedance lines.

(c) Remember that the device needs to be mounted and bonded (or soldered) to the circuit, which means that certain small pad areas would be needed close to the device terminals. These can be included in the complete circuit as short lengths of transmission line.

## Design of the Matching Networks

Input network. Taking, for this example, the value of $\Gamma_s$ as equal to $\Gamma_o$ (0.48 $\angle 168^o$), we firstly plot this value on the Smith chart, and read off the corresponding admittance value, which is

$$Y_o = (0.052 - j0.014) \text{ mho}$$

A shorted shunt stub can be used to tune out the susceptance component; a shorted stub of length $\ell$ and characteristic impedance $Z_o$ has an impedance given by

$$Z = jZ_o \tan\beta\ell \qquad (5)$$

and if $\ell = \lambda/8$, then $\beta\ell = \pi/4$, so

$$Z = jZ_o \qquad (6)$$

Hence the required stub impedance is $Z_o$=1/.014, or 71 Ω. This brings the conductance component round to correspond to a resistive component of 19.25 Ω, which can be matched to the 50 Ω system by using a quarter wave transformer of impedance

$$Z_o = \sqrt{50 \times 19.25} = 31\ \Omega$$

Reference to figure 2 (which shows the complete circuit, including the output matching network to be discussed in the following section) shows how the circuit is arranged. A decoupling capacitor is included at the 'shorted' end of the stub, to feed in the d.c. bias, and the low impedance of the $\lambda/4$ transformer provides an area large enough for the attachment of transistor leads.

Output network. In this case, the value of $\Gamma_L$ (from equation (1)) is $0.841\lfloor 74$ . If we start with a small 'soldering area', consisting of a 0.6 cm length of 50 Ω line, this will rotate $S_{22}'$ round to a value $0.846\lfloor -131^o$, corresponding to an impedance of $(5.0-j22.5)\Omega$. We then add a shorted shunt stub of tune out most of the susceptance, followed by a series line to match the real partofthe resulting admittance to 50 Ω. We then add a short length of open-circuited stub to tune out the remaining admittance, and also to provide a limited tuning capability, since its length and impedance can be varied in situ. Finally, a short length of series 50 Ω line is added as a pad to hold the output blocking capacitor; this does not affect the vswr. The complete circuit is shown in figure 2. (Note: since the foregoing derivation is mainly a matter of Smith-chart manipulation, it is not intended to give all the detail in these notes).

Bandwidth of the final amplifier. This can be estimated from the output Q of the transistor, which is the dominant factor; the normalised impedance corresponding to $S_{22}' = 0.846\lfloor -72^o$ is 0.23-j1.35, and hence

$$Q = \frac{\text{Reactance}}{\text{Resistance}} = \frac{1.35}{0.23} = 5.88$$

The bandwidth to the 3 dB points is given in terms of the centre frequency, $f_o$, by

$$\Delta f = \frac{f_o}{Q} = 680 \text{ MHz for } f_o = 4 \text{ GHz}$$

Theoretical trimming of the circuit by computer optimisation methods outlined in the Section Computer Optimisation, using the calculated elements as starting values, will enable this amplifier to maintain a 'flat' gain over the required bandwidth; since this is only 500 MHz, the problem of constant gain is relatively simple, and the design can be carried out by what is essentially a single-frequency approach.

## Wide-band Amplifiers

The design of amplifiers with bandwidths of an octave or more presents a number of problems, which can briefly be summarised as follows.

(a) The MAG of BJTs decreases with frequency at the rate of 5.5 to 6 dB per octave. Since wide-band specifications require a nominally constant gain over the band, this means that the matching networks must be designed to have insertion losses which vary in sympathy with the gain of the transistor (this is known as 'gain compensation').

(b) The variations of $S_{11}$ and $S_{22}$ over a wide band can be quite large, which causes difficulties in the matching network design.

(c) The process of broad-banding causes a degradation of the noise figure, and it would not be possible to obtain F = 2.6 dB (as in the previous example) over a band covering 2 to 4 GHz, even though the device itself is capable of this performance at the upper frequency.

(d) It is not possible to design wide-band single stage amplifiers to have flat gain and low terminal vswr's simultaneously, because of the gain compensation; this leads to quite severe mismatches at the lower frequencies. It is possible to use negative feedback (see Section Negative feedback) to provide constant gain and easier matching, but this does degrade both the gain and the noise figure.

Interstage gain compensation. For 'single-ended' amplifiers (as distinct from the balanced form described in Section Balanced amplifiers), it is best to arrange that the gain compensation is provided by the interstage networks, and the input and output networks can be designed on band-pass filter principles to provide good terminal matches. Considering, for example, a 3 stage 2 to 4 GHz amplifier, the total MAG of the three devices will fall off at about 18 dB/octave; this means that each interstage network would be designed to have an insertion loss at 2 GHz which is 9 dB greater than that at 4 GHz, and the 4 GHz value would naturally be made as low as possible.

In recent years, much effort has been given to the problem of designing what are essentially 'sloped response' band-pass filters (3), and analytical solutions have been obtained. The use of a computer is normally required, because of the complexity of the equations, and a comprehensive programme is now commercially available (4). This will synthesise a network for a given sloped response, and then calculate the impedance transformation required between the output of one transistor and the input of the next. It is not really possible to deal in detail with the procedure in these notes, and the student is advised to consult the given references.

Terminal matching. This problem is somewhat simpler, since sloped insertion loss is not required, and the circuit elements can be calculated in the form of Tschebycheff filters from fairly straightforward and closed-form equations (5). The procedure is to match the 50 Ω source impedance to the input impedance of the first transistor, and the output impedance of the last transistor to the 50 Ω load, using the simplest LRC equivalent networks for these impedances which can be derived to give a reasonably accurate representation.

The most accurate method is to calculate the circuit performance of the 3-stage amplifier with the two interstage networks connected, and obtain the input and output impedances taking into account the feedback ($S_{12}$) through the transistors. These impedances are then plotted on a Smith chart, from which approximate equivalent circuits of the forms shown in figures 4(a) and 4(b) can be derived. Note that, in most cases, the elements of these circuits will not be completely independent of frequency, but they should be constant enough over an octave band to represent them by the values at, say, mid-band. These values can then be used as the impedances to be matched, using the methods of reference 5.

Computer optimisation. The use of the foregoing procedures will yield a circuit having a reasonable approach to the specified gain and vswr's, which can be used as a good starting point for optimisation by computer methods. A full discussion of these methods is outside the scope of these notes, and it is suggested that reference 4 is consulted; this describes a commercially-available optimisation programme capable of dealing with noise and gain parameters. Basically, the method is to take the complete circuit and vary certain selected elements (decided by the design) within given constraints, so as to make the performance approach the specified requirements as closely as possible. The data required for this type of programme generally consists of:

(a) The s-parameters of the device(s) used, at various equally-spaced points in the given frequency band (an 'interpolation' routine is provided to give closer-spaced values when required).

(b) The noise parameters as defined in Section Fundamental BJT properties, as functions of frequency. Interpolation is usually provided in this case.

(c) The complete circuit description, with the element values obtained as described in Section Interstage gain compensation and Section Terminal matching.

(d) The specification (i.e. gain, terminal vswr's, and noise figure). For example, a gain of 24 dB with a flatness of ±1 dB could be specified, along with maximum vswr's of 2:1 and a maximum noise figure of 6 dB. It is also possible to place 'weightings' on each requirement, so that the most important feature can be optimised at the expense of others if necessary.

Once a satisfactory design is obtained, it will usually be necessary to re-optimise, including in the total circuit all extra elements that are required (e.g. quarter-wave bias chokes, d.c. blocking capacitors, etc) but do not form part of the main matching circuits. It is also necessary to take into account such 'parasitics' as wire bonds, stray capacitances, and losses in series and decoupling capacitors, and also to make as much allowance as possible for discontinuities in the microstrip circuit.

Practical realisation. These circuits are normally realised in open microstrip on an m.i.c. substrate such as alumina (dielectric constant $\approx$ 9.6), and should be laid out to occupy the smallest possible substrate area. Suitable earthing pads must be provided for all points which are intended to be directly earthed, such as the emitters of the transistors, and these can be connected to the ground plane in the following ways:

(a) By placing the pads at the edges of the substrate, and plating round the edge to the ground plane.

(b) If (a) is not possible, a hole can be drilled in the substrate at the centre of the pad, which can be plated through to the ground plane, or connected by gold tape or rivets.



Transistors encapsulated in coplanar headers can be attached to the circuit using solder or gold-loaded epoxy; chip transistors can be eutectically mounted on the collector pad, with bond-wires connecting the base and emitter to the appropriate places.

Capacitors should preferably be of the ceramic chip type, which can be attached using solder or epoxy on one side with tape or bond-wire connections to the other side. MOS capacitors can also be used, and these can be eutectically mounted.

Balanced amplifiers. This configuration is one way of solving the problem of designing wide-band single-stage amplifiers with flat gain and good terminal matches. The method, originally proposed by Kurokawa (6), is to use two identical amplifier circuits connected together using 3 dB hybrid quadrature couplers, and is shown schematically in figure 5. The couplers used almost universally in this type of circuit are of the interdigitated structure due to Lange (7), which divides the input power equally between the two 'sides' of the circuit, thus providing the gain, and recombines the two outputs in the second coupler.

The properties of the coupler are such that any reflected signals at the input or output (due to high transistor vswr's) are 'dumped' in the 50 Ω resistive loads connected to the isolated port of each coupler, and so the matches seen at the input and output via the couplers can be perfect 50 Ω matches, provided:

(a) the two sides of the circuit have identical input and output vswr, in magnitude and phase.

(b) The couplers are free of imperfections.

The gain of the system is equal to the gain of one side; it can be shown that the overall s-matrix of the balanced unit is given by the following equations

$$|S_{11}| = |t^2 S_{11a} - (1-t^2) S_{11b}| \tag{7}$$

$$|S_{21}| = t\sqrt{1-t^2}\,|S_{21a} + S_{21b}| \tag{8}$$

$$|S_{12}| = t\sqrt{1-t^2}\,|S_{12a} + S_{12b}| \tag{9}$$

$$|S_{22}| = |t^2 S_{22a} - (1-t^2) S_{22b}| \tag{10}$$

Subscripts a and b apply to each single-ended side, including the matching networks (i.e. the portions shown in dotted lines in figure 5, which at this point are not assumed to be identical). The quantity t is a function of the electrical length (θ) of the coupler, and is given by

$$t = 1/\sqrt{2 - \cos^2\theta} \tag{11}$$

Hence, if the coupler is made λ/4 long, then $\theta = 90^o$, and $t^2 = 0.5 = (1-t^2)$, so the above equations reduce to

$$|S_{11}| = 0.5|S_{11a} - S_{11b}| \tag{12}$$

$$|S_{21}| = 0.5|S_{21a} + S_{21b}| \tag{13}$$

$$|S_{12}| = 0.5|S_{12a} + S_{12b}| \tag{14}$$

$$|S_{22}| = 0.5|S_{22a} - S_{22b}| \qquad (15)$$

If the two amplifier circuits are made identical, so that $S_{11a} = S_{11b}$, etc, then we have

$$|S_{11}| = |S_{22}| = 0 \qquad \text{(perfect matches)}$$

And the gain, $|S_{21}|$ equals the gain of one side.

The advantages of this configuration are:

(i) That good terminal matches and flat grain response can be obtained simultaneously, even though the actual amplifying circuits may have high vswr's.

(ii) The stages can be readily cascaded with very little interaction (this is known as the 'modular' approach to amplifier design).

(iii) The amplifier circuits can be designed for optimum noise source impedance and still maintain a good match.

(iv) The output power is twice that obtainable from a single transistor, and the distortion performance is correspondingly improved.

(v) The balanced unit can still function, although with reduced gain, if one of the devices fail.

To obtain these advantages, it is necessary to use two devices and a somewhat complex circuit, which increases the cost compared to a similar single-ended stage.

Negative feedback. It is possible to apply resistive negative feedback (NFB) to flatten the gain response and reduce the vswr's (8,9); such techniques allow the circuit designer a good measure of control over the performance, and also help to reduce performance variations due to batch-to-batch spreads in device s-parameters. It does, however, degrade the noise figure and reduce the gain available from a given device. The basic method, including both shunt and series feedback, is shown in figure 6 (the capacitor C in the shunt connection is purely for the purpose of d.c. isolation).

The simple low-frequency treatment of Ulrich (9) can be applied to BJTs by modifying his equivalent model to include the finite input resistance of the common-emitter circuit ($r_{b'e}$), as shown in figure 7; otherwise, the circuit is identical to that of Ulrich. In the present case, the transistor base resistance is neglected, and for a representative low-noise device biassed at 4 mA operating current, $r_{b'e}$ is in the region of 300 Ω and $g_m$ is about 150-160 mS. This high value of transconductance makes the bipolar transistor very suitable for the application of NFB; Ulrich points out that FETs, in order to use NFB effectively, would need to exhibit much higher values of $g_m$ than are usual at the present time.

The Y-matrix of the circuit of figure 7 can readily be written, viz.

$$Y = \begin{bmatrix} \frac{1}{R_T} & -\frac{1}{R_2} \\ \frac{g_m}{1+g_m R_1} - \frac{1}{R_2} & \frac{1}{R_2} \end{bmatrix} \qquad (16)$$

$$\text{where } R_T = \frac{R_2 r_{b'e}}{r_{b'e} + R_2} \qquad (17)$$

The equations to convert the y-parameters to s-parameters in a system of impedance $Z_o$ (normally 50 Ω) are as follows:

$$S_{11} = \frac{(1-Z_o y_{11})(1+Z_o y_{22})+y_{12}y_{21}Z_o^2}{\Delta} \qquad (18)$$

$$S_{12} = -\,2y_{12}/\Delta \qquad (19)$$

$$S_{21} = -\,2y_{21}/\Delta \qquad (20)$$

$$S_{22} = \frac{(1+Z_o y_{11})(1-Z_o y_{22})+y_{12}y_{21}Z_o^2}{\Delta} \qquad (21)$$

where

$$\Delta = (1+Z_o y_{11})(1+Z_o y_{22})-y_{12}y_{21}Z_o^2 \qquad (22)$$

If, for example, the series feedback is set to zero ($R_1 = 0$), we find that $S_{11}$ is given by:

$$S_{11} = \frac{1}{\Delta}\left\{1 + Z_o\left(\frac{1}{R_2} - \frac{1}{R_T}\right) + \frac{Z_o^2}{R_2}\left(\frac{1}{R_2} - \frac{1}{R_T}\right) - \frac{g_m Z_o^2}{R_2}\right\} \qquad (23)$$

and it is thus possible to make $S_{11}$ approach zero (i.e. a perfect match) with a suitable choice of $R_2$; similar remarks apply to $S_{22}$. It can also be shown that $|S_{21}|$ can be made relatively insensitive to changes in the transistor parameters,which leads to effective 'flattening' of the gain-frequency response.

The equations given above are merely for general guidance, and apply to low frequencies. At higher frequencies, the following approaches should be used:

(a) Computer calculation using the S-parameters of the device measured without feedback, and adding the feedback network and re-calculating the overall parameters. This is useful for theoretical exploratory purposes, to assess the extent to which feedback is useful in any particular case.

(b) Practical measurement of the overall S-parameters of the transistor mounted in the chosen feedback arrangement estimated from (a) above. This provides good accuracy for a practical design, since any effects of parasitic elements in the circuit are included in the measured values.

The final design of an amplifier using NFB techniques will mainly consist of the provision of matching networks to tune out reactive components of $S_{11}$ and $S_{22}$; quite simple circuits have been used in the design of a multi-octave amplifier with flat gain and low vswr's (8), which is capable of being directly cascaded.

## BJT POWER AMPLIFIERS

This section will give a brief introduction

to power amplifier design. Since such amplifiers often operate in non-linear regions, small-signal parameters are not suitable for design purposes, and it is necessary to define and measure some appropriate 'large-signal' parameters to characterise the device.

Large-signal Source and Load Impedances

A power device can be usefully described in terms of the large-signal source and load impedances required to produce a given power and gain at a given frequency. A basic measurement system is illustrated in figure 8; the device under test is placed in a suitable (microstrip) jig with bias applied, and the input signal level is varied and the tuning stubs adjusted until (a) power meter C indicates the required power level and (b) meter B on the input coupler, which is monitoring the reflected power, gives zero reading. Meter A monitors the incident power, and thus gives a measure of the power gain. Under these conditions, the input is conjugately matched, and the output is presented with the load required to produce the given output power.

The device is now removed from the circuit, and the impedance at planes X and Y are measured on the network analyser, and transferred to the device terminals by taking account of the electrical lengths of the jig from planes X and Y to the device itself. The result of this measurement is to produce two reflection coefficients, viz;

$\Gamma_{SM}$ = source reflection coefficient for a conjugate match.

$\Gamma_{PL}$ = load reflection coefficient for the required output power.

Both of these quantities are functions of frequency and power level, and will also depend on whether the transistor is operated under class A conditions (with the device biassed to a current nearly equal to twice the peak r.f. current) or class B (with the device just biassed to cut-off).

It is usual nowadays for manufacturers to quote a range of these values of $\Gamma_{SM}$ and $\Gamma_{PL}$ on data sheets, and it should be noted that the resistive components are quite low, and broad-band matching is difficult.

Circuit Design

Given the values of $\Gamma_{SM}$ and $\Gamma_{PL}$, the methods of single-frequency design illustrated previously (see Section Input network and Section Output network) can be followed. For instance, a particular power device producing 500 mW of power at 2.3 GHz has the following values:

$$\Gamma_{SM} = 0.56\underline{|95^o}, \quad \Gamma_{PL} = 0.6\underline{|-134^o}$$

On the input side, $\Gamma_S$ plotted as an admittance is represented by 42 Ω in shunt with -j31 Ω, and a simple λ/8 open-circuited stub of impedance $Z_o$, chosen to be 31 Ω, can tune this out completely, and bring the conjugate point almost to the centre of the chart.

On the output side, a λ/4 transformer of 26 Ω impedance followed by a short length of 50 Ω line transforms the load impedance to $\Gamma_{PL}$ at the device terminals. Figure 9 shows the complete circuit, which includes small bias chokes, d.c. blocking capacitors, and decoupling capacitors. The chokes can be made from λ/4 lengths of thin-bond wire wound into 'springs' round a former of about 0.5 mm diameter.

Bias Circuits

As mentioned previously, power amplifiers are operated either in class A or class B, the difference arising as a result of the initial bias conditions without applied signal. In class A, with a fairly high bias current, the transistor is not driven into cut-off or saturation at the peaks of the input swing, and this is basically a 'linear' mode, with low distortion but rather low efficiency. Class B is a nonlinear mode, with very little bias current applied, and is more efficient than class A, but produces higher distortion.

Class B biassing is quite simple, in that only a collector voltage supply is required, but for class A, a fixed bias current must be set up. This can be done (both in power and small-signal amplifiers) by using a series emitter resistor to define the current, but this method requires efficient decoupling of the emitters, which is difficult at these frequencies, and is best avoided. It is always advisable to earth emitters directly and avoid small residual impedances in the common lead, which can seriously degrade the performance.

A very useful circuit is shown in figure 10, which uses a pnp control transistor. This transistor is effectively biassed to the appropriate current (about 100 mA in this case); the emitter resistor ($R_C$) senses any change of collector current in the microwave device (due to changes of transistor, temperature variations, etc) and varies the base current supplied to the microwave device to maintain its operating current at the predetermined value. Exactly the same circuit can be used for small-signal device biassing, with appropriate choice of the resistor values in the control transistor circuit.

Further Comments on Power Amplifiers

This brief introduction to BJT power amplifiers is intended to point out the differences in approach between these and small-signal types, which arise largely as a result of non-linearity in the power types, particularly in class B operation. It is possible, in class A operation, to make initial designs using small-signal S-parameters measured at the required (high) bias currents. It is found that the S-parameters of many devices do not vary greatly over quite large input signal swings, and are in fact constant enough for initial design purposes or feasibility estimates.

It should be appreciated that a power amplifier design is dependent on the power level required, and that an amplifier designed for, say, 1 W at 2 GHz would need quite a different circuit to one designed for 5 W at 2 GHz using the same device. Thus, the design parameters $\Gamma_{SM}$ and $\Gamma_{PL}$ must be measured at the operating power level, and should preferably be measured in a jig of the same type as that envisaged for a completed amplifier, using the same mounting arrangement.

Finally, it must be borne in mind that the power device will need to be mounted in such a way as to dissipate the heat generated in it. For this reason, power transistors are encapsulated in large headers, with flanges or studs for mounting to metal surfaces to provide adequate heat sinking. These large headers present high values of parasitic inductance and capacitance, which degrade the performance and make broad-banding difficult because of high values of Q. To ease this, a number of manufacturers now assemble power devices with a degree of 'internal matching', provided by placing small chip capacitors and extra inductances inside the header as close to the device chip as possible. These components help to tune out the parasitics, and transform the device resistances from their intrinsic low values (about 2 Ω) up to the more easily-matchable levels of 10 Ω or more.

ACKNOWLEDGEMENT

Acknowledgement is given to the GEC Hirst Research Centre for permission to publish.

REFERENCES

1. Snapp, C.P.: 1979, "Microwave Bipolar Transistor Technology - Present and Future", Proc. 9th E.M.C., Brighton, England.

2. Haus, H.A. and Adler, R.B.: 1958, Proc. I.R.E. 46, 8, 1517-1533.

3. Mellor, D.J. and Linvill, J.G.: 1975, IEEE Trans., MTT-23, 1013-1020.

4. Besser, L.: 1979, " 'CADSYN' and 'COMPACT' " User Manuals, Compact Engineering Inc., Palo Alto, CA94303

5. Levy, R.: 1964, Proc. I.E.E., 111, 1099-1106

6. Kurokawa, K.: 1965, B.S.T.J., 44, 1675-1698

7. Lange, J.: 1969, IEEE Trans. MTT-17, 1150-1151

8. Walker, J.L.B. et al.: 1976, "The Application of Negative Feedback to Microwave Bipolar Transistor Amplifiers", Proc. 6th E.M.C., Rome, Italy.

9. Ulrich, E.: 1978, Microwaves, 17, 10, 66-70.

PROBLEMS

1. Referring to equation (1) for the variation of noise figure with source impedance, show that:

   (a) The quantity $\dfrac{|\Gamma_S - \Gamma_O|^2}{(1-|\Gamma_S|^2)}$ is equal to a constant (N), and express N in terms of device noise parameters.

   (b) Show that the equation in (a) above represents a circle with centre at the vector point given by $C_F = \Gamma_O/(1+N)$ and radius $R_F = \{N^2+N(1-|\Gamma_O|^2)\}^{\frac{1}{2}}/(1+N)$

   (Method: Cast the equation into the polar form $r^2-2ar\cos(\theta-\phi)+(a^2-b^2)=0$, which represents a circle of radius b and centre at $a\underline{|\phi}$. In this case, $\Gamma_S$ corresponds to the r,θ radius vector).

2. Show how equation (3) (cascaded noise figure) can be obtained from equation (2) by considering an infinite cascade of identical stages.

3. Taking the S-parameters of the low-noise transistor quoted in Section Selection of suitable device, calculate the input reflection coefficient ($S_{11}'$) when the transistor is terminated in a load reflection coefficient of $0.5\underline{|60}$. Comment on how this load alters the $S_{11}$ value from that when the device is terminated in a 50 Ω load.

4. The input impedance of a particular low-noise transistor at 4 GHz is 6-j20 Ω and its optimum noise source impedance is 20+j10 Ω. Calculate the resulting input vswr with an input circuit designed for minimum noise operation.

5. A Lange coupler, required to cover the octave band 2 to 4 GHz, is designed to be λ/4 long at 3 GHz. Using equation (11) calculate t at 2 and 4 GHz, and show how equation (7), for $|S_{11}|$, is altered at these band edges. Comment briefly on how $|S_{11}|$ alters over the band if the two transistors are exactly matched.

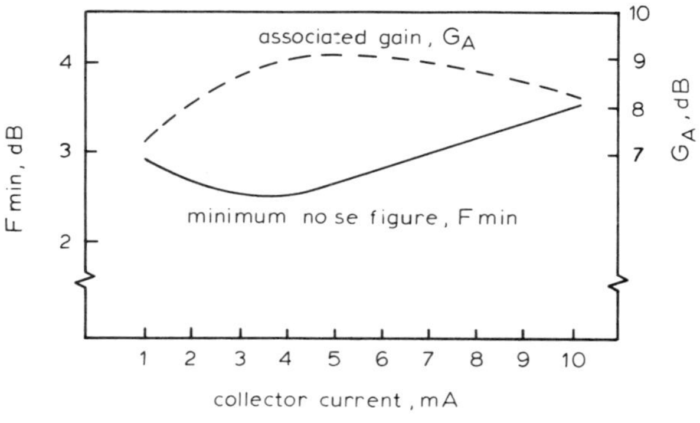


Figure 1  F min and G A against collector current for a typical BJT

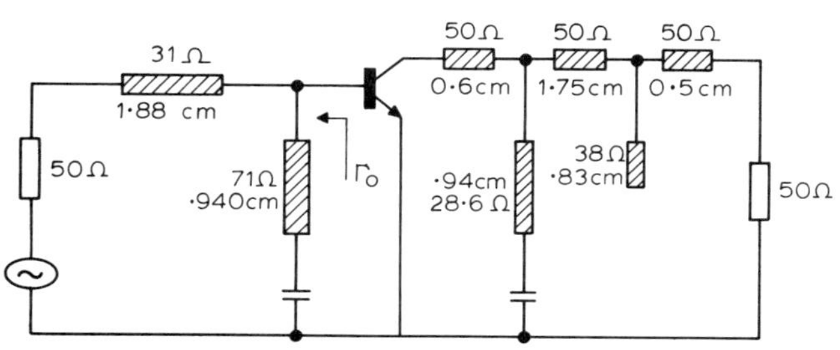


Figure 2  Complete circuit of the 4 GHz narrow-band amplifier

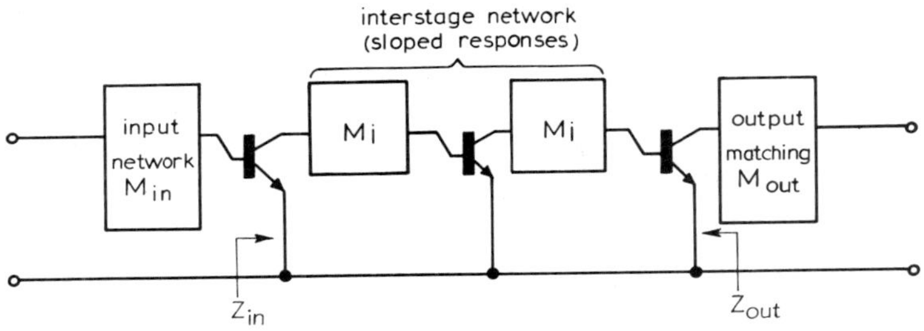


Figure 3 Schematic of 3-stage wide-band amplifier

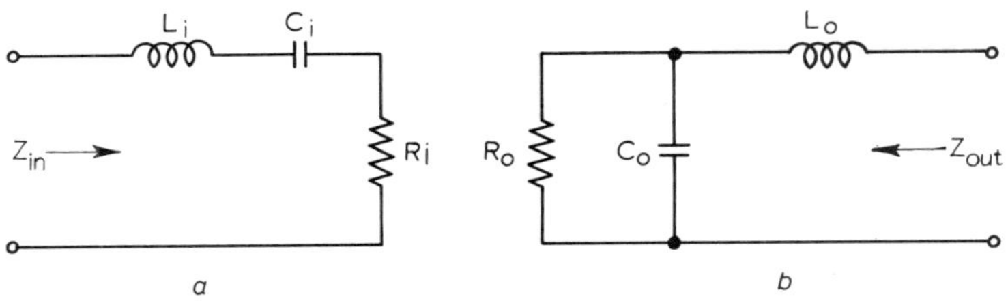


Figure 4 Equivalent circuits of the input (a) and output (b) of the 3-stage amplifier

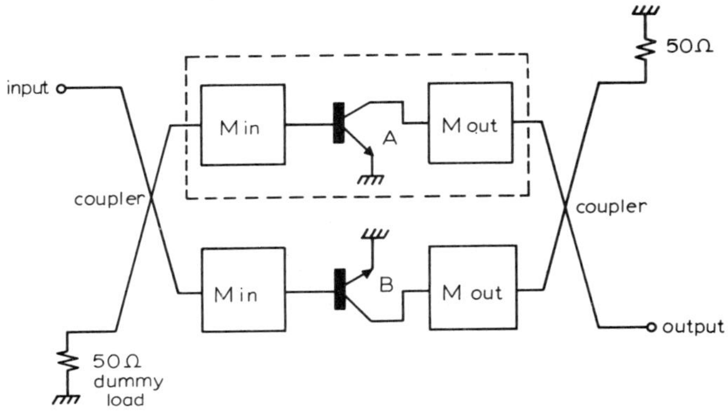


Figure 5 Arrangement of a balanced amplifier using two 3dB couplers

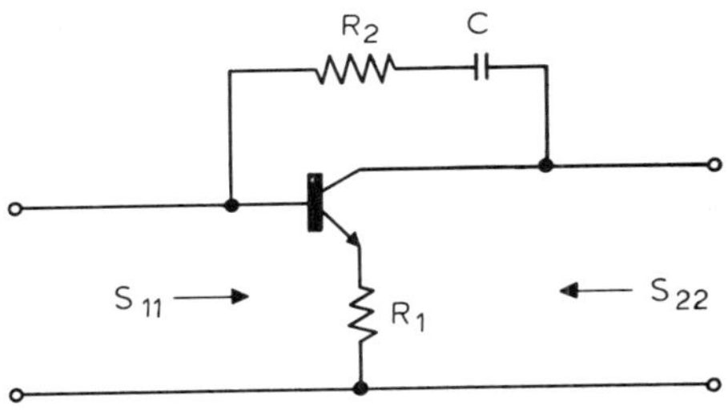


Figure 6 Transistor with series ($R_1$) and shunt ($R_2$) resistive feedback

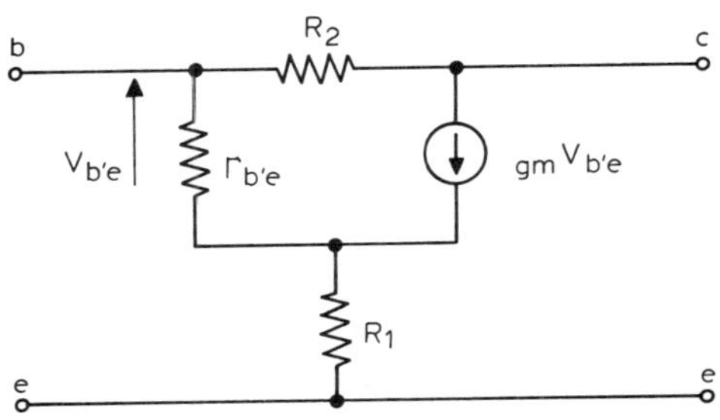


Figure 7 Low-frequency equivalent circuit of a transistor with feedback

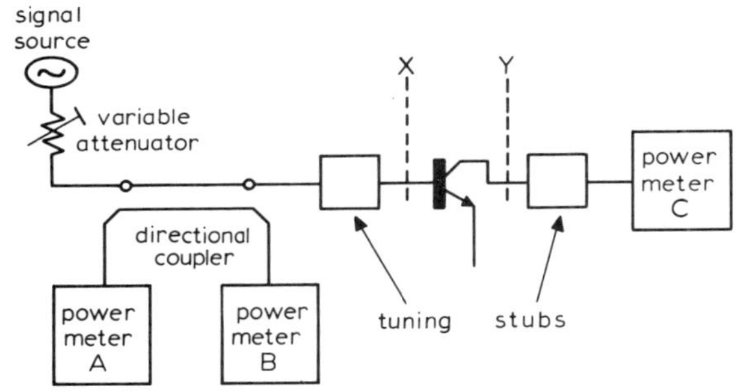


Figure 8 Test system for measuring power transistor performance and impedances

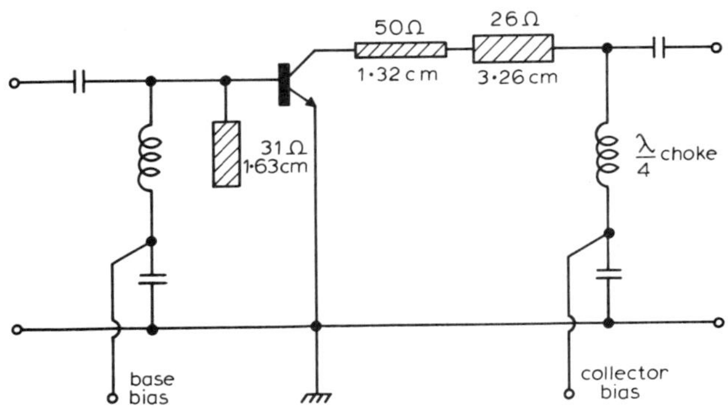


Figure 9 Circuit of the 0.5 W 2.3 GHz amplifier (line lengths given in air)

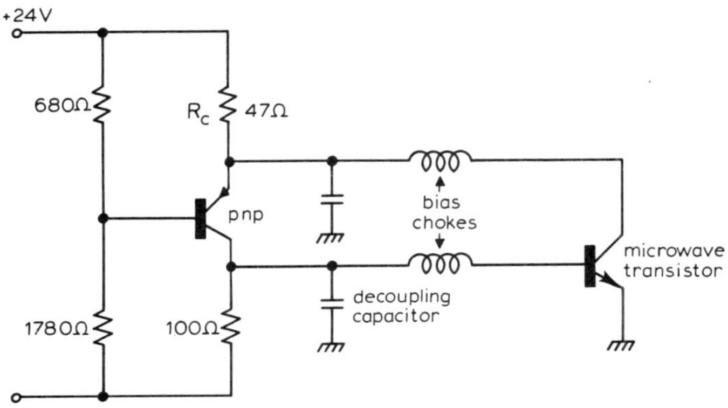


Figure 10 Bias current control circuit for class-A power or small-signal operation

# GaAs FET AMPLIFIER REALISATION

S.C. Cripps

Watkins-Johnson Limited (UK)

## INTRODUCTION

In any branch of engineering design the step from "drawing board" to practical and measurable hardware is recognised as a difficult and often precarious one, often more so than the engineer responsible is prepared to admit. This effect is particularly noticeable in the realisation of electronic hardware as the operating frequency increases, to the extent that the microwave man has often been regarded as a rather special breed of Homo Sapiens, combining green fingers with patience beyond the normal call of duty to coax his small black (or blue?) boxes into meeting their specifications. Whilst this view is clearly extreme, it is certainly true that microwave circuit design as presented in textbooks and journals often lays a heavy emphasis on the paper-work and computer-aided paperwork aspects of achieving a design without considering the practical problems of realisation. It is the purpose of this course segment to present a series of topics on the more practical side of amplifier design at microwave frequencies, other parts of the course having dealt very adequately with the more academic process of reaching a paper design. The treatment is deliberately light-weight although it is hoped to put over an important general message to the effect that the key to efficient realisation of microwave circuit designs is to steer a mid-course between the too-academic approach of expecting everything to work exactly as the computer predicts and the equally unproductive opposite extreme of assuming that there is no limit to what can be achieved if you 'tweak it for long enough'.

## CIRCUIT DESIGN

The process of microwave amplifier circuit design is often regarded as that of matching the input and output impedances of the active device to 50 ohm over its operating bandwidth. Such matching can be achieved for up to 10% bandwidths by simply using a Smith Chart to design a two-element network which achieves a perfect match at the centre band, and "acceptable" (usually interpretted as 2:1 VSWR) match at the band edges. For larger bandwidths it is necessary to use many more elements, and a "synthesis" approach has to be used. Matching by filter synthesis is a useful approach in which the source and load impedances are the device impedance and 50 ohm respectively and the matching network is regarded as a band pass filter. The filter is designed using tables for minimum passband insertion loss in order to achieve the best in-band match. All these techniques are well-known and readily accessible in published literature (10)

Merely matching a device, however, is not enough to realise a typical amplifier specification requirement. In order to achieve a flat gain over a significant bandwidth, the gain must be supressed at the L.F. end where the device gain is higher. This can be achieved either by deliberate reactive mis-matching, or by the incorporation of resistive gain compensation circuitry. Resistive absorption often has a disastrous effect on noise figure and consequently reactive mismatch has to be used. In addition to gain equalisation, an amplifier design usually has to meet a noise figure requirement in which case the device has to be presented with a mismatch on its input in order to give its minimum noise figure. The net effect is that a suitable single stage amplifier will, in most practical cases, have poor VSWRs at input and output over most of its frequency range. In order to cascade several stages it is therefore necessary either to use ferrite isolators or a balanced approach (Fig.11). Both approaches have advantages and disadvantages but in general the balanced approach is favoured in broadband, high gain designs. A typical paper design, obtained using a combination of Smith Chart, synthesis and computer-aided design, results in a circuit with transmission lines of various lengths and impedances. At lower frequencies the physical lengths of each distributed element are much longer than the width or electrode spacing of the chosen transmission line medium and physical realisation of the design can usually be achieved with quite good agreement to the theoretical predictions. The first major problems of distributed circuit realisation arises as the frequency gets higher, and typical lengths which come out of the design procedure start becoming comparable to the width of the transmission line conductors. This causes uncertainties both in the behaviour of the element itself, and in the end-effects where the adjacent elements join. Fig12 shows a typical situation at 10 GHz, where a quarter-wave transformer of 20 ohm impedance is wider than its length on an alumina dielectric of commonly used thickness. Whether or not the designer decides, in this particular case, that the configuration actually looks more like a central line with two parallel stubs, the inescapable fact is that one-dimensional transmission line theory cannot be used to predict the behaviour of a structure which is clearly three-dimensional.

There is no simple solution to this problem, and in practice the designer has to use a combination of practical and theoretical aids in an attempt to reduce the uncertainty of distributed designs. Fortunately a large amount of data has been published on the behaviour of microstrip discontinuities at X-band frequencies. This data is usually presented in the form of equivalent circuits which can be built-in to the original design. A complementary approach is to reduce the discontinuities by introducing gaps and cut-outs in the original structure (Fig. 13). In any event, the final design is not exact and some degree of variability is usually built into the circuit for final trimming purposes (Fig.13.). More recently it has become clear that as even higher frequency designs are tackled by designers, a return to lumped element concepts can often result in fewer uncertainties than in a distributed approach. Provided the physical dimensions of a component are small compared to a wavelength, it can be characterised over a wide frequency range by an equivalent lumped circuit. Short lengths of bond wire and small beam lead or momolithic capacitors may not behave as perfect indicators and capacitors, but the discontinuities between them are small and designs above 20 GHz have been reported using this approach. (Fig 4.)

## CIRCUIT CONFIGURATION & REALISATION

Of the various forms of transmission line realisation available to the designer (see Fig 5) by far the most commonly used is microstrip. Stripline (5b) is easier to characterise analytically but is inconvenient practically in that the entire circuit is confined within the ground planes making adjustment and modifications difficult. Coplanar waveguide (5d) has some attractive features. A slightly wider range of characteristic impedances are achievable in comparison to microstrip and both series and parallel lumped elements can be easily incorporated. Suitable launchers for connectors pose problems, however, and little data is available to date on discontinuities.

Choice of substrate material is broadly between the two classes known as "softboard" which are low-loss plastic type materials, and the ceramic or crystalline materials such as alumina, sapphire and quartz. The latter materials have the properties of mechanical rigidity at high temperatures, enabling conducting metallic layers to be deposited onto them by evaporation or sputtering. The resulting coated substrate can be masked and chemically etched to give patterns with very high resolution (typically 10 microns). Softboard material is usually clad with copper foil, and circuit patterns are of much lower resolution quality. The lower dielectric constant does, however, reduce the resolution quality required and circuits can be fabricated using normal PC board techniques. In general, softboard is used extensively up to about 6 GHz, where the simplicity and low cost are considerable advantages. At higher frequencies the so-called 'thin-film' technology is almost universal due to the higher tolerances required and the need to bond device chips directly into the microstrip circuit, a process usually requiring the substrate to be heated to a high temperature.

Alumina is the commonest material used for thin-film microwave circuits. It is available in large size sheets of varying thicknesses and can be polished to a high surface finish with accurate dimensions. Its high dielectric constant (9) is useful at lower frequencies in enabling very small size circuits to be designed, although at frequencies below 5 GHz its loss is slightly higher than a good quality softboard material. Sapphire has similar electrical properties to alumina, but can be polished to a much higher quality of surface finish which can significantly reduce the loss at X-Band and higher frequencies. Sapphire is not such a good material mechanically as alumina, being more brittle due to its crystalline structure. From an electrical point of view, quartz (fused silica) represents an excellent compromise as a microwave thin film material due to its fairly low dielectric constant (4) coupled with very low losses in suitably processed material. The need for a lower dielectric constant becomes evident at X-band and higher frequencies due to the relieving of the microstrip 'aspect ratio' problem shown in figure 2. Unfortunately, quartz is a very fragile material and requires great care in its handling, but has become widely used at frequencies higher than 10 GHz.

A widely used substrate material is 0.025" (0.6mm) thick alumina. The alumina is polished and sputtered with a 'seed' layer of chromium or nichrome. This is then covered with a sputtered layer of gold or copper. Substrates can be purchased readily sputtered, so that the required circuit can be processes during a straight forward procedure of photoresist exposure and chemical etching/plating.

On alumina of this thickness, transmission lines of impedances between 20 ohm and 90 ohm are readily realisable, those representing the practical upper and lower width limits. Significantly higher impedances are often desirable, but are difficult to realise in this medium (a 25 micron bond wire laid across the surface still looks like a line of 140 ohm impedance). In a move towards lumped circuit realisation, inductive spirals are sometimes used at lower frequencies (Fig 6(a)). These elements are quite complicated in their behaviour, but can be useful in reducing the space required for lower frequency designs. Capacitors are always required in some form, either as D.C. blocks or lumped matching elements. Beam lead types can be thermocompression bonded directly across a gap in the microstrip line, although below 5 GHz either ceramic or monolithic types are usually soldered or bonded into the circuit using rather larger gaps (Fig 6(b)). At higher frequencies, interdigital capacitors can be used (Fig 6(c)) although effective capacitance is under 1pF maximum and somewhat higher tolorances must be maintained to achieve consistent performance.

Resistors can either be chip types bonded into place or more effectively can be formed by etching away the conducting layer completely to leave an area of the chromium seed layer (Fig 7). This use of the seed layer as a basis for resistors has great advantages in terms of reduced parastics and higher reliability, especially at higher (ie X band) frequencies, but clearly requires a much closer control of the seed layer composition and thickness. Several studies have also shown that chromium and Nichrome films are rather unstable under high temperature cycling conditions. Tantalum nitride is therefore becoming more widely used as a seed layer, despite the greater complexity of its deposition.

The manufacture of 3 dB hybrid couplers is an important feature in the practical realisation of microwave FET amplifiers, since as explained in the previous section the balanced amplifier approach is necessary to cascade two or more stages without resorting to the use of ferrite isolators. Microstrip lines laid side by side have directional coupler properties, but to achieve 3dB coupline the line spacing has to be of the order of a few microns and is beyond the resolution limits for normal thin film processing. An ingenious method of realising 3dB coupling using standard thin film technology was originally proposed by X Lange(11). The two coupling lines are effectively split up into a series of interleaving fingers (interdigital) to increase the coupling whilst keeping the spacing realistically wide. Correct odd and even mode velocities are maintained by bridging band wires (Fig 8). Couplers of this type can be designed to cover more than an octave bandwidth and can give losses well under 0.5 dB at 10GHZ. The vast majority of commercially available GaAs FET amplifiers consist of several cascaded substrates, each having two lange interdigital couplers and a pair of identical amplifier stages, all metallised in one operation from a single photographic mask.

## TUNING & ALIGNMENT METHODS

It probably cannot be denied that if a particular device is carefully and accurately enough making full use of the most expensive modern aids, and its proposed amplifier circuit is designed with meticulous allowance for every possible source of uncertainty using published data the final result will be a fairly close agreement between reality and prediction. This statement is of little interest, however to the commercial amplifier manufacturer who has to deal with large numbers of devices with varying parameters, varying substrate thickness and dilectric constant, limited mechanical tolerances on bondwires, connectors and housings not to mention the often all important effects of temperature range. Clearly the impact of all these effects is a strong function of frequency, but certainly by the time X-band frequencies are reached, the combined uncertainties introduced from one set of parts to another would be catastrophic if some form of fine tuning were not employed. This actually applies also in a less dramatic way at much lower frequencies; an amplifier design which gives 1.5 dB noise figure at 3 GHz with one set of parts can easily give 2dB using a different set purely through the typical statistical variations in device and circuit parameters.

Microstrip in some respects is not well suited to tuning techniques. Waveguide components can be very effectively fine-tuned by the use of screws, and there is no direct substitute in microstrip. The use of thin metal discs or pieces soldered or epoxied onto the microstrip transmission line elements is quite common, however the process of fixing these elements down such that the final performance is exactly as desired can be an irksome and skilful process. Several refinements have been used to case this problem (see Fig19). Usually it is found that particular circuit has certain sensitive areas where the metal pieces always have to be fixed. It is therefore possible to simulate the metal piece by bonding together several small 'islands' which can form part of the circuit metallisation pattern(19b). The usual approach in a production environmentis to 'overbond' the tuning area, ie interconnect all the islands in the knowledge that wires only need to be removed by the alignment operator in order to bring the circuit into specification. These islands can be used in numerous ways, eg to extend the lengths of a low impedance transformer, or to introduce parallel capacitance. An alternative approach is to use variable capacitance diodes, however these introduce loss and do not give the same range of tuning as the moveable metal element, due totheirfixed position. A fairly large number of diodes is necessary to ac hieve a good tuning range and the difficulties of providing independent bias become too great for this technieque to have achieved widespread use.

Whatever tuning method is employed, the knowledge that tuning will inevitably end up being used on a particular circuit design should be taken into account by the designer at the outset. Computer circuit, design programmes usually offer a sensitivity analysis which should be used to determine which areas of the circuit are most sensitive and consequently need the flexibility of tuning arrangements. In addition to this, it should be borne in mind that it is, for instance, much more easy to 'tweak' the length of an open circuited stub than a short circuit element, and the design should be based as far as possible on elements that offer the greatest possibility of fine tuning.

## ENVIRONMENTAL CONSIDERATIONS

Amplifiers which use bare semiconductor chips usually have to be enclosed in hermetic packages to avoid degradation in hostile atmospheres particularly where high humidity or very low temperatures are likely. Different manufactures take many different approaches to this problem, which is a difficult one due to the incompatibility of properties between good microwave materials and materials which lend themselves well to hermetic sealing. The use of gaskets can be a relatively cheap solution but usually cannot offer the degree of long term hermeticity provided by a welded seal The hermetic problem also causes difficulties with microwave connectors, since at higher frequencies the loss and discontinuities caused by glass to metal seals can become intolerable.

Wide temperature ranges ($-40^{o}C$ to $70^{o}C$ or $+100^{o}C$) are a common requirement in military equipment and the gain of an amplifier usually has to be compensated. This is usually achieved using a pin diode attenuator placed in the output stage of the amplifier. Forward bias,derived from a temperature sensitive circuit, reduces the gain as the temperature decreases. In a tupical case, a GaAs FET can be expected to give 0.002 dB of gain variation per dB of gain per $^{o}C$. Thus an amplifier with a nominal gain of 30 dB varies by 7dB over a range of $-40^{o}$ to $+70^{o}$. This level of variation can only effectively be dealt with by using a pin diode attenuator, since variation of gate bias modifies the FET impedance too much and degrades the overall frequency response of the amplifier.

A wide temperature range also caused problems inside the amplifier box. In particular the expansion of the substrates and carriers can fracture the bonds between connector pins and intersubstrate joins, and many techniques have been used to make allowance for this effect, but they usually result in yet another uncertainty factor for the designer.

## REFERENCES

10) Matthaei, Young, Jones; Microwave Filters impedance - matching networks and coupling structures, McGraw-Hill 1964.

11) Lange, J.;'Interdigitated quadrature hybrid', I.E.E.E. Trans. MTT-17(12) December 1969.

## PROBLEMS

(1b) Taking values of S11 from the data at 3 GHz derive a circuit model for the impedance of the GaAs FET consisting of a series RC network. Use the model to predict the values of S11 at 2 GHz and 4 GHz. Compare the results with the data sheet.

(2b) Using the values of S11 given at 3 GHz, design a network to match the device input impedance at 3 GHz using a 50 ohm series line and a parallel stub. In a practical case, the magnitude of S11 may be accurately known, but the phase angle can be uncertain. Calculate the VSWR of the same amplifier network if the phase angle of the device increases (negatively) by 10 degrees.

(3b) The amplifier of question (2b), with the 10 degree phase angle error is "tweaked" using a tuning disc. Assuming the disc can be treated as a parallel lumped capacitor, estimate the size and position of the disc required. (Assume that the dielectric used has a dielectric constant of 10 and a thickness of 0.6 mm.)

(4b) Typically the effective capacitance of a tuning disc such as in question (3b) can be varied between about 0.5 pF and 1.5 pF by altering its transverse orientation on the microstrip line. Sketch the range of impedances which are presented to a device by such a tuning disc and a 50 ohm line.

(5b) Representing two discs as moveable lumped capacitances of 1 pF at 3 GHz, sketch the full tuning range provided by the discs on a 50 ohm microstrip line with remote termination.

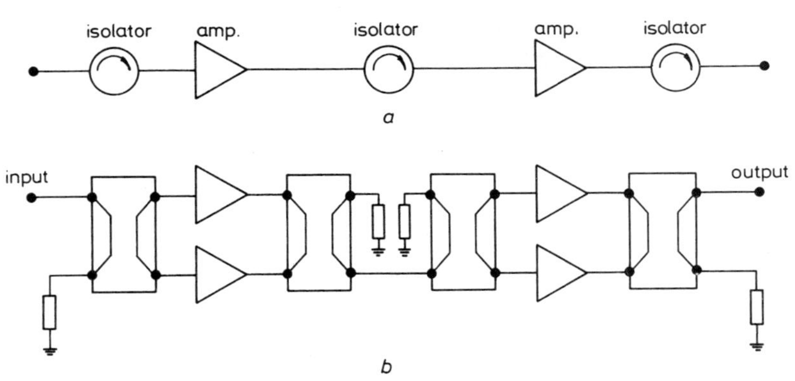


Fig.11 Amplifier configurations

(a) Single-ended design (b) Balanced design

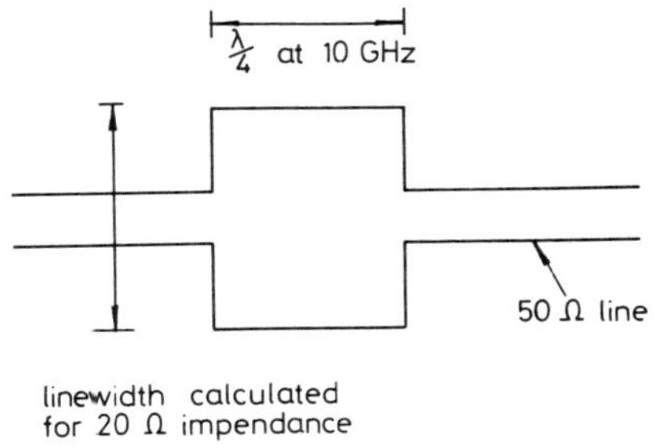


Fig.12 Aspect ratio problem in microstrip

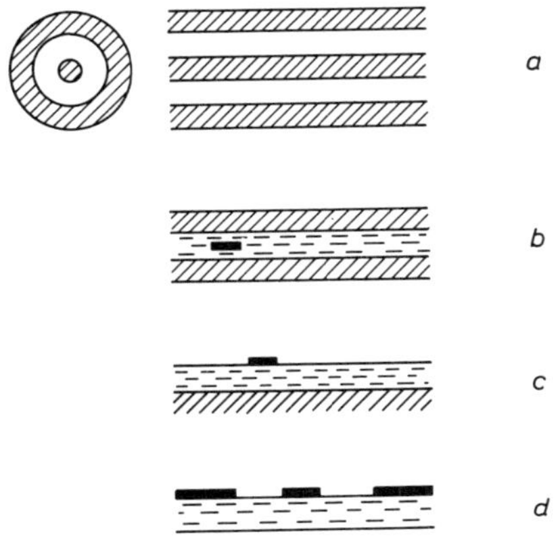


Fig.15 Practical transmission lines

(a) Coaxial line
(b) Stripline
(c) Microstrip
(d) Coplanar waveguide

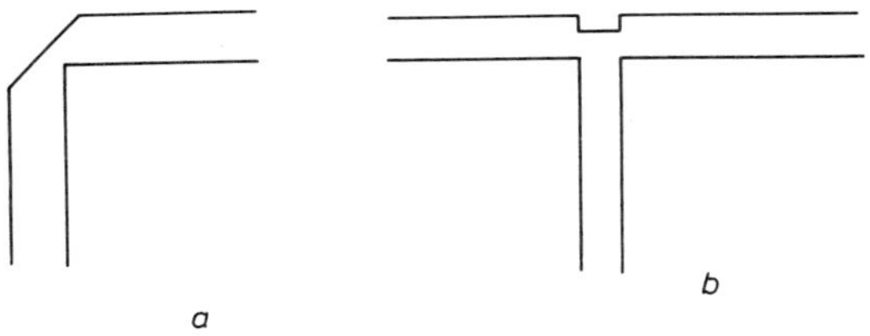


Fig.13 Examples of discontinuity compensation

(a) Mitred bend

(b) Recessed stub

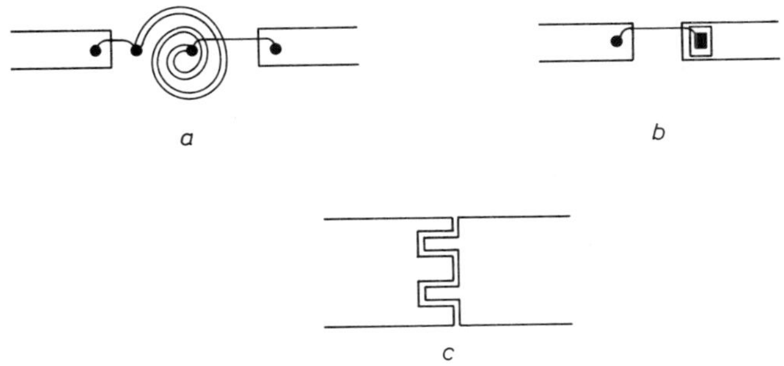


Fig.16 Microwave quasi-lumped elements

(a) Spiral inductor
(b) Chip capacitor
(c) Interdigital capacitor

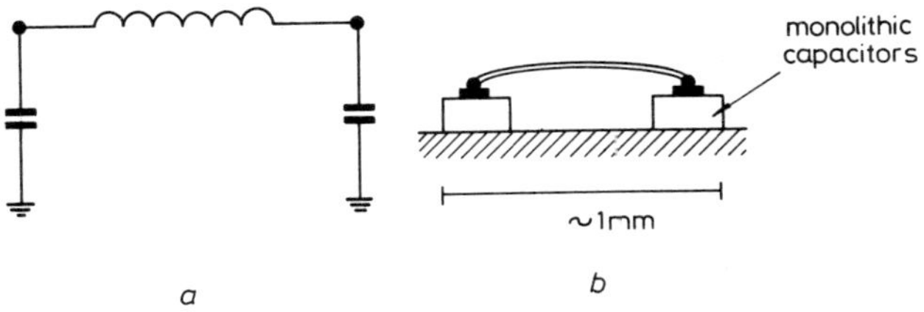


Fig.14 Lumped circuit approach

(a) Intended network

(b) Realisation

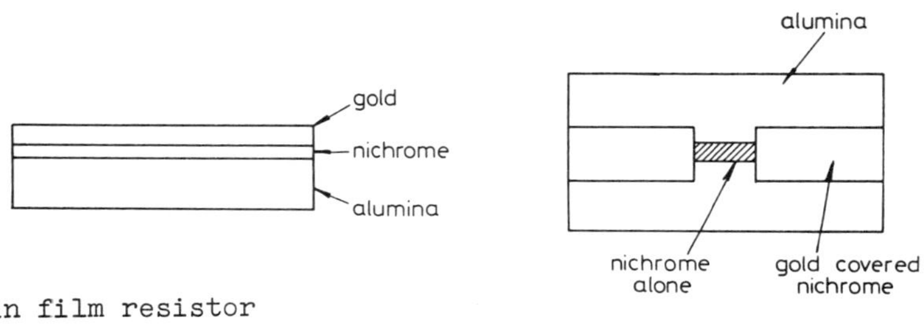


Fig.17 Integrated thin film resistor

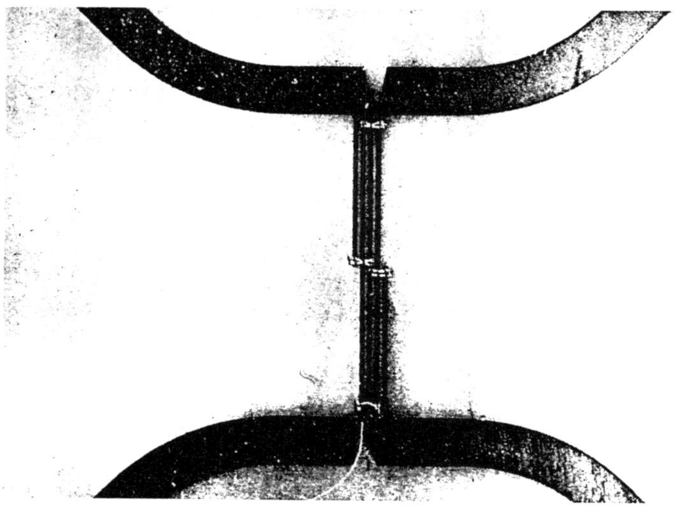

Fig.18 Lange coupler

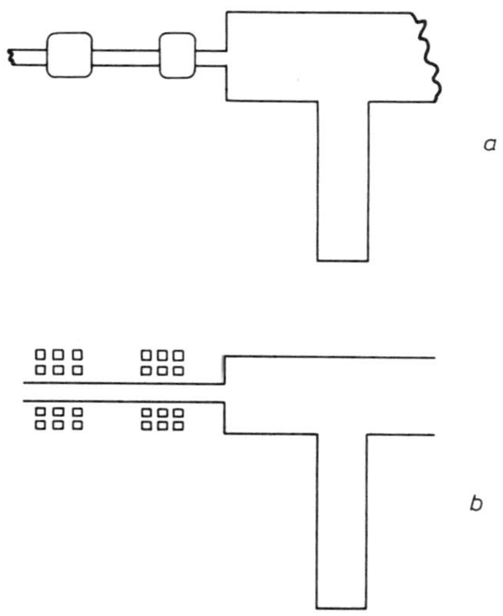


Fig.19 Tuning techniques

(a) Metal pieces

(b) Bonded islands

# CHAPTER 10

# Design of reflection amplifiers

Dr. R. Ranson/ Dr. P. Huish

# DESIGN OF REFLECTION AMPLIFIERS (PART I)

R. G. Ranson

MEL (A Division of Philips Electronic and Associated Industries Ltd.)

## INTRODUCTION

An amplifier may be regarded as a black box with input and output terminals, where signals applied to the input appear magnified at the output. The increase in signal energy being derived from some other source which is usually, but not always, DC (c.f. transistor amplifiers and parametric amplifiers). Such a definition can readily be related to active devices such as transistors since they possess obvious input and output terminals which are isolated to some extent. However, consider the simple circuit illustrated by figure 1 (a). The circuit consists of a source and load with the amplifier "black box" containing a negative conductance. In this circuit both the input and output terminals are directly connected and only two nodes can be identified.

Nevertheless the circuit does provide power gain since

$$\text{Power in load} = V^2 g_1 = \frac{I^2 g_1}{(g_s - g + g_1)^2} \quad \ldots(1)$$

$$\text{Max. available power (without amp.)} = \frac{I^2}{4g_s} \quad \ldots(2)$$

$$\text{Transducer gain} = \frac{4g_s g_1}{(g_s - g + g_1)^2} \quad \ldots(3)$$

Thus, the transducer gain of a negative conductance amplifier can be set to any level by a suitable choice of source and load conductances. Also, as the input and output terminals are electrically identical, the amplifier circuit could be redrawn as shown in figure 1(b). In this case the amplifier has only one port and the power gain is the ratio of the power incident to the power reflected from the negative conductance amplifier.

Reflection amplifiers of this form have use at high frequencies where conventional devices such as transistors and FETs cannot as yet operate. However, the much simplified circuits presented in figure 1 are not practical and only serve as an introduction since no pure negative conductance can exist.

## PRACTICAL NEGATIVE RESISTANCE DEVICES(NRDs)

A pure negative resistance is physically impossible as such a device could generate infinite power over an infinite bandwidth i.e. when $|g| = g_s + g_1$. On the other hand devices with limited negative resistance properties do exist. Some common examples are tunnel diodes, transferred electron devices (Gunn diodes), IMPATT and TRAPATT diodes. The physical mechanisms responsible for the negative resistance properties are not important in the understanding of the use of such devices; but it should be noted that all practical NRDs exhibit a terminal immittance which has a negative real part over only a finite bandwidth and dynamic range. Thus, the design of circuits for these devices must consider these basic limitations and in particular the circuit model on which designs are based must also include these restrictions.

### Circuit models for NRDs.

It is common practice in active circuit modelling to consider the device as defined by its operating point, which is fixed by external influence, e.g. bias circuits and power supplies. Then, the applied signal represents small changes around this operating point and the device parameters are not significantly affected by these changes. This small signal theory is implicit to all the circuits to be discussed in this chapter as the large signal case is beyond the scope of these notes. Thus, in any device model the elements are assumed constant for a given operating point and for small signal operation.

The bandwidth limitations of NRDs may be represented in circuit form by some reactance associated with the negative resistance element. Such a model, in its simplest form, is a shunt combination of a negative resistor and a capacitor. This model has certain physical justifications but it should be emphasised that in terms of circuit design the only justification of a model is that it accurately represents the electrical characteristics of the network in the frequency range of interest.

Finally, the stability of NRDs is most important. As the input and output signals appear at the same port, excessive gain can result in large signal oscillations. In addition high gain levels are only possible as $|g|$ approaches $(g_s+g_1)$ and so fluctuations in device parameters have large effects on the amplifier performance. Thus, circuit designs usually limit the gain to prevent oscillations, give improved circuit parameter stability and maximise the amplifier bandwidth.

### Amplifier circuits using NRDs

Most practical reflection amplifiers provide some means of isolating the input and output signals in order to improve stability and increase gain bandwidth performance. Two examples of such circuits are the;

a) balanced 90° hybrid circuit and
b) circulator coupled circuit.

The 90° hybrid circuit, illustrated by figure 2(a), provides input/output isolation by the phase relation of the reflected signals from the two identical NRD modules combining in the hybrid. The circulator coupled circuit (figure 2(b)) achieves isolation by means of the directional properties of the ferro-magnetic junction in the circulator. In both cases the transducer gain is the ratio of power incident to power reflected from the NRD. Thus, for a circulator circuit the power gain is the square of the modulus of the reflection coefficient (Scanlan (1))

$$G_T = \left| \frac{Z_D - Z_L^*}{Z_D + Z_L} \right|^2 \qquad \ldots(4)$$

Similarly, in the hybrid circuit the two identical halves each amplify half of the input signal with the relative phase relationships providing addition at the output and subtraction at the input. Thus, the power gain is identical to the circulator coupled case.

The bandwidth of the two circuits is also identical if the NRD modules are the same. For example, if the NRD is assumed to be equivalent to a shunt combination of a negative resistor and capacitor as discussed previously, then, using an inductance to resonate the device at the desired centre frequency, it can be shown that the midband gain and 3dB bandwidth (in Hz) of the circuits are: (1)

$$G_T\Big|_{max.} = \left( \frac{R_o + R_D}{R_o - R_D} \right)^2 \qquad \ldots (5)$$

$$(G_T^{\frac{1}{2}} - 1)(1 - \frac{2}{G_T})^{\frac{1}{2}} B = \frac{1}{\pi C R_D} \qquad \ldots (6)$$

Thus, although it is possible to choose the midband gain arbitrarily by adjusting the source resistance, the gain bandwidth relation is limited by the reactance associated with the negative resistance of the device.

## A FUNDAMENTAL THEOREM FOR NEGATIVE RESISTANCE CIRCUITS

The design of circuits employing negative resistance elements is not usually encountered in network theory. However, there is no basic limitation in this respect and it is possible to extend the usual passive network theory to encompass such circuits. The fundamental theorem relating circuits with negative resistances to those with positive resistance has been derived by Aron (2) and is stated here without proof.

Theorem. If p is the input reflection coefficient of a linear passive network terminated in a positive resistance R then p′, the input reflection coefficient of the same network terminated in a negative resistance of equal magnitude, is related to p by

$$|p'| = \frac{1}{|p|} \qquad \ldots (7)$$

The usefulness of this result is demonstrated in the circuit synthesis section, where standard passive circuit design tables can be used to synthesis negative resistance circuits. It can also be used to relate the general passive matching circuit constraint of Bode(3) to reflection amplifiers.

Bode (3) states that for a passive RC load, whatever the form of the lossless matching network then:

$$\int_0^\infty \log_e \frac{1}{|p'|} \cdot dw \leqslant \frac{\pi}{RC} \qquad \ldots(8)$$

That is, that a perfect match $|p'| = 0$ is only possible at one frequency and a broadband match is only possible at the cost of increasing $|p'|$. This may be related to the negative resistance load case using equation (7) in conjunction with gain of a reflection amplifier equation (4) to yield

$$\int_0^\infty \log_e G_T^{\frac{1}{2}} \cdot dw \leqslant \frac{\pi}{RC} \qquad \ldots(9)$$

From this it can be seen that the optimum bandpass response must consist of $G_T = 0$ out of band and a constant finite value within the pass band. In this case equation (9) reduces to: (see figure 3)

$$B \log_e G_T \leqslant \frac{1}{RC} \qquad \ldots(10)$$

This represents a limit for the gain bandwidth relation of any lossless matching network and so all real networks which approximate to the optimum response can only approach this limit.

## SYNTHESIS OF BROADBAND REFLECTION AMPLIFIERS

The fundamental theorem of Aron (2) and the modified Bode relation (equation (10)) provide the basis of the synthesis of negative resistance circuits. The problem of designing an amplifier with a specific gain response is first reduced to realising a lossless matching circuit, terminated in an equal positive resistance, which produces the inverse response. Then, the synthesis of this passive circuit can be related to standard design tables by suitable magnitude and frequency transformations.

For example, figure 4(a) shows a circuit for a reflection amplifier where the negative resistance device is represented by a series-RLC circuit. If the desired reflection gain response, $|S_{11}|^2$, is as shown in figure 4(b) then using equation (7) the same circuit terminated in +R must have the inverse response, $|S_{11}'|^2$, shown in figure 4(c). This can be realised by first scaling the magnitude of all elements by 1/R and scaling the frequency by $(w_2 - w_1)/w_o$. The resulting design can then be synthesised using standard low pass filter prototypes and then rescaled in magnitude and frequency. This yields a ladder network which must be chosen to have the same final element resonance as the device (series in this case). The resulting design for $|S_{11}'|^2$ is shown in figure 4(d). Finally, the reflection amplifier is made by substituting the final element and load by the device and adding "remainder components" to make up the difference between the diode and filter immittance.

The general design procedure outlined above is for an arbitrary choice of gain and bandwidth. Hence, the matching circuit design is independent of the load presented by the negative resistance device. However, it should be noted that for a physically realisable circuit the "remainder components" must have a positive value. Thus, there must be a restriction on the choice of gain and bandwidth for a particular load. In fact the exact relation between gain and bandwidth for a particular load depends on the type of matching circuit used and its circuit order. But, since the example of figure 4 is for the ideal response, the relation for this case is modified Bode limit (equation (10)) adjusted for the dual configuration used.

$$B \log_e G_T \leqslant \frac{R}{L} \qquad \ldots(11)$$

Equation 11 shows that the maximum gain bandwidth is only achieved when L is a minimum and this corresponds to the "remainder components" being zero. Thus for a given device it is usual to specify either the minimum gain or bandwidth that is desired and calculate the matching network element values such that the final elements are equal to the device. This is then called an optimum design.

In practice an approximation to the ideal response must be used for circuit design, but this does not alter the basic principles discussed above. Thus, Scanlan (4) has deduced the relations between gain, bandwidth, circuit order and load for the Chebyshev approximation. In this case the gain ripple must also be chosen for a truly optimum design. However, as the ripple calculated in this way is rather excessive, it is usual to specify the ripple as desired and suffer the slight degradation in performance.

Design example

This lumped circuit design example is based on the relations deduced by Scanlan and the general method discussed previously. It also illustrates the method of Getsinger (5) for deriving a simple circuit model for a device using measured data (figure 5). The example shows the design of an amplifier with 5.2dB gain and 0.2dB ripple but does not consider the circulator or hybrid which is needed for final realisation.

Device model. The measured gain, with reference to a 50 ohm system, of a suitable stabilised transferred electron device is shown in figure 5(a). This may be modelled by a series tuned circuit, where the peak gain at 14.8GHz is 5.2dB. Hence using equation (5) $Ro/R = 3.4399$. Then the normalised slope parameter $X_1/R = 18.04$ is calculated as described in figure 5(b)(5). For a 50 ohm system, this yields R = 14.54, $L = X_1/wo = 2.82$nH and a resonating capacitor C = 0.041 pF as the device model.

Comparison of the calculated response of this model with the measured data in figure 6 shows close agreement over a 10% band centred on the resonant frequency.

Low pass prototype. The element values for a low pass prototype of this design can be obtained from standard tables (6) or calculated using the equations given by Scanlan (1).

$$g_1 = 2.59089$$
$$g_2 = 0.13770$$
$$g_3 = 3.43991$$

Transformation to final design. Equating these prototype values to the circuit model yields the magnitude scaling factor R = 50/3.43991 and the bandwidth scaling factor $w = g_1R/X = 0.1436$. Thus, scaling the remaining prototype element gives

$$b_2 = \frac{g_2}{Rw} = 0.06595$$

$$C_2 = \frac{b_2}{wo} = 0.709\text{pF}$$

$$L_2 = 163\text{pH}$$

The final network for the N=2 case is shown in figure 7, together with the basic N=1 circuit and the corresponding gain responses.

Comparison of results. The normalised bandwidth of the device (N=1 case) is 0.182 and with the addition of one extra resonator (N=2 case) this is expanded to 0.2872. The improvement in performance of the N=2 case is significant and it is interesting to compare this performance with the maximum possible according to Bode, equation (11). This is 0.582 for 5.2dB gain which means that the N=1 and 2 cases represent 31.3% and 49.3% respectively of the theoretical maximum.

The use of a more complex matching circuit enables the Bode limit to be approached but never exceeded. In fact the law of diminishing marginal returns is particularly relevant to this problem. The greatest single improvement in gain bandwidth is achieved by the first additional resonator. Subsequent increases in the number of resonators have a progressively lesser effect. For example, a matching circuit with 8 resonators can achieve approximately 90% of the gain bandwidth predicted by Bode and, in practice the small theoretical improvement due to further resonators is not realised because of practical tolerances and the difficulty of alignment. Finally, since the maximum performance of a reflection amplifier is inversely proportional to the load immittance, minimisation of the reactance associated with the negative resistance device is the most important circuit design consideration.

REFERENCES


1. Scanlan, J.O., 1966, "Analysis and synthesis of tunnel diode circuits", p.76 ff, John Wiley & Sons, London.

2. Aron, R., 1961, "Gain bandwidth relations in negative resistance amplifiers", Proc. IRE, Vol.49, pp 355-356.

3. Bode, H.W., 1945, "Network analysis and feedback amplifier design", p.281 ff, D. Van Nostrand Co. Inc., New York.

4. Scanlan, J.O., and Lim, J.T., 1964, "A design theory for optimum broadband reflection amplifiers", IEEE Trans., MTT 12, pp 504-511.

5. Getsinger, W.J., 1963 "Prototypes for use in broadbanding reflection amplifiers" IEEE Trans., MTT-11, pp 486-497

6. Matthaei, G.L., Young, L. and Jones, E.M.T. 1964, "Microwave filters, impedance-matching networks, and coupling structures", McGraw Hill, New York.

TUTORIAL EXAMPLES

a1. Prove the gain bandwidth relation (equation 6) for a single tuned negative resistance circuit of the form of figure 2.

Hint: assume a parallel LCR tuned equivalent circuit for the NRD so that

$$G_T(w) = \left| \frac{1 - Y_D'(w)}{1 + Y_D'(w)} \right|^2$$

where $Y_D'(w) = \dfrac{Y_D(w)}{G_o} = -g' + jb'(w)$

and $b'(w) = \dfrac{Cw_o}{R_o}\left(\dfrac{w}{wo} - \dfrac{wo}{w}\right)$

Finally, note that the frequency dependent term is that of a bandpass response so that the 3dB bandwidth (in radians) is:-

$$B = wo\left(\frac{w}{wo} - \frac{wo}{w}\right)$$

a2. Show that if the simple negative resistance amplifier of figure 1 is replaced by a parallel LCR device then the 3dB bandwidth of the circuit is

$$(G_T^{\frac{1}{2}} - 1)\,B = \frac{1}{2\pi RC}$$

Hence show that for high gain circuits the circulator or hybrid coupled amplifiers provide twice the bandwidth of the circuit in figure 1.

a3. How could the design example be modified to produce a greater gain than 5.2dB and what other amplifier characteristics would such a change affect?

a4. Assume a 10dB design for the example in the text and calculate the bandwidth possible from the device for the N=1, N=2 and N=$\infty$ cases.

Hint: the load determines the fundamental limit to the gain bandwidth but the bandwidth scaling factor represents the bandwidth for a given prototype circuit.

a5. Continue the design example in the text for N=3 using the prototype values

$$g_1 = 3.86485$$
$$g_2 = 0.33381$$
$$g_3 = 1.98602$$
$$g_4 = 0.29070$$

where for a series load each element is defined by

$$\frac{woLi}{R} = \frac{gi}{w}$$

for i odd and

$$woC_iR = \frac{gi}{w}$$

for i even.

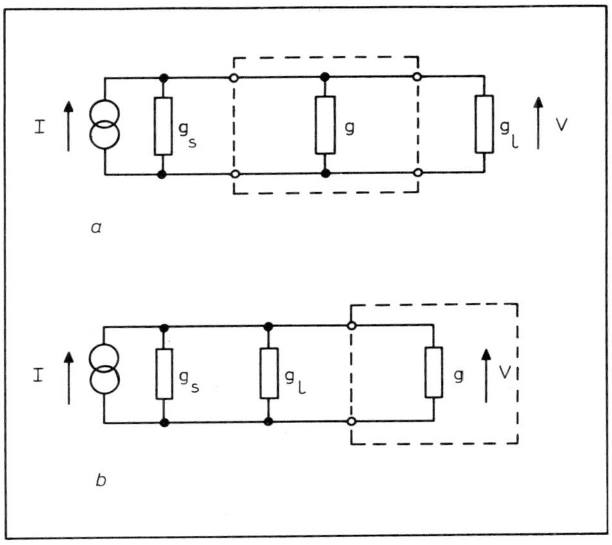


Figure 1 Simple negative resistance amplifier

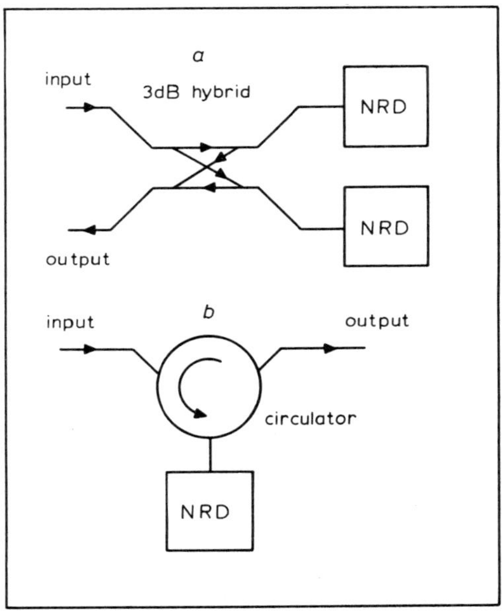


Figure 2 Practical negative resistance amplifiers

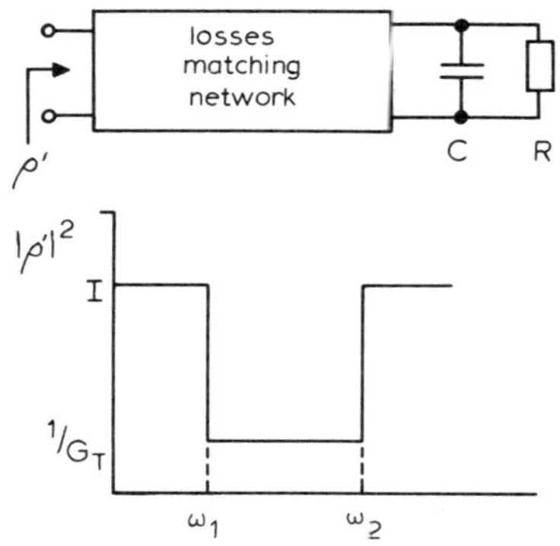


Figure 3 Optimum matching circuit performance

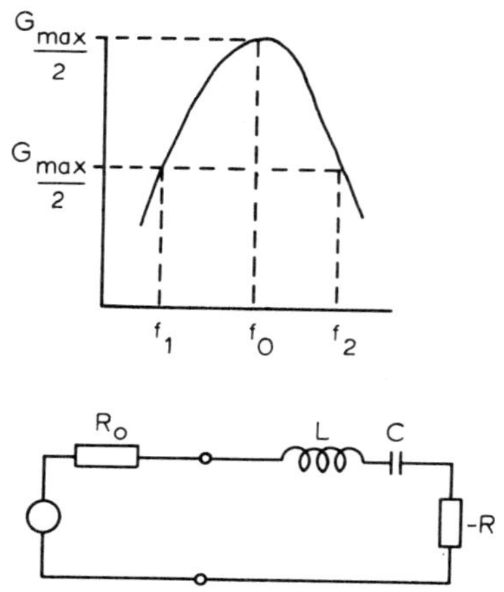


$$u = \frac{f_2 - f_1}{f_0}$$

$$k = 10\,\frac{G_{max}}{Z_o}$$

$$\frac{R_o}{R} = \frac{1 + k}{1 - k}$$

$$\frac{X_1}{R} = \frac{2\sqrt{k}}{u\,(k-1)}$$

$$L = \frac{X_1}{\omega_o}$$

$$C = \frac{1}{L\,\omega_o^2}$$

Figure 5 Derivation of device model

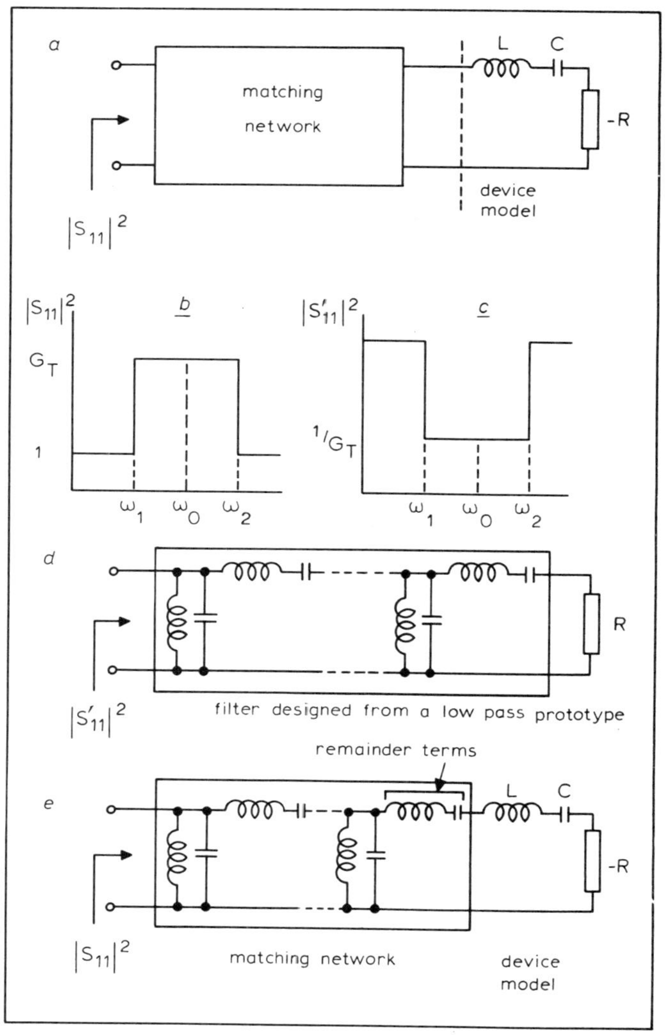


Figure 4 Design steps for matching negative resistance elements

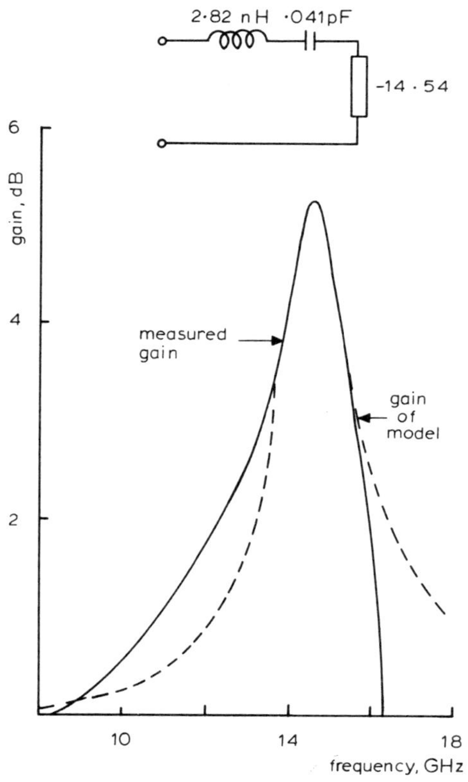


Figure 6 Data for design example

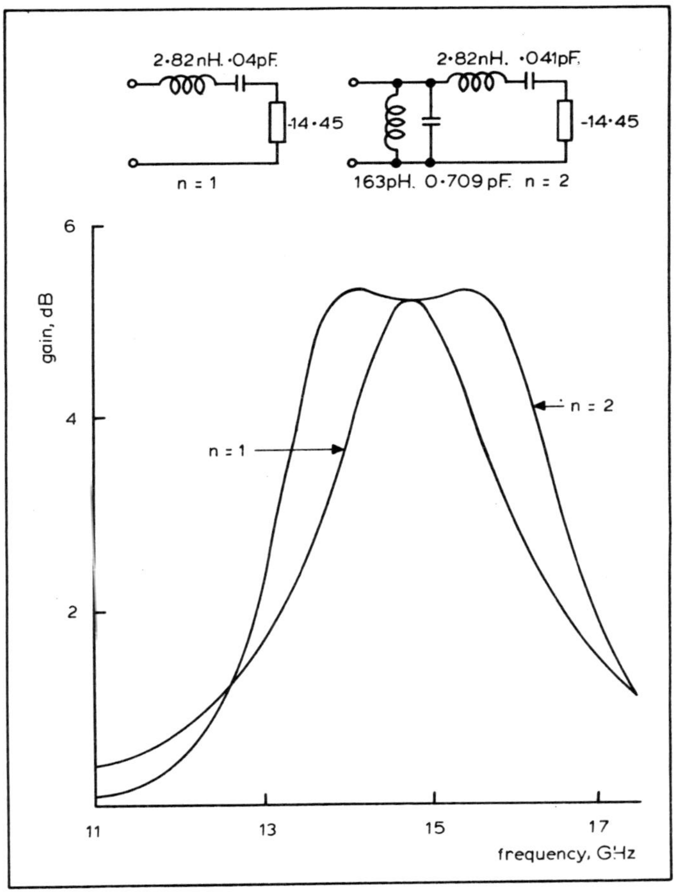


Figure 7 Lumped element design example

## b) PRACTICAL REALISATION

P W Huish

Post Office Research Centre, Martlesham Heath, Ipswich, Suffolk IP5 7RE

### 1 INTRODUCTION

This section is concerned with the realisation of microwave negative resistance amplifiers using GUNN or IMPATT diodes. The problem to be resolved in designing such amplifiers is the achievement of a suitable transformation between input and output impedances and device impedance allowing the desired gain, centre frequency and bandwidth to be obtained whilst avoiding spurious outputs at other frequencies. The solution to these problems is described in the following paragraphs by way of practical example. Most of these are applicable to both GUNN and IMPATT diodes, but there are differences between the two devices which will be outlined below.

Generally speaking the GUNN diode exhibits a larger value of negative resistance over a broader bandwidth than the IMPATT. This means that it is possible to obtain amplifier operation over a broader band with the GUNN diode but that the circuit design must ensure that the load impedance presented to the diode does not permit out of band oscillations. On the other hand the IMPATT diode exhibits only a few ohms of negative resistance over a relatively narrow frequency range. Low circuit losses are important if efficient amplification is to be obtained. In addition GUNN diodes operate in a low constant voltage, high current mode, whereas IMPATTs operate in a high voltage, low constant current mode. The IMPATT has advantages of high output power, typically several watts at X-band, whereas the GUNN can generate about 0.5W cw at this frequency but has lower noise and lower cost.

### 2 DEVICE MEASUREMENT

GUNN and IMPATT devices are rarely supplied fully characterised except in terms of operating voltage, current output power and frequency. Before amplifier design can commence it is necessary that the diode impedance at the desired power level be determined.

A suitable method is to mount the device in the short circuit plane of a 50 Ω 7 mm air line connected to a microwave network analyser as in fig 8. Also shown in fig 8 is a travelling wavetube for generating enough power for large signal measurements. Bias can sometimes be applied through the analyser or through a bias tee placed between the analyser and diode. High power attenuators may also be necessary to limit the power incident on the harmonic frequency converter. When large signal measurements are undertaken it is essential that all components are capable of withstanding the high power levels involved.

It is convenient to display the measurements on a compressed polar display, fig 9. This gives fair accuracy for the reactive impedance but can become inaccurate for the low values of negative resistance associated with large signal IMPATT measurements. Alternatively a phase-gain measurement can be used for the results, but the same comments about accuracy apply.

A similar method uses a slotted coaxial line to measure the VSWR due to the diode which can then be used to calculate the impedance. Such a measurement is however tedious and time consuming and is not extensively used.

In both slotted line and network analyser measurements, the measured impedance is representative of the diode mounted in a 50 Ω 7 mm coaxial air line. The field distribution in the immediate vicinity of the diode can have a profound effect on the measured impedance, particularly the reactive impedance. This means that any difference in the mounting configuration of the diode in the measurement rig and the amplifier will cause change in impedance and therefore gain and centre frequency.

Such effects can, however, be turned to advantage to modify the diode impedance and a few examples are given in fig 10. Typically, recessing the diode into either the holder or the transformer lowers the reactance, whereas post-mounting raises the reactance.

Network analyser and slotted line measurements are useful up to about 20 GHz, above which the coaxial line dimensions start to become impractically small. An alternative technique, used by the author, and also applicable to lower frequencies uses an oscillator circuit of well defined characteristics. This technique is useable up to at least 60 GHz and probably 100 GHz, and has the advantage of determining the output power of the diode as a function of bias voltage and current. The circuit used by the author is the hybrid coaxial waveguide circuit, described by Kenyon [7].

Kenyon shows that the equivalent load resistance presented at a voltage node in the coaxial line is given by:

$$R = \frac{Z_o\, Q_{ex}}{Q_{ex} + Q_L}$$

where $Z_o$ is the impedance of the line
$Q_L$ is the loaded Q of the oscillator
$Q_{ex}$ represents the coupling between coax line and cavity

The position of the diode relative to the voltage node can be determined and thus the load impedance seen by the diode obtained. The diodes impedance is simply the negative of the load impedance.

This method has been used to confirm large

signal measurements of high efficiency X-band IMPATT diodes made on the network analyser. An advantage is that the diode mounting configuration can often be very similar to that used in the amplifier, allowing accurate estimation of the diode reactance.

The diode impedance is calculated from the load resistance presented to the diode for optimum oscillator output power. The resistance at a voltage node in the coaxial line may be determined using the equivalent circuit and the loaded Q of the oscillator. The actual load impedance may then be determined knowing the position of the diode in the line with respect to the voltage node.

## 3 CIRCUIT CONFIGURATION

Negative resistance amplifiers may take either the reflection or transmission forms shown in fig 11. The transmission form fig 11a is confined mainly to waveguide circuits and, due to its symmetry can operate in either direction of operation. A typical characteristic, taken from ref 8 is shown in fig 12. This shows that the transmission amplifier can also exhibit reflection gain just outside the transmission pass band. If this is not to prove troublesome, input and output isolators are required, adding considerably to the cost of the complete circuit.

The transmission amplifier does however possess advantages for broadband applications where the limited bandwidth of isolators and circulators becomes an embarassment. But this is only true if the out of band reflection gain can be tolerated or nullified by suitable filters.

The reflection amplifier, fig 11b, is by far the most common form appearing in coaxial, stripline and waveguide media. Since it has a combined input and output port, these signals are separated in a circulator, although for experimental purposes a directional coupler often suffices.

The reflection amplifier circuit bears a superficial resemblance to an oscillator but a simple experiment carried out with an oscillator circuit soon shows several distinct differences.

An oscillator circuit with near independent control of the reactive and resistive match is particularly useful for this experiment, two such oscillators being the hybrid coaxial waveguide circuit mentioned earlier and the capacitively coupled coaxial circuit. If the oscillator is first adjusted for maximum power, then, as the load resistance is increased, it will be found that the output power falls until finally oscillation ceases. If the oscillator's reflection coefficient is viewed on a swept reflectometer display, a tall gain spike will be observed at a frequency corresponding to the oscillation frequency of the circuit. Further increases in the load resistance will reduce the magnitude of the gain spike.

There are, however, several important differences between a true amplifier circuit and this readjusted oscillator circuit. Fixed tuned oscillators are usually designed for high Q, to give low noise and low pushing and pulling characteristics. When adjusted for amplifier operation such a circuit will exhibit narrow bandwidth, often too narrow for any use. Another effect is noticed at high levels of RF drive and is often associated with an abrupt saturation of RF output. This is the appearance of spurious emissions at or near the first subharmonic frequency ie $f_o/2$. These emissions are parametrically related and occur due to the non linear depletion capacitance being pumped by the RF signal. Sometimes low frequency oscillations (up to a few hundred MHz) may be observed due to low frequency variations of the charge stored in the device. These are often referred to as bias oscillations and have been considered by Brackett [9].

Most amplifiers are destined for communications use where spurious emissions cannot be tolerated and where a wide transmission bandwidth is necessary. The following sections will illustrate, by example, how useful amplifiers can be constructed in various media.

## 4 WAVEGUIDE AMPLIFIERS

The use of waveguide as a circuit medium is most prevalent at high frequencies where the dimensions of coaxial and similar structures become impossibly small. Transformation between the waveguide impedance and the diode can be performed by reducing the waveguide height, linearly or in steps, or by using a ridge transition. The diode is often mounted in a coaxial section coupled to the waveguide. Two examples of waveguide circuits are shown in fig 13. Fig 13a uses a re-entrant coaxial line section to achieve an impedance transformation between the waveguide impedance and the diode [10]. No provision is made for subharmonic attenuation and it may be assumed that with the high gain of this design, the RF drive level is fairly low and does not pump the diode into parametric behaviour.

The reduced height waveguide circuit of fig 13b also uses a coaxial line section to match the diode to the waveguide. This circuit has been used successfully for GaAs and InP GUNN diodes [11] [12]. The length of the coaxial line is chosen to resonate the diode impedance and the iris and reduced height waveguide section serve as an impedance transformer to match the coaxial line section to the waveguide load.

There is a similarlity between this circuit and the single tuned circuits described in [7]. However in the circuit of 13b, the absorber is preceded by a bandstop filter. This is necessary with GUNN diodes since their high negative resistance implies low VSWR in the line. This means that a significant proportion of the power would be dissipated in the absorber. This is minimised by the inclusion of the bandstop filter.

A further form of waveguide amplifier takes the form of a post filter as shown in fig 13c [8]. As mentioned previously, such a circuit is bidirectional and also provides fairly low gains - in the region of 4-6 dB.

## 5 COAXIAL, MICROSTRIP AND SIMILAR CIRCUITS

These seem to comprise the majority of published work on the reflection amplifier and, at low microwave frequencies, appear to offer advantages in both size and construction. A basic coaxial circuit is illustrated in fig 14 and consists of an impedance transforming element. Although bias may be applied to the circuit through a conventional bias tee, a more practical and compact approach is illustrated in fig 14 where the bias is fed to a low impedance point through a high impedance line and RF filter. A DC block is usually necessary at the output and care must be exercised in its design. The introduction of tuning screws as shown allow adjustment of gain and frequency but large scale changes are effected by the length and diameter of the transformer, which can be determined using the chart shown in fig 15.

The use of a multisection impedance transformer gives improved bandwidth although the design is more complicated.

Similar circuits can be constructed in microstrip and other variants and give similar results. Whilst microstrip is compact it offers less scope for tuning and the design must therefore be correct otherwise expensive and time consuming empirical optimisation is necessary.

The problem of subharmonic oscillations are conveniently solved in this class of circuit with the trap circuit shown in fig 15. This consists of a chip resistor or a piece of microwave absorber connected to a half wavelength of line which is open circuited at its far end. At the operating frequency the resistor appears open circuited but at half this frequency it appears in shunt with the diode and is effective in inhibiting subharmonic oscillations.

Greater control over the resistive match to the diode, thereby allowing gain adjustment is given by the use of a capacitive output coupling. A simple method is to provide axial movement of the DC block in fig 14, although more elaborate schemes can be devised. Axial movement of the transformer can also be useful in adjusting the centre frequency of the amplifier.

Similar circuits can be realised in microstrip and a typical example illustrating the use of the DC-block, subharmonic trap and bias filter is shown in fig 16. One advantage of microstrip is that the circulators can be formed on the same substrate which allows very compact circuits to be realised. As an example a 3 stage circulator coupled microstrip amplifier can be built on a substrate 50 mm x 25 mm [13].

## 6 MULTIPLE DIODE CIRCUITS

In cases where the output power required exceeds the capability of a single diode it is often advantageous to combine the outputs of several diodes in one circuit. One circuit which has given particularly satisfactory results was proposed by C T Rucher [14] and is shown in fig 17. A number of diodes are connected by a star of impedance matching transformers. These are connected at their centres by a lossy "hub" and a capacitive tuning plate provides coupling between the transformers and the output. The lossy hub attenuates antisymetric modes and also inhibits spurii at $2f_o$ and $f_o/2$. Adjustment of the coupling allows the load resistance to be adjusted over a large range without significantly affecting the operating frequency. Changes in frequency are obtained by adjusting the transformer lengths or by dielectrically loading the transmission lines.

The authors experience of this circuit has been very favourable with a wide range of diode types. A typical performance obtained using high power GaAs IMPATTs is shown in fig 18, where a bandwidth of 430 MHz was obtained. Calculations indicate a bandwidth in the region of 400 MHz for 2, 3 or 4 low reactance slope diodes at 6 dB gain. With higher values of gain, a reduction in bandwidth occurs.

An alternative to multiple diode circuits is the use of splitters or hybrids to combine a number of single ended amplifiers. These devices do, however exhibit an insertion loss of about 1 dB per pass which can seriously reduce the gain and overall efficiency of a high power amplifier. By way of example an output stage consisting of two 4W added power 6 dB amplifier can be compared with a two diode Rucker circuit containing two 4W diodes. The Rucker circuit can add 8W of power at 6 dB gain giving 10.6W output. The two single ended amplifiers lose 1 dB in the splitter giving a combined output power of 8.46W. In addition the overall gain of the single ended version is reduced to 4 dB due to the loss in the splitter. This means that extra gain is required in either this stage or in preceding stages, thereby reducing the bandwidth of the amplifier.

It was mentioned earlier that if stable amplification is to be obtained, then the load resistance must always be greater than the negative resistance of the device. This means that if the load resistance is adjusted for high large signal gain where $R_L \sim |R_D|$ it is quite likely that under small signal conditions $R_L \leqslant |R_D|$ and oscillation will occur. This effect can be resolved in at least two ways. The first is to limit the large signal gain. The second employs a squelch circuit which adjusts the diode bias to suit the incident power and can be a simple on-off switch or a sophisticated continuously variable circuit.

A variation on this last theme reported by Deadman [13] is applicable to diodes which exhibit a large change in bias voltage (or current) with RF drive such as high efficiency IMPATT diodes. A constant voltage bias supply is used, which effectively allows the bias current to vary with the RF drive. Such an arrangement reduces gain compression and can completely stabilise the diode at the small signal level.

## 7 MULTISTAGE AMPLIFIERS

It was shown in the previous section that the gain of a simple single stage amplifier must be traded against its bandwidth. If insufficient gain is obtained at the required bandwidth then a further stage of amplification is required. Stability requirements may also require a further stage of amplification to achieve the desired gain. The usual configuration is a cascade of circulator coupled stages connected through isolators to improve interstage isolation and stability. To ensure the best overall efficiency it is preferable that the early stages should have high gain whereas the last stage should add the greatest power but may have low gain.

## 8 CONCLUSIONS

Negative resistance amplifiers may be realised in either reflection or transmission forms using GUNN or IMPATT diodes and even field effect transistors. The gain and bandwidth of the amplifier is determined by the load impedance presented to the diode. Thus the design of the matching element between the device and load is the major task in the design of negative resistance amplifiers. The design of the matching element requires knowledge of the device impedance as a function of signal level. Such information may be obtained by network analyser measurements up to about 20 GHz or by oscillator measurements up to at least 80 GHz.

The most common form is the reflection amplifier which can be realised in waveguide, coaxial line or microstrip and its derivatives. Of these, microstrip offers the most compact and economic solution but lacks ease of tuning and adjustment. Waveguide circuits seem to be preferred for high frequency work due to the increasing losses of microstrip and the intricacy of coaxial circuits.

Stability can be achieved by considering the change of device impedance with signal level and adopting a suitable choice of large signal gain. Higher gains require the use of a squelch circuit which monitors the input signal and adjusts the bias to ensure stability.

High gain requirements are met by cascading several amplifiers usually with interstage isolation. High output powers can be obtained by using splitters to combine several amplifiers or by using combining circuits such as the one described by Rucker.

## 9 ACKNOWLEDGEMENTS

Acknowledgement is made to the Director of Research of the British Post Office for permission to publish this work.

## 10 REFERENCES


7 Kenyon, N.D., 1973, "Single tuned solid state microwave oscillators", International Journal of Circuit Theory and Applications, Vol 1, pp 387-393.

8 Bowers H.C., Midford T.A. and Plants S.T., 1970, "IMPATT diode multistage transmission amplifiers, IEEE Trans Microwave Theory & Techniques MTTL8 No 11.

9 Brackett, C.A., 1973, "The elimination of tuning induced burnout and bias circuit oscillations in IMPATT oscillators", Bell System Technical Journal, Vol 52, No 3.

10 Hewlett Packard Ltd, Application Note no 968.

11 Hamilton et al, 1976, "InP GUNN effect devices", IEEE Transactions on micrlwave theory and techniques, Vol MTT 24, No 11.

12 Brookbands, D.M., Cotton, F.J., Howard, A.M., Lang, R.J., 1979, "Q-band InP transferred electron amplifiers", Proc 1979 European Microwave Conference, Brighton England.

13 Deadman, H.A., Clough, L.D., Smith, J.G., Tearle, C.A., Birbeck, J.H.C., 1979, "High efficiency J band diode amplifiers", ibid.

14 Rucker, C.T., 1969, "A multiple diode high average power avalanche diode oscillator", IEEE Trans on Microwave Theory & Techniques, Vol MTT 17, No 12.


Tutorial Questions

b1. The apparatus of fig 1 has a phase accuracy of $\pm 2^{o}$ and an amplitude accuracy of $\pm$ 0.2 dB. If a diode has a large signal impedance of $-2 + j10\Omega$, estimate the limits of uncertainty in the measured impedance.

b2. An amplifier with a gain of 10 dB is connected to a circulator with a VSWR of $\Gamma$. Calculate the maximum value of $\Gamma$ before instability occurs.

b3. A diode has the small and large signal negative resistance characteristics shown in fig 19 as a function of frequency. What is the lowest frequency at which stable amplification can be obtained with large signal gains of 6 and 10 dB.

b4. Show graphically or otherwise how a series cpacitance may be used to tune the real part of the load impedance presented to a diode in an amplifier circuit. A diodes reactance varies linearly with frequency from $-j10\Omega$ at 9 GHz to $+j10\Omega$ at 11 GHz and it has constant negative resistance of $-2\Omega$. Find the change in centre frequency due to a change in gain from 3 to 6 dB by adjusting the capacitive loading. Assumea 30$\Omega$ impedance transformer $\lambda/4$ in length and 50$\Omega$ load impedance.

b5. A diode is tested in a hybrid coaxial waveguide circuit and is found to give 1.5W at 10 GHz when positioned 14.7 mm below a voltage mode in the circuit. The load resistance seen from a voltage node is 1.5$\Omega$. Find the length and impedance of a suitable single step transformer to make a coaxial 10 dB gain amplifier at this frequency. What will be the output power and what factors may require the transformer dimensions to be adjusted in practice. Assume 50$\Omega$ coaxial lines in both oscillator and amplifier circuits.

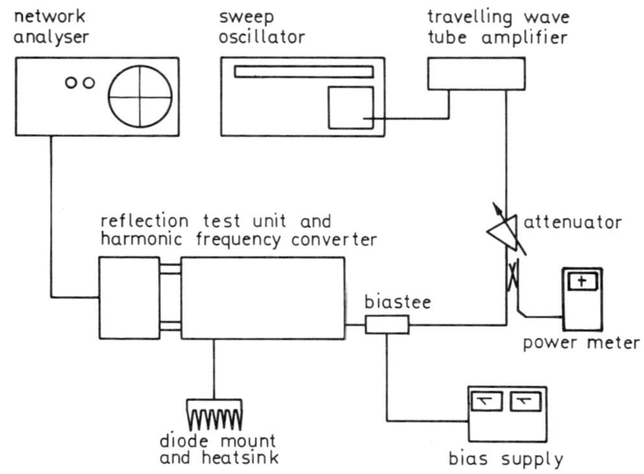


Figure 8 Equipment for determining diode impedance

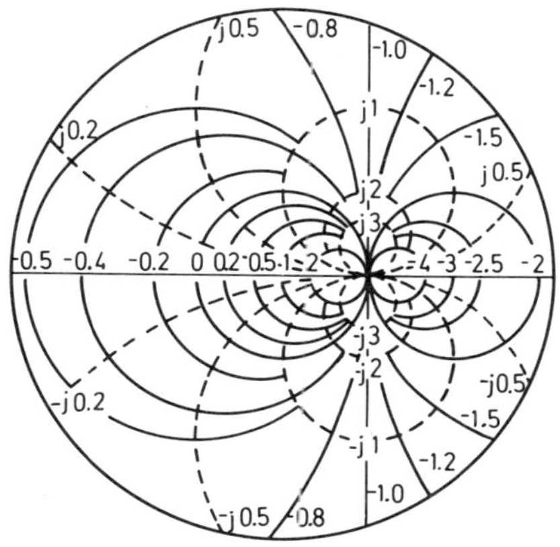


Figure 9 Compressed Smith chart for display of negative resistance

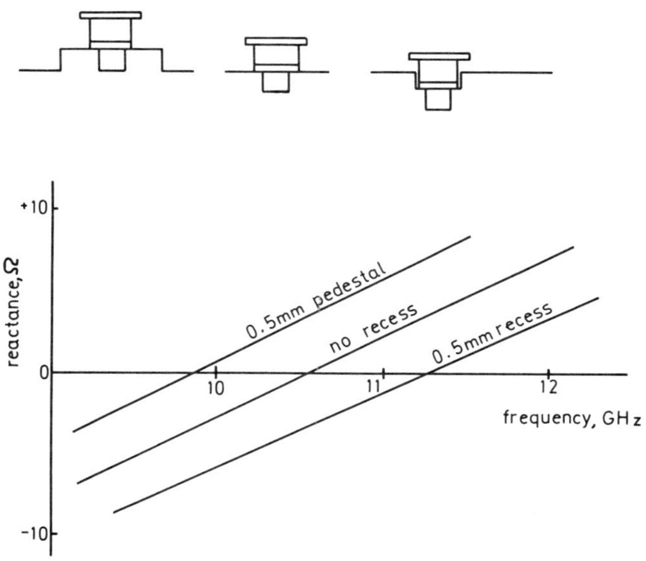


Figure 10 Effect of package mounting configuration on reactance

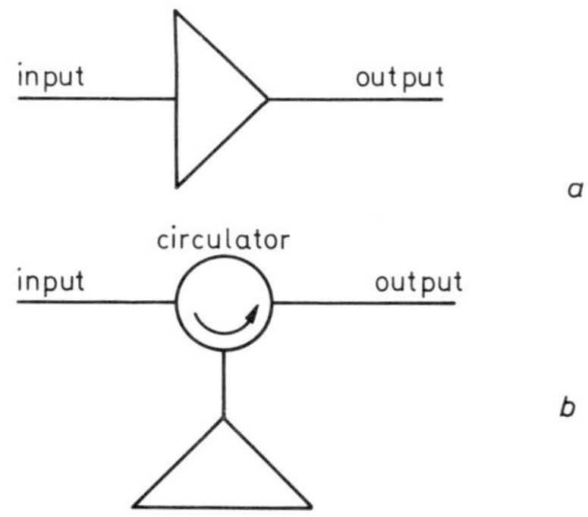


Figure 11 (a) Transmission amplifier
(b) Reflection amplifier

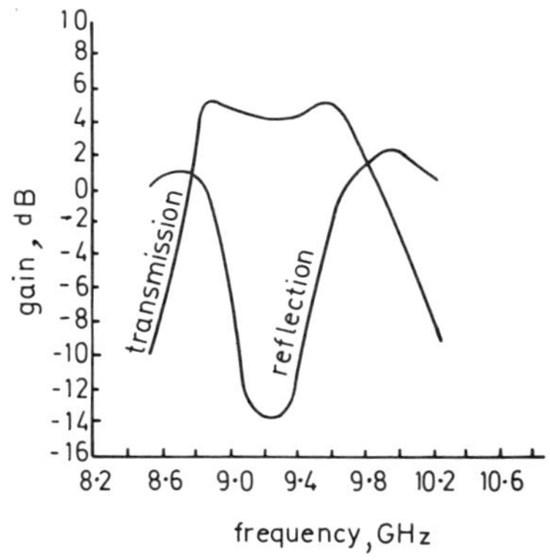


Figure 12 Performance of a transmission amplifier

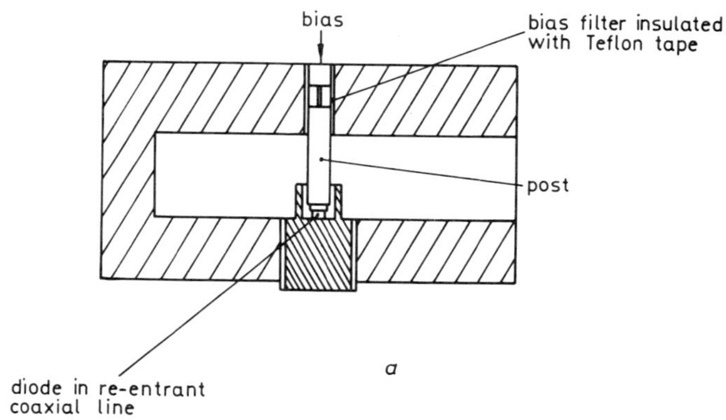


Figure 13 (a) Waveguide amplifier using re-entrant co-axial line

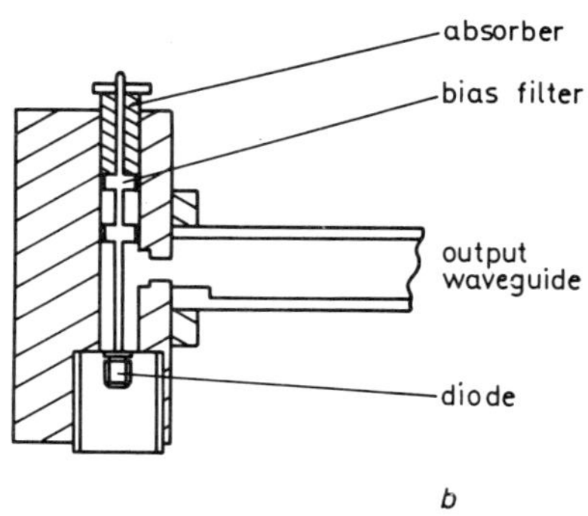


(b) Reduced height co-axial waveguide circuit

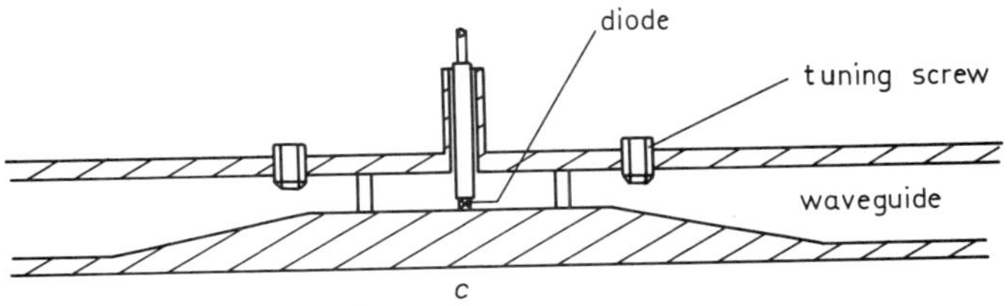


(c) Transmission-type waveguide circuit

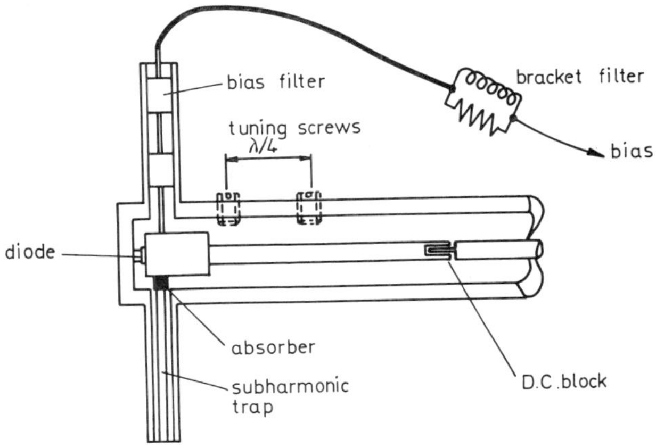


Figure 14 Coaxial reflection amplifier

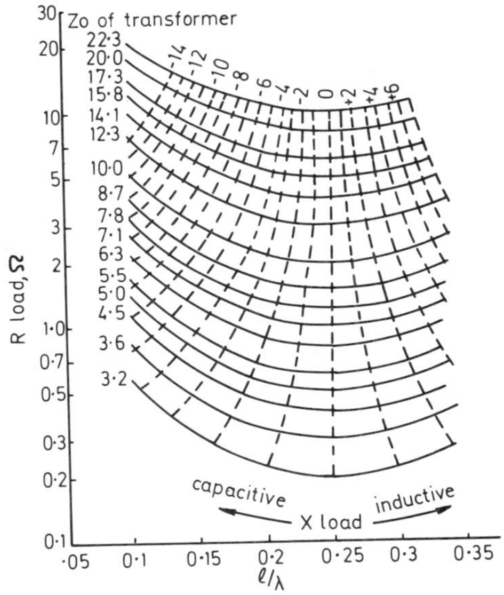


Figure 15 Design aid for single step transformer. Load impedance = 50 Ω

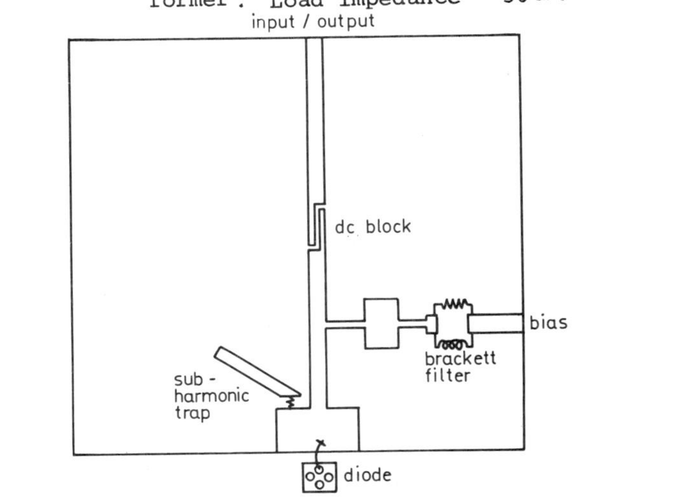


Figure 16 Microstrip reflection amplifier circuit

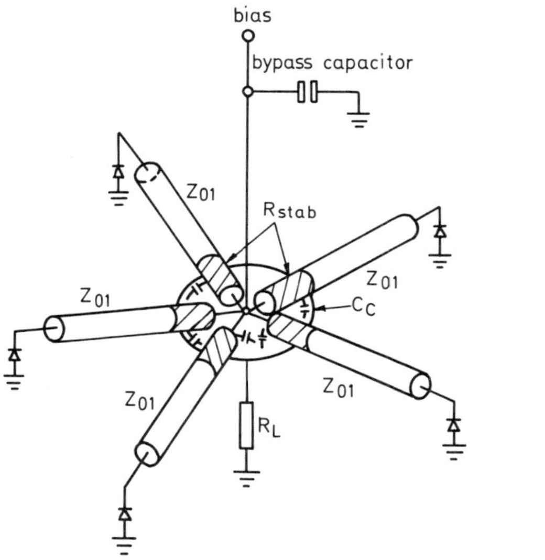


Figure 17 Sketch of five-device 'Rucker' combining circuit

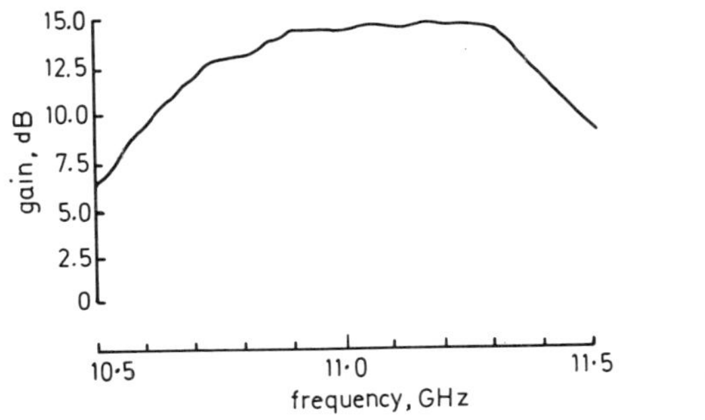


Figure 18 Gain against frequency of a two stage 15 dB gain 7.3 W output power amplifier

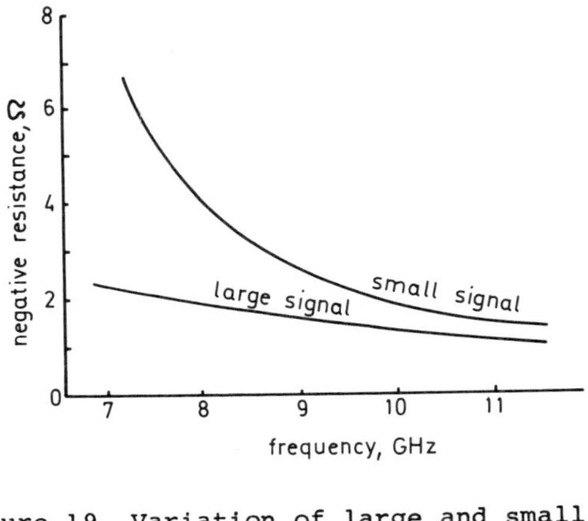


Figure 19 Variation of large and small signal negative resistance for question 3

# CHAPTER 11

# Amplifier evaluation techniques

J. Arnold

# AMPLIFIER EVALUATION TECHNIQUES

J. Arnold

Plessey Research (Caswell) Ltd., Allen Clark Research Centre, Towcester, Northants., U.K.

## INTRODUCTION

Microwave amplifiers are being increasingly used in many modern Microwave Systems in both Civil and Military roles. Applications include radar systems, earth terminals, scientific and communication satellites, microwave links, etc. These systems necessitate amplifiers to operate under extreme environmental conditions, from temperatures of -50°C to +125°C, under hard vacuum with high radiation levels, and in the presence of interfering signals.

With the need to make efficient use of the microwave spectrum, systems are becoming increasingly complex utilising techniques such as quadrature phase shift keying, multiple carrier, frequency reuse etc. These techniques place increasing demands on the performance of microwave amplifiers since the systems ideally require the amplifier to allow the signals to be transmitted without causing degradation to performance. This basically simple requirement has wide ranging implications on gain response, noise performance, phase characteristics, linearity etc. ViFian (1). This chapter reviews techniques to evaluate these parameters for microwave amplifiers without specifically identifying the effects of each parameter on particular systems.

## MEASUREMENT OF AMPLIFIER GAIN AND IMPEDANCE

### Network Analysers up to 18 GHz

All microwave engineers are faced with the problems of measuring the transmission or reflection characteristics of passive or active devices either in component form or when integrated into complete systems, such as in amplifiers. Not so many years ago, the engineer requiring to measure impedances had to resort to the use of slotted lines and to perform point by point measurements. When only the magnitude of reflection coefficients were required, could the use of well matched directional couplers and crystal detectors together with a ratiometer, enable swept measurements to be used to determine the return loss characteristics (return loss = $-20\log_{10}$ (voltage reflection coefficient)). A similar technique allowed gain or loss characteristics to be obtained. What was needed was an instrument capable of providing amplitude and phase information at each frequency as the frequency was swept across the operating band. This data was required in both reflection and transmission modes, so that the eight parameters that define the two port network scattering parameter would completely define the network $\begin{bmatrix} S_{11} & S_{12} \\ S_{21} & S_{22} \end{bmatrix}$.

These requirements were met by the introduction of the network analyser. This instrument allowed in-detail measurements to be made from 0.01 GHz to 2 GHz, and 2 GHz to 18 GHz. Manual versions of this analyser became available which enabled swept measurements to be made, but for high accuracy measurements that corrected for systematic system imperfections, a fully computer controlled system became available. This has subsequently been designated A.N.A. (automatic network analyser), and the early versions were so expensive (requiring several racks of computer hardware) that only a few standards laboratories acquired these equipments. In the UK only 3 or 4 such equipments existed. Recently, however, the computer has been succeeded by the microprocessor that has resulted in equivalent capacity being available in a desk-top computer. As a result, the cost and size of the system has been reduced so that many laboratories have these available. The same basic microwave hardware is utilised. Fig. 1 shows a block diagram of the system. Fig. 2 and Fig. 3 show in detail two vital components of the system and provide an insight into the operation processes necessary to obtain phase and amplitude information. In principle the transmission/reflection RF unit is two high directivity couplers, one providing a reference signal, the other determining the reflected signals. A line stretcher in the reference channel allows off-setting of the phase. The reference and test channels are coherently down-converted to an IF frequency of 278 KHz where samples of each signal are fed to limiter and phase detectors for determining the r.f. phase angle and the other samples of amplitude are compared in a synchronous detector to provide an amplitude output. The outputs can be processed by the computer along with calibration references to provide measured phase and amplitude data.

The basic system suffers from a number of inherent imperfections. The coupler directivities are not infinite, the reference and test couplers need not necessarily track with frequency and the source impedance seen by the test unit may not be exactly 50Ω. These errors can be modelled as shown in Fig. 4, together with the equation that represents the overall measured impedance. By using three known standards, the unknowns can be determined and error correction applied (2).

Directivity errors can be determined as shown in Fig. 5 using a quality sliding load. By making measurements at a number of settings of the load position, the magnitude and phase of the directivity can be determined. The use of a fixed open and fixed short circuit can determine the other two unknowns, but experience has shown that a fixed short circuit and a fixed offset short circuit provide more accurate results. This arises since the open circuit cannot be modelled having fixed fringes capacitance for all frequencies. A similar model can be derived for transmission measurements. Fig. 6 shows the results of a load measurement before and after correction routines have been used. Note that the correction routine removes the large oscillations, from points of in-phase and out-phase combinations of reflections.

As a result of applying these error corrections there still remains some residual non-systemic errors which in themselves give the overall accuracy. Best accuracy is achieved when reference and test channel signals are of nearly equal level.

Transmission accuracy:

Amplitude better than $\pm$ 0.25 dB up to 40 dB
Phase better than $\pm$ 2°

Reflection accuracy:
(worst case for $\Gamma_{LOAD}$ = 1)

Amplitude better than $\pm$ 0.1
Phase better than $\pm$ 5°

Averaging of measurements can reduce these errors. Anderson et al (3).

In measuring the frequency response of small signal amplifiers care must be taken to ensure that the amplifier is not driven into compression. The signal level in to the amplifier can be reduced by lowering the r.f. signal level into the test system, but this results in the system being unable to phase lock. Signal levels down to -30 dBm into the test amplifier can be achieved whilst maintaining approximately -10 dBm into the reference channel sampling coupler. Calibrated attenuators can be used combined with the test amplifier to improve dynamic capabilities.

The signal level from the output of the amplifier must be maintained at less than -10 dBm (into harmonic converter) so fixed attenuators can be added after the amplifier or before the test channel of the harmonic converter. The latter will limit the dynamic range for impedance measurements, if performed at the same time.

For some amplifiers, the minimum input signal level of -30 dBm may still cause compression. For example, if the amplifier gain is 50 dB then the output power of +20 dBm would be close to the compression point. Hence attenuation will be required at the amplifier input to limit the signal level to -40 dBm, say.

As an example consider the signal levels into the reference and test channels shown in Fig. 7. The reference channel has been reduced to -40 dBm at the harmonic converter input, giving a signal level of -30 dBm at the unknown port. This must be reduced to -40 dBm at the amplifier input so a 10 dB attenuator can be added. At the amplifier output a 20 dB attenuator must be added to maintain the test channel at -10 dBm. In fact to satisfy another condition that the test channel-reference channel $\leq$ 20 dB this attenuator should be 30 dB.

As shown a dynamic range of 58 dB exists in the test channel to investigate out of band gain performance, and the attenuator can be calibrated during the initial calibration routine. The use of -30 dBm at the UNKNOWN port will not allow the short circuit calibration through the 10 dB attenuator, since only 28 dB dynamic range exists at the test port when reflection measurements are being performed. Recalibration will be necessary for impedance measurements with the 10 dB attenuator removed. This will give a dynamic range up to -28 dB return loss. For FET amplifiers, higher input signal levels will not significantly change the impedance levels.

This example shows the considerations to be given to any amplifier measurements using the system.

Fig. 8 shows the gain response of a broadband amplifier measured on ANA.

The reproducibility of results achieved by these systems have significantly reduced the discrepancies between measurements performed by different personnel and have become standard test equipment for customer amplifiers.

Most amplifiers utilise SMA connectors, and some problems of reproducibility of mating have been experienced. The use of 3.5 mm connectors should improve the problems.

## Network Analysers Above 18 GHz

Adaptations of the above system using auxiliary coherent down converters have been utilised up to 40 GHz. The instantaneous bandwidth of these systems is 2 GHz, but a computer control of the local oscillator frequency can provide full coverage over the bands 18 to 26.5 GHz and 26.5 GHz to 40 GHz.

## Scalar Network Analysers

These fall basically into two types:

(1) A.C. coupled - AM modulated carrier

(2) D.C. coupled - CW carrier.

Type 1. AC Coupled Systems. This system uses hot carrier detector diodes primarily for their high sensitivity (-50 dBm) and excellent reproducible characteristics. This enables standard instrumentation to be utilised to compensate for the non square law portion of the transfer characteristic and to extend their usable dynamic range to 60 dB. In order to improve the sensitivity, d.c. bias is applied to the diode, and this necessitates the use of an RF modulated signal to detect signals in the presence of the bias.

The development of ultra broadband high directivity dual directional coaxial couplers and directional bridges enable systems to operate over bandwidths of 10 MHz to 18 GHz. These components together with the broadband detectors enable simultaneous measurements of reflection coefficient magnitudes and transmission losses or gains.

Fig. 9 shows a system connected to perform these measurements.

This is a straightforward reflectometer system. The swept frequency r.f. carrier is amplitude modulated at a frequency of 27.8 KHz. The directional coupler samples the r.f. input signal (-20 dB) and is detected at detector A, whilst the main through signal passes almost unattenuated to the output of the coupler. With a short circuit connected to the coupler, the reflected signal is returned to detector B. The outputs from detectors A and B can be ratiod and the level set to 0 dB. The displayed level will not be flat due to frequency tracking errors and due to a non ideal source match. If a true open circuit can be placed at the reference plane of the short circuit and the response noted, then the average of the short circuit and open circuit plot represents the true plot. By connecting the third detector C through an attenuator (to reduce mismatch errors), the ratio of C to A will calibrate the transmission reference. Offset controls of 1 and 10 dB and system sensitivity controls of 0.25, 1, 10 dB (all operating at 27.8 KHz) enable the screen to be calibrated, and a display of ± 40 dB can be shown.

With an amplifier connected to the coupler, both the gain and reflection coefficient performance can be displayed.

Since directivity errors cannot be removed without phase information, these still remain as an error in reflection measurements. However for most amplifiers, the system can resolve to 1.2:1 VSWR's.

The accuracy of the offset attenuator indications of gain are accurate to ± 0.5 dB and the frequency response is flat to typically ± 0.25 dB when ratio measurements are made, Fig. 10.

Because of the dynamic range of the detectors (-50 dBm to +10 dBm) and the high coupling values used, severe limits are imposed on amplifier measurements. The use of attenuators is essential for high gain amplifiers to avoid saturation and it will be necessary to overdrive the amplifier to measure input and output reflection coefficients. Fig. 11 shows the problems of dynamic range for a medium gain amplifier of 20 dB.

Type 2. DC Coupled Systems. Some manufacturers still produce reflectometer/ transmission equipments based on d.c. operated detectors. It is argued that with the development of zero biased diodes with a high sensitivity that it becomes unnecessary to use a modulated carrier system. Hence drift errors due to the changes in the bias requirement are removed. High dynamic range logarithmic amplifiers form the heart of these systems and self balancing is used to ensure stability at low detection levels. These systems also rely on the development of high precision microwave bridges. These are basically Wheatstone bridges with high precision microwave loads. These bridges exhibit excellent bandwidths, 10 MHz to 18 GHz, and directivities of 40 dB. The performance of these systems compare well against Type 1 systems.

Data Normalisers

The introduction recently of data normalisers have simplified the problems facing the amplifier designer in aligning an amplifier so that the flat gain response he obtains is truly the amplifier response and not just the combination of the system response and the amplifier response combined. These normalisers store the system response without the amplifier and remove the additional errors of frequency tracking within couplers and detectors. As a result of normalisation, a reference response can be flattened to less than 0.05 dB. The display with the amplifier included is presented as (MEASURED-STORED) giving a true response. The normalisation capability of these normalisers is a function of the digitation of the display. The memory must be large enough to provide sufficient resolution to reduce the amplitude variations observed during calibration to a negligible size. Using an 8 bit word, the memory system will reduce the vertical deviations by a factor of 256 at each point. If the vertical uncorrected amplitude variations are ± 2 dB, the corrected deviation will be ± $0.00\overline{78}$ dB. In the horizontal direction the required resolution depends on the frequency range covered by the display. For most octave-band measurements, 256 points of correction provide sufficient horizontal resolution, as the rate of change of amplitude with respect to frequency is not great.(4)

Accuracy. Fig. 12 shows many of the limiting factors for accurate measurement of VSWR's for scalar reflectometers.

The directivity (defined as leakage into reflected signal detector) with a perfect load on the system is primarily caused by

(1) limited inherent directivity of the coupler

(2) test port reflection coefficient

(3) test adaptor reflection coefficient.

A further factor effecting the resultant detected reflection signal is the source impedance feeding the reflectometer, and this parameter becomes an increasing error as the test load reflection coefficient increases. This can be readily reduced by using a precision attenuator to define the source impedance. Items (1) and (2) combine to form the inherent reflectometer directivity. Fig. 5 shows the phasor combination of the reflective signal and directivity to produce a ripple error. Fig. 13 shows the results of a worst case analysis. Curves (A) and (B) show the effects of a directivity of 40 dB on measured return losses, whilst curve (C) shows the worst case errors due to an adaptor of return loss 30 dB (VSWR of 1.065) on measured return losses. It is seen that without directivity correction the errors in measured reflection coefficients can be significant.

NOISE MEASUREMENTS

Basic Theoretical Background

In any receiving system, each component should ideally allow signals to traverse the system with no degradation to the signal-to-noise of the received signal. In practice, however, such ideal components do not exist, and it is required to define a figure of merit for the components to enable the system designer to optimise the overall system performance. Such a figure of merit is noise figure defined as,

$$\text{Noise figure, } F = \frac{\text{Input signal-to-noise}}{\text{Output signal-to-noise}} = \frac{S_i/N_i}{S_o/N_o}$$

Now $S_o = G S_i$, therefore

$$F = \frac{N_o}{N_i . G} \qquad (1)$$

This expression is now independent of the signal, Si. By manipulation,

$$F = \frac{N_o/G}{N_i} = \frac{\text{output noise referenced to input port}}{\text{input noise}} \qquad (2)$$

Consider the input to the system terminated in a resistive source impedance. If this resistive termination is at temperature T, then a voltage is generated at the open terminal due to random motions of thermally agitated electrons. This random motion gives rise to an infinitely broadband spectrum of noise (Johnson). Nyquists theorem (5) states that when the resistive source impedance is connected to a matched load, the noise power delivered to the load is given by

$$P = \tfrac{1}{2}hfB + \frac{hfB}{e^{hf/kT} - 1} \qquad (3)$$

where T = termination temperature in $^{o}$K

K = Boltzmann's constant, 1.38 x $10^{-23}$ Joules/$^{o}$K

h = Planck's constant, 6.62 x $10^{-34}$ Joules. seconds

B = bandwidth, Hz

$f$ = frequency, Hz.

This represents the available power from the resistive source (maximum power transferable). If $hf/KT \ll 1$,

$$P = kTB \qquad (4)$$

In the definition of noise figure, equation (1), the input noise, Ni, is that generated from a matched resistive termination maintained at an arbitrary temperature of 290$^{o}$K. ($T_o$).

Thus equation (1) becomes

$$F = \frac{N_o}{kT_oBG} \qquad (5)$$

For the ideal system, the component should cause no additional noise to be generated so that $N_o = kT_oBG$, and the noise figure is 0 dB.

Of the total noise output power of the system $N_o$, a portion is due to the input termination $(kT_oBG)$. The amount of noise added by the system or receiver is,

$$N_s = N_o - kT_oBG = (F-1)\,kT_oBG \qquad (6)$$

This equation gives rise to a very useful conceptual model. Fig. 14 shows the model consisting of the resistive termination $kT_oB$ in parallel with another resistive termination $(F-1)\,kT_oB$ at the input to an ideal receiver of gain G with a 0 dB noise figure $(F=1)$.

From this model, another useful system concept of equivalent noise temperature can be introduced. The additional noise generated by the non-ideal receiver is represented by the term $(F-1)kT_oB$. Since this is a resistive element, this noise power will be generated by a resistor having a temperature, $T_e$, where

$$T_e = (F-1)T_o \qquad (7)$$

Thus, for any noise figure, $F$, we can introduce an equivalent noise temperature, $T_e$, which represents the temperature of a matched termination that gives rise to amount of added noise at the output of the system equivalent to the non-ideal receiver. This temperature is referred to the receiver input terminals. Noise temperatures can be added directly to give system performance. For example, suppose an antenna is connected to a receiver with noise temperature, $T_r$. An antenna temperature, $T_{ANT}$, can be defined that is the temperature of a matched resistive element at the input of the receiver that gives rise to the same noise as that due to the antenna. For the antenna, this temperature is a function of the angle of elevation, the conditions prevailing in the sky, and is a function of the sidelobe performance of the antenna. These sidelobes will generally receive noise generated by the background earth temperature. Thus antenna temperature represents the temperature "seen" by the antenna.

Thus the system equivalent noise temperature, $T_{system}$ is given by

$$T_{system} = T_{ANT} + T_r \qquad (8)$$

Noise Measurements

CW Method. From the definitions so far derived, a very simple measurement of noise figure can be made. The only parameters that require measurement are No, B, G. No and G can readily be determined but B is more difficult, since it is the noise bandwidth (bandwidth of an ideal rectangular response characteristic) that gives rise to the same power output as the actual gain response. This can be determined by measuring the complete gain response then integrating to determine the noise bandwidth. Alternatively a measurement of the 3 dB points can be made and a correction figure applied. For example the ratio of noise bandwidth to 3 dB bandwidth for a single tuned circuit is 1.57 whilst for a transitionally double tuned circuit is 1.11.

This method, besides being tedious, is very inaccurate. Alternative methods should remove the necessity to determine the gain-bandwidth product, and to this end a broadband noise generator can be utilised.

Such a generator should be capable of generating noise power considerably in excess of the input termination reference at 290$^{o}$K when fired.

For such a source,

$$P_{excess\ noise} = k(T_N - T_o)B \qquad (9)$$

where $T_N$ is the noise temperature of the source in the fired condition.

The relative excess noise ratio $(T_N - T_o)/T_o$ is the figure defined by noise source manufacturers. For example if $T_N$ = 10,000$^{o}$K the excess noise power ratio is 15.24 dB.

Consider the noise outputs from a receiver with the noise source operated in the "fired" and "unfired" states.

With the noise source "unfired" or OFF, the noise output from the receiver, $N_{OFF}$, consists of the receiver added noise plus the input termination noise multiplied by the receiver gain. Fig. 15(a).

Thus,
$$N_{OFF} = kT_oBG + (F-1)kT_oBG \quad (10)$$

With the noise source "fired" or ON, the noise output will rise by the excess noise ratio of the noise source multiplied by the receiver gain. Fig. 15(b).

Thus,

$$N_{ON} = kT_oBG + (F-1)kT_oBG + k(T_N - T_o)BG \quad (11)$$

From equations (10) and (11)

$$\frac{N_{ON}}{N_{OFF}} = \frac{FT_o + (T_N - T_o)}{FT_o}$$

or

$$F = \left(\frac{T_N - T_o}{T_o}\right) \times \frac{1}{\left(\frac{N_{ON}}{N_{OFF}} - 1\right)}$$

thus

$$F_{dB} = 10\log\left(\frac{T_N - T_o}{T_o}\right) - 10\log\left(\frac{N_{ON}}{N_{OFF}} - 1\right) \quad (12)$$

The first term, is defined as the excess noise ratio, for the noise source, equation (9). In the second term, the $N_{ON}$ to $N_{OFF}$ ratio is called the $Y$ factor, and represents the ratio in decibels between the receiver output power with the source in the ON and OFF states.

Thus

$$F_{dB} = ENR_{dB} - 10\log(y-1) \quad (13)$$

Double Noise Power Method. This method is still used in industry for manual systems measurements, and consists in setting $N_{ON} = 2 \times N_{OFF}$. The method requires a means of controlling the excess noise from the noise source, and this is provided by a precision attenuator. By utilising a 3 dB attenuator to calibrate the x2 level, and using the precision attenuator to reset the level, the noise figure is calculated as ENR dB minus the attenuator setting.

$Y$ Factor Method (RF attenuator method). This method ensures that the detector or power meter used for monitoring the noise output is maintained at the same working point on the characteristic when the source is ON and OFF, thus eliminating errors due to non-linearities in the system.

With the precision attenuator set to 0 dB and the noise source OFF, a reference level is noted. With the noise source ON, the precision attenuator is adjusted to reset to the reference level. The noise figure is then calculated from equation (13).

IF Substitution. Since r.f. attenuators are usually calibrated against i.f. standards attenuators, improved accuracy can be achieved by down-converting from r.f. to i.f. Over the dynamic range from ON to OFF the mixer will remain linear, and the above procedure can be repeated using the i.f. attenuator to resolve the $Y$ factor reading

Automatic Noise Figure Method. The methods so far discussed do not give direct reading of noise figure, so electronic processing techniques have been developed to perform automatically the steps performed manually above. The speed of operation of this system has been designed to enable direct indication of noise figure whilst optimising the system (or device evaluation for amplifier design).

Fig. 16 shows the basic diagram for a popular version of an automatic noise figure meter. The noise source is gated ON and OFF through a modulator that is operated from a master gate generator at 400 Hz. The output from the receiver unit consists of 400 Hz modulated noise. This signal is amplified and detected in a square-low detector. A sample of this modulated noise is detected in a second detector which is gated from the master generator to provide an output when the noise generator is OFF. When the noise source is ON, the detector is gated OFF. Its output is converted to a d.c. signal proportion to the noise OFF signal. This d.c. level is amplified with a suitable time constant and applied to AGC in the IF amplifier to maintain a constant (noise OFF) output to the following detector.

This detector provides an output voltage proportion input power. A delay of 200$\mu$sec is introduced to ensure that the noise source is allowed to settle before making a measurement of the ON power. A synchronous detector determines the peak to peak levels. It can be shown that the output is given by

$$V_{OUT} \propto N_{OFF} \cdot \frac{T_N}{F} \quad (14)$$

This output, when displayed on a meter gives a logarithm indication of noise figure.

During calibration, $T_N$ is set by providing an offset to $N_{OFF}$ via the AGC control.

Automatic Swept Frequency Method. With modern system requirements it has become essential to measure noise performance of amplifiers quickly and accurately across greater than octave bandwidths. With the introduction of desk-top computer controlled equipment these measurements have become feasible at low cost. The basic system is shown in Fig. 17. This in principle is a high resolution swept frequency $Y$ factor analyser.

The desk top computer controls the switching of the noise source, the frequency of the local oscillator source, and receives inputs from the power sensor. Initially the system is calibrated at the mixer interface, so that power readings are obtained with the source ON and OFF at each frequency point. The measurements are repeated with the amplifier added into the system, when a second set of power readings are obtained with the source ON and OFF. From these readings, the total noise figure can be obtained, and from a simple arithmetic manipulation by the computer, the correction factor due to the mixer can be extracted, and the gain of the amplifier can be determined.

The enhanced accuracy achievable over previous swept noise figure systems is primarily due to the fact that the computer corrects at each frequency for the noise source ENR, and the accuracy and repeatability of the power meter.

Noise Sources

For the majority of noise measurements some form of calibrated noise source is essential. Over the years, a number of different sources have been developed, each dependent for its operation on different physical principles.

Gas Sources. These are of two types (1) coaxial with a tightly wound coupling helix around the gas tube, (2) waveguide with the gas tube mounted across the waveguide. Type (1) have fallen from favour since the tightly coupled helix results in large transient spikes on the noise output, and have been known to damage small geometry solid state devices.

Type (2), especially through the efforts of workers like Halford (7) and Olson (6), have been produced to such a high quality and reproducibility that they still form the basic noise standards for calibration purposes. These sources consist of argon filled tubes mounted across a waveguide in the E plane at angles of 10 to 15°. In the early days of tube manufacture, the ENR was based on the effective electron temperature of the plasma for the particular tube radius pressure product, but the measurements by Olson (6) showed these results to be approximate. Special attention was paid to dimensional properties of the glass envelope to ensure consistent wall thickness, concentricity and centralisation within the waveguide. Quartz was initially chosen for the glass envelope because of its low loss properties, but eventually boronsilicate glass was chosen to provide a low change of ENR with frequency.

Tube diameters and pressure were optimised for each frequency range, to ensure excellent VSWR in both the on and off conditions (less than 1.05). Tube currents were selected so that operating current was above a critical current limit below which unstable plasma operation occurred causing impedance modulation at low frequencies. The operation above this limit also resulted in small changes of ENR with tube current.

All these precautions and controls resulted in gas sources with excellent stability, less than 0.005 dB over 2 years, and excellent correlation to NBS standards, less than 0.02 dB.

The use of a "hot" plasma does result in considerable heating of the mount, especially since typical maintaining voltages and currents are 86 volts and 150 mA. This dissipation elevates the ambient temperature of the cold load, causing errors in to 290°K reference.

Solid State Sources. These sources consist of a semiconductor diode driven into avalanche by the application of a low voltage (typically 20 to 28 volts) pulse and maintained from a constant current source of typically 5 mA. These diodes are capable of high ENR's (40 dB), and by providing high quality attenuating elements the ENR can be reduced to 15.5 dB and at the same time provide excellent ON to OFF VSWRs, typically less than 1.1:1. Little has been written about the stability of these sources, and, due to lack of evidence the solid state source has received a poor reputation for accurate noise figure measurements. However, in the authors experience over many years of having these sources checked against gas standards, the variations have been typically less than 0.05 dB (includes operator measurement repeatability). In addition these sources do not exhibit large transient overshoots.

Hot/Cold Standards. These hot/cold standards form the reference for all noise standards and with the availability of good quality commercial hot/cold terminations they can now be seen in many noise laboratories. With the requirement to measure ultra low noise amplifiers (approximately 1 dB) they have become an essential piece of test equipment.

A typical thermal noise standard, Fig. 18, consists of a matched resistive element thermally isolated by a uniform transmission line to the interface adaptor. These resistive loads may be thin film resistive elements, or for higher frequencies, microwave load elements. The transmission lines for the isolating sections are usually fabricated from copper plated stainless steel and have distributed temperature and transmission line factors. The construction techniques for hot or cold loads are similar. The main difference is the method used to obtain temperature stability of the load element. Hot loads usually use accurate thermostatically controlled heating elements (373.1°K) whilst cold loads normally use a boiling liquid such as liquid nitrogen (77.36°K). For the cold loads a constant thermal differential is maintained between the output connector and the cryogen by heating the output end of the isolation section to provide a barrier that prevents moisture condensation.

Extensive computer modelling is utilised to established the distribution of temperatures and losses along the waveguide sections thermal with differentials. Losses have to be measured to better than 0.002 dB. For accurate measurement corrections are necessary for changes of parametric pressure and ambient temperature. Typically a typical cold load at 18 GHz,

$$\text{Noise temperature} = 85.674 + 0.01053\,\Delta P + 0.0372\,\Delta T_{01} + 0.00674\,\Delta T_{02}$$

where P = pressure differential from 760 mm of Hg

$\Delta T_{01}$ = mounting plate temperature (typically 36.5°C)

$\Delta T_{02}$ = ambient temperature (22.5°C)

For these conditions, noise temperature = 87.13°K with a VSWR of 1.03.

For the measurement of low noise temperatures, it is necessary to consider the errors introduced by the noise theory introduced by Nyquist (5). This is less than 2% at 10 GHz.

For a commercial source the accuracy of the calibration noise temperature is better than ± 1.5°K.

Basic Pitfalls of Noise Measurements

(1) Never rely on the general calibration provided by a noise source manufacturer. Experience has shown that variations of 1 dB can be obtained easily whilst using noise source with "on the body calibrations".

(2) Mismatch effects can cause very significant errors. In particular, the calibrated noise source is calibrated to give an ENR when matched into a conjugately matched termination. In practice most sources having VSWR's less than 1.05 will provide minimal errors when operated into good VSWR's. However with a mismatched terminating load (for example, a receiver with a VSWR of 2.0:1) the error could be ± 0.2 dB. See Fig. 19.

For device measurements where VSWR's of greater than 5.0:1 can be obtained, it is essential to provide isolation between the noise source and the device being measured. This effect of interactive VSWRs is further complicated where the noise source VSWR's vary between the ON and OFF states.

Where mixers are used, care should be taken to ensure that the mixer sees a constant source impedance, since the mixer image noise is dependent on the image termination. The use of an isolator or broadband attenuator will aid to maintain a constant source impedance. This is particularly important if filters are included in the system.

(3) Spurious responses in amplifiers can lead to erroneous noise figures.

(4) Never rely on sweeper sources as local oscillator sources without checking the measured noise performance against a known stable source. Sweeper sources invariably have poor f.m. performance which cannot be rejected by the use of balanced mixers (latter provides AM rejection only).

(5) Many arguments can be initiated by discussing single sideband and double sideband noise figures. These arguments are more related to systems performance. For amplifiers, in isolation the same noise figures are measured provided sensible use is made of image rejecting filters, or image rejecting mixers are used. Fig. 20 shows a typical example where 3 dB differences can be measured (having corrected for filter losses). For many systems (other than radiometers) only a single response of the receiver is used. Providing the amplifier gain is adequate a filter will ensure s.s.b. operation.

(6) For d.s.b. measurements choose an i.f. frequency that ensures the gain at both upper and lower sidebands are equal. For most microwave systems, 30 MHz i.f. is adequate.

(7) Ensure that the noise measuring system has adequate dynamic range. Calculate the maximum output signal level with the noise source ON and compare this level with the CW 1 dB compression point of the system. Ensure that the operational level is at least 10 dB below the compression point. This is to prevent noise clipping with consequent reductions in measured Y factor.

## Measurement Accuracies

(1) VSWR errors can be minimised, but still remain a significant source of uncertainty. Tedious measurements of the magnitude and phase of the reflection coefficients can reduce this error further (8) (9) (10). In practice error typically ± 0.15 dB.

(2) Noise source calibrations traceable to noise standards ± 0.1 dB.

(3) Instrumentation errors include ability to reset to a reference level, system backlash, drift, and reference attenuator accuracy. Typically, for a normal measurement this can be ± 0.05 dB. Averaging can reduce this error.

(4) Large errors can result in measuring noise figure by Y factor method using cold load reference, due to low value of Y factor.

(5) Conventional automatic noise figure equipment (not computer controlled) have errors in the automatic mode, due to non square law characteristics, and duty cycles not equal to 0.5. These can be reduced by Y factor calibration.

(6) Correction should be made for ambient temperature deviations from 290°K. See Fig. 21.

(7) Second stage correction should be made using Friis (11) correction, but error is dependent on level of correction. Error could be ± 0.1 dB.

Table 1 shows accuracy for a number of commercial measurement systems.

Fig. 22 shows the results obtained for a 12 GHz amplifier using solid state and cold load measurements.

## AMPLIFIER STABILITY

Amongst the many demands made on the microwave amplifier designer to meet exacting specifications for noise figure, gain responses, bandwidth, linearity etc. the one parameter most easily overlooked, both from the designers point of view and from the customers point of view is amplifier stability. Even the best amplifier can be rendered useless to the system integrator if, as the result of combining the amplifier with filters or waveguide sections the whole amplifier bursts into oscillation, or even generates low levels of spurious signals. Although these signals may be generated well away from the operating frequency, intermodulation within the amplifier can result in signals appearing within the operational bandwidth.

The problem of instability results from the device inherent characteristics. The input and outputs of active devices are not isolated, and reverse feedback exists both within and outside the transistor. The latter may be caused by package parasitics or even by the mounting of the device into a poorly designed environment.

Traditionally, the measure of a devices stability has been defined by the Rolletts stability factor (K), or Inverse Linvill factor, C, where $K = \frac{1}{C}$.

This is defined as

$$K = \frac{1 + |S_{11}S_{22} - S_{12}S_{21}|^2 - |S_{11}|^2 - |S_{22}|^2}{2\,|S_{12}|\,|S_{21}|} \qquad (15)$$

When K < 1 the device is potentially unstable, when a value of passive load placed at one or both of the device terminals will cause oscillation. The definition of K does not indicate whether the source or load or both is the potentially unstable port. However, the plotting of stability circles will establish which ports are unstable. When $K \geqslant 1$, the device is unconditionally stable, and any passive load placed at either of the device terminals will not cause oscillation.

Most active devices exhibit instability at the lower frequency end of the spectrum, but it should not be assumed that instabilities cannot occur at higher frequencies. The latter occurs due to reducing isolation both of the device and the enclosures. Changes to d.c. operating points can cause the marginally stable device to become unstable, and reducing temperature also aggrevates the device stability.

Whilst it is essential to ensure $K \geqslant 1$ for an amplifier, experience has shown that amplifiers satisfying this condition can be made oscillatory if a second factor, KK, is less than zero.

Here

$$KK = 1 + |S_{11}|^2 - |S_{22}|^2 - |S_{11}S_{22} - S_{12}S_{21}|^2 \qquad (16)$$

When $K < 1$, the device may exhibit an input or output that is of negative impedance, but may not be oscillatory since the combination of the termination resistance and device resistance is still positive. This may be particularly true when the amplifier is embedded between 50Ω terminations.

Alternatively the device may exhibit a real input impedance when terminated in a 50Ω load, but may become a negative impedance when the load is replaced by a variable short circuit.

These combinations demonstrate the necessity to satisfy further conditions, namely

$$|\Gamma_{SOURCE}| . |\Gamma_{IN}| < 1 \qquad (17)$$

$$|\Gamma_{LOAD}| . |\Gamma_{OUT}| < 1 \qquad (18)$$

These expressions are familiar to those who have attempted to design oscillator circuits.

These expressions form the basis of a simple test to establish amplifier stability. Since the worst case values for $|\Gamma_{SOURCE}|$ and $|\Gamma_{LOAD}|$ are unity for passive loads, then $|\Gamma_{IN}| < 1$ and $|\Gamma_{OUT}| < 1$ must be satisfied for the complete range of phases of short circuits at the load and source ports respectively.

Any wide band reflectometer system can be utilised to perform these measurements. Firstly observe $\Gamma_{IN}$ across the frequency band whilst adjusting the phase of a short circuit connected to the amplifier output. Ensure that $\Gamma_{IN}$ remains less than the level of the short circuit calibration level ($\Gamma = 1$). Reverse the amplifier, and observe $\Gamma_{OUT}$ whilst adjusting the phase of a short circuit connected to the amplifier input. Again $\Gamma_{OUT}$ should remain less than unity.

Fig. 23(a,b) shows these test configurations and Fig. 24 shows the results of tests on a low noise 2.1 GHz amplifier showing potential instability within the operating band.

An alternative test configuration, that can be a useful indicator of oscillation, but which may not define the offending frequencies is shown in Fig. 25(a,b).

In the context of stability it is interesting to consider the additional benefits to be obtained from balanced amplifier configurations in terms of stability.

For such an amplifier the overall S parameters can be represented by,

$$S_{11} = e^{-2j\phi}\left(t^2 S_{11}(a) - (1-t^2)\, S_{11}(b)\right)$$

$$S_{22} = e^{-2j\phi}\left(t^2 S_{22}(a) - (1-t^2)\, S_{22}(b)\right)$$

$$S_{21} = j e^{-j2\phi}\, t\sqrt{1-t^2}\,\left(S_{21}(a) + S_{21}(b)\right)$$

$$S_{12} = j e^{-j2\phi}\, t\sqrt{1-t^2}\,\left(S_{12}(a) + S_{12}(b)\right)$$

where subscripts a, b represent the amplifiers within the balanced configuration, t represents the coupling, $\phi$ is an arbitrary transmission phase angle.

For 3 dB couplers, $t = 1/\sqrt{2}$, so these equations reduce to

$$S_{11} = e^{-2j\phi}.\tfrac{1}{2}.\left(S_{11}(a) - S_{11}(b)\right)$$

$$S_{22} = e^{-2j\phi}.\tfrac{1}{2}.\left(S_{22}(a) - S_{22}(b)\right)$$

$$S_{21} = j e^{-2j\phi}.\tfrac{1}{2}\left(S_{21}(a) + S_{21}(b)\right)$$

$$S_{12} = j e^{-2j\phi}.\tfrac{1}{2}.\left(S_{12}(a) + S_{12}(b)\right)$$

When a = b, for identical amplifiers,

$$\left.\begin{array}{l} |S_{11}| = 0 \\ |S_{22}| = 0 \end{array}\right\} \text{matched}$$

$$|S_{21}| = |S_{21}| \quad \text{a or b}$$

$$|S_{12}| = |S_{12}| \quad \text{a or b}$$

Therefore,

$$K = \frac{1 + (|S_{12}|.|S_{21}|)^2}{2\,|S_{12}|.|S_{21}|}$$

Let $P = |S_{12}|.|S_{21}|$.

Therefore $K = \dfrac{1+P^2}{2P}$

This expression has a minima of 1 when P = 1, that is when $|S_{21}.S_{12}| = 1$, and $K > 1$ for all other values of $|S_{12}|.|S_{21}|$.

Table 2 shows the S parameters at 2.1 GHz for a typical single ended amplifier.

| | Real | Imaginary |
|---|---|---|
| $S_{11}$ | -0.699 | -0.348 |
| $S_{22}$ | 0.309 | 0.459 |
| $S_{12}$ | 0.009 | 0.015 |
| $S_{21}$ | -10.907 | 7.895 |

Table 2

From this data, K = 0.48 and this amplifier is potentially unstable. Table 3 shows the overall amplifier S-parameters for a balanced module.

Table 3

| | Real | Imaginary |
|---|---|---|
| $S_{11}$ | 0 | 0 |
| $S_{22}$ | 0 | 0 |
| $S_{12}$ | 0.009 | 0.015 |
| $S_{21}$ | -10.907 | 7.895 |

From this data, K = 2.24, and the amplifier has become unconditionally stable. This remains true within the bandwidth of the quadrature couplers.

## GROUP DELAY MEASUREMENTS

Ideally a network should be capable of transmitting a signal with zero distortion. This necessitates that over the frequency interval of interest, taken to be the band containing the signal energy, that the amplitude of the transfer function must remain constant and that the phase of the transfer function must be linear with frequency (non dispersive). Such a media could be a well terminated air coaxial line, where the phase of a single frequency passing along the line is directly related to the signal frequency and the propagation velocity. A plot of frequency versus phase would be a straight line dependent on the propagation velocity and the effective electrical length of the path. Equation (19) determines the steady state relationship between the input and output signal (phase delay).

$$t_o = -\frac{\phi}{\omega} \qquad (19)$$

A convenient measure of a networks phase linearity is the group delay, $t_g$ , equation (20) which determines the propagation delay of signal energy (envelope delay).

$$t_g = -\frac{\partial\phi}{\partial\omega} \qquad (20)$$

For a linear phase network, equation (19) and (20) become identical, and a constant group delay results.

For a practical network, such as an amplifier or filter, with a band limited response, the transfer characteristic will exhibit deviations from linear phase (due to internal VSWR's), or alternatively deviations from constant group delay. These deviations from linear phase result in different frequency components traversing the unit with different phase delays and hence differential time delays resulting in intersymbol interference due to broadening of pulses in digital transmission or co-channel interference in systems carrying a number of FM or television channels.

Consider an amplitude modulated signal being applied to a network.

Then

$$V_{IN}(t) = A(1 - m \sin W_s t) \sin W_c t.$$

where $m$ = modulation index

$f_s$ = modulation frequency, $\omega_s = 2\pi f_s$.

$f_c$ = carrier frequency, $\omega_c = 2\pi f_c$.

Expanding,

$$V_{IN}(t) = A \sin \omega_c t + \frac{mA}{2} \cos(W_c - W_s)t - \frac{mA}{2}\cos(W_c + W_s)t.$$

that is, carrier plus two sidebands.

Assuming constant group delay ($t_g$) from $(W_c - W_s)$ to $(W_c + W_s)$

$$V_{OUT}(t) = B \sin(W_c t + W_c t_g) + \frac{mB}{2}\{\cos\{(W_c - W_s)t + (W_c - W_s)t_g\}\} - \frac{mB}{2}\{\cos\{(W_c + W_s)t + (W_c + W_s)t_g\}\}.$$

This reduces by manipulation to

$$V_{OUT}(t) = B\left[1 + m \sin(W_s t + W_s t_g)\right] \times \sin(W_c t + W_c t_g)$$

therefore,

envelope phase shift, $\theta_s = W_s t_g$

or $t_g = \theta_s / \omega_s$.

or
$$t_g = \frac{\phi_s}{360 \times f_s} \qquad (21)$$

where $t_g$ = group delay in seconds

$\phi_s$ = modulation envelope phase shift in degrees

$f_s$ = modulation frequency in Hz.

As a physical interpretation of these expressions, again consider the carrier with angular velocity $\omega_c$ and modulated at frequency $f_s = \omega_s / 2\pi$ . This modulated signal possesses sidebands at $\omega_c + \omega_s$ and $\omega_c - \omega_s$ . When passing through a network with an envelope delay of $t_g$ seconds, the upper and lower sidebands are symmetrically shifted in phase in opposite directions with respect to the carrier by $\omega_s t_g$ radians. This produces a phase shift in the modulation envelope of $\omega_s t_g$ radians, where these radians are now radians at the modulation frequency. Thus, if the carrier frequency sidebands of an amplitude modulated wave are shifted symmetrically by $10^{\circ}$ with respect to the carrier frequency, then the modulation envelope of the wave at the output of the network is likewise shifted in phase by $10^{\circ}$. After detection this envelope delay appears at a $10^{\circ}$ phase delay of the modulation output from the detector.

Hence from the fundamental definition of group delay and the corresponding modulation delay, two basic measurement methods are possible.

### Carrier Increment Method

By incrementing the carrier frequency and measuring the corresponding phase shift, the incremental delay can be derived. The group delay so determined represents the average linear delay between the frequency steps. The size of the frequency steps chosen is dependent on the delay resolution required.

This in turn is limited by the phase resolution capabilities of the measuring equipment at the carrier frequency. Fig. 26 shows the trade offs. These measurements are easily performed on a network analyser using a desktop computer when many frequency points can be measured and the computer used to calculate and plot the corresponding group delay. For 10 MHz steps system resolution should be better than ± 0.05 nseconds, but in practice residual variations of approximately ± 0.15 nseconds can be achieved. This may be due to phase jitter caused by the use of a sweeper source. The use of a synthesiser could reduce this to ± 0.1 nseconds. Fig. 27 shows the group delay response for an amplifier, and for comparison the group delay for a coaxial air line shows the limits of the measuring system.

Amplitude Modulation Method

This method consists of measuring the envelope phase shift of an amplitude modulated signal when transmitted through the network. This is an extension of a method suggested by Nyquist and Bland (12). Using this technique the carrier frequency can be simultaneously swept across the operating bandwidth of the network. Fig. 28 shows a typical test configuration.

The carrier modulation is derived from a low distortion sinusoidal source (with less than 0.5% distortion), which together with a d.c. bias is used to modulate a PIN diode. A sample of the input modulation and the detected modulation is fed to a vector voltmeter. The latter enables the use of modulation frequencies from 1 MHz to 1 GHz, or for larger group delays 20 KHz can be used and the detected signals fed directly to the IF phase comparator within the voltmeter. Initial calibration is performed without the amplifier using the calibrated phase offset. The choice of modulation frequency is again determined from Fig. 26. This is a measure of the smallest bandwidth within which fluctuations in delay can be measured. For reference, a phase shift of $1^{o}$ with 2.78 MHz modulation results in a 1 nsecond delay. With care this method can resolve to ± 0.10 nsec.

Frequency Modulation Method

Some systems are operated with hard limiting, which eliminates a.m. modulation. This can be overcome by the use of a low modulation index frequency modulation giving rise to a frequency spectrum similar to amplitude modulation. The a.m. detector is then replaced by a discriminator.

AM and FM methods can be used to measure group delay in frequency translating systems.

Some specifications for amplifiers actually specify a maximum phase deviation from linearity instead of specifying a group delay deviation. These requirements can be readily demonstrated by using a high quality airline to offset the nominal linear phase variation with frequency. The resultant display shows the deviation from linear phase. This procedure can be readily implemented on a network analyser by offsetting the phase in the reference channel. Fig. 29 shows a typical response.

The variation from constant group delay can be represented in a quadratic expression which will include a number of coefficients. These will include a linear component (nS/MHz), a parabolic component (nS/MHz$^2$) and a ripple component (nS peak-to-peak) after extracting the other components.

Accuracy Considerations

(1) It is essential to ensure that VSWRs are reduced to a low level, and to this end the amplifier should be enclosed between high quality attenuators.

(2) The modulation frequencies should be accurately measured since this accuracy relates directly to the group delay accuracy.

(3) Spurious signal from the signal generator should be maintained at least 40 dB below the fundamental carrier level.

(4) Noise within the system limits accuracy. This noise can arise from signal source residual FM, the noise performance of the amplifier under test, or even the phase detector noise. As an example, for a signal-to-noise of 60 dB, phase error is ± 0.06$^{o}$ which for a 2.78 MHz modulation corresponds to a group delay error of ± 0.06 nseconds.

(5) Averaging due to the choice of too high a modulation frequency. This is dependent on the ripple in the true group delay ripple characteristic. For example if the sampling width is 5% of the ripple periodicity the error in the group delay measured is 1.2%.

MEASUREMENT OF AMPLIFIER NON-LINEAR TRANSFER CHARACTERISTIC

When a Class A or AB operated amplifier is driven with an increasing input signal level, a level will be reached above which the transfer characteristic becomes non-linear. Continued increase of signal will result in the output level becoming saturated. The amplifier is now said to be operating under large signal conditions. This non-linear transfer characteristic gives rise to a number of new parameters.

COMPRESSION POINT

This parameter is a measure of the deviation from linearity. It is defined as the point at which the output level power is 1 dB below the extrapolated linear characteristic. A comparison of the 1 dB compression point for different amplifiers enables the system designer to optimise a complete system for optimum linear performance. Thus the output stages will exhibit the highest compression point. The compression point is dependent on frequency and will vary within the amplifier bandwidth, especially for wideband amplifiers.

This parameter can be readily measured by a number of methods:

(a) Using a ramping waveform to increment the output power from a source, the input power to the amplifier and the output power can be monitored on power meters and a recording produced of Power Input versus Power Output.

(b) By incorporating a desk top computer into the system, the measurement can be automated to increase the power level by known amounts whilst monitoring the output levels and calculating the point at which the deviation from linearity is 1 dB.

(c) By adapting a swept display system used for reflectometer applications, and calibrating the detector for absolute power levels against power meters, a system can be produced that can provide a single frequency indication of the power transfer characteristic or with small modifications can produce a swept frequency display of the compression characteristic. Care must be taken to correct for power levels if a modulation system is used. This method has the advantage that the display is in real time, and hence enables amplifiers to be optimised. The inclusion of a data normaliser will enable the small signal characteristic to be stored, and the true gain compression can be determined as the input power is increased. Fig. 30(a) and (b) show a CW system and swept system, respectively.

## INTERMODULATION PRODUCTS

Modern satellite systems are utilising multiple carrier systems to system capacity. These systems place increasing demands on system linearity. Berretta et al (13).

When two or more carriers are fed through a non-linear amplifier, interaction of the carriers and/or their harmonics create additional frequency components at the output. These components will appear in adjacent channels, thus reducing the signal-to-noise performance in that channel. These products are known as intermodulation. With a non-linear system the classical approach is to provide "back-off" to improve intermodulation performance. This requires the output power to be reduced below the 1 dB compression point to improve the linearity and hence reduce the intermodulation products. In general, these intermodulation products are given by

$$f_{OUT} = \mp n_1 f_1 \mp n_2 f_2 \mp \cdots n_i f_i$$

where $n_i$ = integer constant

$f_i$ = input frequency components

$\Sigma n$ = order of intermodulation product

An example of a third order product may be

$$f_{OUT} = 2f_2 - f_1 \text{ or } f_{OUT} = \mp f_1 \mp f_2 \mp f_3.$$

The most common product is the third order. When signals $f_1$, $f_2$ are applied to the input of the amplifier, output components appear predominantly at frequencies $2f_1 - f_2$ and $2f_2 - f_1$, due to intermixing with the second harmonic products. These third order products are measured with respect to the fundamental input signals at the output of the amplifier. Fig. 31(a).

Consider a single carrier input causing just detectable peak distortion. When two carriers are present, there will be periods of time when the signals phases reinforce, and periods when they cancel. This period occurs at the difference frequency. At points of reinforcing, the voltage is doubled, and in order to avoid clipping each signal must be reduced by 6 dB. Hence there is a total power reduction of 3 dB (3 dB back-off). When more carriers are introduced the probability of the phases causing total reinforcement are reduced and the rate of back-off diminishes until more than six carriers causes little change.

For multiple carrier inputs, the intermodulation levels are determined by arranging for a window with no fundamental signal level, and then determining the sum of the products in this channel at the output of the amplifier. Fig. 31(b). As the input signal level is varied, so the intermodulation products vary, and the concept of an intercept point is introduced. At this power level the fundamental and intermodulation products are equal in amplitude. This is a theoretical level, since gain compression and saturation occurs, but the principle is useful to determine intermodulation levels at different drive levels. The intercept point is different for each order. For the third order the slope of the intermodulation product is 3 dB/1 dB, so that a differential rate of 2 dB/1 dB occurs in the product levels relative to the carrier.

Fig. 31(c) shows a typical system for two carrier measurements. The following cautions should be noted:-

(1) Sources should have low harmonic content.

(2) Isolation between sources is essential to avoid intermodulation products between the sources themselves. The use of a high directivity coupler and isolators are sufficient. Note their positions in the test configuration.

(3) Lack of isolation can cause one source to lock to the other source through the ALC loop when the carrier spacings are low.

(4) Non-linearities in the spectrum analyser cause their own intermodulation products. These can be observed by switching RF attenuation levels on the spectrum analyser.

(5) For high dynamic range measurements, the spectrum analyser may not have sufficient range. Considerations of the analysers spurious free range must be considered. When the measured intermodulation products are close to the noise baseline, the product levels will be added to the noise and cause erroneous levels. Fig. 32 shows the results of multicarrier tests on an amplifier.

## AM TO PM CONVERSION

For many systems, especially multicarrier systems, operated near to the saturation level of the amplifier to maintain system efficiency, another equally important parameter is AM to PM conversion. As the amplifier is driven into saturation, not only does the gain reduce but the transfer phase is changed. As a result the AM-PM can be defined as the change of transfer phase for an increment in the input power of 1 dB, and expressed in degrees/dB. This is a function of the input power.

A simple static method of measurement is to measure the amplifier change of phase on the network analyser as the drive level is changed, but the method suffers from changes of phase in the test equipment as the drive levels are changed. An alternative approach is to use a dynamic method. Here the carrier is amplitude modulated with a peak to peak amplitude of 1 dB and a phase noise meter used as a discriminator to determine the phase change. Fig. 33 shows such a test schematic used for AM-PM measurements on a 12 GHz amplifier. Fig. 34(a,b) shows results from static and dynamic measurements.

## E.M.C. (ELECTROMAGNETIC COMPATIBILITY)

This is becoming an increasingly important consideration, not only for systems, but also for the amplifiers within the system. The primary items that require consideration are

(1) Conducted susceptibility.

(2) Conducted emission.

(3) Radiated susceptibility.

(4) Radiated emission.

These items are self explanatory. (1) examines the performance of the amplifier whilst being exposed to ripple or transient signals injected into its primary supply lines, whilst (3) examines the amplifier performance when it is situated in an r.f. field. Items (2) and (4) establish the levels of emissions along the primary supply lines and radiated from the unit. These, may in turn, cause susceptibility problems to adjacent equipments. The methods adopted for these tests are well documented in MIL-Specifications, but a revision of terminologies is considered.

### Field Strength

The measurement of field strength is the measurement of the electric field in volts/meters as given in equation (22):

$$E = \sqrt{120\pi P} \qquad (22)$$

where $E$ = rms value of field strength in volts/meter

$P$ = power density in watts/meter$^2$

$120\pi$ = impedance of free space in ohms

### Power Density

The field strength is proportional to the square root of the power density. The latter represents the power flowing through a unit area on the surface of a sphere surrounding the radiating point source of power.

$$P = \frac{P_T}{4\pi r^2} \qquad (23)$$

where $P$ = power density in watt/meter$^2$

$P_T$ = transmitted power in watts

$r$ = distance in meters

### Effective Aperture

This represents the effective "catching" area of an antenna. For an isotropic antenna (receives equally from all directions), the effective aperture is $\lambda^2/4\pi$, where $\lambda$ is the wavelength, and by the definition of antenna gain the effective aperture for an antenna of gain $G$ is

$$A_r = \frac{G\lambda^2}{4\pi} \qquad (24)$$

where $A_r$ = effective aperture, metre$^2$.

$\lambda$ = wavelength in meters

### Antenna Factor

To convert the voltage reading at the terminals of the antenna (the point at which the receiver is connected) to field strength a conversion factor is added (in dB) to the terminal voltage. This is called the antenna factor, and is supplied by the antenna manufacturer for EMC applications. The units may be dB/meter.

For example, $E = KV$

where K = antenna factor, dB/meter

V = voltage at terminals, volts

or E dB volts/meter = K dB/meter + V dB volts (25)

Alternatively

E dBμV/meter = K dB/meter + VμdB V

represent more realistic measured parameters. For a 50 Ω system, a conversion factor is required from μV to the more conventional power level in dBm.

Thus, dBμV = dBm + 107 dB

For a practical measurement of emission from a unit the antenna is positioned a defined distance from the unit, and the antenna is connected to a spectrum analyser. The latter indicates power levels in dBm, and with the above conversion factors will enable the field strength to be determined. Conversely, this system can be utilised to establish the field strength for radiated susceptibility tests.

For conducted susceptibility measurements a low frequency power source is connected to the supply lines either via a capacitor or via a toroidal transformer. The power level is adjusted to establish the required peak to peak voltage on the unit supply lines. A spectrum analyser connected to the output of the amplifier being tested will monitor the level of spurious signals generated along with the wanted signals.

Careful considerations should be given to the design of filters, regulators and box screening to meet the specified test conditions.

## CONCLUSIONS

This chapter has dealt with the problems faced by amplifier designers and systems engineers to establish compliance with specifications. Some theoretical background has been introduced to simplify understanding of the evaluation procedures. The subject is too extensive to be covered adequately in the notes, and references are provided for additional reading.

## REFERENCES

1. Vifian, Hugo, "Introduction to distortion parameters in linear networks", H.P. Seminar London, January 8, 1980.

2. Fitzpatrick, J., "Error models for system measurements", Microwave Journal, May 1978.

3. Anderson, Doub, Ross, Taubman, "Signal averaging improves dynamic range, resolution and accuracy of R.F. Measurements", Microwave Journal, October 1969.

4. Britton, R.H., "Memory improves swept frequency measurements", Microwave Journal, November 1973.

5. Nyquist, H., "Thermal agitation of electrical charge in conductors", Phys. Rev. Vol. 32, p.110, July 28, 1928.

6. Olson, K.W., "Measured noise temperature versus theoretical electron temperature for gas discharge noise sources", IEEE MTT-16, Vol. 9, p.640-645, Sept. 1968.

7. Halford, G.J., Denson, C.I., "Plasma noise sources of improved accuracy", IEEE MTT-16, Vol. 9, p.655-663, Sept. 1968.

8. Otoshi, T.Y., "The effect of mismatched components on microwave noise temperature calibrations", IEEE MTT-16, Vol. 9, p.675-686, Sept. 1968.

9. Woods, D., "Effect of some component tolerances and measurement errors on noise measurements", Electronic Letters, Vol. 2, No. 4, p.130-131, January 1966.

10. Halford, Merlo, Houghton, "Effect of some component tolerances and measuring errors on noise measurements", Electronic Letters, Vol. 1, No. 9, November 1965.

11. Friis, H.T., "Noise figure of radio receivers", Proc. IRE, Vol. 32, No. 7, p.419-422, July 1944.

12. Nyquist, H., Bland, S., "Measurements of phase distortion", Bell Syst. Tech. Journ., Vol. 9, p.522-549, July 1930.

13. Berretta, G., Gough, R.A., Guijarro, J., "Predicting amplifier performance for multicarriers using computer simulation" Microwave Journal, July 1975.

QUESTIONS

(1) The general expression for the phase response of a single tuned circuit is

$$\phi = \tan^{-1}\left(-2Q\frac{\Delta f}{f_o}\right).$$

where $Q$ = quality factor

$f_o$ = centre frequency

$\Delta f$ = deviation from centre frequency

By defining a fractional detuning, $x$, as the ratio of $\Delta f$ to the 3 dB bandwidth, show that the fractional change in group delay, $\Delta$, relative to the group delay at $f_o$, for small values of $x$ is given by $\Delta = -4x^2$.

What effects does cascading of a number of single tuned circuits have on the group delay response.

(2) An amateur radio operator wishes to determine the noise figure of his 10 GHz amplifier. It is noted that pointing the antenna/receiver combination firstly towards the sun, then towards quiet sky results in a 1 dB change in the output power from the receiver. When pointing the 1.2 metre diameter antenna at quiet sky, the antenna temperature is 30°K. Given that the equivalent noise temperature of the sun (referenced to the antenna terminals) is given by

$$T_{SUN} = \frac{FG\lambda^2}{3.468}.$$

where $\lambda$ is the wavelength in metres

$F$ is numerically equal to 270

$G$ is the numerical antenna gain

Assuming the antenna efficiency factor is 0.54, calculate the noise figure of the receiver.

(3) Calculate, using the concept of noise temperatures, the figure of merit ($G/T$ = antenna gain/system noise temperature) for a receiving system consisting of a 40 foot diameter antenna with a feeder loss of 0.15 dB operating with a receiver of noise figure, 0.6 dB. The antenna efficiency factor is 0.54 and the frequency of operation is 4 GHz. Assume that the antenna noise temperature is 40°K. If the bandwidth of the system is 30 MHz calculate the level of the minimum detectable signal level.

(4) In balanced amplifiers, one of the essential building elements is the quadrature coupler. Derive, by means of a phasor diagram, a general expression for the input VSWR and the power delivered to the isolated termination for an ideal, lossless coupler when terminated in loads with reflection coefficients ( $\rho_1, \theta_1$ ) and ( $\rho_2, \theta_2$ ).

Show that, when identical terminations are presented as loads to the coupled and direct ports, that the input VSWR is unity, and all the reflected power is returned to the isolated termination.

Calculate the input VSWR when the magnitude of the load reflection coefficients differ by 0.1, and the phase of the reflection coefficients differ by $10^o$. (Take $\rho$ mean = 0.7).

(5) In a balanced amplifier consisting of two modules with gains $G_1$, $G_2$ and noise figures $F_1$, $F_2$, it is required to determine the overall performance. Using ideal, lossless couplers, determine the total amplifier gain and noise figure.

(6) Derive from first principles the following expression for antenna factor, K, for an antenna of gain G operating at frequency f into a receiver of impedance $Z_o$:-

$$K = \frac{f\pi}{75}\sqrt{\frac{30}{Z_o G}}$$

where f is in MHz.

(7) An amplifier tested without a regulator exhibits a gain change of 1 dB/volt change of the supply line. When integrated into the system, the ripple voltage will be 100 mV peak to peak on the supply line. It is required that the sideband levels corresponding to the modulation frequency (ignore added noise components) should be less than -80 dBc. What rejection is required in the amplifier regulator.

(8) An amplifier exhibits third order intermodulation products of -30 dBc at 3 dB back off from the 1 dB compression point (+10 dBm). Determine the level of the third order products when the output power is -20 dBm.

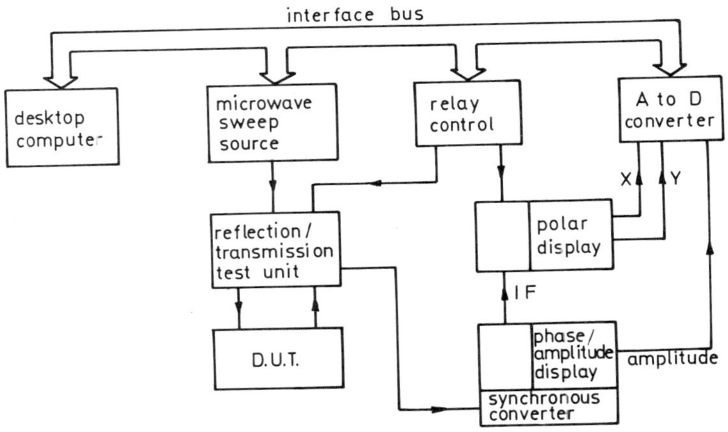


Figure 1 Block Diagram of network analyser system

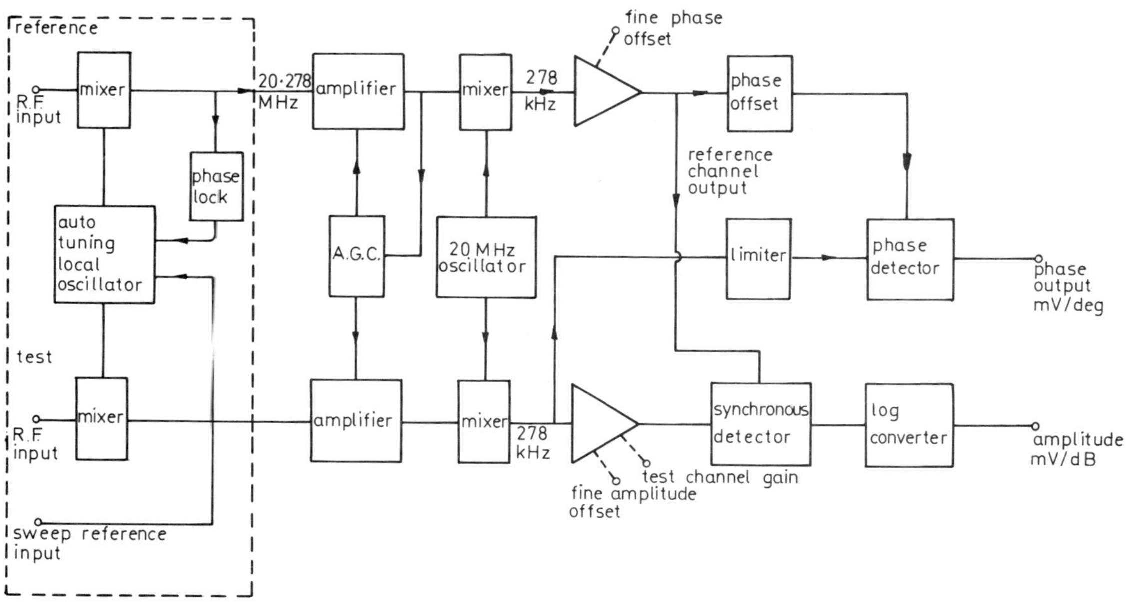


Figure 2 Block Diagram of frequency converter/amplitude/phase conditioner

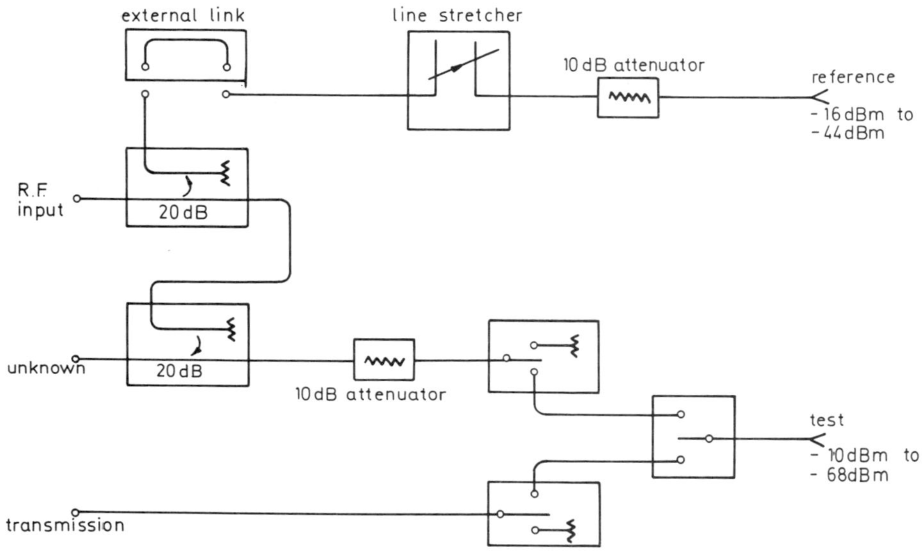


Figure 3 RF circuit layout for transmission and reflection measurements

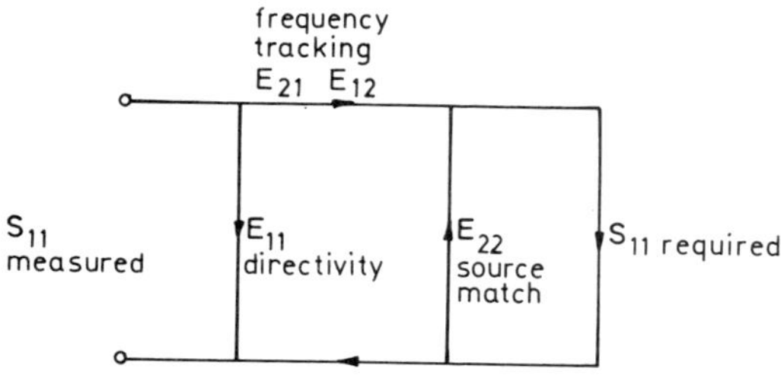


$$S_{11}\,\text{measured} = E_{11} + \frac{S_{11}\,\text{required}\ \ (E_{21}\ E_{12})}{1 - E_{22}\ S_{11}\,\text{required}}$$

Figure 4 Error model for reflection measurements

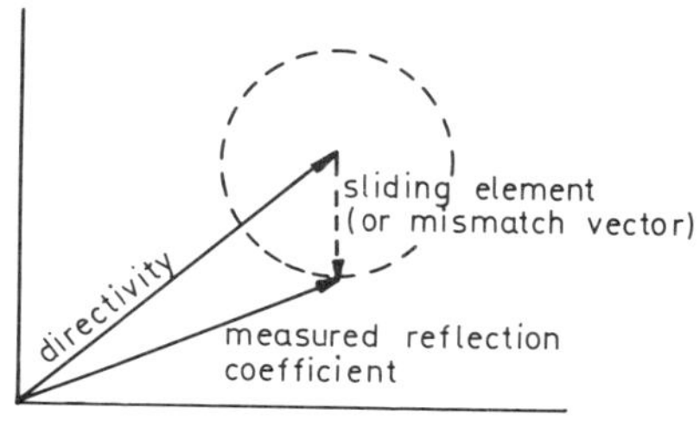


Figure 5 Phasor diagram for directivity

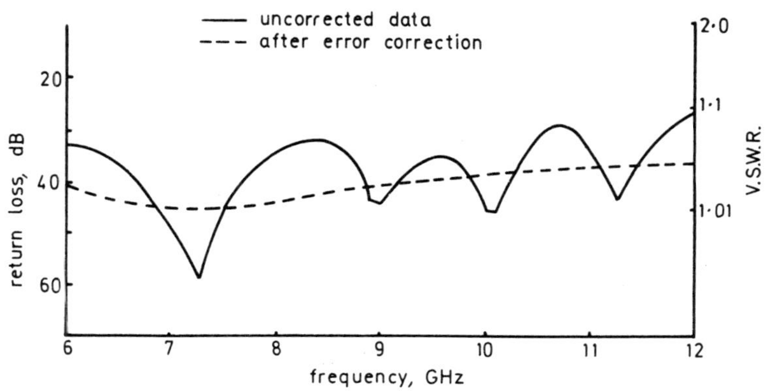


Figure 6 Measured response of a termination before and after error correction

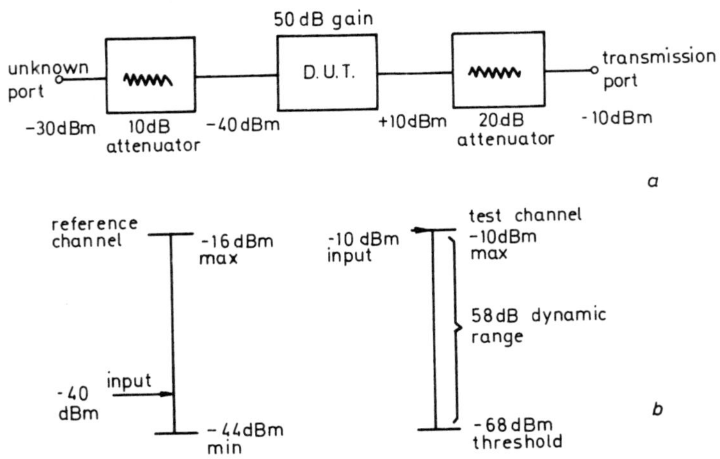


Figure 7 Dynamic range considerations in network analyser

(a) At test ports
(b) Harmonic converter operating levels

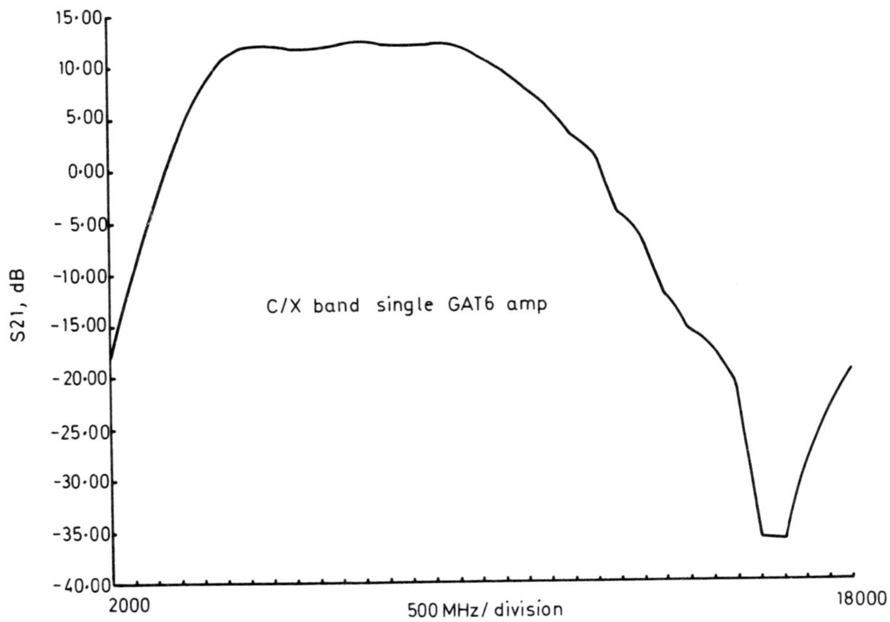


Figure 8 A.N.A. gain response of a broadband amplifier

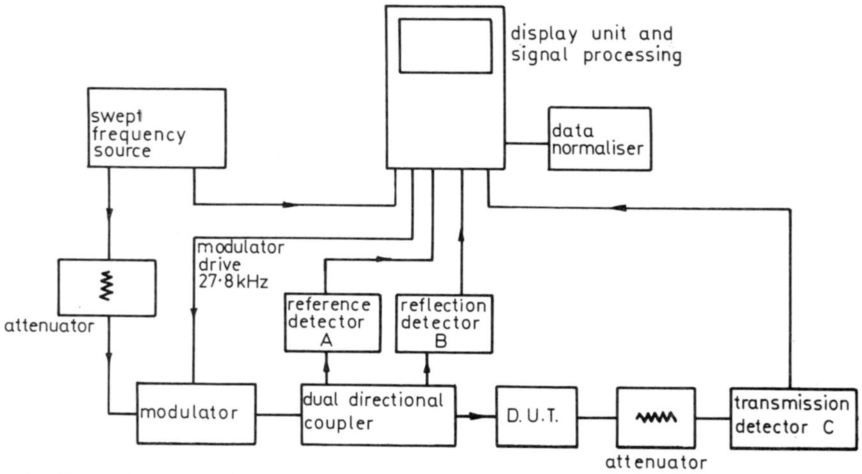


Figure 9 Swept gain/impedance schematic

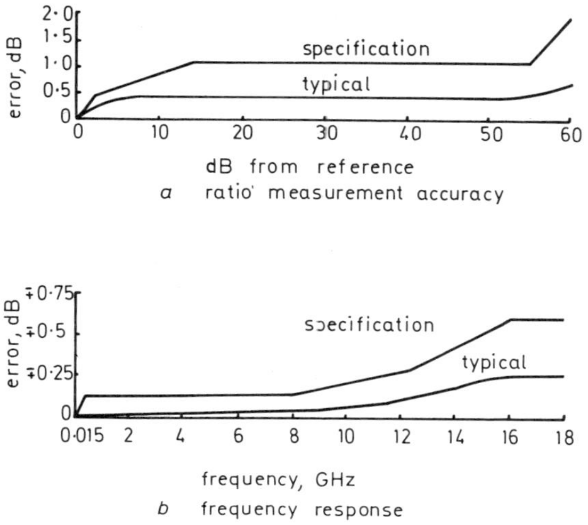


Figure 10 Accuracy of ratio measurements for gain

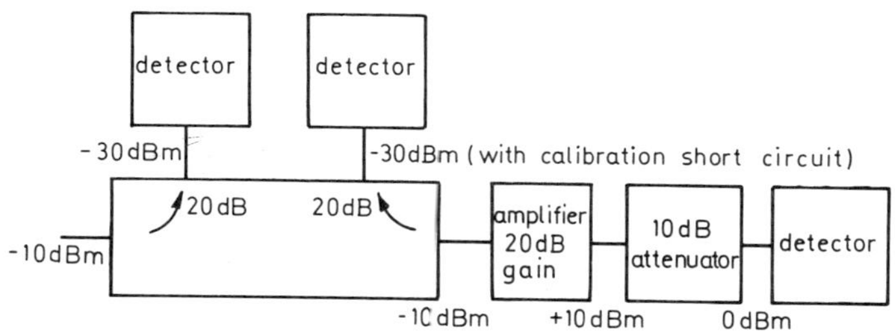


Figure 11 Dynamic range consideration in scalar analysers

(a) For a 20dB gain amplifier with compression point ≯ + 10 dBmgain can be measured but detector dynamic range is only 20dB (VSWR = 1.2)

(b) For higher gain input attenuator; required when simulataneous gain and VSWR cannot be measured.

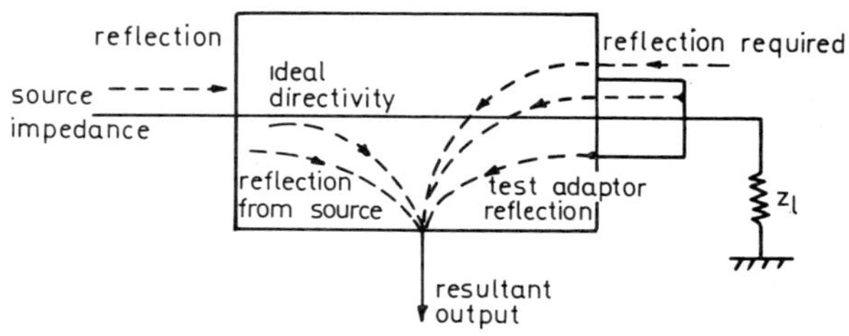


Figure 12 / Multiple reflection errors in scalar analysers

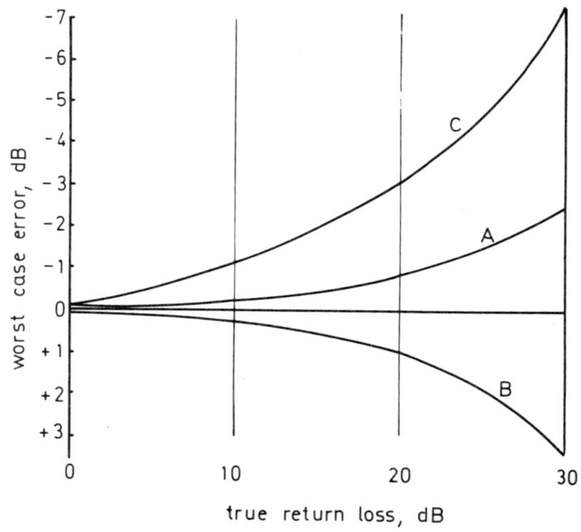


Figure 13 Limitations of measurements due to directivity and connectors on reflection coefficient

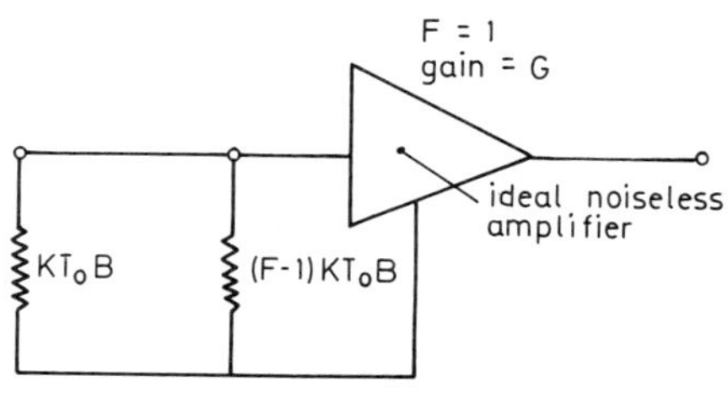


Figure 14 Noise model of amplifier

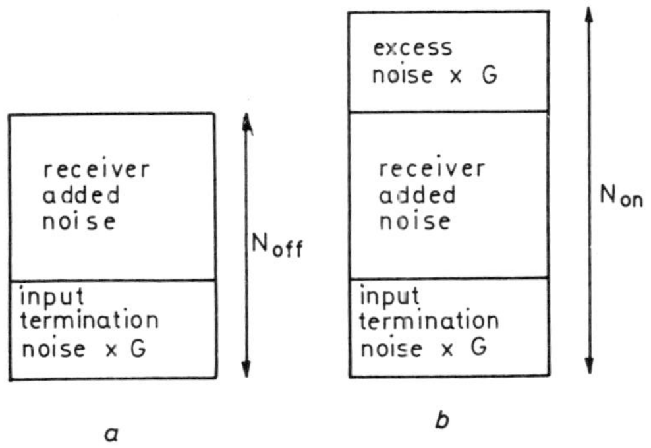


Figure 15 Added noise through amplifier

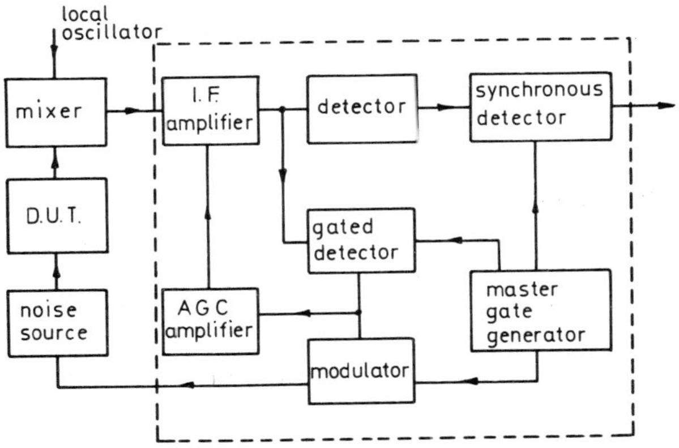


Figure 16 Block diagram of automatic noise meter

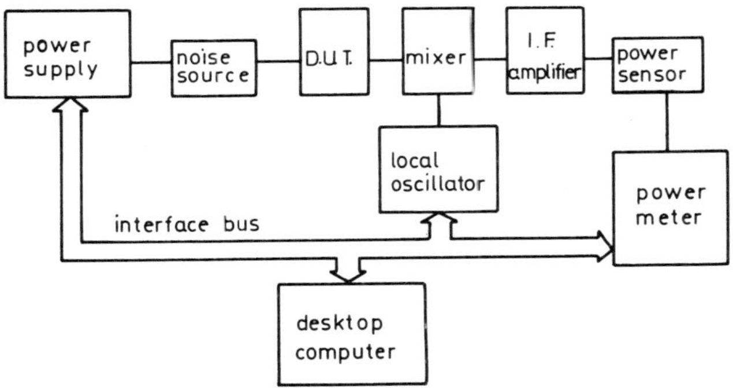


Figure 17 Block diagram of automatic swept noise system

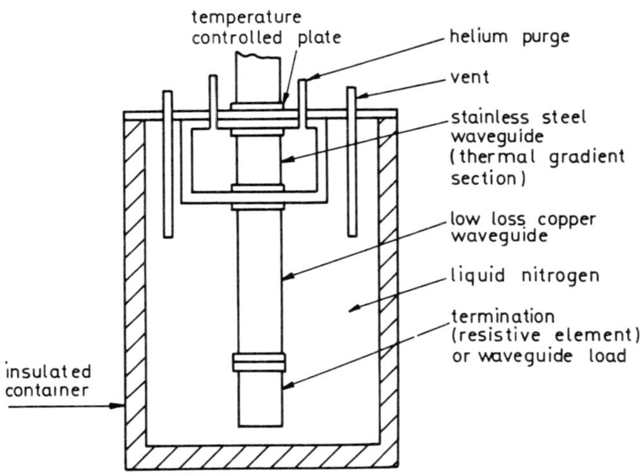


Figure 18 Schematic of cold standard

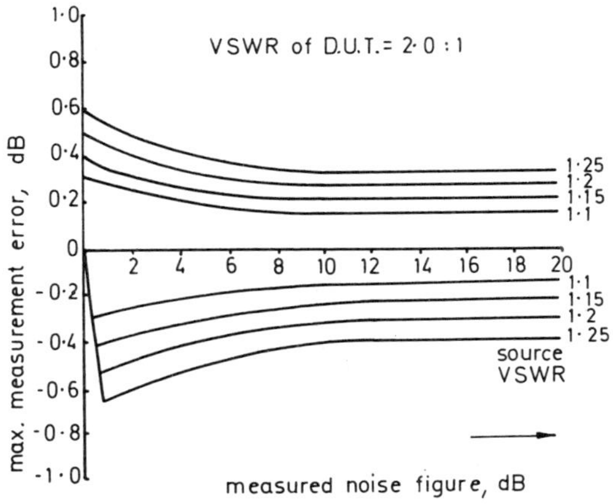


Figure 19 Noise figure errors with interactive VSWRs

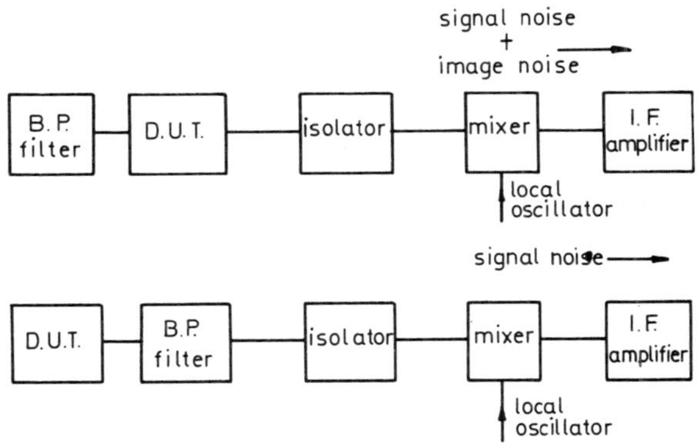


Figure 20 System optimised for single channel operation

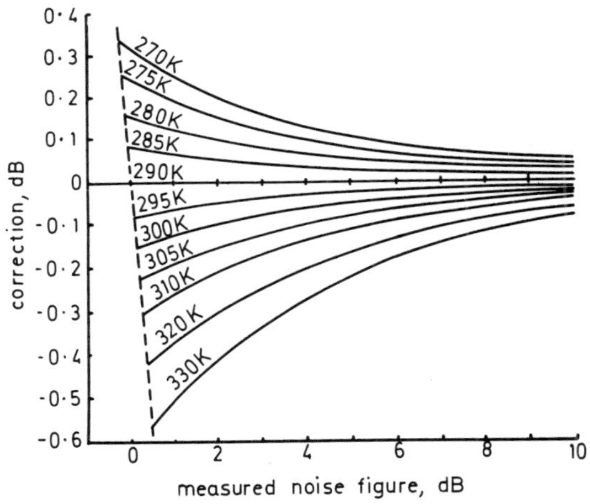


Figure 21 Noise figure correction due to ambient temperature

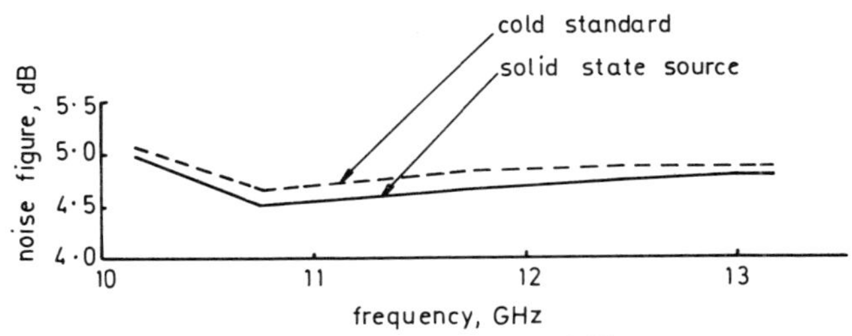


Figure 22 Noise figure for amplifier comparing cold standard versus solid state source

35dB gain amplifier using packaged devices.

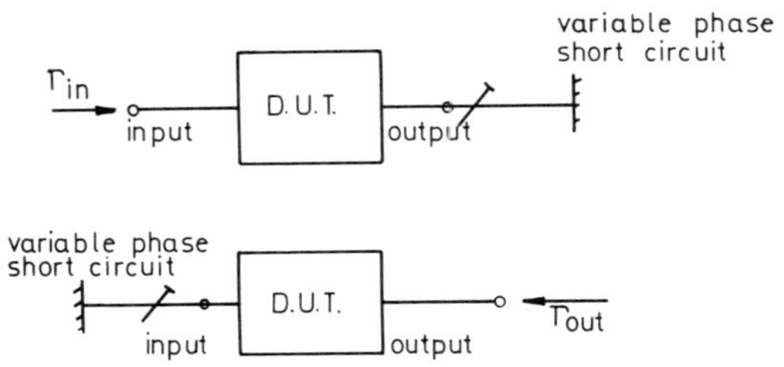


Figure 23 Stability test configurations using reflectometers

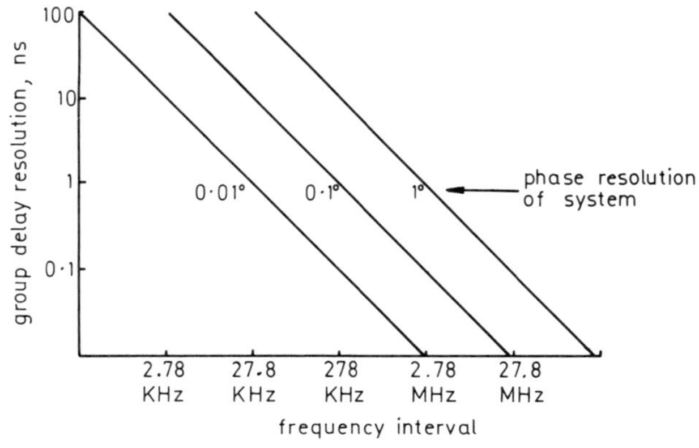


Figure 26 Trade offs for group delay measurements

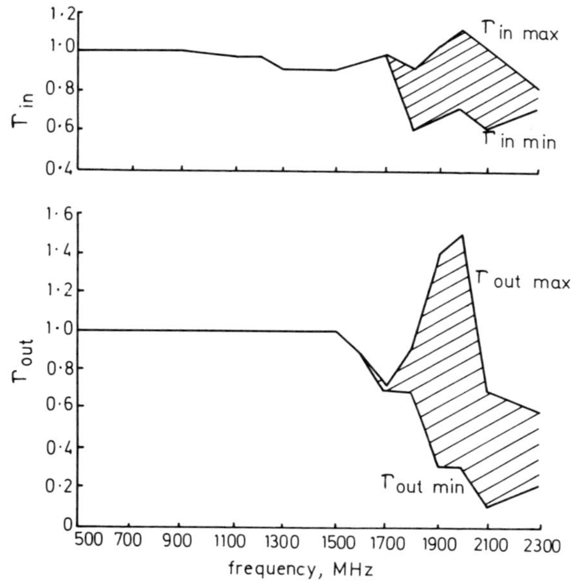


Figure 24 Plot of stability test on 2.1 GHz amplifier

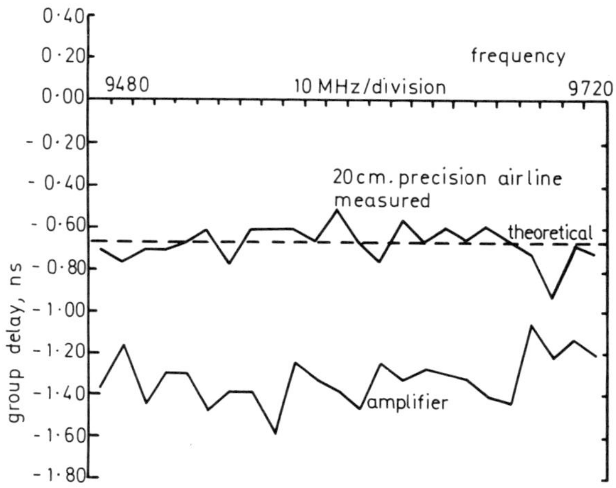


Figure 27 Group delay measurement of an amplifier

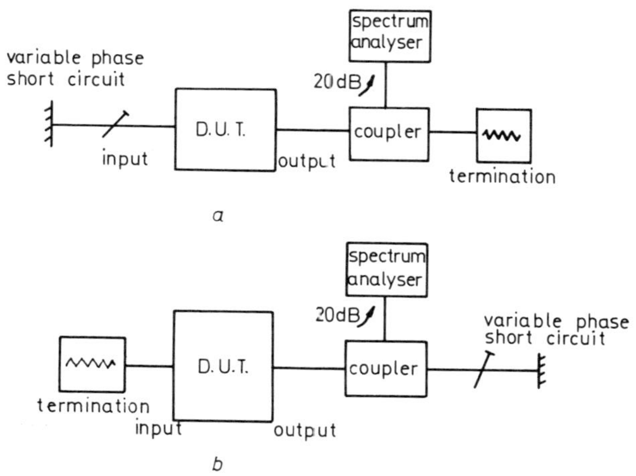


Figure 25 Stability test configuration using spectrum analysers

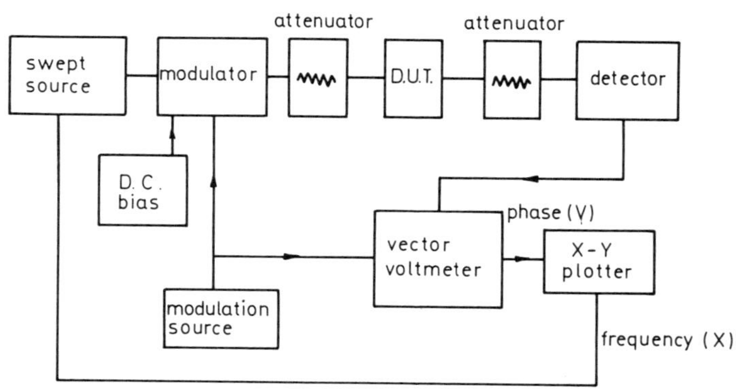


Figure 28 Group delay measurement with envelope detection

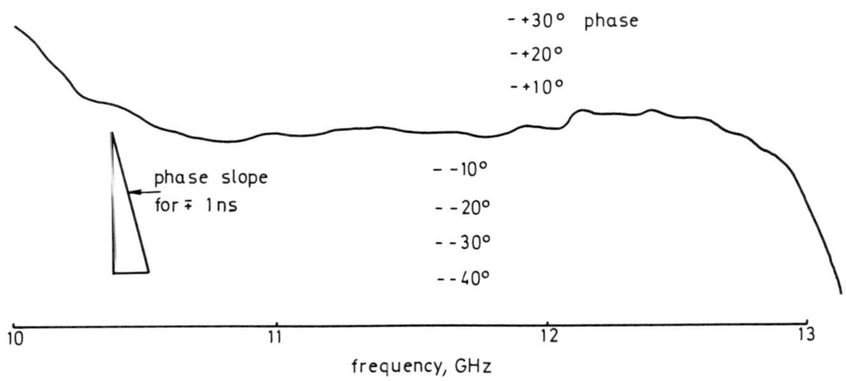


Figure 29 Phase deviation response of an amplifier

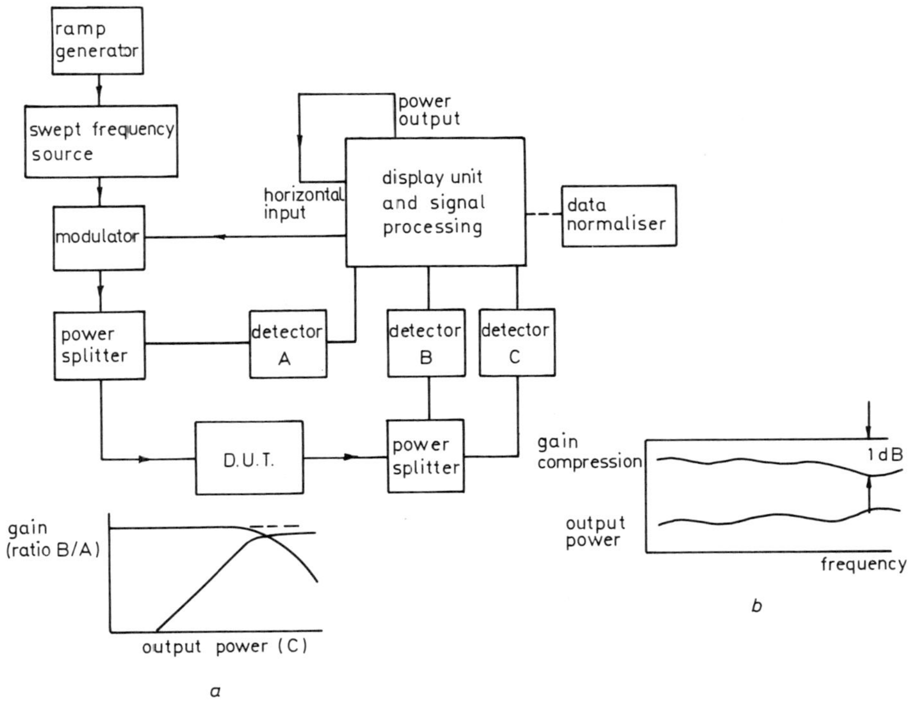


Figure 30 Schematic for compression measurements (CW and swept frequency). For swept frequency response disconnect ramp generator, add power control attenuator and normaliser.

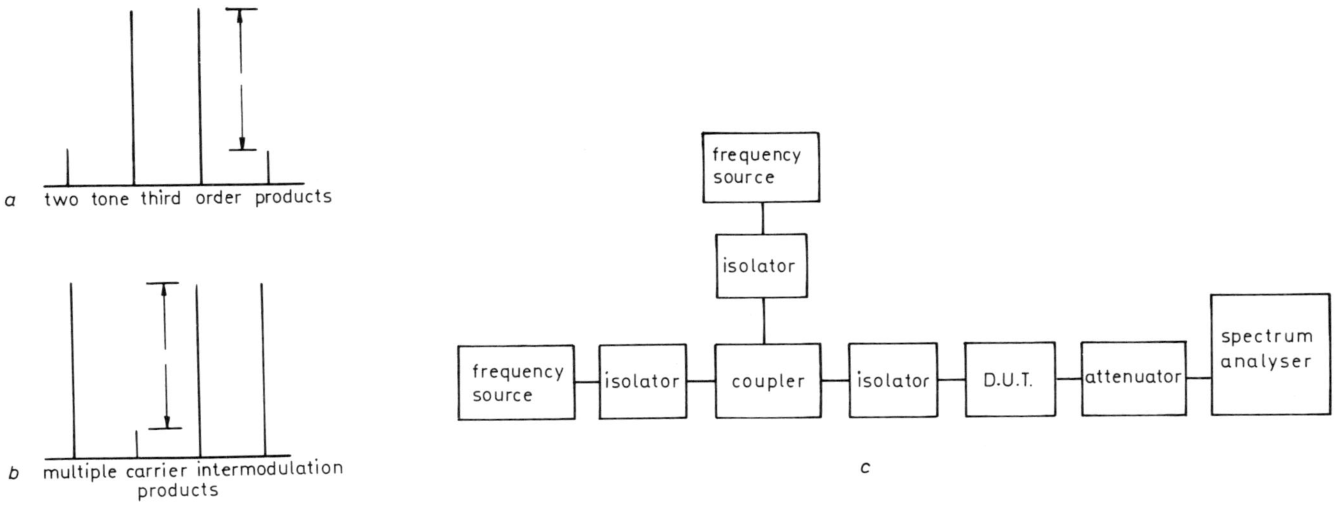


Figure 31 Schematic for intermodulation products (third order)

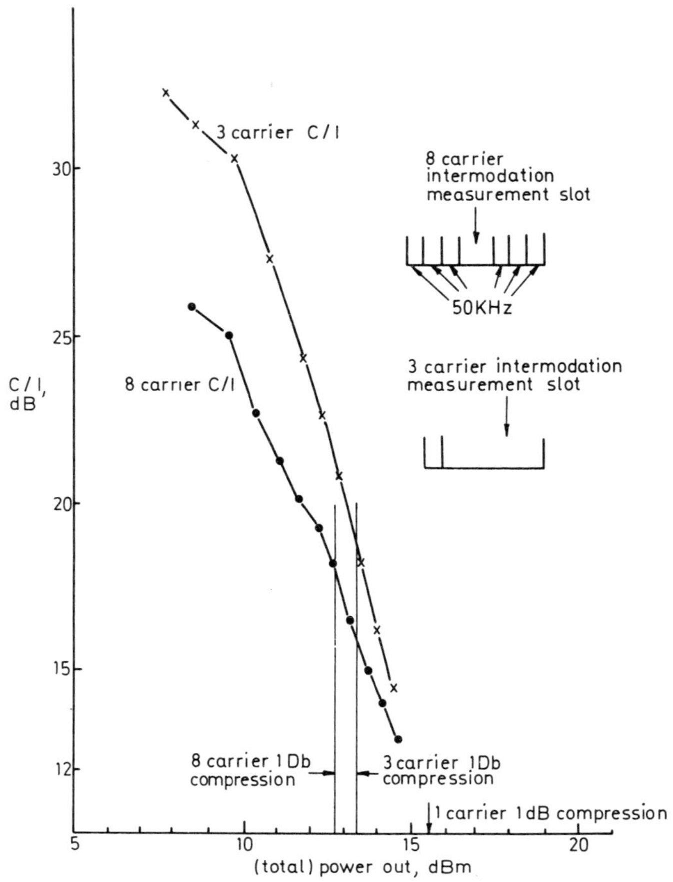


Figure 32 Intermodulation characteristics for multicarrier operation

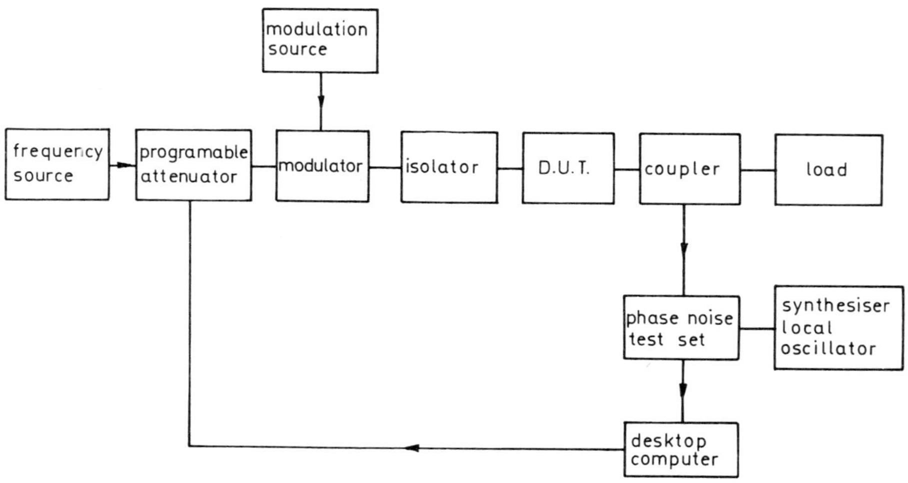


Figure 33 Schematic for am-pm measurements

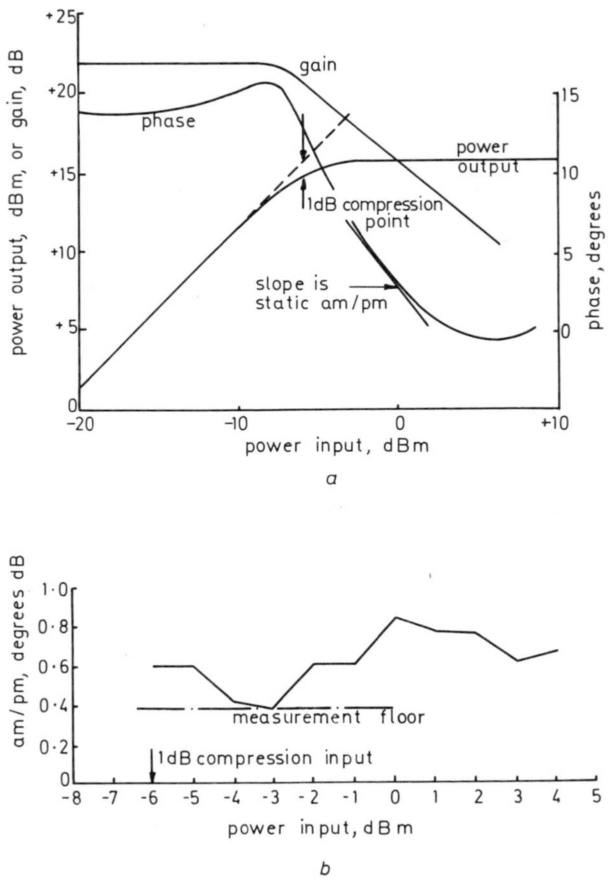


Figure 34 Static and dynamic results for am-pm conversation.
(a) Static
(b) Dynamic

TABLE 1

COMPARISON OF BEST AVAILABLE NOISE FIGURE SYSTEMS

| | Conventional Automatic Noise Meter | Computer controlled Noise meter | Y factor measurement using IF substitution with hot/cold load |
|---|---|---|---|
| Δ ENR calibrated against standard | ± 0.1 | ± 0.1 | ± 0.01 |
| Mismatch uncertainty | ± 0.25 | ± 0.15 | ± 0.01 |
| Instrumentation or operator errors | ± 0.15 | ± 0.04 | ± 0.05 |
| Second stage contribution | ± 0.1 | ± 0.1 | ± 0.10 |
| Ambient temperature if corrected | 0 | 0 | 0 |
| System nonlinearities (mixers) + assymetrically image gains | ± 0.05 | ± 0.05 | ± 0.05 |
| Total worst case | ± 0.65 | ± 0.44 | ± 0.31 |
| RRS | ± 0.33 dB | ± 0.22 dB | ± 0.16 dB |

# CHAPTER 12

# Design of microwave osillators

Dr. N. Kenyon/Dr. F. Myers/Dr. M. J. Howes

# CHAPTER 12 - DESIGN OF MICROWAVE OSCILLATORS
## (a) Basic Principles

N. D. Kenyon

Post Office Research Centre

### 1 GENERAL PROPERTIES OF NEGATIVE RESISTANCE OSCILLATORS

#### 1.1 Oscillation Condition

Fig 1 shows a simple equivalent circuit for a negative-resistance oscillator; it has for convenience been divided into two halves:

- a linear impedance $Z(\omega) = R(\omega) + jX(\omega)$, including most of the microwave circuit, cavities, tuners, and so on;
- a non-linear negative impedance $-\bar{Z}(i) = -\bar{R}(i) - j\bar{X}(i)$ dependent on the rf current amplitude (i) but independent of the frequency to a first approximation, representing the diode chip and also the parasitics of its package.

For equilibrium oscillation the total circuit impedance is zero, ie

$$Z(\omega) = \bar{Z}(i)$$

These two impedances, both having positive real part, may be plotted as loci on the complex plane, and can be referred to as the "load line" and "device line" respectively (Fig 2) and it is clear that any intersection of these loci represents a potential oscillation point. However for the oscillation to remain stable it is also necessary that the angle $\theta$ defined in Fig 2 lie in the range $0 < \theta < \pi$ [1].

#### 1.2 Starting

Oscillations build up rapidly from thermal noise, but only if the small signal total impedance $Z(\omega) - \bar{Z}(0)$ encloses the origin, in the usual way. If there should be more than one stable intersection, there is no simple way of predetermining which frequency will be generated.

#### 1.3 Stability

Two types may be distinguished: drift stability and perturbation stability.

a. Drift refers to basic changes in the load and device lines, leading to changes in operating amplitude and frequency. The load line is likely to be affected mainly by temperature, while the device line may move with change of temperature or bias current and with ageing of the semiconductor. If the load line is marked with uniformly spaced frequencies, then the further apart these are (ie the higher the circuit Q) the smaller will be the influence of device-line drift on frequency, but there is no corresponding remedy for amplitude changes.

b. Perturbation refers to the inclusion in the circuit of an additional small voltage (v) into the circuit, such as noise, an injected signal for locking purposes, or a reflected wave from a distant mismatch to the oscillator output port. Dividing v by the current i, and normalizing frequency to that of the oscillator, a complex perturbing impedance $Z_p$ is derived, and (in first approximation) the operating points on device and load lines are displaced from the original intersection until their vectorial difference is $Z_p$, as shown in Fig 3. It is clear that high circuit Q is resistant to FM noise and to output mismatch, while the opposite is true of a circuit whose Q is kept low so that rapid changes to the operating point can be made for modulation purposes (by injection locking, bias variation, or electronic circuit-tuning). Clearly too, for a given circuit the stability is greatest when $\theta = \pi/2$; as $\theta$ approaches 0 or $\pi$ the noise and perturbation sensitivity become very large. When $\theta = \pi/2$ the "oscillator external Q" as defined by locking or pulling criteria is equal to the loaded-Q ($Q_L$) of the circuit (load-line) which may be taken as $\frac{\omega}{2R}\cdot\frac{\partial X}{\partial \omega}$; for other values of $\theta$ the locking-Q is lower by the factor $\sin\theta$.

#### 1.4 Power

The power delivered by the negative resistance is $i^2\bar{R}(i)$, which has a maximum for some value $\bar{R} = \bar{R}_{opt}$ (Fig 4). The circuit should be designed to provide a $Z(\omega)$ having these properties:

a. real part close to $\bar{R}_{opt}$ and imaginary part at the desired frequency equal to $\bar{X}_{opt}$, preferably with independent adjustments to these two;

b. only one intersection;

c. appropriate $\partial X/\partial\omega$ and $\theta$.

#### 1.5 Efficiency

The real component of load $R(\omega)$ is the sum of two parts, a circuit loss resistance $R_c$ and a useful load $R_L$. The optimised oscillator has a circuit efficiency

$$\eta_c = \frac{R_L}{R_c + R_L} = 1 - \frac{R_c}{R_{opt}}$$

(This is not to be confused with the electronic efficiency represented by the ratio of $\frac{1}{2}i^2\bar{R}_{opt}$ to the bias input power). Clearly, for good efficiency, $R_c$ must first be made as low as possible, and then $\bar{R}_{opt}$ must be considerably greater.

#### 1.6 Diode Size

It is this $R_{opt}$ restriction which prevents the generation of ever-larger power simply by increasing the active area; although the diode may deliver power proportional to the area ($\bar{R}_{opt} \propto 1/\text{Area}$, $i_{opt} \propto \text{Area}$) the proportion of it reaching the useful load declines with increasing area.

### 1.7 Thermal Impedance

This is another difficulty which arises when the diode size is increased; assuming a fixed electronic efficiency, the input power at a fixed current density increases proportionally to the area, while the thermal conductance into the heat sink increases with proportionality closer to the diameter than the area, so that temperature of the active layer increases. However for present purposes this problem may be regarded as a practical one, since the temperature can in principle be held almost constant if the diode area is increased in other ways - such as by forming an annular cross-section or by stitching several small diodes together.

### 1.8 Power-Impedance Product

The generated power $P = \frac{1}{2}i^2\bar{R}_{opt}$ can be rewritten as $\frac{1}{2}J^2\bar{\rho}_{opt}/\bar{R}_{opt}$, where J is (rf) current density and $\bar{\rho}_{opt}$ is the optimum negative resistance per unit area. Thus for a given structure and current density the power-impedance product is independent of the diode area, and is therefore a useful means of comparing different devices.

### 1.9 Quality Factor

If in the neighbourhood of the oscillation frequency the load impedance is similar to a single-tuned tank circuit, the unloaded and loaded Q's are defined respectively by

$$Q_o = \frac{\omega}{2R_c} \frac{\partial X}{\partial \omega} \quad \text{and} \quad Q_L = \frac{\omega}{2(R_c + R_L)} \cdot \frac{\partial X}{\partial \omega}$$

If follows that $\eta_c = 1 - Q_L/Q_o$ ... (3)

Thus regardless of the diode impedance there is a basic compromise between efficiency and loaded-Q of an oscillator. If $Q_L$ is to be in the hundreds then $Q_o$ must be in thousands. If $Q_L$ is itself to be in the thousands, sufficiently high $Q_o$ is difficult to obtain directly in the oscillator primary resonant circuit, and it is necessary to provide an auxiliary resonator suitably coupled (Fig 5); even so the efficiency penalty exacted to achieve a given oscillator Q is still that of equn. (3).

Where it is desired to achieve low $\partial X/\partial\omega$ (for modulation purposes usually) there are two approaches:

a. Energy storage is kept low; then the energy changes or exchanges required to alter the oscillation frequency are correspondingly low.

b. A second resonator (or rather, antiresonator) is added; then when the frequency is deviated the bulk of the exchange energy passes between the resonators, and demands on the modulation source are relatively light. Such an approach is particularly suitable where only a narrow band of deviation is required (eg for AFC).

## 2 THE NEGATIVE IMPEDANCE DEVICE

### 2.1 Microwave Diode Impedance

In purely resistive circuits, a negative conductance device would be recognized as a function $\bar{G}(V)$ tending to some maximum value $\bar{G}(0)$ for small-signal perturbations, and tending to zero as oscillations become large and approach some physical limit. The real part of the rf admittance of an Impatt for example, behaves in this way; however there is a displacement current which is most often considerably greater than the particle current, and as Fig 6 shows the combination is better regarded as a negative resistance with some series capacitive reactance. The important fact is that this negative resistance may be only a couple of ohms, particularly if the diode area is large to get high power. Alternatively, the same fact can be expressed differently by considering the standing waves in transmission line of "standard proportions (eg 50 ohm coaxial line) with the diode at one end and a load at the other:-

a. Looking towards the diode, the VSWR has a high negative value, particularly if the power travelling away from the diode is close to the maximum the diode can deliver.

b. Looking towards the load, a high VSWR (suitably phased) must be attained before the system will oscillate at maximum load power. Other types of microwave diode, such as transferred electron devices, may not reflect such a high VSWR at the end of a 50 ohm line, and matching for maximum power is rather easier.

### 2.2 Effective Negative Resistance

When the diode is mounted at the end of a two-conductor uniform transmission line, it is useful to characterise the "active device" by means of the complex VSWR, which is a quantity that can be easily and accurately measured (eg para 3.2h below). At a suficient distance (usually quite small) from the diode only one electromagnetic mode is present in the line: the SWR and the position of the E-field minimum, as a function of absolute wave amplitude, define a "device line" which includes not only the semiconductor chip but also its package, the surrounding irregular volume and the stored energy of evanescent modes in the transmission line. Of course the device line will not be exactly the same if the diode is transferred to another transmission line of the same impedance but different geometry.

Thus when this modified-device impedance is a slowly varying function of frequency, compared with the rest of the circuit (which may not be the case if the package is resonant), the diode can conceptually be replaced by a pure negative resistance $\bar{r}Z_o$ situated at the E-field minimum.

### 2.3 Chip Terminal Impedance

If the circuit between the terminals of the active device and the E-field minimum is lossless, it can be represented[2] for a single frequency by the circuit of Fig 7; the values of $B_o$ and $n^2$ can be determined by a simple technique in which the position of the voltage node is measured as a function of capacitance below avalanche breakdown. The chip impedance can then be determined[3] for any other bias condition by the position and magnitude of $\bar{r}$, together with $B_o$ and $n^2$.

### 2.4 Irregular Behaviour of Device Impedance

The behaviour of the microwave diode impedance at the desired frequency is often not as simple as implied in para 1.1: for example $\bar{Z}(i)$ generally has some dependence on $Z(2\omega)$, at least for larger amplitudes of oscillation[4]. Impatt diodes under large signal conditions experience appreciable susceptance modulation[5], and if the circuit impedance in the region of $\frac{1}{3}$ to $\frac{1}{2}$ $\omega$ is appropriate they can break into self-pumped parametric oscillation; since circuits of rectangular waveguide output do not radiate such low frequencies, the only reliable external indication is the irregularity in the power/bias current characteristic (Fig 8).

## 3 OSCILLATOR CIRCUITS [10]

### 3.1 Simple Oscillator

From section 2.1 it is easy to envisage a high-VSWR load comprising a matched output transmission line $Z_o$ and a transformer section of some different impedance. For example, such a section could be roughly $\sqrt{Z_o \bar{R}_{opt}}$ in impedance and rather more than $\lambda/4$ long, reaching right to the diode. Of this type of circuit the following remarks may be made:

a. The transmission line must be of two-conductor type (at least), and is very likely to require 50 ohm output, with a split into high-pass/low-pass for biassing.

b. Adjustments, for optimization purposes, to the phase and magnitude of the VSWR are not independent and are difficult to perform in a low-loss manner.

c. Non-adjustment circuit can have very low circuit loss.

d. Stored energy is low, so the circuit is appropriate where modulation is to be applied. Higher Q can be achieved by making the line sections longer, but there is then great danger of violating the condition (b) of section 1.4.

e. The circuit can be used for characterisation of the active device.

f. It is difficult to design with rectangular waveguide output, particularly at millimetre-wave frequencies.

### 3.2 Resistively Stabilized Circuit[6] (Fig 9a)

The diode is biassed through a two-conductor transmission line, but instead of splitting into high-pass and low-pass paths the line simply passes through a region of high microwave loss, such as iron-loaded foam or resin. The diode being of low $\bar{r}$ there is no question of any oscillation. Next a high VSWR is introduced at one frequency only by coupling a resonator to the line: the magnitude of VSWR is governed by the resonator Q and the coupling coefficient, while the phase is determined by its position. Note that:

a. Resonator Q is controlled by its coupling to the useful load; using a rotary iris as the output to waveguide, very little frequency change occurs.

b. This in turn controls the resistive or power-matching to the diode, while ...

c. ... resonator position controls the phase-matching to get the desired frequency.

d. The fraction of power lost in the microwave absorber is only about $\bar{r}$.

e. High VSWR is only possible over a narrow range of frequency so no oscillation can occur elsewhere.

f. If the resonator has higher order resonances, the phase-matching is unlikely to be suitable; no intersection with the device line occurs, so there is no unwanted oscillation (Fig 9b).

g. The resonator need not be a two-conductor type, and the output can be rectangular waveguide; oscillators from 8 to 115 GHz have been made in this way.

h. The resonator affords an excellent means of determining the VSWR on the diode transmission line (DTL); when it is located at an antinode the resonant frequency, as determined by reflection measurements at the output, or by the oscillation frequency itself, is exactly equal to the unloaded resonant frequency of the resonator; the magnitude of the VSWR is determined from the Q of the resonance.

i. The resonator acts as a narrow-band filter and suppresses am noise components away from the carrier - useful for local oscillators.

In the given equivalent circuit[6], the definition of the coupling parameter $Q_{ex}$ is the same as the normal one for an external loading to a resonator; the resonator appears to the DTL as a normalized series impedance

$$\frac{Q_L}{Q_{ex}\ (1 + j.2Q_L\delta)}\ ,$$

where $2\delta = (\omega/\omega_o - \omega_o/\omega)$. Therefore at resonance the VSWR in the line is $(1 + Q_L/Q_{ex})$ and must be adjusted for equality with $1/\bar{r}_{opt}$. We have seen that for good efficiency $Q_L \ll Q_o$: for small $\bar{r}_{opt}$ (large diode) we must also impose $Q_{ex} \ll Q_L$; fortunately it is possible in some geometries to secure exceedingly tight coupling such that $Q_{ex} \sim 5$, enabling diodes of 1 ohm or even less to be accommodated.

To a good approximation, the circuit resistance seen by the diode is $Z_o Q_{ex}/Q_o$ which can be as low as one or two tenths of an ohm. It is possible to share even this between two or more diodes, each mounted at the end of a DTL which is coupled to the same cavity resonator; however this process is limited by the loss within the DTL's.

### 3.3 The multiple-diode oscillator[7] is a good solution to the problem of obtaining very high powers:-

i. Diodes are well separated, facilitating heat removal.

ii. Diodes are rf-wise in series, giving a lower $Q_L$, and better efficiency.

iii. Diodes are dc-wise in parallel, thus requiring no greater voltage, yet can be current-controlled independently from outside the rf circuit.

iv. The presence of the absorbers in each DTL inhibits push-pull modes between the various diodes.

### 3.4 Low-Q Circuits

Low stored energy is not easy to achieve especially with Impatt diodes: if the tuning were achieved with a simple lumped inductance, the Q would be that of the diode itself, usually of order 10. To get a lower effective Q than this, double tuning would be necessary (see section 1.9(b)). However since the Impatt $\bar{R}_{opt}$ differs greatly from standard output impedances (50 ohms or more), a transformer is necessary, and this inevitably stores additional microwave energy. For example, if a negative resistance $-\bar{R}$ is tuned by a $\lambda/4$ length of transmission line $\sqrt{Z_o \bar{R}}$ loaded by the output line $Z_o$ then the extra

Q is $\frac{\pi}{4}\sqrt{Z_o/\bar{R}}$.

A successful low-Q circuit of waveguide output [8] is shown in Fig 10; volume is minimized by using greatly-reduced-height waveguide and very close-in tuning elements. A circuit incorporating stabilizing absorber and rotary output coupling is shown in Fig 11.

3.5 Cap Circuit[9]

This circuit is not easy to design or optimize for power and frequency simultaneously (Fig 12); however it is very popular with those working on diode evaluation, because of its simplicity of construction (even up to 200 GHz) and potentially low loss in the optimum power condition.

REFERENCES


1 KUROKAWA, K., 1969, "Theory of Microwave Circuits", Chapter 9. Academic Press.

2 KUROKAWA, K., ibid, Chapter 1..

3 GROVES, I. S., and KENYON, N. D., 1973 Electronics Letters 9, 331.

4 OKAMURA, S. and SHINDO, Y., 1977, URSI Colloq. Measurements in Telecomms., Lannion, Paper K4, p 468.

5 GOEDBLOED, J. J and TJASSENS, H., 1974, Proc. 4th Eu.M. C., Montreux, p 328.

6 KENYON, N. D., 1973, Int J Circuit Theory and Applications, 387-393.

7 KUROKAWA, K., 1971, IEEE Trans MTT-19, 793-81.

8 KUNO, H. J., "High Data Rata MM-Wave Exciter-Modulator Amplifier", Tech Report AFAL-TR-72-243, Hughes Research Labs., Sept 1972.

9 KENYON, N. D., 1973, Proc 3rd Eu.M. C, Brussels, paper A1.1.

10 KUROKAWA, K., 1976, Chapter 5 of "Microwave Devices", Howes and Morgan (Ed), Wiley & Sons.


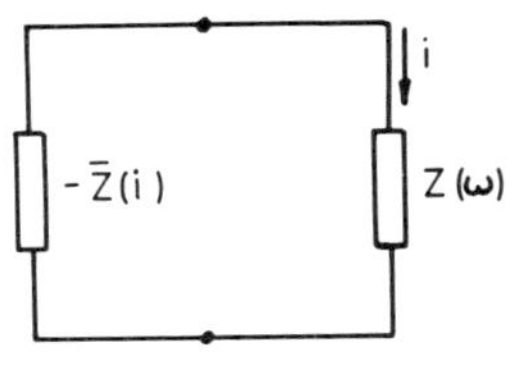


Figure 1 Equivalent circuit of negative resistance oscillator

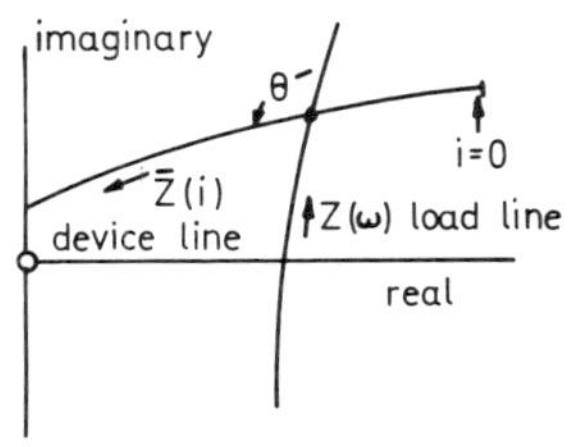


Figure 2 Oscillation condition in impedance plane

Questions on: Microwave Oscillators - (a)

a1 An Impatt chip has electronic efficiency $\eta$ and delivers at its terminals a power per unit area of P'; it is to be mounted in a circuit of unloaded quality factor $Q_o$ and its area is selected so as to maximize the useful power. What is then the overall efficiency?

a2 An oscillator is built according to the circuit of Fig 9, in which $Z_o$ = 50 ohms, $Q_o$ = 2000, and $Q_{ex}$ = 10. Impatt diodes are available having a power-impedance product of 0.2 watt-ohms. If the circuit efficiency is to benot less than 80%, what is the maximum useful power, and the corresponding Q? How much power is taken up in the absorber element, and how might this be reduced?

a3 In the oscillator of question 2, the diode is biassed just at breakdown; the swept reflection coefficient (magnitude) at the oscillator output port is displayed on an oscilloscope. Describe how this display changes:-

a. as the distance between diode and resonator is varied.

b. as the output coupling is varied.

c. as the reverse bias current is increased.

How can the quantity $Q_{ex}$ be determined? [b]

a4 Considering the simple circuit of Fig 13, show that no stable oscillation is possible for $\theta$ in the range $\pi$ to $2\pi$. Show also that, for optimum power out of the negative resistance device, $\partial i/\partial\bar{R} = -\frac{1}{2}i/R$.

a5 A cap-circuit oscillator has the (slightly simplified) equivalent circuit of Fig 12b; show, using the Smith chart or otherwise, that there are two positions of the short circuit where maximum power is obtained, unless the diode is too small. How does the oscillation frequency vary with short-circuit position?

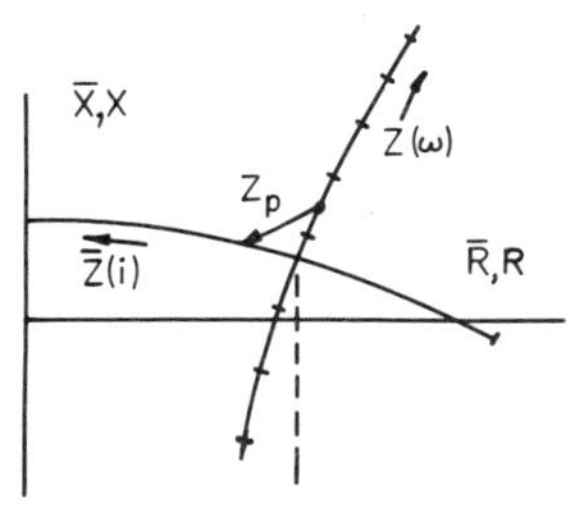


Figure 3 Perturbation of oscillation conditions

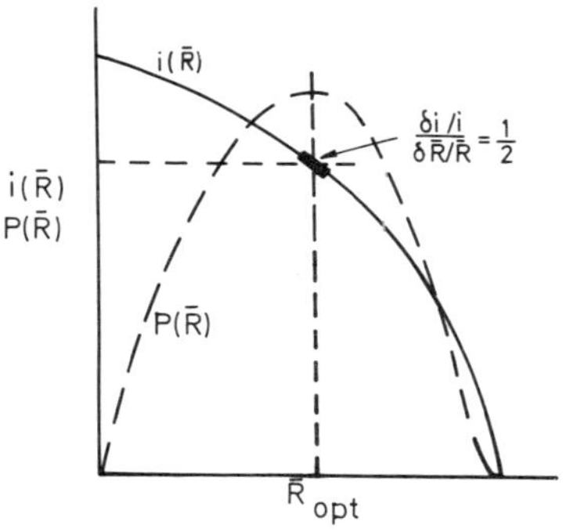


Figure 4  Maximization of power from the diode

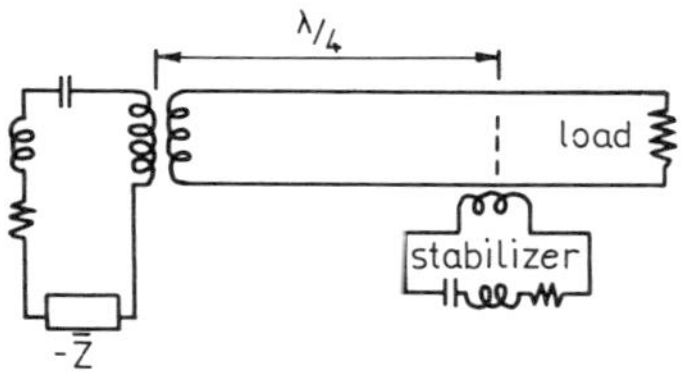


Figure 5  Stabilization of an auxiliary resonator

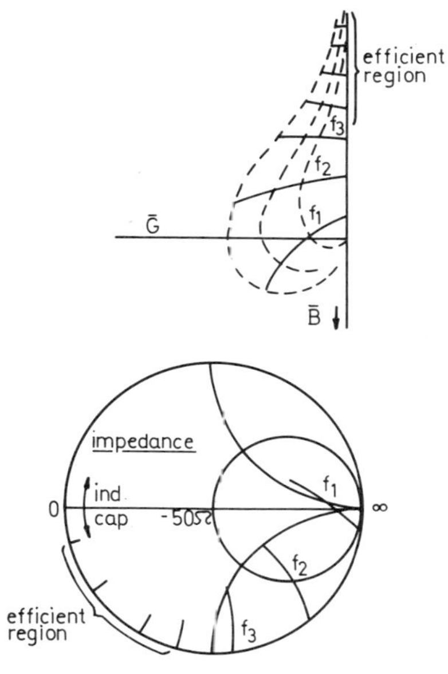


Figure 6  IMPATT diode immittance

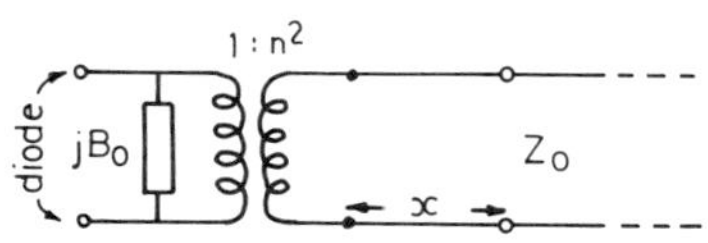


Figure 7  Bilinear transformation between chip and effective impedances

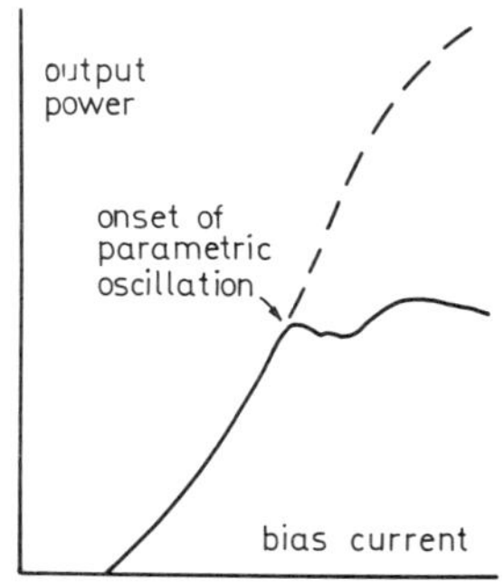


Figure 8  Parametric oscillation indication

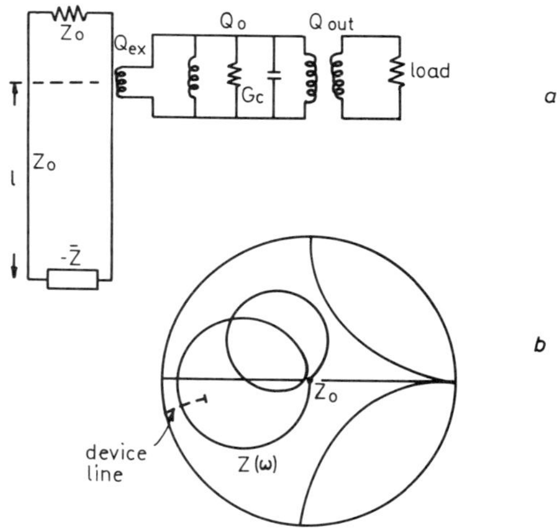


Figure 9  Resistively stabilized circuit

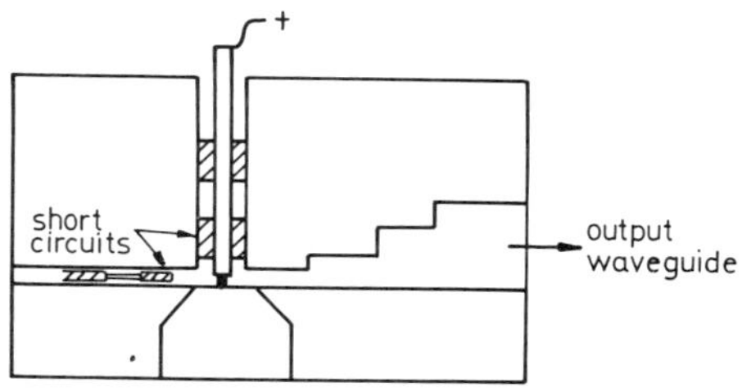


Figure 10 Hughes low-Q circuit

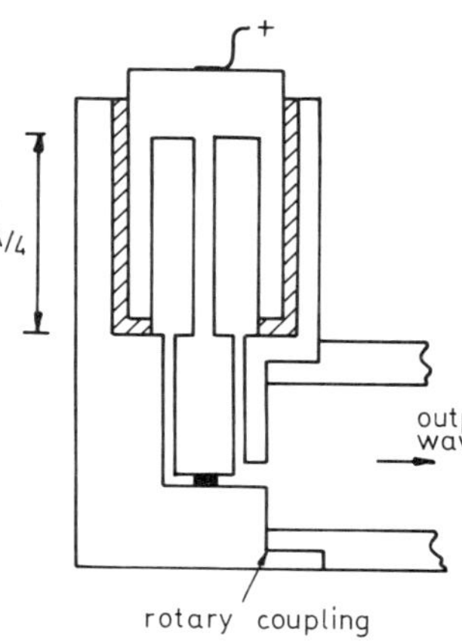


Figure 11 Resistively stabilized low-Q circuit

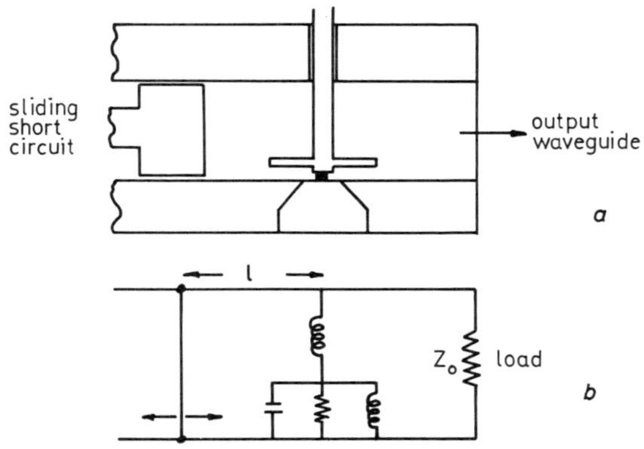


Figure 12 Cap circuit in waveguide

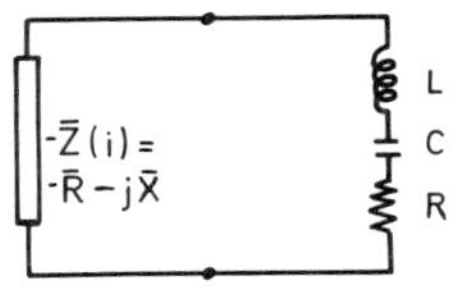


Figure 13 Oscillator circuit

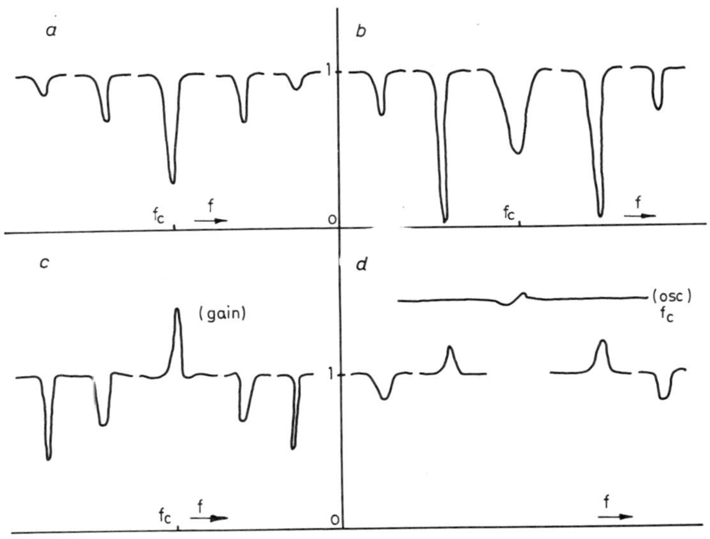


Figure 14 Swept reflection coefficient for five successive diode positions, and increasing bias (a) - (d)

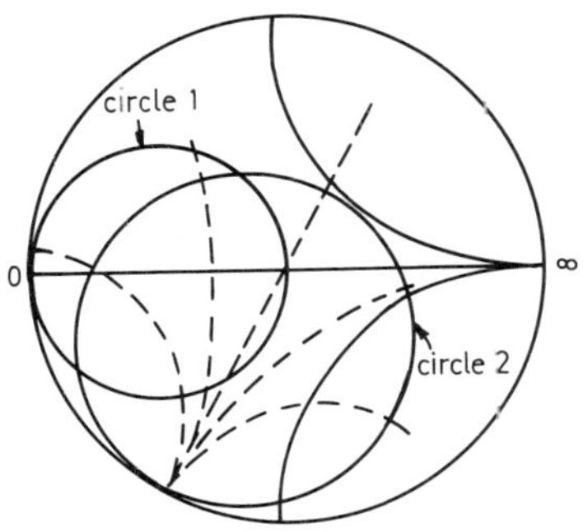


Figure 15 Impedances in a cap circuit oscillator

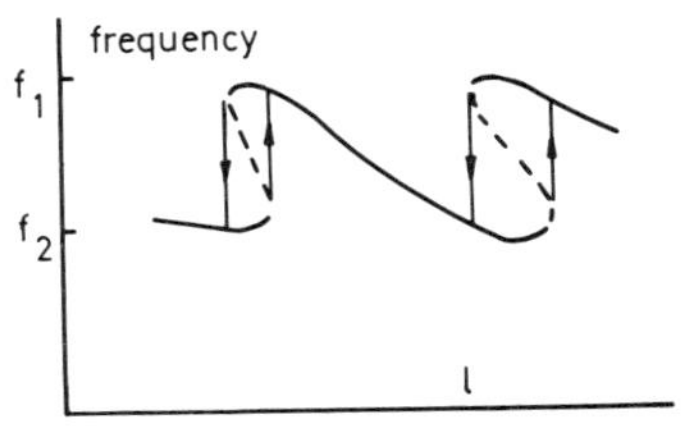


Figure 16 Cap circuit oscillator, variation of frequency with short-circuit position

# NARROW BAND MODULES

F.A. Myers

Plessey Research (Caswell) Ltd.

## INTRODUCTION

Ever since the advent of true microwave systems, about 40 years ago, there has been the requirement for tunable oscillators, such as local oscillators in radar systems. Until recently, the requirements have usually been satisfied by tunable klystrons with the occasional transistor multiplier but, since the commercial exploitation of the Gunn effect began 10 years ago, these applications are now dominated by solid state devices.

Relatively narrow band units ($\leqslant$10% bandwidth) are required for a variety of applications including local oscillators with AFC, FMCW radars, communications transmitters and local oscillators, tellurometers (range measurement) etc. Broad band oscillators, which are not the subject of this discussion, are utilised for sweep generators, frequency agile oscillators for ECM/ECCM etc., and require somewhat different design approaches with emphasis on aspects such as bandwidth and tuning speed rather than on stability etc.

## AVAILABLE MICROWAVE GENERATORS

The state of the art in terms of power and efficiency of CW solid state generators is outlined in Fig.17 a,17 b with each device, apart from that laboratory curio the CW TRAPATT, having some advantage for particular systems.

### Gunn Device

The Gunn device fabricated from GaAs is the oldest, most common and to date most useful of the devices available. Although limited commercially to about 300 mW CW at X-band at 5% efficiency its ease of biasing, reliability and general ease of operation make it ideal for most narrow band applications, which udually do not demand too much power.

It does not produce enough power for communication systems unless device combining is utilised, and becomes very inefficient at mm wave frequencies (~1% at 100 GHz) although its low noise compared to Impatts still make it a useful device. The more recently investigated material, InP, offers particular promise at the higher frequencies and should enable a considerable improvement in efficiency to be achieved. Although threatened by FET's at lower frequencies this device will be around for some years to come.

### Impatt Devices

Impatts (Impact Avalanche Transit Time), or avalanche diodes, are generally high power, high efficiency devices with relatively poor noise performance compared to Gunn devices. As such, they are not commonly encountered in tunable oscillators for local oscillator application but are utilised for communication transmitters and mm wave oscillators.

Generally, GaAs is the preferred material at frequencies up to 20 GHz with Si above this frequency. The ultimate operating frequency of this material is not definitely established but is likely to be around 300 GHz.

### FET's

Although not shown in Fig.17 , as they are more commonly thought of as amplifying devices, the FET is capable of being used as an oscillator with particularly good conversion efficiency (~30%, similar to a GaAs READ Impatt). To date, however, in conflict with their low noise amplifier properties the oscillator modes appear relatively noisy. They are generally encountered in systems not requiring low noise, such as broad band frequency agile systems. To date these devices are limited to about 20 GHz but recent developments on narrow gate technologies will gradually extend this frequency.

### Tuning Mechanisms

In order to electronically tune an oscillator some circuit element with a variable reactance must be employed. For broad band oscillators the magnetic resonance of Yttrium Iron Garnet (YIG) is commonly used. This gives extremely broad band capability (about one octave) with good linearity and stability but with the disadvantage of slow sweep rate and high power consumption. It is not, therefore, commonly employed for narrow band oscillators.

The solid state generators themselves, such as the Gunn or Impatt diode, have a voltage or current dependent reactance which can be used to tune a circuit over a limited frequency range but this technique suffers from so many problems that it is only used for special applications. Bias tuning also obviously affects the resistive as well as the reactive component of the device causing a.m. distortion and power variation over the tuning range. The change in the operating point can also seriously affect the noise and stability of the oscillator but a very serious drawback is that the dF/dV for a Gunn device (or dF/dI for an Impatt) can vary from device to device making oscillator reproducibility very difficult to achieve.

The circuit element most commonly encountered in tunable oscillators is the varactor diode. This can take the form of a p-n junction but the most common varactor is the metal-n type Schottky barrier, usually fabricated from GaAs. In this device reverse biasing the barrier causes a depletion layer to extend into the semiconductor, this depletion layer acting as the dielectric in a capacitor.

The relationship between applied bias, V, and capacitance, C, is given, for a flat profile varactor, by:-

$$C = \left[\frac{q \mathcal{E} N_D}{2(V_b - V - kT/q)}\right]^{\frac{1}{2}} m^{-2} \quad \cdots\cdots (4)$$

q = electronic charge

$\mathcal{E}$ = semiconductor permittivity

$N_D$ = donor impurity density

k = Boltzmann's constant

T = absolute temperature

$V_b$ = barrier height

V = applied bias

This relationship gives a very non-linear relationship between frequency and voltage, a factor that will influence the microwave design.

Several requirements exist in the design of the varactor, some of them mutually incompatible, as always! These requirements, not always essential but desirable, are as follows:-

(a) Large forward turn on voltage. This is usually around 0.5 to 1 volt and is required as large as possible to stop the varactor rectifying the r.f. voltage at low bias causing excessive power swings and tuning anomalies.

(b) Large reverse breakdown voltage. This enables the varactor to be operated at large reverse bias with high powered devices without running the risk of breakdown at some point in the r.f. cycle.

These two points are particularly important for high power CW oscillators and even more so for pulsed oscillators. A varactor in a pulsed oscillator will have several tens of volts of r.f. across its terminals resulting in perhaps only one or two volts of reverse bias during which a part of the cycle will not push the varactor into reverse breakdown or forward bias. The tuning range is then severely limited.

(c) Low series resistance (or high "Q", high cut-off frequency). The unavoidable series resistance obviously absorbs power and is undesirable for this reason. The resistance also modifies the reactive component of the varactor when transformed through the circuit and can therefore limit the tuning range.

(d) Large capacitance ratio. This is obviously required for large bandwidths. The tuning ratio can be modified, as can the tuning law, by modifying the carrier concentration profile. This is more frequently encountered in varactors for broad band applications.

MICROWAVE CIRCUITS FOR TUNABLE OSCILLATORS

Having discussed the r.f. generator and tuning element, it remains to incorporate both of these into some circuit to obtain optimum performance from both.

The equivalent circuit for the ideal electronically tunable oscillator is shown in Fig.18a. It can be shown by considering the simplified circuit of Fig.18b (see questions) that the maximum available tuning range can be given by the expression

$$\Delta F = -\frac{\Delta C_j V^2 \pi F^2}{Q_L P_T} \quad \cdots\cdots (5)$$

$\Delta C_j$ = capacitance variation

V = r.f. voltage swing across capacitor

F = frequency

$Q_L$ = circuit loaded Q-factor

$P_T$ = total power dissipated in varactor and circuit

It can be seen immediately that several compromises now have to be made. In order to obtain the largest possible bandwidth $Q_L$ should be minimised. This, however, severely reduces the stability of the oscillator resulting in poor noise, temperature stability etc. Thus a fine line has to be drawn between allowable noise degradation and minimum usable bandwidth. In addition, a large r.f. voltage would seem desirable but increasing the swing also reduces the time averaged value of varactor capacitance, reducing $\Delta C_j$. Parasitic elements of the package in which the varactor chip is mounted, as well as the parasitic series resistance, will also reduce the tuning range.

The actual microwave circuits described below represent practical approximation and compromises towards the ideal shown in Fig.18.

Microstrip Resonators

At first sight microstrip circuits on a ceramic or quartz substrate offers considerable advantages of size, weight and ease of manufacture. In practice the Q factor is so low that the stability, even for narrow band applications, makes it unsuitable. The low Q-factor is used to advantage in the design and construction of broad band circuits. In fact, most broad band, frequency agile circuits utilise either a microstrip resonator or lumped circuits on a ceramic substrate.

Further problems arise in microstrip, particularly at frequencies above 12 GHz, in the packaging of devices. The conventional microwave packages, suitable for coaxial or waveguide circuits, presents too high a value of parasitic reactance to match the device.

Coaxial Resonator

A circuit that offers a reasonable Q factor is the coaxial resonator (see section on high power pulsed sources for theory). Shown in Fig.19 is the schematic diagram of the earliest form of varactor tuned oscillator. In this case the varactor is loop coupled into a high magnetic field and the degree of coupling, and hence tuning, is experimentally optimised by adjusing the orientation and insertion of the probe.

Although useful in the early days for producing relatively unsophisticated local oscillators the units are difficult and time consuming to set up and mechanically complex. More modern oscillators mount the Gunn and varactor diodes directly in the same cavity in one of the configurations shown in Fig.20. A schematic of a practical circuit utilising devices mounted in the same centre conductor approximately $\lambda_o/2$ apart is shown in Fig.21.

This oscillator is readily capable of producing 5% - 10% bandwidth with up to 100 mW output power with a frequency temperature coefficient of -1 MHz/°C at J-band, with an external Q-factor of 30. The structure can be ruggedised and is readily capable of withstanding missile type environments.

Waveguide Resonator

The high Q-factor of a waveguide resonator (typically 500 - 1000) makes its stability very good and, providing the coupling is light, makes it ideal for low noise, narrow band applications such as narrow band local oscillators.

A schematic diagram of a typical waveguide resonator is shown in Fig.22. In this case the Gunn and varactor diode are mounted approximately $\lambda g/2$ apart in waveguide. Typical bandwidths required, and achieved, are ½ to 1% with a temperature stability of -0.3 MHz/°C at J-band at up to 300 mW power output.

It should be noted that it is possible in a waveguide circuit to increase the coupling between the varactor and the circuit to such an extent that identical bandwidths to the coaxial case can be achieved. It may appear at first sight advantageous as waveguide is normally considered a high stability medium. On further examination, however, the Q-factor and stability are found to have degraded to values at least as low as the coaxial case but without that circuits device tolerance and flexibility.

One further disadvantage of waveguide circuits over coaxial is the inability to satisfactorily model the circuit and derive useful theoretical information. The relatively straightforward nature of the TEM coax modes enable at least first order approximation to be carried out.

MISCELLANEOUS OPERATING PARAMETERS

Noise

Most narrow band oscillators have a very tight (often impossible!!) noise specification. The a.m. noise of a CW oscillator is not affected by the cavity Q-factor (that of a pulsed oscillator is), and is thus not dependent upon the cavity used. The f.m. noise, as mentioned above, is very strongly dependent upon the Q-factor and hence the bandwidth.

F.M. noise is readily measured (1) using the equipment shown schematically in Fig. 23. The f.m. noise measured on three oscillators is shown in Fig.24. This shows the unavoidable degradation observed in going from a high Q fixed frequency waveguide oscillator, through a narrow band waveguide oscillator to a low Q, medium bandwidth coaxial resonator.

Post Tuning Drift

It is often required to switch the frequency rapidly over a considerable bandwidth in the minimum possible time. Applying a step function voltage to the varactor obviously causes the frequency to change rapidly but the frequency may over or order shoot the ultimate value and take a considerable time to settle to the final value.

Of prime importance in determining the tuning speed and settling time are thermal effects in both the Gunn and varactor diode. As the frequency is changed the power output changes, resulting in a change in current through the Gunn diode and hence different power dissipation. Similarly the r.f. power dissipated in the varactor diode changes causing small, but significant, thermal effects.

Somewhat more complex effects can take place in the varactor if trapping centres or surface states exist in/on the device. In this case electronic time constants caused by the filling or emptying of these sites can cause limitations.

In practice, the frequency can be switched from one band end to the other in time scales of a few nanoseconds but the frequency may then drift by a few MHz over the next few 100$\mu$ secs. The actual values are difficult to specify as they depend critically upon the oscillator type.

Linearity

Apart from the large change in tuning slope caused by the varactor, fine grain non-linearities can occur over small voltage ranges. These may be caused by load pulling effects and the fitting of an isolator is often essential to overcome these. Of more significance, and far more difficult to overcome, are non-linearities caused by spurious resonances in the circuit itself. An example of this effect is shown in Fig.25. These effects must be minimised in the initial design or they will for ever cause problems in production.

LINEARISERS

The basic frequency/voltage characteristic of an electronically tuned oscillator is highly non-linear (owing to the varactor capacitance/voltage law), the non-linearity being expressed as either deviation from the best straight line fit (+/- 15% typically) or as the ratio of maximum/minimum tuning slopes (8:1 typically). This non-linearity usually presents difficulties to a system designer and it is quite common to interface the oscillator with a lineariser to approximate the tuning characteristics to a straight line. A YIG tuned oscillator, incidentally, usually has extremely good linearity making it ideal for broad band sweeper type applications.

Several aspects must be considered in the design of a lineariser. It must obviously achieve some specified linearity value but of equal importance the unavoidable degradation in microwave performance must be minimised for the particular system. This usually means comprising tuning speed, f.m. noise, post tuning drift etc.

There are many possible techniques of linearisation but two fairly common ones will be described that illustrate the various compromises.

Break-Point Lineariser

This type of lineariser is described in Fig.26 . Basically, the tuning slope is approximated to a series of straight lines and then a variable gain amplifier is designed to produce a mirror image transfer function. This can use diodes biased to a particular voltage, as shown, or transistors. The resultant frequency/voltage curve is still non-linear but it can usually be made to approach an adequate value by correct choice of the number and position of the break points.

The modulation characteristics depend very much upon the amplifier used but large signal bandwidths of a few 100 KHz with small signal bandwidths of several 10's of MHz can be achieved for an f.m. noise degradation of 3 dB's.

Continuous Approximation Lineariser

This is outlined in Fig. 27 . It can be seen that over a particular voltage range the frequency/voltage curve and hence the desired function can be approximated by a power series of a particular length. This function can be generated electrically by a series of analogue multipliers with the coefficients determined by fixed resistors. Excellent linearity can be obtained, over the design voltage range, but usually at the expense of other parameters. With present day multipliers the bandwidth is limited to a small signal value of 100 KHz and the f.m. noise degradation can be 10 to 20 dB. However, for systems where linearity is the vital parameter, then this may prove acceptable.

AUTOMATIC TEST EQUIPMENT

Electronically tuned oscillators required for sophisticated systems require extremely complex measurements to be carried out. In some cases this would be tedious by manual methods and the measurement interval so coarse that some information, such as localised non-linearities, would not be observed.

With modern microwave equipment usually being obtainable with interface bus capability, enabling information to be read and instructions given to the instrument, it is very attractive and sometimes essential to use calculator/computer controlled test equipment (Automatic Test Equipment or A.T.E.) to gather the data. The block diagram of a system based upon an HP9830A calculator is shown in Fig. 28 .

The A.T.E. described can readily measure the following parameters:-

Frequency/voltage
Power/voltage
Frequency/temperature coefficient
Pushing/pulling
Start up time/drift
Post tuning drift
Burn-in/ageing effects

To date, vibration and noise performance are not easily measured by the A.T.E.

An example of a typical print out of the characteristics of a medium bandwidth coaxial oscillator are shown in Fig. 29 .

In addition to the measurement role, the computational capability can be used, for instance, to curve fit the characteristics and determine values of select on test resistors for lineariser design. With the advent of lower cost microprocessors this type of equipment will become much more common.

REFERENCES

11. Ondria, J.G., 1968, IEEE Trans. MTT 16(9) pp. 767-781.

QUESTIONS

b1 Given the general expression for the capacitance voltage law of the varactor (see text), determine the general expression for depletion layer width.

b2. Calculate the zero bias capacitance, and capacitance at -30 volts for a 60 micron diameter varactor fabricated on GaAs with a donor density of $10^{22}$ $m^{-3}$. ($\rho \simeq 0.001$ m).

$q = 1.602 \times 10^{-19}$ coulombs

$\varepsilon$ $10^{-10}$ F/m

$k = 1.4 \times 10^{-23}$ W/°K

$V_b = 0.5$ volts

$T = 300$°K

b3. If the semiconductor layer of the varactor above is 5 micron wide calculate (V=-30)

(a) the amount of undepleted epitaxial material

(b) the parasitic resistance of the undepleted layer.

b4. Show that the tuning range of the simple parallel tuned circuit of Fig. 18b is given by

$$\Delta F = -\frac{\Delta C_j V^2 \pi F^2}{Q_L P_T}$$

b5. Show that the Q-factor of a parallel tuned circuit is given by

$$Q = \frac{\omega_0}{2G}\left(\frac{\partial B}{\partial \omega}\right)_{\omega = \omega_0}$$

$\omega_0$ = resonant frequency

B = susceptance

G = conductance

Show further that if susceptance B is a capacitance C is parallel with an inductance L that

$$C = \frac{1}{2}\left(\frac{\partial B}{\partial \omega}\right)_{\omega = \omega_0}$$

b6 . By assuming the equivalent circuit of Fig. 18c show that the frequency temperature stability is given by

$$\frac{dF}{dT} = \frac{\left(\frac{\partial B_c}{\partial T} + \frac{\partial B_0}{\partial T}\right)}{\left(\frac{\partial B_c}{\partial F} + \frac{\partial B_0}{\partial F}\right)}$$

$$= \frac{Q_0}{Q_L} \cdot \frac{dF_0}{dT} + \left(\frac{G_c}{G_c + G_L}\right)\frac{Q_c}{Q_L} \cdot \frac{dF_c}{dT}$$

$Q_L$ = loaded Q of oscillator.

$Q_D$ = Q of device

$Q_C$ = Q of circuit.

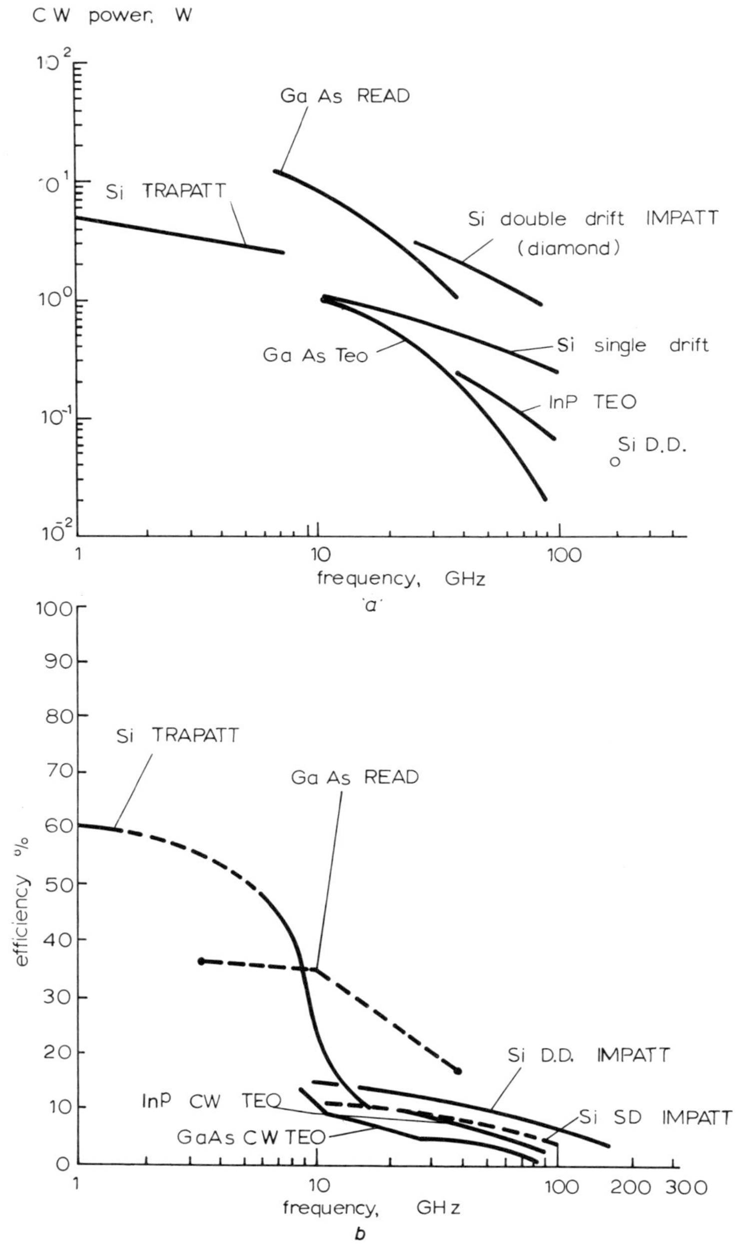


Figure 17 (a) Laboratory CW power performance of microwave diodes vs frequency
(b) Efficiency of microwave diodes vs frequency

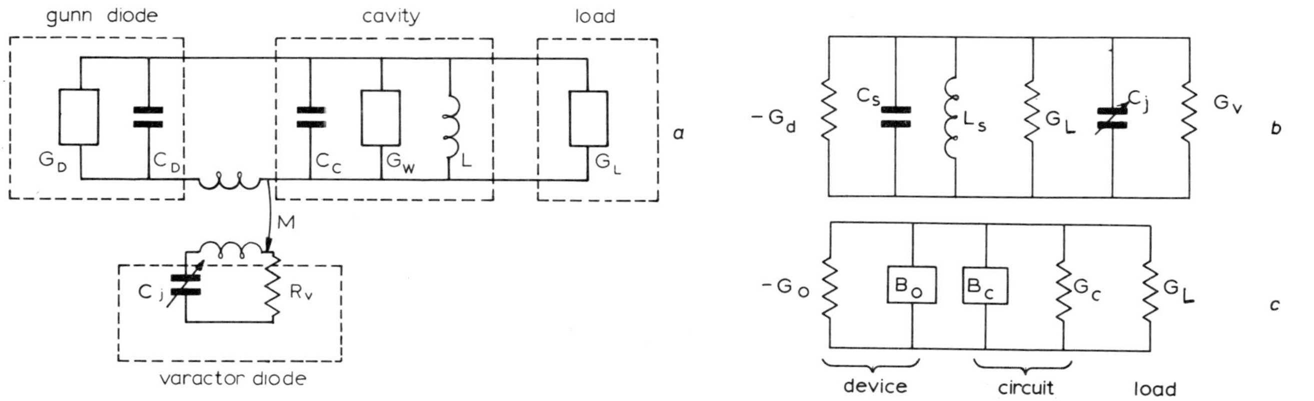


Figure 18 (a) Equivalent circuit of electronically tunable oscillator (b) Simplified equivalent circuit $C_j$ = varactor capacitance, $G_V$ = varactor loss conductance, $G_l$ = load conductance, $C_s$ = circuit + device capacitance, $L_s$ = circuit indictance, $-G_D$ = device conductance
(c) Basic resonant circuit

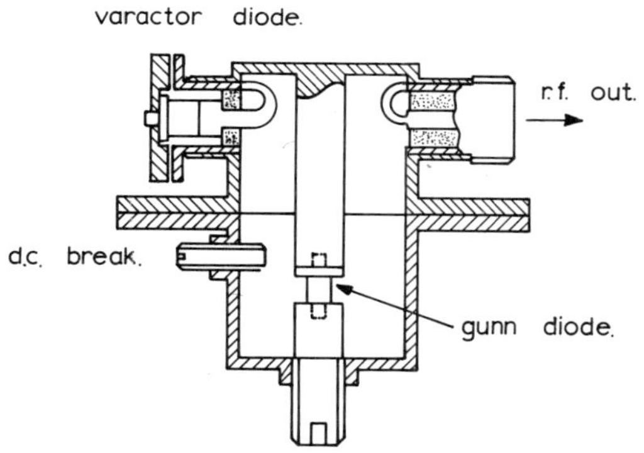


Figure 19 Schematic diagram of loop coupled coaxial cavity

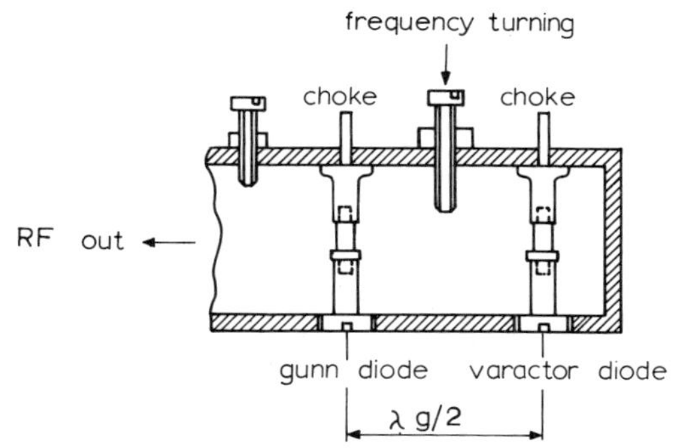


Figure 22 Typical varactor tuned waveguide oscillator

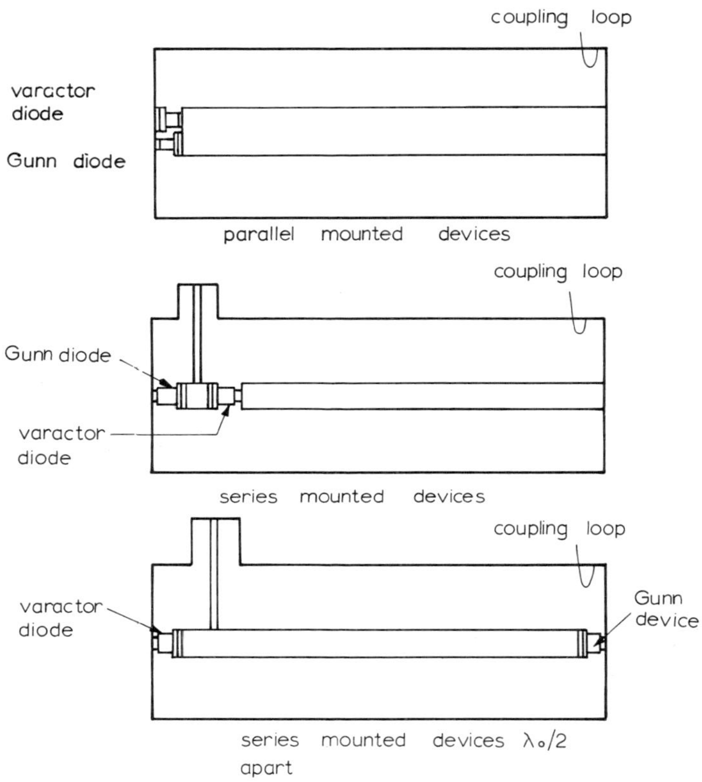


Figure 20 Schematic diagram of directly coupled cavities

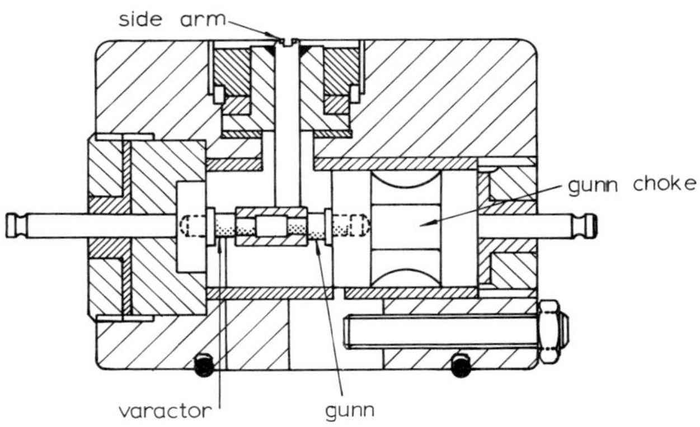


Figure 21 Practical circuit employing devices in same cavity

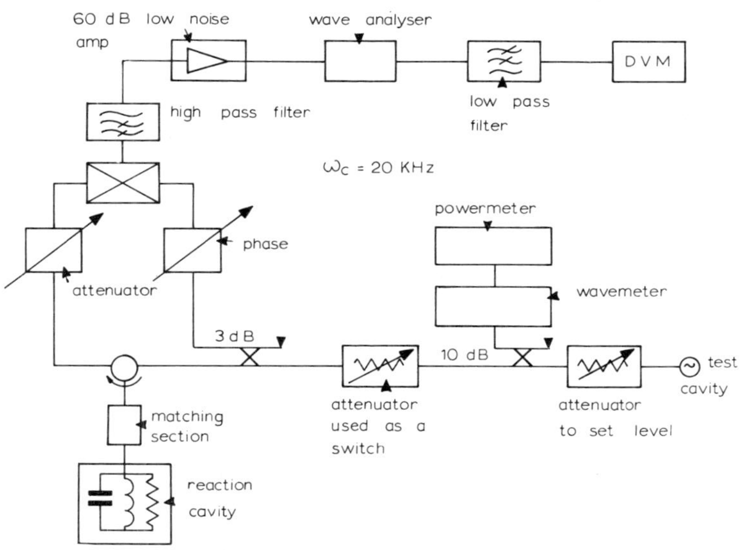


Figure 23 F.m. noise measurement equipment

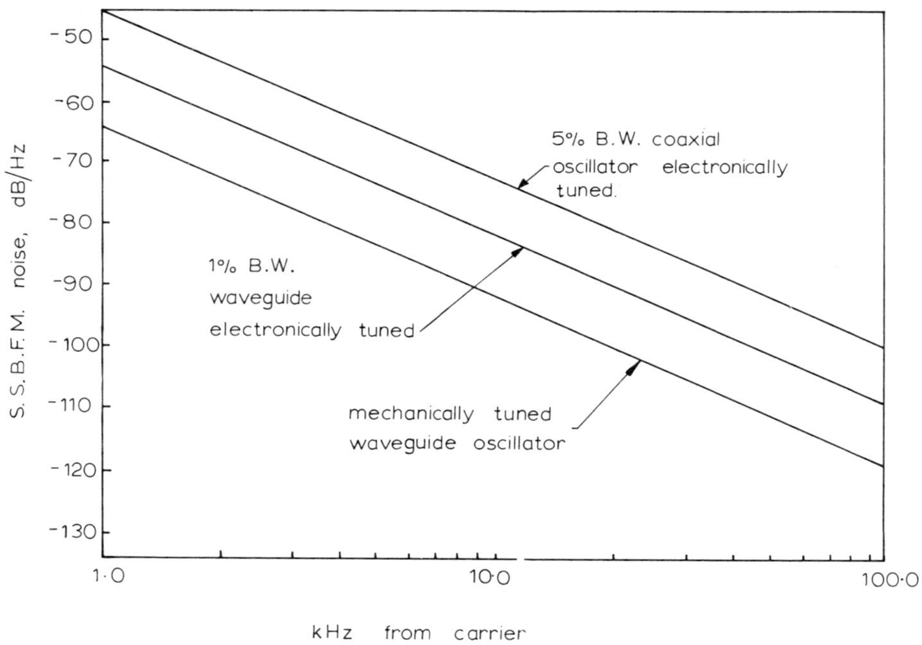


Figure 24 F.m. noise for various oscillators

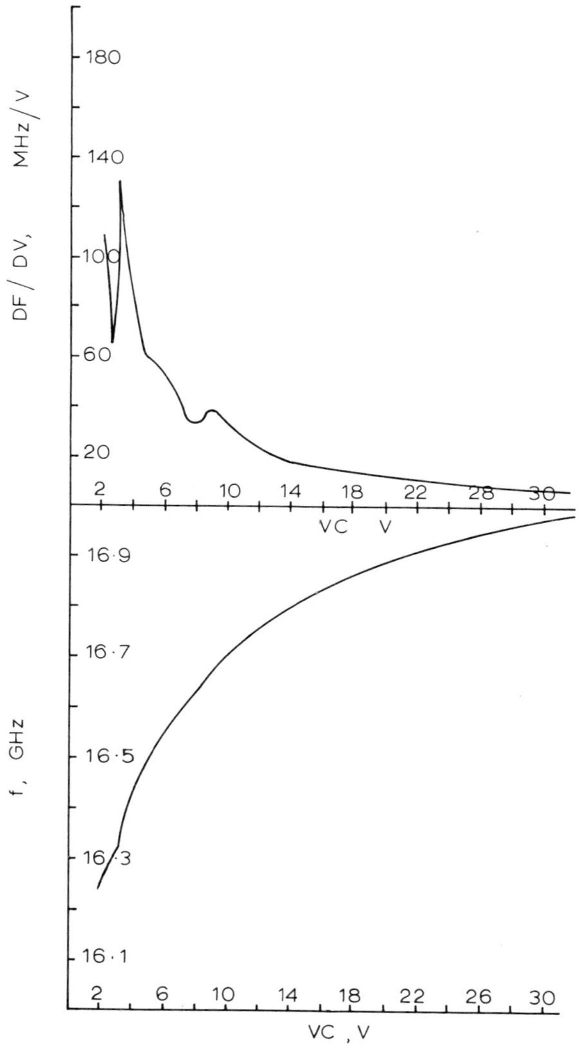


Figure 25 Fine grain non-linearities

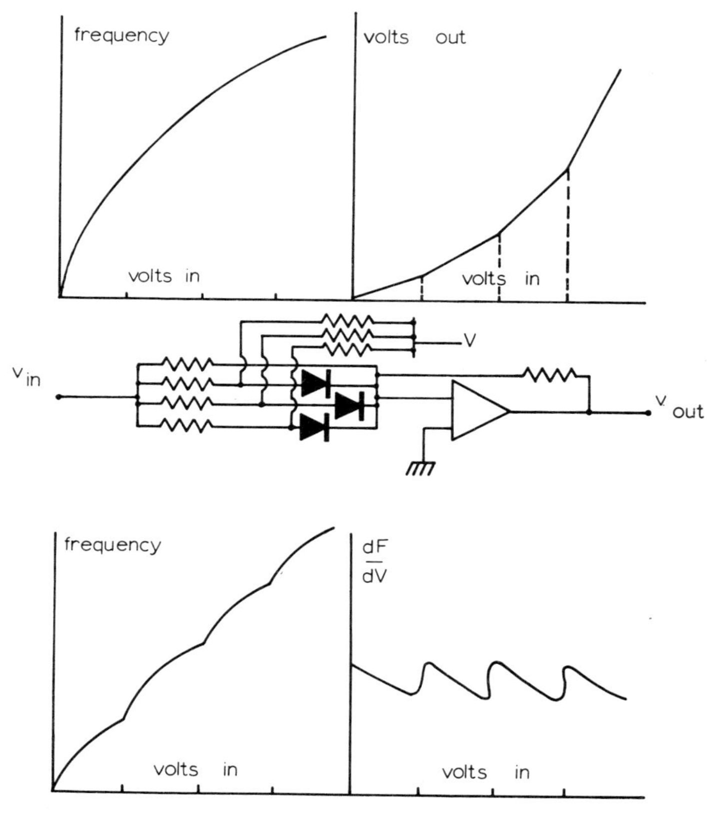


Figure 26 Break point lineariser

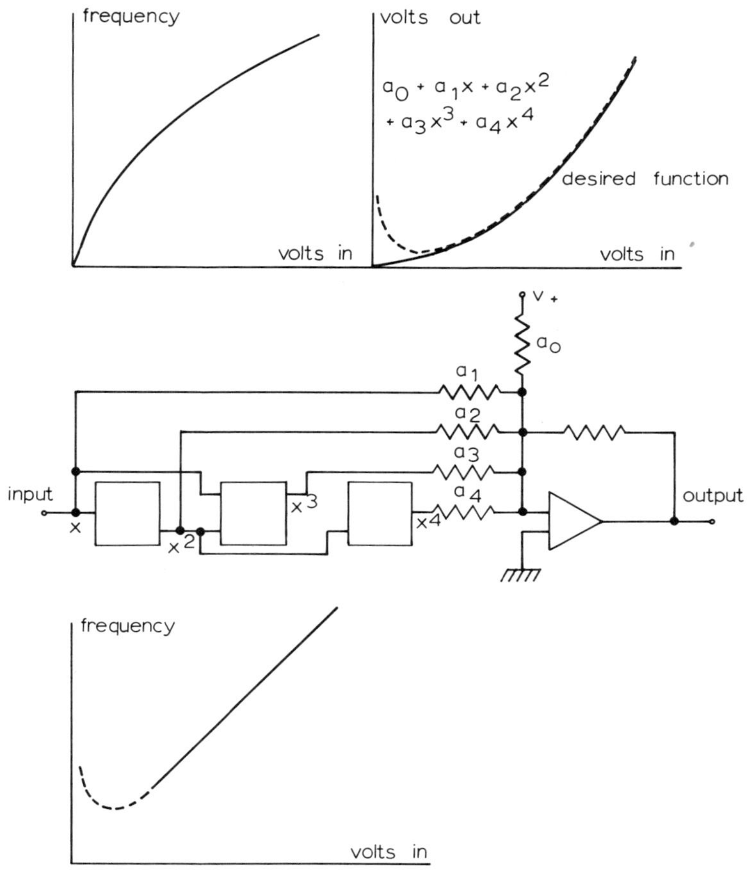


Figure 27 Continuous approximation lineariser

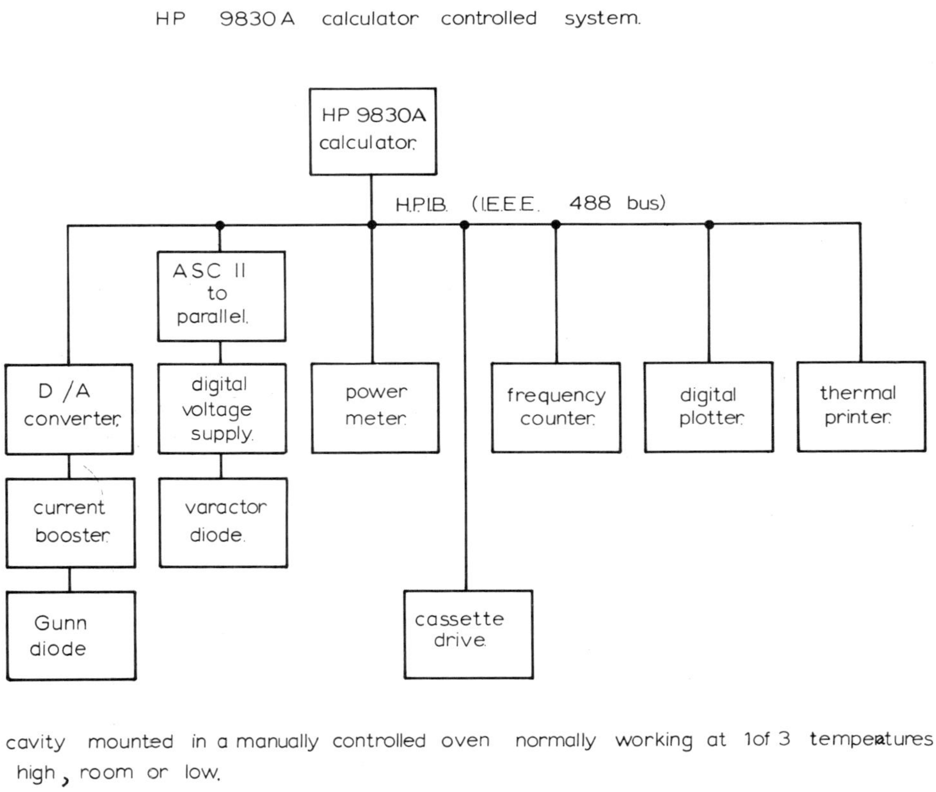


Figure 28 A.T.E. Block diagram

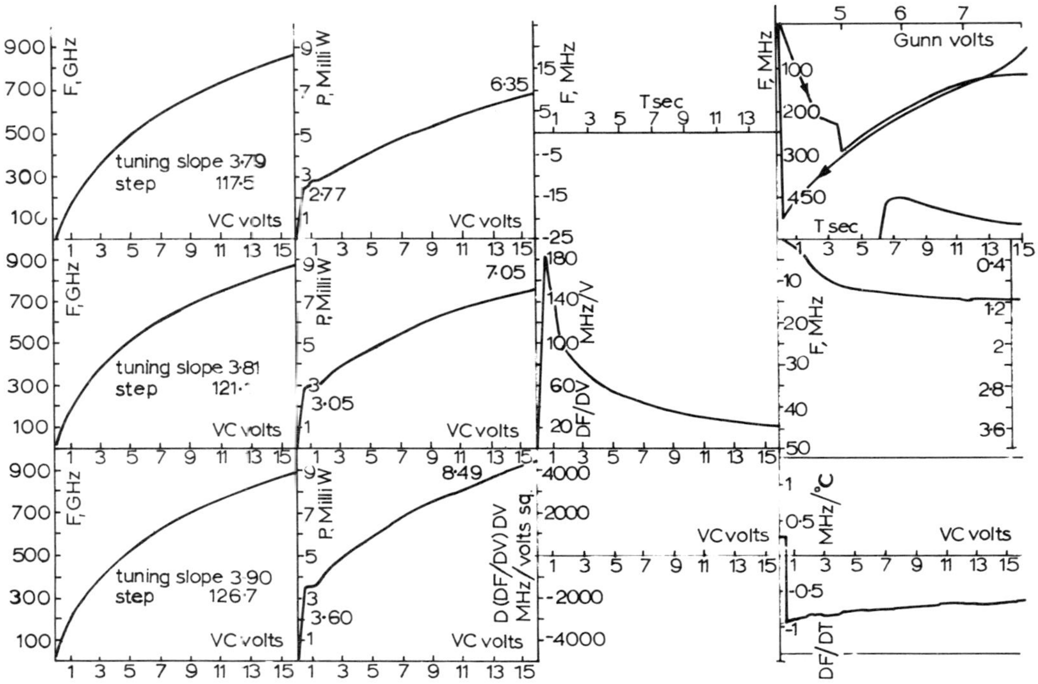


Figure 29 A.T.E. print out cf coaxial oscillator

# HIGH POWER PULSED OSCILLATORS

F.A. Myers

Plessey Research (Caswell) Ltd.

## INTRODUCTION

With the advent of practical pulsed solid state microwave generators about 10 years ago (the Gunn diode or GaAs Transferred Electron Oscillator), the designer of microwave systems found himself dealing with a novel device requiring completely different handling to the conventional devices then available, e.g. magnetrons, klystrons. In fact, so novel were the device's operational parameters that it has taken many years for them to enter true systems use and in the meantime many other devices have been developed which can complement the Gunn diode or in some cases supercede it.

In the discussions that follow, a brief summary of available pulsed sources will be given and then the operational parameters and design guidelines for the pulsed Gunn oscillator will be discussed. This oscillator is the most common one encountered in systems to date (e.g. transponders, portable radars, radar altimeters etc.) and illustrates most of the advantages/problems encountered.

## AVAILABLE SOURCES

Referring to Fig.30(a,b) it can be seen that over the frequency range 4 to >100 GHz that there are potentially many available devices, some being "record" holders for power/efficiency and others, like the Gunn diode, apparently not offering any obvious advantages although in terms of their second order parameters they do.

### Trapatt

The Trapatt device (12) (Trapped Plasma Avalanche Triggered Transit) is a particular type of avalanche device offering reasonable peak power at very high efficiency up to about 10 GHz. Above this frequency, material growth difficulties limit the device capability but it is enjoying limited systems use at the low microwave frequencies, primarily in relatively straightforward pulsed applications.

### GaAs Read, or High Efficiency Impatt

The pulsed modes of operation of the high efficiency Impatt (13) are relatively new and extremely encouraging. By multiple chip combining techniques 10's of watts of peak power have been achieved at very high duty cycles (25%). For applications requiring modest peak powers but high mean powers (e.g. phased array systems, missile homing heads) then this device is highly attractive. Although historically the Impatt has not been thought to have a good rise time performance this is certainly not the case for this device.

### Si Impatt

The Si avalanche device (14) has been mostly used as a CW generator although power levels of up to 50 watts peak have been achieved at X-band. The main area of application for this device, however, is likely to be at mm wavelengths where very dramatic power levels (∼10W at 94 GHz) have been achieved. As these frequencies become exploited this device will mature and the silicon 'chip' will no longer be the preserve of the microprocessor!

### Transistors

The Si bipolar transistor is a very powerful generator of r.f. energy up to the microwave region but to date does not extend beyond about 4 GHz. The power device types of the GaAs FET (15), although being extremely useful generators of CW power, to date do not have a very high pulse-up capability (i.e. pulsed power ∼ CW power). Although this situation may change they are therefore not presently used for pulsed applications.

### Gunn Device

Although producing high peak powers this device is the lowest efficiency (at least in GaAs) and is limited in frequency to about 40 or 50 GHz. However, it is a fairly simple, rugged device and relatively straightforward to handle and is the one most commonly encountered and will be discussed below in more detail. Alternative materials such as InP are improving the efficiency by at least a factor of 2 and are likely to offer the same improvement for frequency and this material should start to enter systems in a few years time.

## PULSED GUNN OSCILLATORS

Pulsed Gunn oscillators are produced in a wide variety of circuits (e.g. waveguide, coaxial or microstrip) each with its own peculiar operating parameters and design problems. Each circuit type will be dealt with below and it should be noted that there can be a read-across of techniques from one circuit to the other although for brevity these will not be discussed. For example high Q locking resonators will be discussed with waveguide oscillators but they are equally applicable to microstrip where the waveguide resonator could be replaced by a barium titanate cylindrical resonator.

One aspect that should be stressed is that in the case of the active two terminal generators there are no equivalents to, say, a transistor 'S' parameter. Although equivalent circuit models exist and, with difficulty, actual device impedances can be measured, the values obtained do not usually enable theoretical design to take place. It is hoped that this situation will change.

### Coaxial Oscillator

The schematic diagram of a typical coaxial oscillator is shown in Fig.31, for operation in either X or J-band depending upon the cavity dimensions and the device used.

In order not to compromise the oscillator efficiency the cavity is designed for maximum unloaded Q-factor. This is given by

$$Q_o = \frac{\lambda_o}{\delta(4 + \frac{2L}{D}(1 + \frac{D}{d})/(\ln \frac{D}{d})}$$

where $\delta$ is skin depth, L is cavity length, D and d are diameters of outer and inner conductor and $\lambda_o$ is the free space wavelength. $Q_o$ reaches a maximum when $D/d = 3.6$ and is given by

$$Q_{o\,max} = \frac{\lambda_o}{\delta(4 + \frac{7.2L}{D})}$$

In order not to support higher order modes the dimensions must be such that

$$d + D < \frac{\lambda_c}{1.47} \quad (\text{for } D \sim 3 \times d)$$

where $\lambda_c$ is the cut off frequency of the $TE_{11}$ mode. The characteristic impedance is then given by

$$Z_o = \frac{130}{\sqrt{\epsilon_r \mu_r}} \log_{10} D/d \quad (\sim 77 \text{ ohm for optimum Q air line})$$

Having designed the cavity with the above constraints in mind the operating frequency is simply set by the cavity length:- ideally $\lambda_o/2$ but in practice foreshortened to allow for the loading effect of the device. Power is extracted, and the device matched, by either an electric field probe $\lambda/4$ from a short circuit or by a magnetic field iris, at or close to, the short circuit plane.

In practice, the coaxial oscillator is very flexible in matching devices and forms a very good basic test circuit. However the external Q-factor is low (typically 10 to 50) and this results in very bad stability, limiting its use for the majority of systems. The frequency drift during the pulse (chirp) is high, typically 50 to 100 MHz/0.5μsec for a 10W peak device and the frequency temperature coefficient is large, typically -1.5 MHz/°C at X-band.

However, because of the low Q-factor and small linear dimensions this circuit does excel in one parameter, namely r.f. rise time. Shown in Fig.32 is the resultant r.f. output for a bias pulse input length of 1 nsec. It should be pointed out, however, that this can only be achieved by the use of a specially optimised pulsed device with minimum parasitic resistance in order to maximise the small negative resistance.

## Waveguide Oscillators

Waveguide is the most commonly used medium for constructing the resonators, owing to its reasonable stability, but unfortunately the designs are based to a large extent on experience and microwave 'green fingers'. This situation is slowly being rectified (16), (17), but design principles at the moment remain outside of day to day use.

However, some basic principles can be discussed. Two main approaches have been made to the design of waveguide oscillators. The first consists of mounting the diode across the guide approximately $\lambda g/2$ away from a short circuit as shown schematically in Figure 33a. The diode is usually mounted on posts often at mid-guide height to deter coaxial modes and coupling and frequency tuning achieved by inserting screws between the diode and the load and the diode and the short circuit, respectively. The second, (Figure 33b) consists of mounting the diode close to a short circuit, approximately $\lambda g/2$ from a thin iris which couples power into the load. Output coupling is set by the iris dimensions and frequency tuning is achieved by insertion of a screw, usually dielectric, between the diode and the iris. Both arrangements which are electrically very similar, lead to cheap production methods such as castings, pressings etc., and both have high stability. In order to raise $Q_L$ the diode can be mounted in a smaller guide than "normal" and, if required, the impedance matched by moving the diode off-centre into a lower impedance plane. The frequency of operation is closer to guide cut off consequently with a higher $\partial B/\partial \omega$, and hence $Q_L$. By suitable choice of step length between guide sizes the mismatch can be corrected. To allow for diode to diode variations, the length between the diode and the short circuit or iris is set to be slightly shorter than that for the required operating frequency and the frequency screw inserted a small amount. See Figure 34. Too much screw penetration lowers the cavity $Q_o$. The diode is mounted on posts which often have to be determined experimentally. Larger posts will give a lower $Q_o$ but better switch on performance. Thin posts give higher stability but can give low temperature switch-on problems.

The position of the output tuning screw is determined by matching, using a moving stub tuner and transforming back an integral number of half guide-wavelengths. The closer the screw is to the diode, the greater the bandwidth. With a large diode/tuning screw separation, switch-on problems can occur.

Both the cavity styles shown in Fig.33 are used in practical oscillators, although the system shown in Fig.33a is more commonly employed owing to its somewhat simpler mechanical configuration.

The waveguide oscillators described above form the basis for most of the designs in use. They have a reasonable external Q-factor (50 to 200 depending on whether the diode is high or low powered) with reasonable chirp (~20-50 MHz/0.5μsecs) and temperature stability (~-0.5 MHz/°C at X-band). The emphasis, however is on "reasonable". For some systems, such as a radar altimeter operating at short pulse lengths the stability is adequate, for others the combination of chirp and temperature drift may take the frequency outside of the receiver bandwidth, or compromise some other systems parameter. It is therefore often necessary to further improve stability.

Temperature Stability. The temperature stability is related to the external Q-factor as shown in Fig.35. As the Q-factor exceeds about 500 the dF/dT approaches a limit defined by the expansion coefficient of the material used. The Q-factor can be increased (see cavity locking below) but even then some other technique must be used to ensure that the basic active oscillator temperature stability does not exceed that of the locking resonator.

Several methods have been described and a commonly used one is that involving a reactive stub, which is arranged to minimise the change in cavity resonant frequency as the temperature varies (18). A schematic is shown in Figure 36. The tuning rod is attached at the top to a support, e.g. nylon, mounted on the waveguide. The tuner can be a metal such as Invar, arranged to be non-contacting, or a dielectric. As the temperature increases, the support expands and withdraws the tuning rod, thus raising the frequency. This compensates for the expansion of metal, or changes in the device, which would normally lower the frequency by making the cavity longer. Compensation down to $\pm$ 50 kHz/$^{o}$C is easily achievable at X-band and down to $\pm$ 20 kHz/$^{o}$C with patience. The tuning rate for metal rods can be changed by varying the initial penetration of the tuner. For dielectric tuning, materials such as Lucalox offer greater compensation for the same length of rod over PTFE, quartz, etc. To design a metal rod compensated cavity, a plot is first made of tuner penetration versus oscillator frequency. Hence the change in tuner length for a given frequency change is found. To a first order, only the expansion of the support is considered and this then is used to define support lengths and hence tuner length. By moving the tuner to a region of high tuning sensitivity, overcompensation may be achieved.

Despite adding extra mechanical complexity to the oscillator this approach is quite simple and fairly rugged w.r.t. shock, vibration etc.

Chirp Reduction. Various techniques are available for minimising chirp, such as varactor compensation, cavity locking, frequency or phase locking etc. The first two techniques are discussed here:-

Varactor Compensation. By inserting a varactor diode into the waveguide resonator, either in the same E-plane or $\lambda g/2$ removed from the Gunn diode (see Fig.37a), a small but useful amount of electronic tuning can be obtained.

This varactor can be used in a phase lock system for general frequency control but can be used also for an open loop chirp compensation system (19). The chirp in solid state oscillators is caused by a temperature rise in the active device chip and has an almost exponential dependence with time. By simply integrating the bias voltage to the active device and applying the resultant non linear ramp to the varactor the frequency of the resonator can be adjusted to increase during the pulse by almost the same magnitude as the decrease caused by device heating. Chirp values of $\sim$5 MHz/0.5$\mu$sec at 17 GHz can be fairly readily obtained by this technique. Lower values can be obtained but at the expense of added complexity and setting up time. Fig.37b shows a temperature compensated integrating circuit that can produce, with patience, chirp values of 2 to 3 MHz/ 0.5$\mu$secs over the tempature range -40 to +70$^{o}$C.

The technique is only suitable for reducing chirp down to the order discussed above. Lower values would need complex linearising circuitry (and infinite patience). It is also limited to low power devices. Even at the 5W level the varactor is still conducting forward current at a reverse bias of 20 volts implying a 40 volt peak to peak voltage for the particular coupling employed. A fine compromise between chirp, coupling and diode forward burn out, reverse breakdown etc. is implied.

Cavity Locking. The second method commonly used to reduce chirp, and improve temperature stability, insensitivity to load mismatch etc. is to use a high Q locking resonator (20). Chirp values $<$ 1 MHz/0.5$\mu$sec can be achieved by this technique.

Three basic formats for cavity stabilisation can be used. These are shown schematically in Fig.38a,b,c. For a pulsed oscillator 9a is the most usable as this one allows the generation of r.f. power at full magnitude before the locking resonator fields are established and can therefore be used for fast rise time locked oscillators. For the case of Fig. b, c the output gradually rises to maximum at a rate defined by the high Q resonator Q-limited rise time.

Shown in Fig. 39 a and b are the schematic diagram and equivalent circuit of a practical pulsed oscillator with external resonator. The condition for stable oscillation is that the susceptances of the basic oscillator and high Q resonator should sum to zero. This can be seen more clearly in Fig. 40. The lower graph represents a circuit with external Q-factor of 50 and it can be seen that a given change of susceptance $\Delta$ B will result in a change in frequency $\Delta$f. However, if this oscillator is stabilised with a resonant cavity of unloaded Q-factor 3000 with an insertion loss of 2 dB then the same change in suspectance $\Delta$ B will result in a vastly reduced change in frequency. This is the basic mechanism of cavity stabilisation.

The degree of stability obtained is primarily a function of the unloaded cavity Q-factor. In Fig. 41 are shown B($\omega$) curves for three cavities with various Q-factors. For a given susceptance change the resulting frequency change decreases with increasing Q-factor.

However, the oscillator does not become stabilised until the high Q cavity fields are established and hence the oscillator exhibits free running chirp for a time equivalent to the Q-limited rise time, and stabilised chirp thereafter. It can thus be seen that there is an optimum Q-factor for the locking cavity for any particular pulse length to obtain minimum overall chirp. This is illustrated in Fig.42 . Figs.43a and 43b show the spectrum of a high power pulsed InP Gunn oscillator with and without the external cavity.

## Microstrip Resonators

Microstrip resonators appear at first sight very attractive as they are small, lightweight and ought to be readily manufacturable and have relatively low cost. In practice, however, the external Q-factor of a microstrip resonator is extremely low ($\sim$10) and the stability very poor. By the time techniques are applied to improve the stability the complexity of the approach defeats the object. They are therefore seldom used for free running pulsed oscillators.

However, if injection locking is required for a particular system, such as an agile transmitter with a bandwidth that cannot be obtained by a directly coupled varactor, then the low Q-factor can be turned to advantage. In Fig. 44 is shown schematically the approach taken to produce a 10W pulsed source in J-band with a 6% bandwidth driven from a low power oscillator. Shown in Fig. 45 is a close up of the final power output stage illustrating the general principles employed. Because of the low impedance of the pulsed devices and the limited range of impedances that can be presented by a microstrip matching circuit residual parasitics caused by wire bonding, stray capacitances etc. are critical, Incidentally, the frequency chirp is reduced to zero and locking can be used for this purpose.

## Miscellaneous Design Points

**Power Combination.** Higher output powers can be achieved in practical oscillators by incorporating several discretely packaged devices in a waveguide (or other) array. Shown in Fig. 46 is a 4 diode cavity designed to produce in excess of 40W peak at 15 GHz. Chip combining can be, and is, carried out but requires very complicated bonding and the utilisation of diamond heat sinks etc. and at present is a less attractive alternative.

**Phase Priming.** Phase priming, the injection of a low level signal outside of the locking range of the pulsed source, can be used in systems to advantage. This can have two effects. The first of these is to reduce any jitter and/or delay that may occur during the leading edge of the r.f. pulse. This is particularly important in the case of an InP pulsed TEO which may otherwise have a delay and jitter of 50 - 70 nsecs at -40°C. The effect seems to be to drive the device through the noise build up time and establish the fields somewhat more rapidly. The effect on delay as a function of priming power and offset from the oscillator is shown in Fig. 47.

The second effect is to produce coherency in the spectrum of the pulsed source providing (a) the pulse modulator is crystal referenced and hence free from pulse to pulse jitter and (b) the signal source used for priming is stable. Under these conditions the individual Fourier p.r.f. lines in the spectrum can be observed and utilised for signal processing.

**Power Supplies.** It should be obvious that to obtain most or all of the above parameters from a practical oscillator requires very careful design for the modulator. Each device type will require its own type of modulator (e.g. voltage drive for a Gunn, current drive for an Impatt), but it should be emphasised that the design of these is by no means trivial. As an example to obtain fast rise times from a high power pulsed GaAs TEO the circuit used in Fig. 48 has been developed. This uses a relatively slow transistor to drive the device close to threshold and hence supply about 80% of the operating current (up to 10 Amps) and then an avalanche transistor switching a delay line to define the fast pulse. Rise times of the order of 1 nsecs can be readily generated by this technique.

## REFERENCES


12 Oxley, C.H., Howard, A.M. and Purcell, J.J., IEE SSED Vol. 1, pp.24-30.

13 Gewartowski, J.W., IEEE Trans. MTT, Vol. 27, pp.434-442.

14 Midford, T.A., Bernick, R.L., IEEE Trans. MTT, Vol. 27, pp.483-492.

15 DiLorenzo, J.V., Wisseman, W.R., IEEE Trans. MTT, Vol. 27, pp.367-378.

16 White, J.F., IEEE Trans. MTT, Vol. 20, pp.372-378.

17 Eddison, I., Howes, M.J., IEE M.O.A. Vol. 1, pp.103-109.

18 Stevens, R., Myers, F.A., IEE Elec. Lett. Vol. 10, pp.463-464.

19 Bullimore, E.D., Myers, F.A., IEE Elec. Lett., Vol. 10, pp.220-221.

20 Smith, D.C., Tebbenham, R.L., Proceedings 9th European Microwave Conference, pp.538-542.


## QUESTIONS

C1 It is required to temperature compensate a pulsed oscillator with a frequency/temperature coefficient of -0.5 MHz/°C, operating from 0 to +100°C.

Using a nylon/invar temperature compensator, as described in the lecture notes calculate the length of compensator required to fully stabilise the oscillator. The temperature coefficient of nylon can be taken to be 100 ppm/°C and the tuning rate of the invar stub as 0.25 MHz/micron.

C2 If the same compensator as above is used on a non-linear oscillator with a frequency temperature coefficient varying linearly between -0.3 MHz/°C at 0°C and -0.7 MHz/°C at +100°C calculate the maximum frequency deviation over temperature and the temperature at which this occurs.

C3 Using the equivalent circuit of Fig. 39 (b) show that the normalised susceptance presented to the active oscillator by the absorbtion cavity is given by

$$B_{1-1} = \frac{Q_o V \varepsilon}{(1+V)^2 + (Q_o V \varepsilon)^2}$$

where $V = Q_e/Q_o$ (for the absorbtion cavity);
$\varepsilon = 2 \times 2\Delta\omega/\omega_o$

Note It can be assumed that the susceptance at the reaction cavity terminals is given by

$$Y_{2-2} = V(1 + jQ_o\varepsilon)$$

C4 Using the result obtained above, show that the bandwidth of the locked oscillator is given by

$$\Delta f_{TOTAL} = \frac{f_o}{2Q_{em}(1+V)}$$

$Q_{em}$ is the active oscillator external Q-factor.

C5 A 4 stage locked pulsed oscillator generating 10W peak output with 48 dBs gain at 15 GHz satisfies the Adler locking equation

$$\Delta f = \frac{2 \times f_o}{Q} \left(\frac{P_{in}}{P_{out}}\right)^{\frac{1}{2}}$$

($\Delta f$ = bandwidth, $f_o$ = centre frequency).

Assuming (1) A Q-factor of 10

(2) Circuit losses can be neglected

(3) Locking power does not add to the output power

Calculate

(1) Overall bandwidth

(2) Power output/stage and locking power.

Explain why, in a practical system, that the bandwidth calculated will not all be available for systems use.

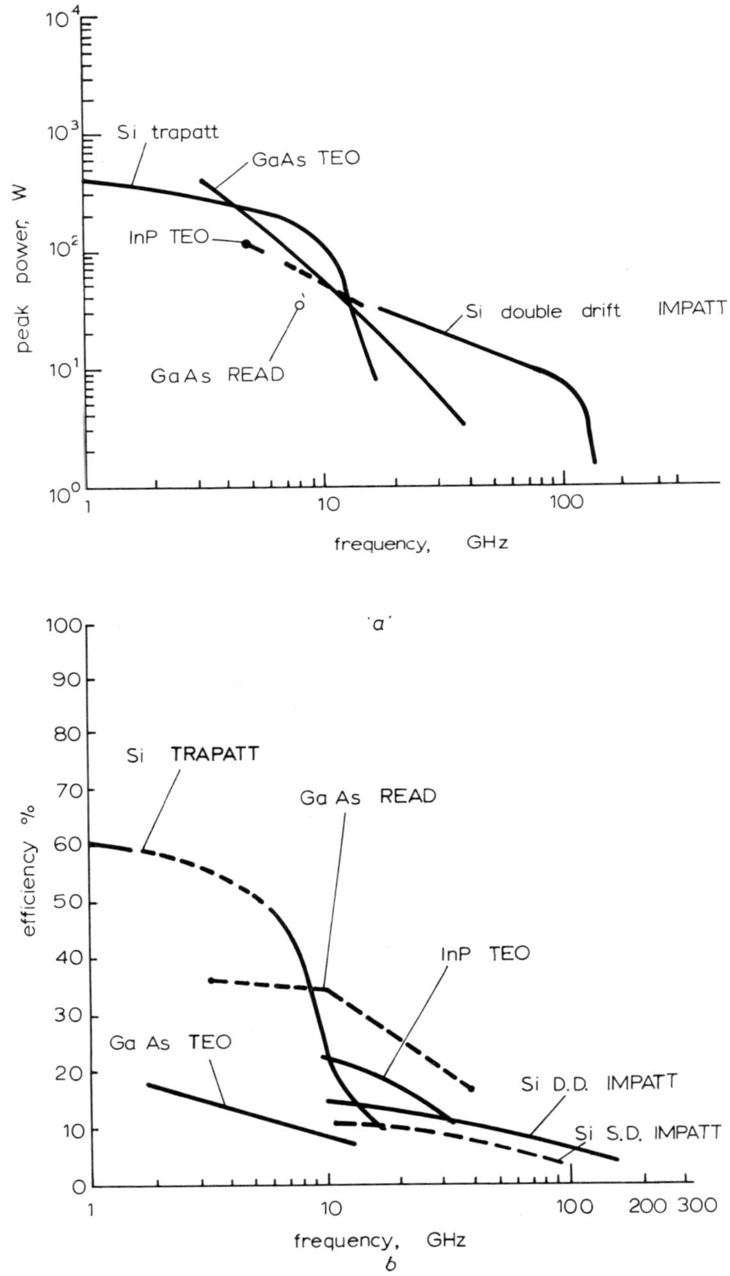


Figure 30 (a) Peak power performance of microwave diodes vs frequency

(b) Efficiences of microwave diodes vs frequency

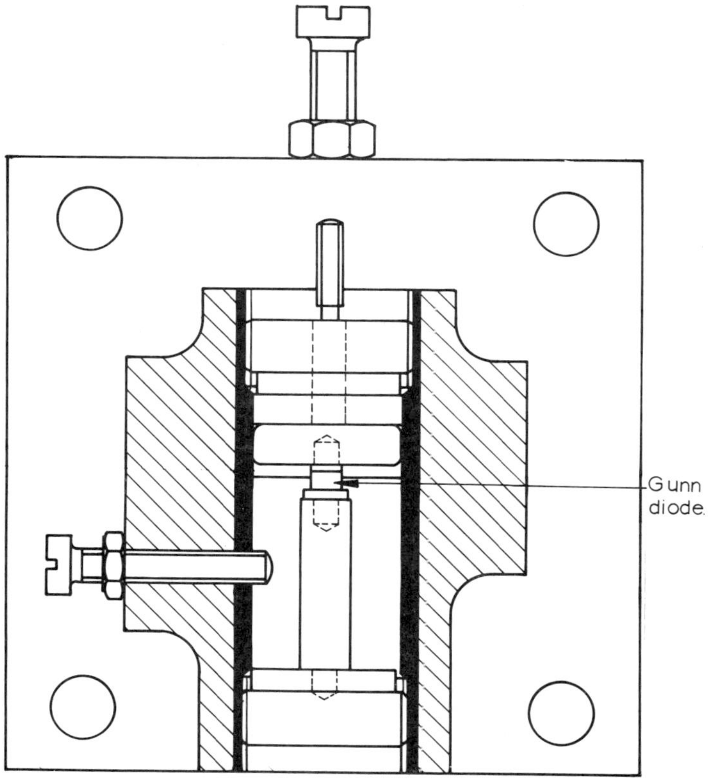


Figure 31 Schematic of coaxial cavity

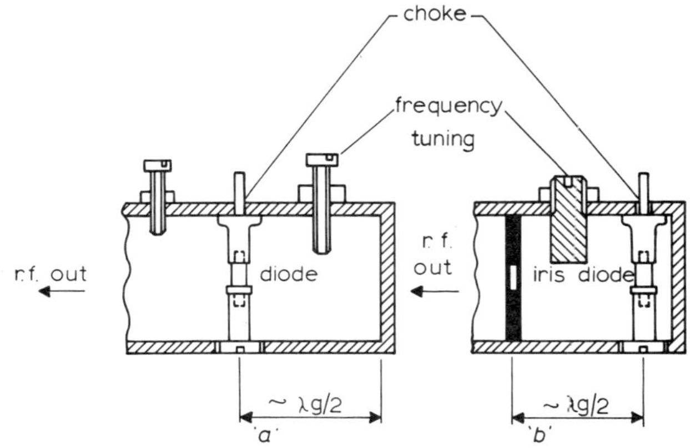


Figure 33 Waveguide cavities
(a) Directly coupled
(b) Iris-coupled

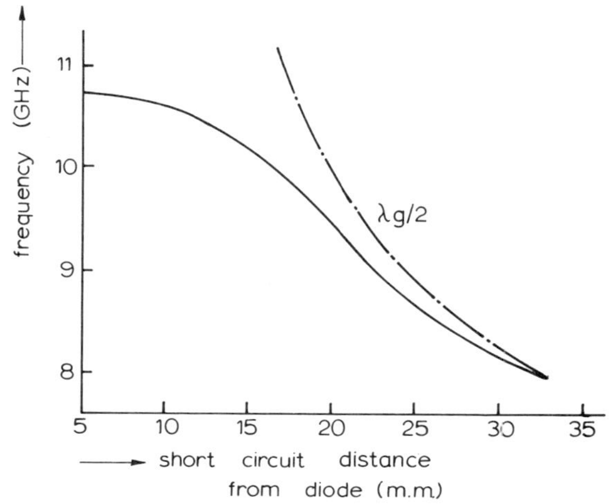


Figure 34 Typical relation between frequency and short circuit distance from diode in WG16 at X-band

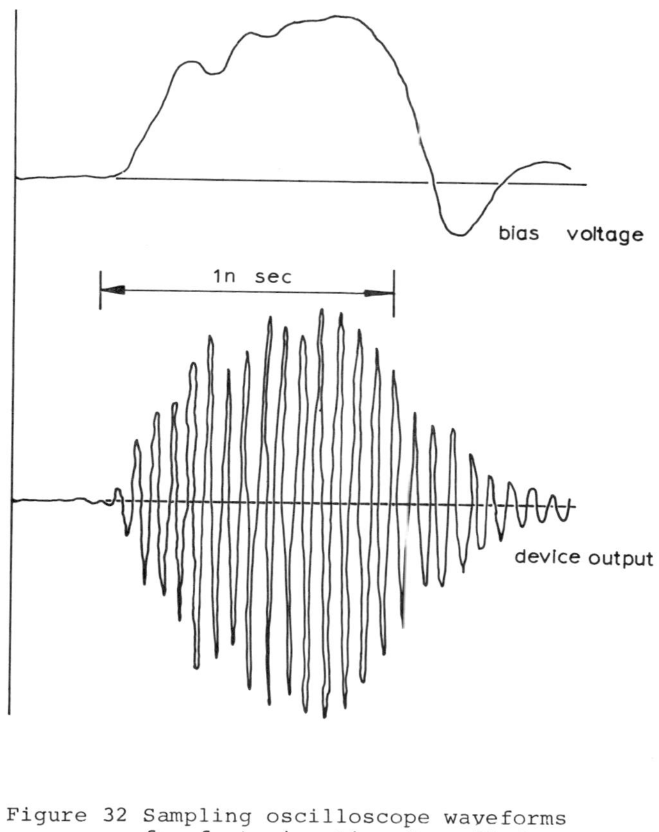


Figure 32 Sampling oscilloscope waveforms for fast rise time Gunn diode

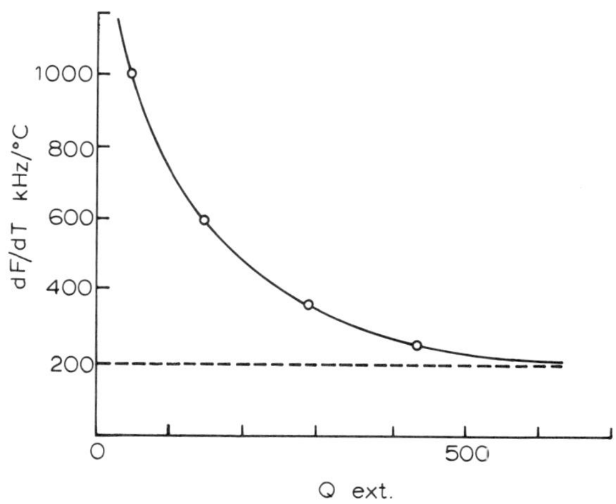


Figure 35 Frequency/temperature coefficient as a function of external Q

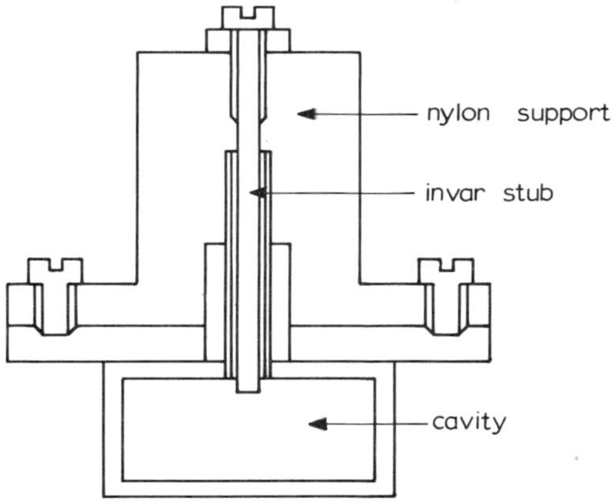


Figure 36 Temperature compensator

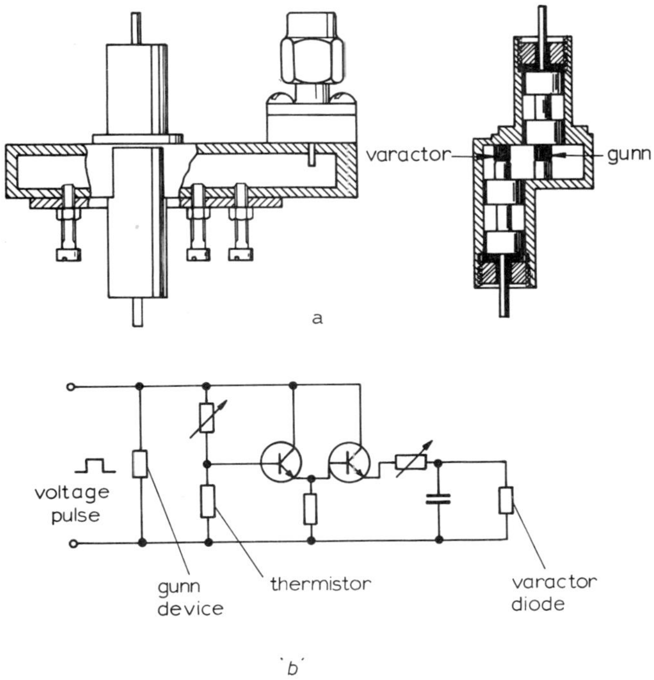


Figure 37 Chirp compensated cavity (a) and circuit (b)

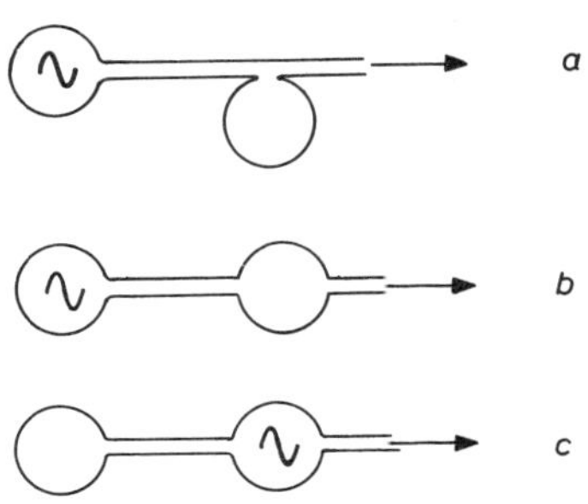


Figure 38 Schematic diagrams of cavity stabilisation

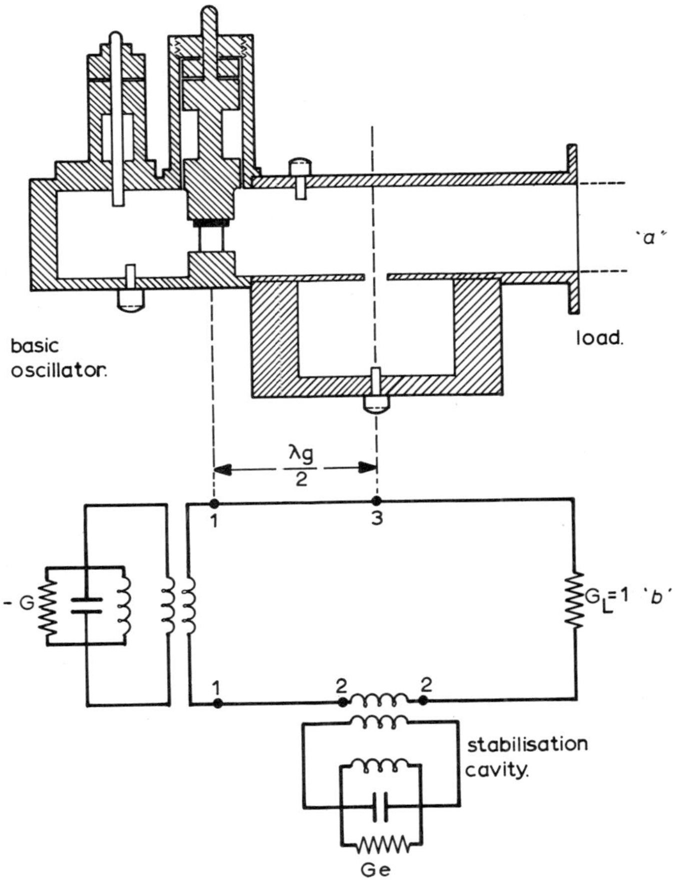


Figure 39 (a) Schematic diagram of a practical stabilised pulsed oscillator

(b) Equivalent circuit of stabilised oscillator

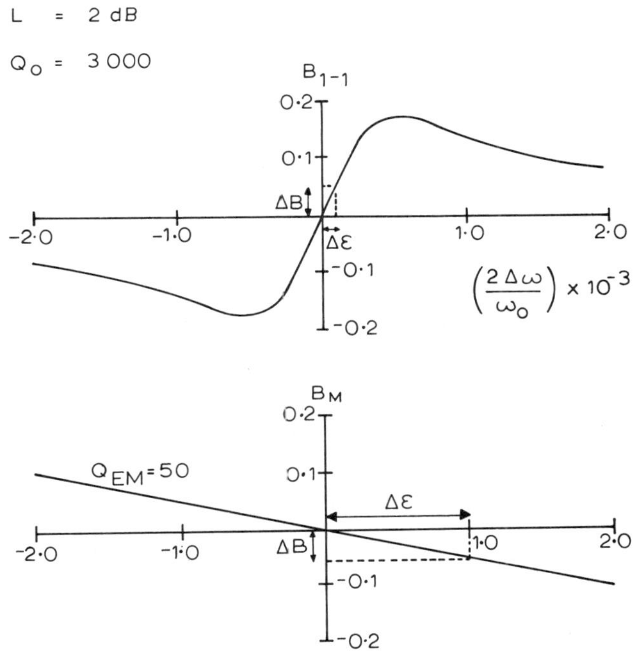


Figure 40 Cavity susceptance as a function of frequency

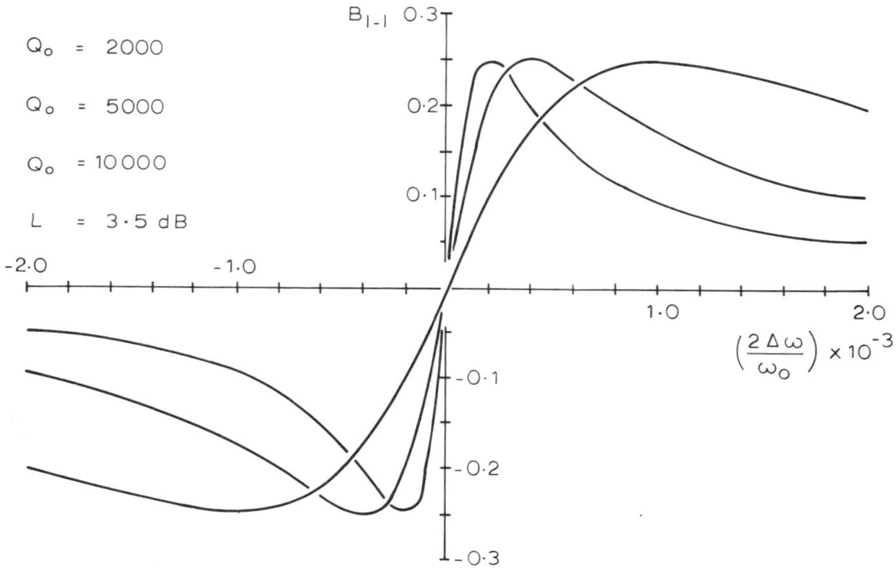


Figure 41 Stabilising cavity susceptance as a function of frequency for various Q values

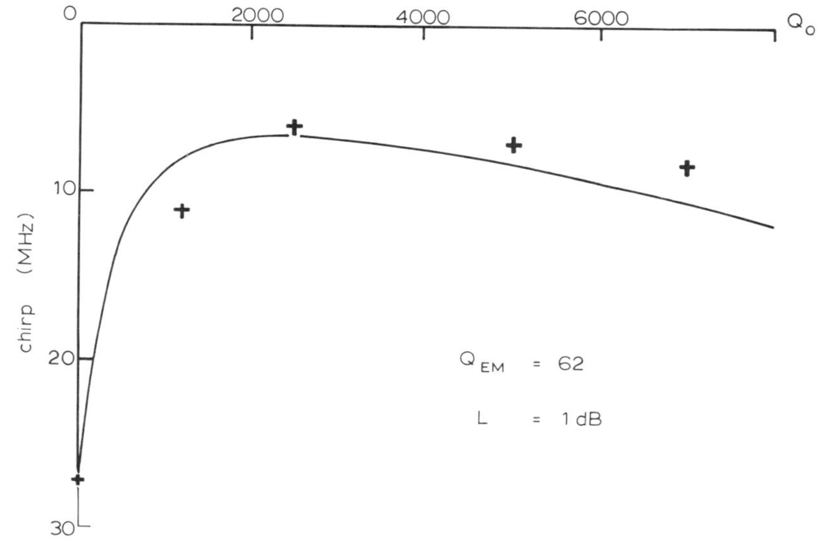


Figure 42 Chirp as a function of stabilising cavity Q

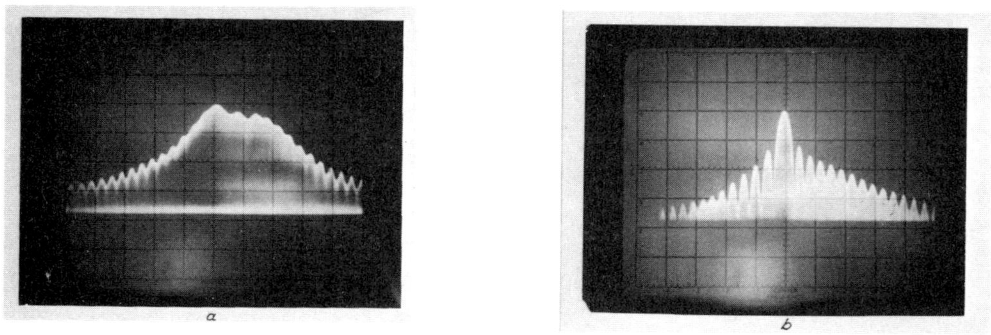


Figure 43 Photograph of spectrum of high power pulsed source
(a) Unstabilised unstabilised F = 28MHz; (b)
10MHz/division, L = 1dB. (a) F = 8MHz, $Q_o$ = 7000

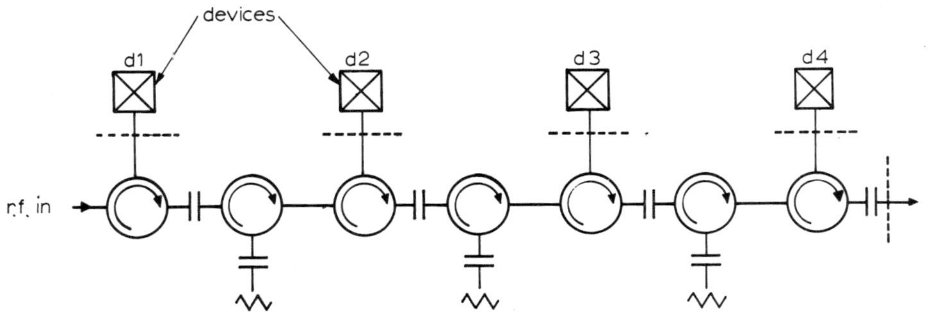


Figure 44  Four diode locked oscillator array

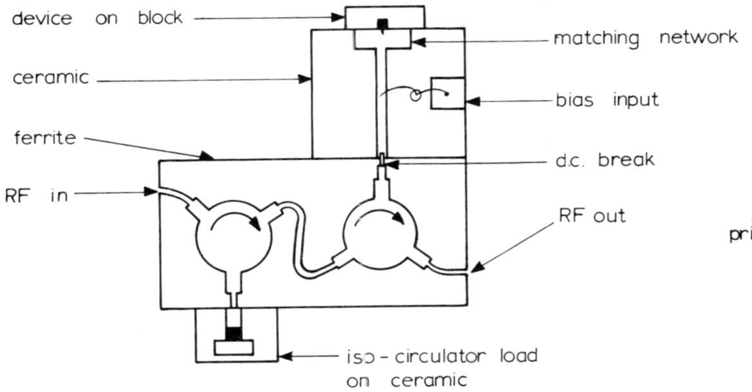


Figure 45  Detail of final power output stage

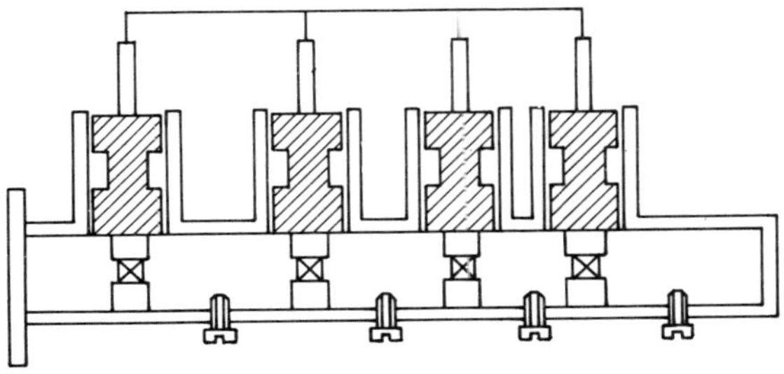

Figure 46  Schematic of four diode array

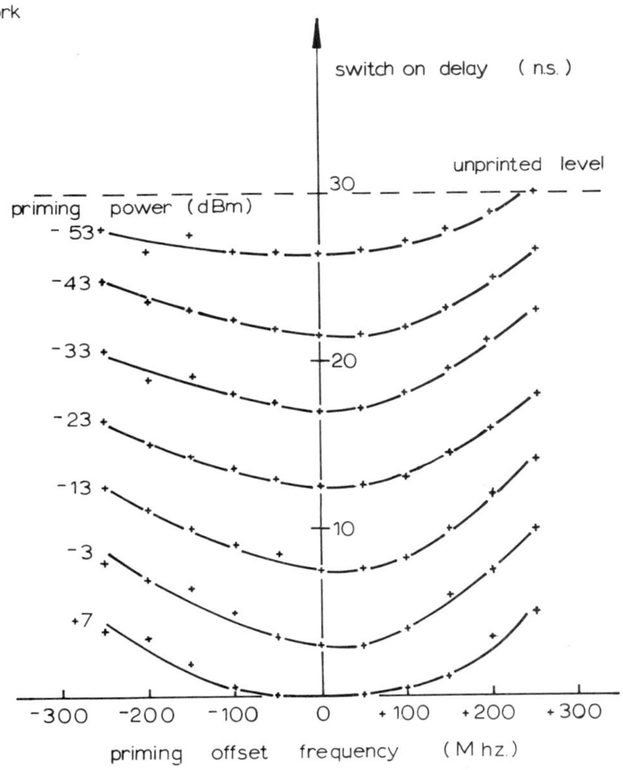


Figure 47  Switch on delay as a function of priming power and frequency

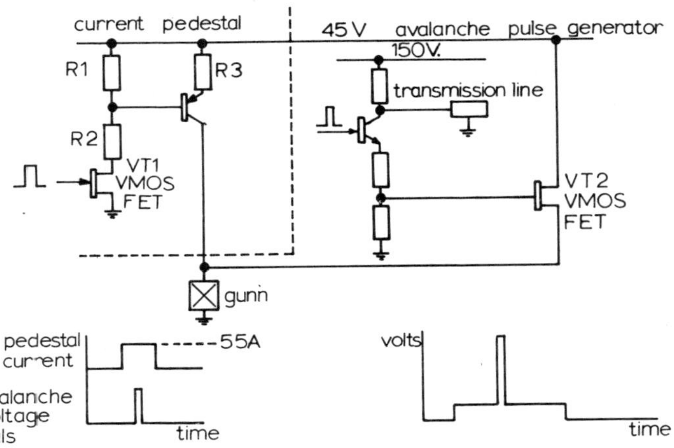


Figure 48  Fast rise time pedestal pulse generator

# ELECTRONICALLY TUNED OSCILLATORS

Dr M J Howes

Dept of Electrical and Electronic Engineering, University of Leeds

## INTRODUCTION

The increasing use of electronically controlled oscillators in a variety of microwave applications has focussed attention on the steady state and transient behaviour of such modules. The demands on oscillator designers are particularly stringent in areas of high sophistication such as electronic warfare systems, where broad band tuning and high source modulation characteristics conflict with high thermal stability and low noise requirements. The important parameters which describe the performance of electronically tuned oscillators are shown schematically in figure

Slew Rate(s). The rate at which the frequency of the oscillator can be changed from one end of the tuning range to the other in response to a step change in the control parameter. Values of slew rate are primarily determined by external shaping circuitry (see 49 'linearity' below) but rates greater than 1 GHz/μsec may be achieved in practice.

Settling time ($t_s$) Settling time is a term stolen from linear amplifier terminology and is defined as the maximum time interval between the application of a step-input tuning voltage and the time the frequency settles within some small frequency band centred on the frequency predicted theoretically from the input tuning voltage.

Post Tuning Drift ($f_p$) Post tuning drift is the largest value of frequency drift measured from the settling time. Such measurements must be carried out over a time interval which is large compared with the settling time.

Tuning Linearity ($f_d$) For many applications the oscillator frequency should be linearly related to the control parameter. A useful measure of this linearity is the maximum frequency deviation between the experimental tuning curve and the best linear fit.

Fine Grain Linearity (g) Fine grain linearity may be quantified by means of the tuning slope ratio which is defined as the maximum value of the tuning curve slope to the minimum value.

These parameters along with spectral purity, which implies low a.m. and f.m. noise and freedom from harmonic and spurious responses, largely specify the broadband electronically tuned oscillator. In particular, the slew rate and settling time determine the frequency agility of the oscillator. Only two basic concepts are available at present in order to realise such oscillators: YIG and varactor tuning.

## ELECTRONIC TUNING

The equivalent circuit of any microwave oscillator is of the form shown in figure 50 . Here Y(w) is the circuit admittance seen by the active device which has an admittance $\overline{Y}(A,w)$. In order that the oscillator should be self starting and deliver power under steady state conditions, the real part of $\overline{Y}$ should be negative over the required frequency band under small signal and large signal conditions. The steady state oscillation conditions are given by

$$Y(\omega) + \overline{Y}(A,\omega) = 0$$

$$\frac{dB}{d\omega}\cdot\frac{\partial\overline{G}}{\partial A} + \frac{dG}{d\omega}\cdot\frac{\partial\overline{B}}{\partial A} > 0 \qquad \ldots(6)$$

Equation 6 determines the frequency and amplitude of oscillation from which it is clear that the frequency may be varied by changing the circuit admittance. For example, mechanical tuning may be achieved by mounting the device in a waveguide cavity as shown in figure 51a with the circuit admittance given by $Y_M - jY_0 \cot\beta l$ where $Y_M$ is determined by the device mounting and encapsulation. In this case, the circuit admittance and hence the oscillation frequency can be varied by means of the cavity length l and a typical variation is shown in the figure. Alternatively, we may realise the circuit admittance using a ferrimagnetic sphere as shown in figure 51b Under certain conditions the sphere behaves as a low loss parallel resonant circuit, the admittance of which is given by $G(H_0) + j\left[wC(H_0) - \frac{1}{wL(H_0)}\right]$ where $H_0$ is an applied magnetic field. Thus, the frequency of such an oscillator can be varied by means of the control parameter $H_0$. The ferrimagnetic material used for the sphere is usually Yttrium Iron Garnet and oscillators based on this concept are commonly referred to as YIG tuned oscillators. The final practical method of realising an electronically controlled admittance is to use the properties of Schottky diodes. Suitably biased, the Schottky diode behaves as a voltage dependent capacitance which may be embedded (figure 51 c) in the oscillator module to control the admittance seen by the active device and hence (according to equation 6 ) the oscillation frequency.

## YIG TUNED OSCILLATORS

The basic principle of YIG tuned oscillators is shown schematically in figure 52 . The YIG sphere is simply strongly coupled via a transmission line to the active element (represented here as a two terminal negative resistance) and less strongly coupled to the external load by an orthogonal coupling loop. Other methods of output coupling are possible, but orthogonal loop coupling possesses the unique advantage in that the YIG element behaves as a narrow bandpass filter with low coupling off resonance and the output spectrum is virtually free from spurii and harmonics. The centre frequency of the filter is given by

$$f_0 = \gamma H_0 \qquad \ldots(7)$$

where γ is the gyromagnetic ratio (2.8 MHz/oersted) and $H_0$ is an applied d.c. magnetic field. It is immediately apparent from equation 7 that the desirable linear relationship between output frequency and control parameter (magnet coil current) should be easy to achieve if leakage flux in the magnet is kept to a minimum.

One method of realising the basic oscillator is shown in figure 53 a[1]. The microwave circuit is mounted in an enclosure of dimensions too small to support unwanted cavity modes. The active element used in this case in a packaged (S4) transferred electron device and the YIG sphere is closely coupled into the circuit by a semi-loop in order to enhance the r.f. field in the vicinity of the sphere. Note that the transmission line is terminated by an r.f. short circuit to maximise the r.f current in the line near the sphere and the bias to the active device is applied via r.f. choke sections. Such an oscillator can be tuned from 6-12 GHz using magnet currents of ~0.5 to 1A. An approximate equivalent circuit of the oscillator shown in figure 53 a is shown in figure 53 b The YIG sphere is modelled by the parallel resonant circuit, the elements of which are given by

$$G_o = (\mu_o . V . d^{-2} . \omega_m . Q_u)^{-1} + G_L$$

$$L_o = \mu_o . V . \omega_m / \omega_o . d^2 \qquad \ldots(8)$$

$$C_o = (L_o . \omega_o^2)^{-1}$$

where $4\pi M_s$ is the saturation magnetisation of the sphere, $w_m = \gamma 8\pi^2 M_s$, V is the sphere volume and d is the coupling loop diameter. The unloaded Q ($Q_u$) is related to the saturation magnetisation and resonance linewidth $\Delta H$ (~0.2 oersted) by

$$Q_u = (H_o - 4\pi M_s/3)/\Delta H \qquad \ldots(9)$$

The basic assumptions on which the equivalent circuit of the sphere is based are that the sphere is always magnetically saturated, the frequency range in which the premature decline phenomenon occurs is avoided and that the sphere diameter $<<\lambda/4$. For YIG spheres ($4\pi M_s$ = 1750 gauss) this implies a minimum frequency of 3.5 GHz but lower frequencies may be accommodated using suitably doped YIG crystals. The size restriction is of little importance since practical spheres are less than 1 mm in diameter.

The linearity, slew rate and other dynamic response characteristics ofa YIG tuned oscillator are dominated by the magnetic circuit design. The magnets used for YIG tuning are simple electromagnets with a single gap to provide a uniform magnetic field. The gap design is critical since any non-uniformity of magnetic field will result in tuning hysteresis and unwanted spurious responses due to the excitation of magnetostatic modes.

Although the resonant frequency of a YIG sphere is not strongly dependent on orientation the change of resonant frequency with temperature is critically dependent upon its orientation and varies monotonically through positive values to negative values. It turns out that in the case of transferred electron devices the inherent df/dT of the active device can be compensated for by that of the YIG sphere and a highly temperature stable oscillator realised. Typical characteristics of an oscillator of the type shown in figure 53 are given in 21 .

Broadband YIG tuned FET oscillators may also be realised and the basic circuit of an X-band common gate GaAs FET oscillator is shown in figure 54 (22). The feedback topology has been designed in order to present an inductive admittance to the resonator, in order to prevent spurious oscillations with the coupling loop whilst retaining a broadband negative conductance. The performance of such an oscillator based on an H.P. FET (L = 1μm, W = 500μm) is shown in figure 55 .

VARACTOR TUNED OSCILLATORS (23)

When considering the potential of varactor tuned oscillators it is useful to consider the case where the tuning element is parallel coupled into the oscillator circuit (figure 56 ) and that in which the coupling may be regarded as a series configuration (figure 57 ). It is also helpful to consider the implications on tuning bandwidth of a small signal analysis even though the tuning element is horribly non-linear.

Parallel Coupling The basic circuit of a varactor tuned oscillator based on parallel coupling is shown in figure 56 . The varactor has been replaced by its small signal equivalent circuit where $C_V$ is the varactor capacitance whose voltage dependence in the case of Schottky type devices is given by

$$C_V = C_o . (1 - V/\phi)^{-1/2} \qquad \ldots(10)$$

$R_s$ is the series resistance of the varactor given by

$$R_s = R_o . (1 - k_1(1 - V/\phi)^{1/2}) \ldots(11)$$

where $R_o$ resistance of the epi-layer outside the depletion region at zero bias and $k_1$ is a function of the device material parameters. The depletion layer conductance G is given by

$$G = eI_s/kT \cdot \exp(eV_B/kT) \ldots(12)$$

where $I_s$ is the reverse bias saturation current and $V_B$ is the terminal bias voltage. Note that the parameter V in equations 10 and 11 is equal to the bias voltage minus the voltage drop across $R_s$. However, under reverse bias ($V_B<0$) and small signal conditions $V \simeq V_B$.

If the varactor capacitance is changed from its value at zero bias ($C_o$) to its value at maximum reverse bias (breakdown) the tuning bandwidth may be deduced from

$$\frac{f_B^2 - f_o^2}{f_o^2} = \left(\frac{C_o - C_B}{C + C_o}\right) \Big/ \left(m^2.(1 + R_s G)^2 + \frac{C_B}{C + C_o}\right) \ldots(13)$$

which implies that octave bandwidths are available for capacitance ratios $C_o/C_B > 4$. The ratio of the power dissipated in the varactor to that delivered to the load at zero bias is given by

$$\frac{P_V}{P_L} = \omega_o^2 . C_o / m^2 . G_L . \omega_{co} \qquad \ldots(14)$$

where $w_{co}$ is the zero bias cut off frequency of the varactor $(2\pi R_s C_o)^{-1}$. Equations 13 and 14 indicate an important characteristic of varactor tuned oscillators. Firstly, increased varactor coupling (achieved by reducing the value of m) increases the electronic tuning range, but only at the expense of output power (see figure 58 ) and the loaded Q factor of the oscillator. Since the f.m. noise performance is closely linked with oscillator loaded Q, there is an inherent trade off between bandwidth and spectral purity.

Series Coupling For the series configuration the tuning bandwidth and power loss characteristics are contained in equations 15 and 16

$$\frac{f_B^2 - f_o^2}{f_o^2} = \left(\frac{C_o - C_B}{C_B}\right) \Big/ \left(m^2 C_o \left(\frac{C + C_D}{C C_D}\right) + 1\right) \quad \ldots (15)$$

$$\frac{P_v}{P_L} = \left(m^2 . \omega_{co} . C_o . R_L\right)^{-1} \quad \ldots (16)$$

It appears that the series oscillator has very similar properties to the parallel case in that octave tuning is, in principle, available and broadband tuning can only be achieved at the expense of output power. However, there is one important difference which makes the series coupled circuit the obvious choice for low noise broadband oscillators. This can be seen by comparing the loaded Q of the oscillator at zero varactor bias normalised to the loaded Q with zero varactor coupling in both cases.

$$Q_L(p) = \frac{1 + 2F/(1 - 1/r)}{1 + 2f_o Q_L F / f_{co}(1 - 1/r)} \quad \ldots (17)$$

$$Q_L(s) = \frac{1 + 2F/(1 - r)}{1 + 2f_o Q_L F / f_{co}(r - 1)} \quad \ldots (18)$$

where $r = C_o/C_{MIN}$ and F is the fractional tuning bandwidth. Thus, for a given varactor and tuning range a larger loaded Q and hence lower noise is available in the series tuned case.

Practical Varactor Tuned Oscillators In practice, the previous analysis only gives a qualitative assessment of varactor tuned oscillator performance. For example, for broadband operation the varactor must be closely coupled into the circuit. This immediately implies a large r.f. voltage across the device and the small signal assumption becomes invalid. In general, the active device and varactor will be embedded in a microwave network which may be of a complex distributed/lumped nature. Consequently the simple equivalent circuits used above will not be sufficient to enable a complete characterisation of the oscillator.

The effects of power dissipation experienced by the varactor under large signal operation are particularly important in agile oscillators where changes in varactor dissipation when switching from one frequency to another result in post tuning drift.

The trade-off between tuning-bandwidth, power output, spectral purity and post-tuning drift is a complex subject which is the subject of increasing interest. However, despite the lack of complete understanding broadband oscillators characterised by acceptable spectral purity have been demonstrated but an octave tuned, low noise oscillator awaits a large signal analysis of a realistic circuit topology. The performance of typical varactor tuned oscillators available at the present time reveals that varactor tuning is to be preferred to YIG tuning on all counts apart from spectral purity and tuning bandwidth.

REFERENCES

21. Easson, R.M: 'Design and Performance of YIG-tuned Gunn Oscillators', Microwave Journal, p 53, Feb., 1971.

22. Trew, R.J:'Design Theory for Broad-Band YIG Tuned FET Oscillators', IEEE Trans. on MTT, Vol.MTT-277, No.1. p 8, 1979.

23. Greenhalgh, S. J: 'Device-Circuit Interactions in Varactor Tuned Transferred Electron Effect Oscillators, PhD Thesis, Leeds University, 1980.

PROBLEMS

d1. The conductance of a two terminal negative resistance device used in a YIG tuned oscillator has a dependence on r.f. voltage amplitude (A<7.5V) given by $-G=(15-2A)$mS. Assuming an equivalent circuit of the form shown in figure 53b plot the power variation of the oscillator when the d.c. magnetic field is increased from 3,000 to 4,500 oersted. ($V=0.75.10^{-6}m^3$, d=2mm, $4\pi M_S=1750$ gauss, $G_L=10$mS, $\ell=10$mm, $Z_0=200\Omega$, $\Delta H=0.5$ oersted).

(d2). The equivalent circuit of a parallel coupled varactor tuned oscillator is shown in figure 56 The varactor is assumed to be of the Schottky barrier type fabricated from epitaxial GaAs (epi-layer thickness = 0.3μm). If the active device is identical to that described in problem (i) determine the power dissipated in the load ($P_L$) and the power dissipated in the varactor ($P_V$) under zero bias conditions. (Varactor structure and material parameters: $\mu=0.45m^2/Vs$, area = $1.5.10^{-9}m^2$, $n_0=10^{22}m^{-3}$. Circuit element values $G_L=10$mS, C=0.5pF, L=0.5nH, m=1).

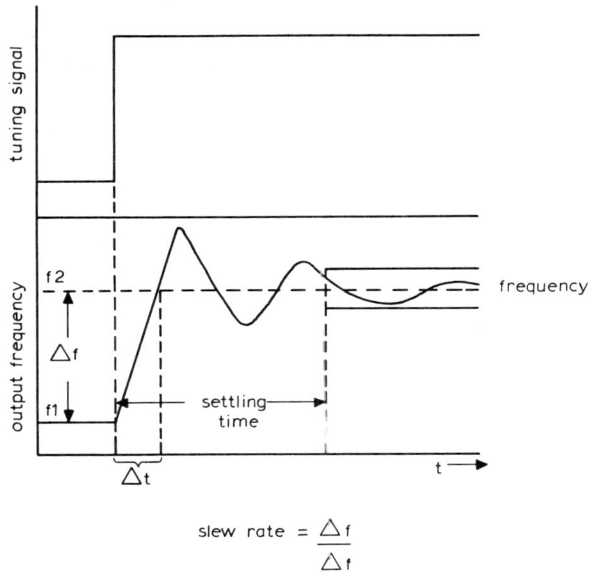


Figure 49 Schematic illustration of parameters which characterise electronically tuned oscillators

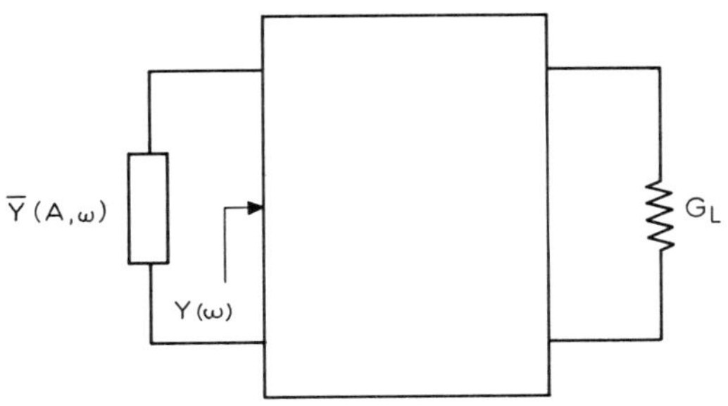


Figure 50 Basis equivalent circuit of microwave oscillator module

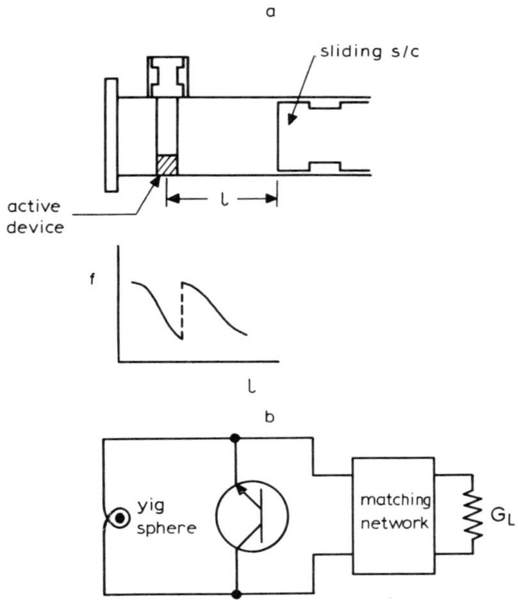


Figure 51 Methods of realising tunable microwave oscillators

(a) Mechanical
(b) YIG - contrcl parameter magnetic coil current

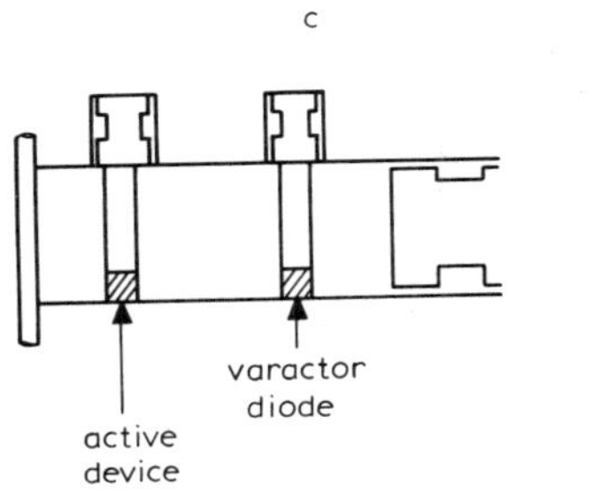


(c) Varactor-control parameter device bias voltage

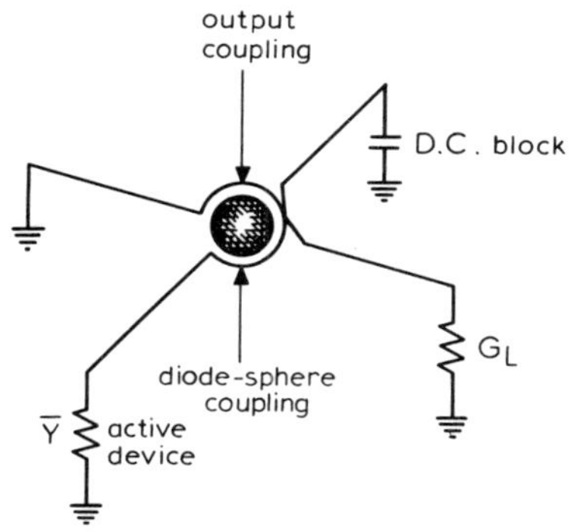


Figure 52 Principle of YIG tuned oscillator

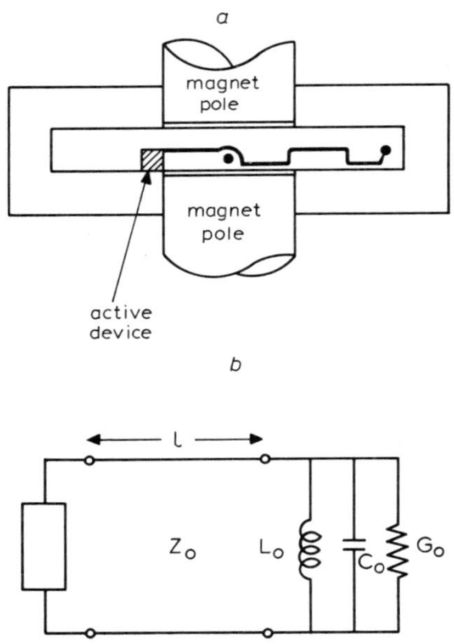


Figure 53 (a) Paractical realisation of YIG tuned transferred electron oscillator

(b) Equivalent circuit of oscillator shown in Figure 53(a)

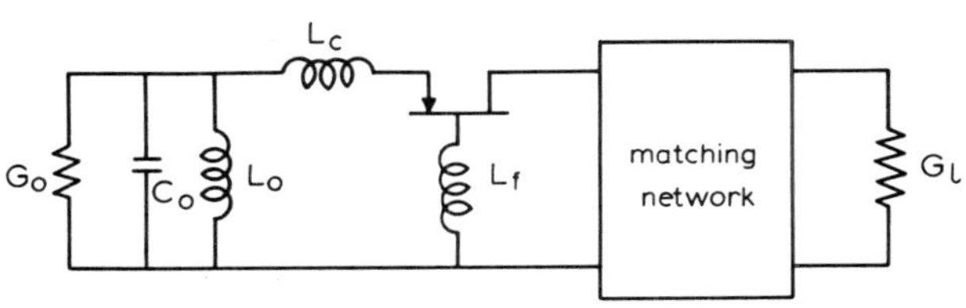


Figure 54 Basic circuit topology of a YIG tuned X-band, common gate MESFET oscillator

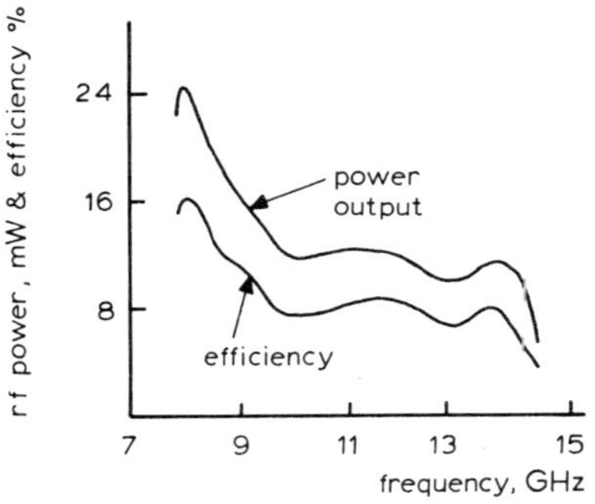


Figure 55

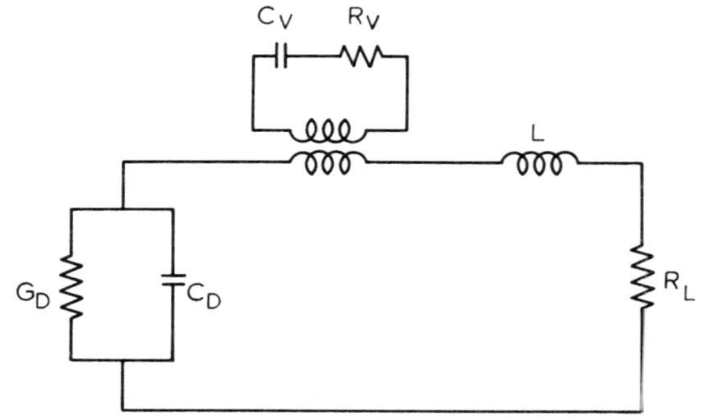


Figure 57 Series configuration for varactor tuned oscillator module

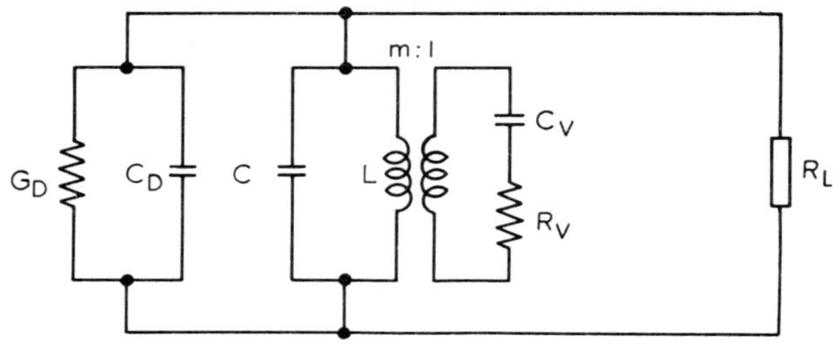


Figure 56 Parallel configuration for varactor tuned oscillator module

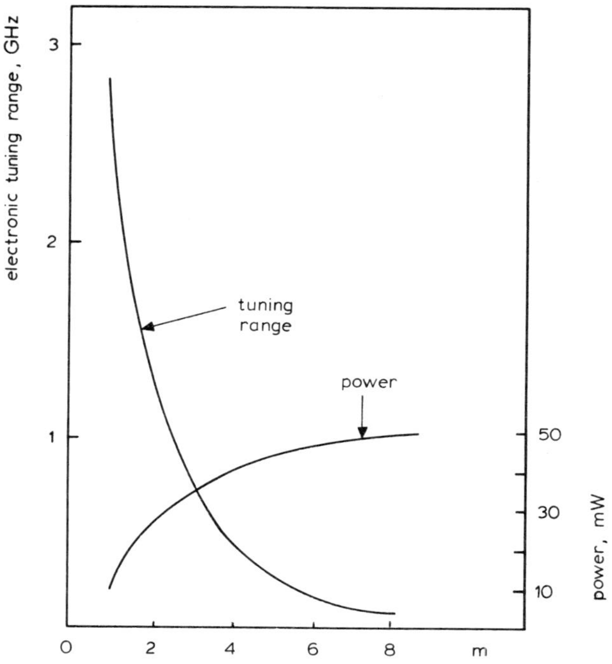


Figure 58 Effect of varactor coupling on the performance of a typical varactor tuned oscillator

# CHAPTER 13

# Osillator evaluation techniques

A.T.M.Vaughan/Dr.B.Mazonas

# CHAPTER 13. OSCILLATOR EVALUATION TECHNIQUES. (a) C. W. OSCILLATORS. CLOSE TO CARRIER NOISE MEASUREMENT

A T M Vaughan

Royal Signals and Radar Establishment, Malvern, Worcs

## INTRODUCTION

The process of equipment development is one of continuing refinement of the performance of systems and associated components to meet the needs of more sophisticated applications. An example that is relevant to this paper occurs in the development of present day coherent radar processing techniques such as doppler frequency radar systems which are designed to extract from a background of clutter, very small radar return signals the magnitudes of which are comparable to the inherent noise modulation associated with the oscillators or amplifiers used in the system.

The unwanted noise generated by these devices defines the ultimate limit of system performance.

This paper describes means of measuring the unwanted noise components that are naturally or artificially impressed upon the wanted signal generated by the microwave source.

## THE PROBLEM

Practical CW Oscillators do not generate a pure monotonic signal but produce a spectrum of very low magnitude and frequency modulation components of a systematic or random nature distributed about the desired carrier frequency ($\Omega c$) as shown in figure 1.

Continuing development of more sophisticated applications places increasingly more stringent limits on the acceptable level of noise which in turn leads to a continuing study of the cause of the generated noise and possible means of reducing it.

The study requires a means of identification and quantification of the noise signals and this paper describes measuring techniques developed for this purpose at RSRE. The power level of the noise signals, distributed about the carrier as shown in Fig 1 varies from several tens of dB/Hz below carrier within a few KHz of the carrier to the order of 160dB/Hz below carrier at frequencies of few MHz from carrier. The measurement task is concerned with isolating and quantifying these very small noise signals in the presence of the much larger carrier signal.

The noise output consists of a random content due to random effects in the active device itself upon which there may be superimposed signals of a systematic nature which result from power supply or drive circuit modulation.

The systematic noise content may be reduced by careful attention to the associated power supply and drive circuits. The random noise is dependent upon the oscillation or amplification mechanism and on the materials and processing used in the active device and the design of the active circuits and whilst some degree of selection and control may be possible in this area such choice is limited and often leads to a compromise between noise and other aspects of the performance eg. output power and efficiency or cost.

Although the random noise is of a continuous nature it is convenient to describe it in terms of equivalent power, single tone frequency and amplitude modulation coefficients as a series of narrow sidebands as illustrated in Fig 2 (Ref 1).

A comprehensive record of the noise performance of a device is made by measuring the equivalent AM and FM signal level in a series of contiguous narrow sidebands covering the frequency range of interest. A typical record is illustrated in Fig 3 which shows that, in addition to the random content, the level of which usually falls as frequency from carrier increases, there are also systematic noise fluctuations related to the power supply frequency-in this case 50 Hz.

## SPECIFICATION OF CLOSE TO CARRIER NOISE

As stated above the random noise modulation is measured in terms of equivalent single tone amplitude and frequency modulation signals in a series of contiguous narrow sidebands covering the frequency range of interest.

At low sideband frequencies (<2 KHz) a narrow bandwidth filter is selected to resolve fine grain structure detail whilst at higher frequencies it is usually acceptable to use a wider analyser bandwidth. Whilst fine grain detail may be desirable, the time taken to traverse a desired sideband frequency range must also be considered. This is governed by the filter bandpass shape which determines its response time. For a Gaussian filter the maximum sweep rate is given by -

$$\text{Sweep rate (Hz/second)} = 0.5\ (\text{Bandwidth(Hz)})^2 \quad \ldots(1)$$

(Ashley et al (2))

It is therefore common practice to change the bandwidth as sideband frequency changes viz:-

| Sideband Frequency Range | Filter Bandwidth (B) |
|---|---|
| Up to 200 Hz | 3 Hz |
| 200 Hz to 2 KHz | 10 Hz |
| 2 KHz to 2 MHz | 100 Hz |
| 2 MHz and above | 1 KHz |

For specification purposes the noise signals are expressed as dB below carrier per Hz bandwidth (-dBc/Hz). The indicated noise voltage levels when using practical wave analyser bandwidths are therefore scaled by

$$\frac{1}{\sqrt{B(\text{Hz})}}. \quad \ldots (2)$$

## ANALYSIS OF THE AM AND FM COMPONENTS

Consider an isolated narrow sideband centred on a frequency $\Omega c + \omega_n$ as illustrated in Fig 4. This may be represented by the vector diagram Fig 5 in which the instantaneous positions 1 and 3 of a pair of contra-rotating vectors yield a result which represents the minimum and maximum values of amplitude modulation. Positions 2 and 4 of a similar pair of vectors yield the resultant frequency modulation limits produced by a single interfering signal at frequency $\omega_n$ above the carrier. There is also a similar interfering signal at $-\omega_n$ below the carrier.

For small signals $V_n << V_c$ the instantaneous signal $V_t$ due to a carrier $V_c$ simultaneously amplitude and phase modulated by a noise signal $\omega_n$ can be represented as (Ref 1)

$$V(t) \approx V_c\left[1+\frac{V_n}{V_c}\cos(\omega nt+\theta_n)\right] \cdot \text{Cos}\left[\Omega_c t+\frac{V_n}{V_c}\sin(\omega nt+\theta_n)\right] \quad (3)$$

MEASUREMENT OF AM AND FM COMPONENTS

Measurement of the AM Components employs a direct detection system in a balanced mixer/detector configuration operating in the homodyne mode. Fig 6 is a block diagram of the system. The essential components are a balanced mixer with switchable detectors, a low noise, wideband video amplifier and a wave analyser. Other monitoring facilities include a microwave spectrum analyser, a video spectrum analyser and a microwave power meter.

The carrier component provides the effective Local Oscillator signal at the detectors which mixes with the noise signals to produce signals at baseband frequencies which are separable into sidebands offset from the carrier frequency $\Omega_c$.

A tuneable wave analyser with switchable bandwidth is used to indicate the level of noise signal in a selected bandwidth. The DC term produced by detection of the carrier is filtered by using capacitive coupling at the input of the amplifier.

Measurement of AM noise signal is made as follows:-

The balanced mixer/detector diodes are connected in the additive configuration. The wave analyser is tuned to a required sideband frequency and the wave analyser bandwidth B is selected as appropriate. The signal indicated by the wave analyser under these conditions is the RMS sum of the noise generated by the unit under test and the noise generated in the diodes and the amplifier. The optimum diode balance is achieved by separately adjustable bias to each diode.

By connecting the detectors in the balanced configuration the noise generated by the unit under test is cancelled leaving only the noise due to the detectors and amplifier. From the wave analyser readings for the two detector configurations the noise due to the source alone can be calculated as follows:-

With the diodes connected in the additive configuration (unbalanced) an equivalent noise voltage due to the oscillator under test plus the system noise is indicated. This is recorded as $V_1$.

With the diodes connected in the balanced configuration a second reading of the noise due to the system only is indicated. This is recorded as $V_2$.

The noise due to the oscillator under test $V_n$ is then found from the relationship

$$V_n = \sqrt{V_1^2 - V_2^2} \quad \ldots. (4)$$

This figure $V_n$ is then factored to take into account the bandwidth B(Hz) selected on the wave analyser filter and the resulting noise in a 1 Hz bandwidth is found from:

$$V_n = \frac{1}{\sqrt{B}} \cdot \sqrt{V_1^2 - V_2^2} \cdot /\text{Hz} \quad \ldots. (5)$$

The phasor diagram (Fig 7) shows the relationship between carrier, AM and FM noise sidebands at the diodes in balanced and unbalanced connection.

MEASUREMENT OF FM COMPONENTS

As a diode detector will only detect amplitude modulation components it is necessary to convert the FM components to amplitude modulation in order to produce an FM measuring system. Fig 5 shows that the FM component is orthogonal to the carrier vector. The phase change of $\pi/2$ to convert this to detectable AM may be achieved by applying the signal to a slightly detuned resonant circuit as shown in Fig 8 but this circuit is susceptible to Amplitude variation (hence the more usual inclusion of a limiter in such a discriminator).

The output response is also a function of the slope of the tuned circuit response curve which must be known if accurate measurements are to be made. A further disadvantage is that the limited frequency range over which linear conversion may be expected is determined by the extent of the linear region of the tuned circuit response curve.

An alternative approach used in the RSRE equipment utilizes a phase sensitive microwave bridge incorporating a carrier suppression filter. The carrier suppression filter consists of a high Q microwave cavity ($Q_L$ > 10,000) and either a 3dB coupler or a low loss circulator. The block diagram is shown in Fig 9.

The signal from the oscillator under test is divided into two separate paths round the bridge. The signal in path No 1 contains elements of carrier and AM and FM sidebands. The signal in path No 2 is passed via the carrier suppression filter beyond which only AM and FM sideband signals proceed to the mixer/detector. A phase shifter is adjusted to ensure that the FM signal from path No 2 is in phase with the residual carrier from path No 1 at the detectors.

The phasor diagram Fig 10 shows the FM components beat in phase with the carrier to produce an AM output whilst the resultant of the AM sidebands is orthogonal to the carrier and produces no detectable output.

CALIBRATION

The accuracy of the measuring system is dependent upon the performance of the components used including:

1. The microwave components comprising the bridge circuit.

2. The detector characteristics.

3. The gain vs frequency response of the base-band amplifier.

4. The bandpass characteristic of the wave analyser.

Whilst it is possible to establish the calibration of the noise measuring system in absolute terms by calculation involving the many factors noted (1 to 4) above, a more satisfactory calibration can be achieved by impressing artificial sidebands of known modulation index upon the carrier from a source (probably an oscillator submitted for test) and carrying out an overall calibration.

To meet this requirement a family of calibrators has been designed by Dunsmore of RSRE for use with noise measuring equipments designed at RSRE. A note on the basic requirements of a calibrator by M R B Dunsmore states the requirements as:-

BASIC PRINCIPLES OF CALIBRATORS FOR CARRIER NOISE ANALYSERS

The accurate calibration of carrier noise analysers (CNA's) relies on microwave systems which are capable of generating modulated signals where the AM and FM contents are known precisely. In addition, it is desirable to be able to vary the amplitude and frequency modulation indices accurately and at will.

A system capable of providing accurate calibrations of a CNA at any microwave frequency within its operating range, and for any form of amplitude and/or phase modulation, must be able to meet the following specification:-

(i) It must be possible to superimpose AM and/or FM on the microwave output of any source whose noise is to be measured on the CNA without such modulation causing any discernible change in the (noise) spectrum of that source eg. by reflecting AM sidebands back into it.

(ii) The accuracy with which the amplitudes of the sidebands are set must be traceable to some passive standard, such as a rotary vane attenuator, without relying on the accuracy or linearity of some intermediate device.

(iii) It must be possible to produce either pure AM or pure FM and ideally, also any mixture of the two.

(iv) The sidebands must be able to be reduced in amplitude by accurately known amounts from their initial setting without altering the proportions of AM and FM.

(v) Ideally, it should be possible to set up known sidebands without having to rely on a microwave spectrum analyser.

(vi) The complete calibration system should be easy to align and use by unskilled operators.

THE CALIBRATOR

This is essentially a double sideband suppressed carrier modulator producing pure single tone modulation of a carrier over a wide sideband frequency range. The level and phase of the resulting sidebands with respect to carrier are independently adjustable to an accuracy determined only by the accuracy of the high grade attenuators and phase shifters built into the calibrator bridge.

A block diagram of the Dunsmore calibrator is shown in Fig 11. The principle of operation is as follows:-

A sample of the CW carrier signal at frequency $\Omega_c$ is diverted through a modulating bridge in which a PIN diode modulator, driven from a signal generator, produces amplitude modulation at the sideband frequencies at which calibration is desired.

The drive and bias to the PIN diode is adjusted to produce a single pair of sidebands.

The spectrum produced is of the form:

$$V_{mod} = \cos\Omega_c t + \left[\frac{M}{2}\cos(\Omega_c+\omega_m)t + \cos(\Omega_c-\omega_m)t\right] \ldots . \quad (6)$$

This signal is combined with a phase inverted sample of the carrier $\cos\Omega_c t$ which is adjusted to be equal in amplitude to that emerging from the modulator. This results in cancellation of the carrier and thus produces a double sideband, suppressed carrier signal.

The DSBSC signal is reintroduced to the undiverted, unmodulated carrier signal in the lower path of the bridge to produce either pure AM sidebands if the carrier is in phase with the sidebands or pure FM if the carrier is phase shifted by $90^o$.

The adjustments of amplitude and phase are carried out at microwave frequencies by means of high class components which ensures that the settings will hold over a wide range of modulating frequencies. The bridge also incorporates switches to isolate either the carrier or the sidebands independently.

The level of sidebands produced are measured and compared to the carrier either by measuring both independently on the same power meter or on a spectrum analyser.

The initial settings are made by adjusting the attenuator in the carrier arm of the bridge to a high value (60 dB) and adjusting the sideband "initial amplitude" attenuator to give equal amplitude sideband and carrier signals.

Subsequently returning the attenuator in the carrier arm to zero dB produces a signal with carrier to sideband ratio of 60 dB to within an accuracy determined by the calibration of the high grade microwave attenuator. The sideband to carrier ratio may be further increased by adjustment of the attenuator in the sideband arm to a level of 120 dB.

The phase setting of the sidebands with respect to carrier makes use of the built-in broad-band diode detector. Initial phase setting is achieved by setting the calibrator to produce a pure FM signal by adjustment of the sideband initial phase shifter to produce a null output from the detector. With this setting achieved, subsequently inserting a $90^o$ phase shift in the carrier arm of the calibrator establishes a pure AM signal from the calibrator. The accuracy of setting is again solely dependent upon the calibration accuracy of the high grade phase shifter in the carrier arm. (All RV attenuators and phase shifters have detailed ANA calibration records).

FM calibration may be carried out by selecting the FM mode for the analyser and zero phase shift in the calibrator.

OPERATION

Operation of the equipment varies with the task to be carried out. Requirements for close-to-carrier noise measurement occur both in the development phase of a new type of device and in the inspection and Quality Assurance facilities of manufacturing contractors.

These two areas of application call for a different operational attitude. The RSRE equipments were designed to meet the need of the first category, that is for laboratory use in assessing the performance of new devices under development.

The requirement in this case calls for a series of equipments capable of carrying out any noise measuring requirement over a wide frequency range. The usage would be spasmodic.

The equipments will in general be operated by skilled personnel, often by the engineer concerned with the development of the device to be measured. In this case the tedious nature of the measurement task is acceptable.

The second category of application mentioned calls for a different approach. In a Quality Assurance application the requirement calls for repeated measurements of similar parameters on similar devices on a 100% test basis. This calls for continuous operation of the test equipment preferably by less skilled personnel and here a degree of automation is justified.

To this end RSRE have been concerned with the development of a semi-automated control and processing system using as a basis a desk calculator with IEEE 488 instrument bus (IEC TC66) to command and process parts of the measurement task. A basic flow chart of the process of a measurement is as shown in Fig 12.

Many of the functions tabulated can be handled by the desk calculator with a resulting considerable saving of time and relief from the need for continued concentration on the part of the operator for example:-

1. Repeated reference to ensure that the DUT power output remains constant within the limits over which the calibration of the detectors remains valid.

2. Continual monitoring of the carrier rejection filter to ensure correct tuning.

3. Selection of wave analyser bandwidth appropriate to the sideband frequency being measured.

4. Calibration checks.

5. Computation of the noise signal displayed on the wave analyser.

This latter function is tedious for a manual observer because due to the random nature of the noise signal, visual assessment of the true reading is difficult and calls for continual concentration and judgement. The automated approach offers the facility of the calculator to integrate the random noise signal over a sufficiently long period to establish a mean reading with an acceptable predetermined standard deviation and to carry out the calculation $V_n = \frac{1}{\sqrt{B}} \cdot \sqrt{V_1^2 - V_2^2}$ .... (5) and store the result for print recording or graph plotting.

A programme to carry out this latter routine using a desk top calculator has been written and proved by my colleague Dave Bettinson of RSRE.

Further advances have been made in the application of automatic control of noise measuring equipment by Marconi Research Labs where equipment for use in a QA facility has been developed.

The automated approach shows a worthwhile saving of time and has proved that the investment in automated control equipment is highly cost effective.

It is interesting to note that the measurement time taken to produce a continuous plot under automated control as in Fig 13 is of the order of 40 minutes. (The automated plotter produces its own graph paper and graphics) and that for the duration of an FM measurement the cavity must remain accurately tuned. This, in turn, requires accurate control of the temperature of the cavity and a variation of the order of .001°C has been calculated as the maximum tolerable variation. Cavities currently being made by Clifton-Angus Technical Services Ltd will meet the requirement provided that the ambient temperature is fairly closely controlled but the problem of temperature variation is eased if the time required for a measurement can be reduced.

One means of reducing the measurement time is to use a Fast Fourier Transform analyser to process the data over the lower video frequency range. Such an analyser operating up to 25 KHz has recently become available for a reasonable cost. Whilst not useable at the higher video frequencies its use for the lower video frequency analysis leads to a considerable saving of measurement time of the order of 10 minutes because the FFT analyser will instantaneously process the large number of contiguous narrow sidebands and rapidly integrate to establish a true mean for each sideband and store the data for recording under control of the calculator.

## RECORDING OF RESULTS

For initial evaluation of a device measured on a manual noise measuring equipment measurements at a few spot frequencies may suffice and the result may be recorded in tabulated form similar to Fig 14 or on specially prepared graph paper designed by Dunsmore Fig 15. Automated equipment offers either tabulated recording at pre selected frequencies or graphical presentation of the continuous frequency plot as shown in Fig 13.

Calculation of single sideband noise power ratio is as follows:-

$$\text{dB ratio} = R + \left[20 \text{ Log}_{10}\left(\frac{V_c}{V_N}\right)\right] + 10 \text{ Log}_{10} B \quad .... (7)$$

where R(dB) is the carrier to calibration signal ratio producing a wave analyser reading of $V_c$.

$V_n$ is the Source Noise Signal calculated from

$$V_N = \sqrt{V_{s1}^2 - V_{s2}^2} \quad .... (8)$$

where $V_{s1}$ is the wave analyser reading of total noise (Source + System)

$V_{s2}$ is the wave analyser reading of Residual Noise (System only)

B is the bandwidth of the wave analyser filter(Hz)

## PITFALLS FOR THE UNWARY

The equipment is highly sensitive to very low level signals. (-160 dB on a carrier input of 1/10 watt = 170 dBW = $10^{-4}$ picowatt).

Ambient radiation may well exceed this level and therefore precautions must be taken against leakage of unwanted signals into the equipment. All components and connectors must be of high quality and soundly assembled.

The video circuits are susceptible to ambient interfering signals and must therefore be accurately screened.

Power supplies used to power the equipment must be meticulously filtered and for very sensitive measurements battery operation of the equipment in a screened enclosure is advised.

Phase and frequency measurement is susceptible to interference due to phase changes in components, particularly connectors caused by vibration or even airborne sound such as human speech.

All components must be of the highest quality and care must be taken to ensure that no component inadvertently converts AM to FM or FM to AM.

In fact the precautions may be summarised by noting that when one is measuring very small signals, very small interference becomes highly significant.

## REFERENCES

1. Ondria, J G., IEEE Vol MTT-16 No 9 Sept 1968; 767-781.

2. Ashley, J. R., Barley, T.A., Rast, G. F; IEEE Vol MTT-25 No 4 April 1977; 294-318.

3. Dunsmore, M. R. B., Unpublished communications.

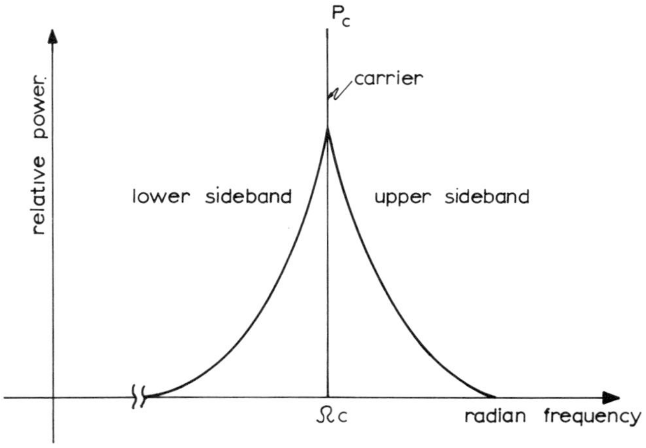


Figure 1 Typical power against frequency spectrum of a microwave source. $\Omega_c$ = carrier radian frequency The carrier power Pc is of much greater magnitude than the mean sideband power

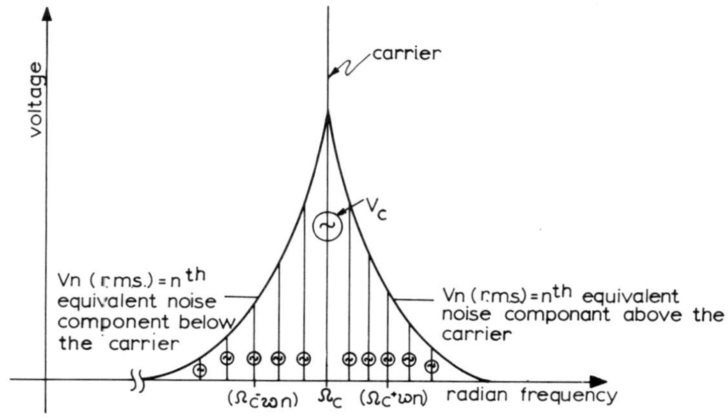


Figure 2 Noise spectrum decomposed into an equivalent spectrum of sinusoidal noise voltages (Reference 1)

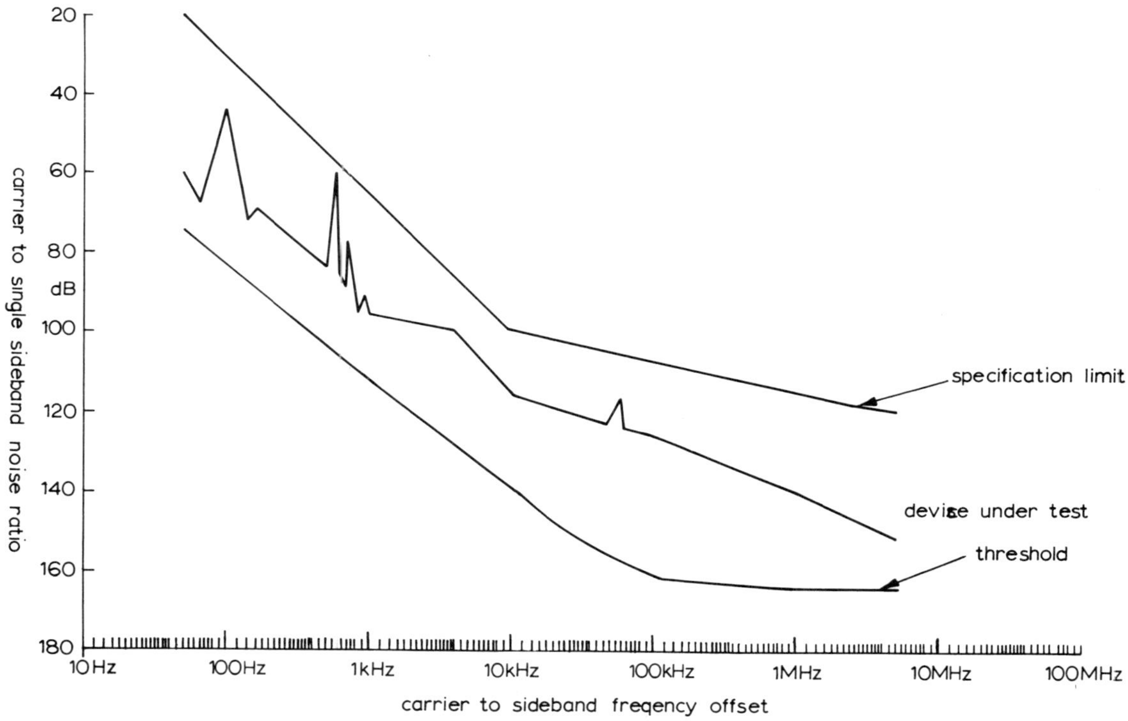


Figure 3 Typical plot of noise against sideband frequency

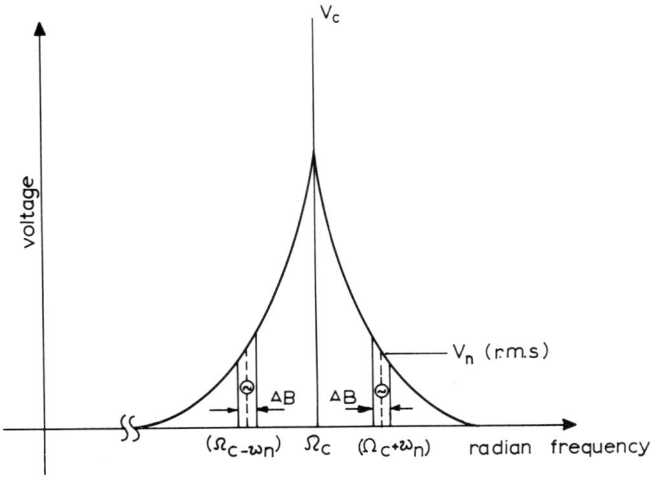


Figure 4 Single pair of sidebands

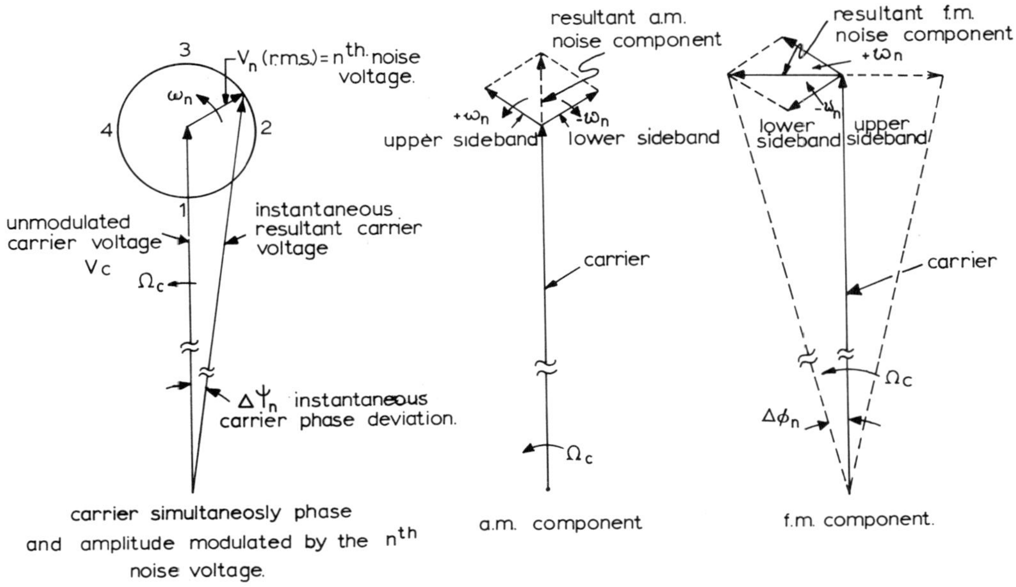


Figure 5 Vector diagrams showing phase relation of a.m. and f.m. sidebands

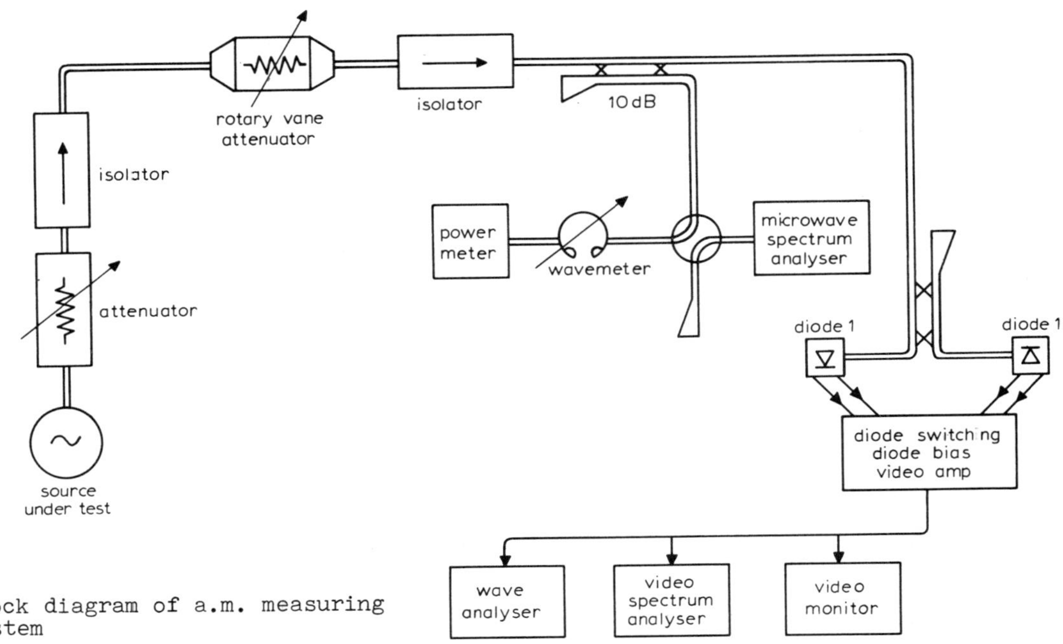


Figure 6 Block diagram of a.m. measuring system

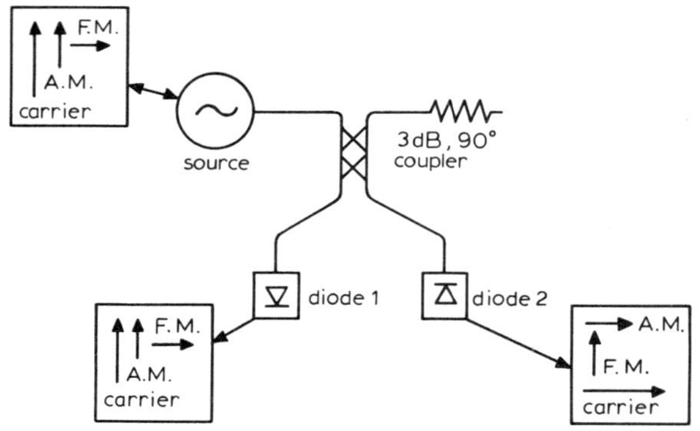


| relative carrier phase ↑ | relative phase of A.M.& F.M.components vs diode switching | | | |
|---|---|---|---|---|
| diode | unbalanced | | balanced | |
| | A.M. | F.M. | A.M. | F.M. |
| D 1 | ↑ | → | ↑ | → |
| D 2 | ↑ | → | ↓ | ← |
| output | A.M. | 0 | 0 | 0 |
| | source noise & system noise | | system noise only | |

Figure 7 Phasor relationship of carrier to a.m. and f.m. sidebands against diode switching

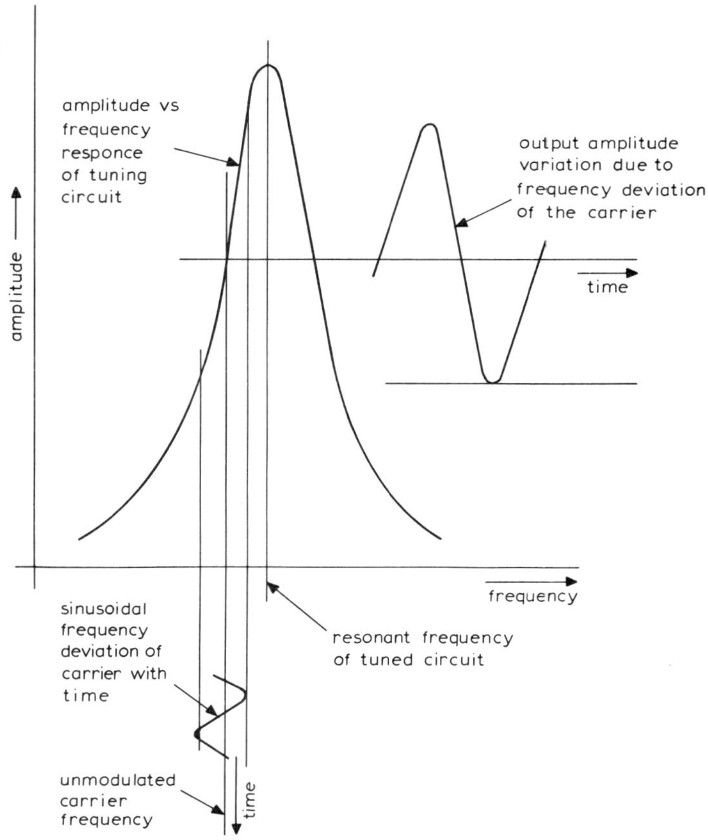


Figure 8 Slightly detuned resonant circuit as an f.m. discriminator

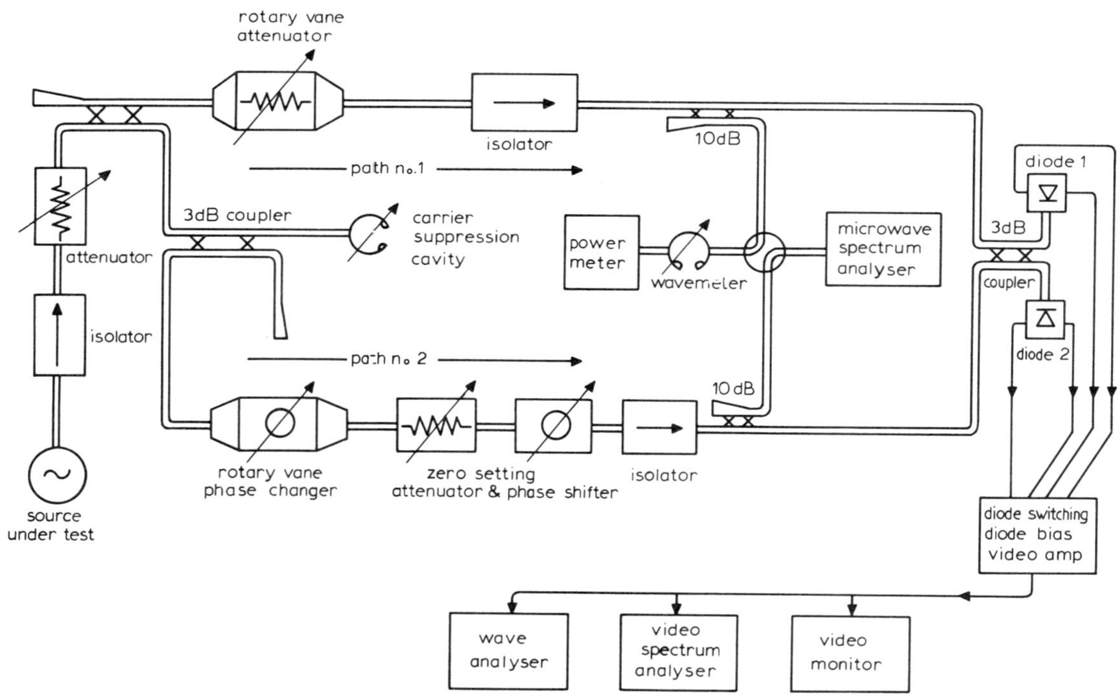


Figure 9 Block diagram of f.m. bridge

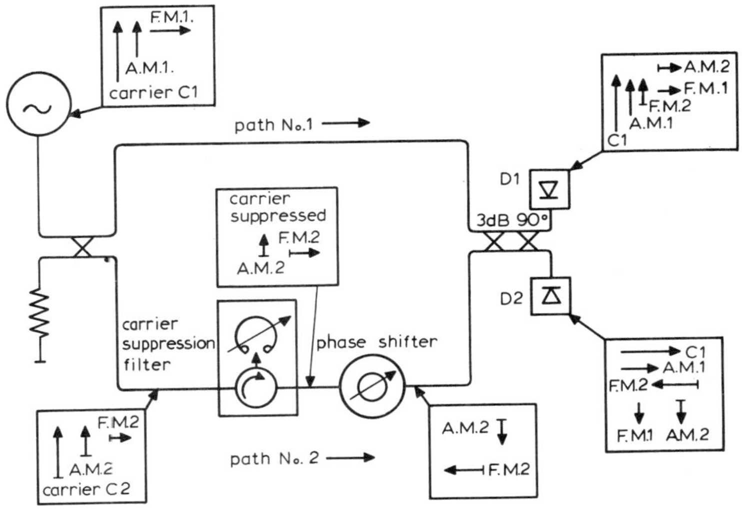


| relative carrier phase → | relative phase of A.M. & F.M. components w.r.t. carrier | | | |
|---|---|---|---|---|
| | path No.1 | | path No.2 | |
| | A.M. | F.M. | A.M. | F.M. |
| D1 | ↑ A.M. 1 | F.M.1 → | A.M. 2 ↦ | ↥ F.M. 2 |
| D2 | ↑ A.M. 1 | F.M.1 → | A.M.2 ↦ | ↧ F.M. 2 |
| I.F. out | 0 | 0 | 0 | 2 F.M. 2. |
| | with diodes balanced, output at video is due only to F.M. from path No. 2. | | | |

Figure 10 Phasor diagram of f.m. bridge

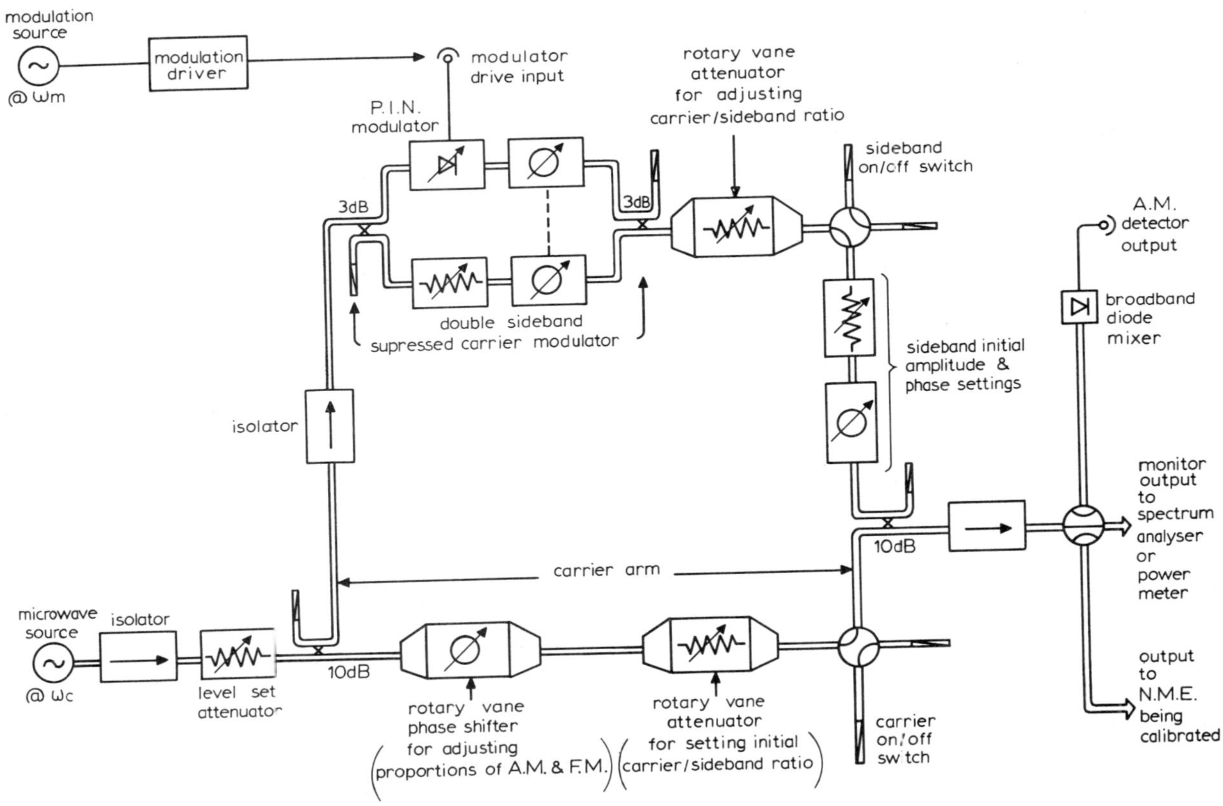


Figure 11 Precision a.m./narrow band f.m. modulation generating bridge

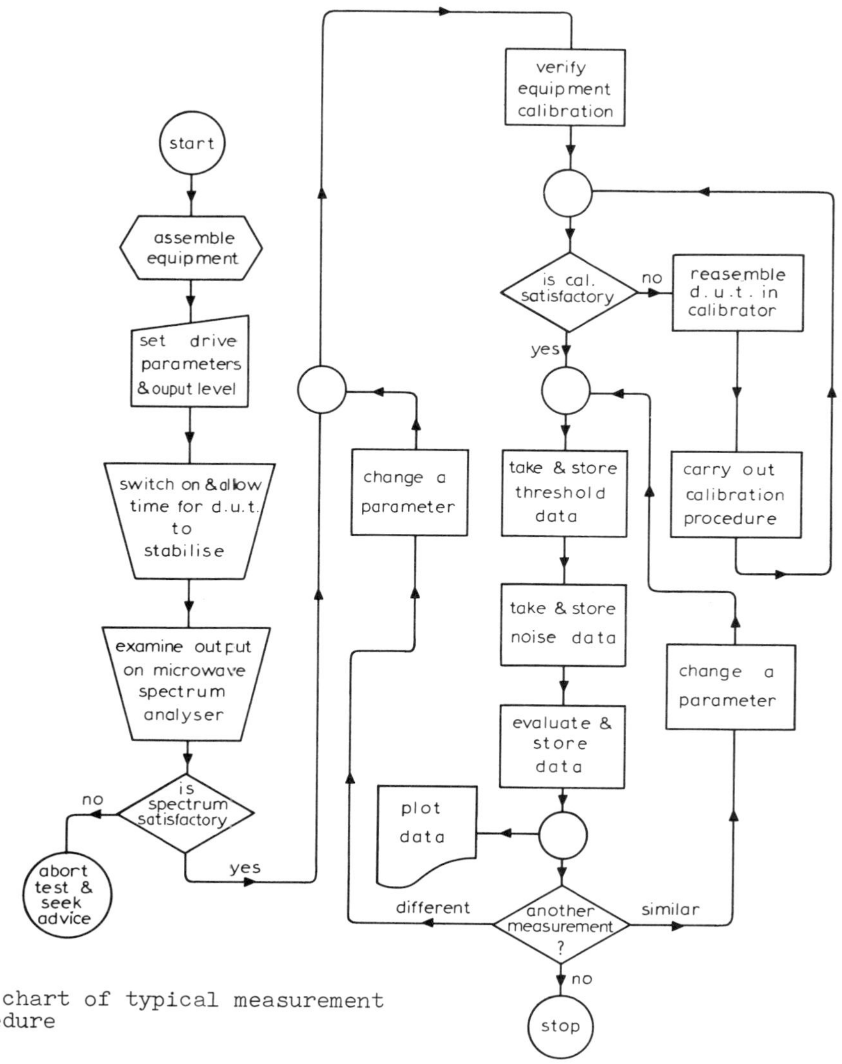


Figure 12 Flow chart of typical measurement procedure

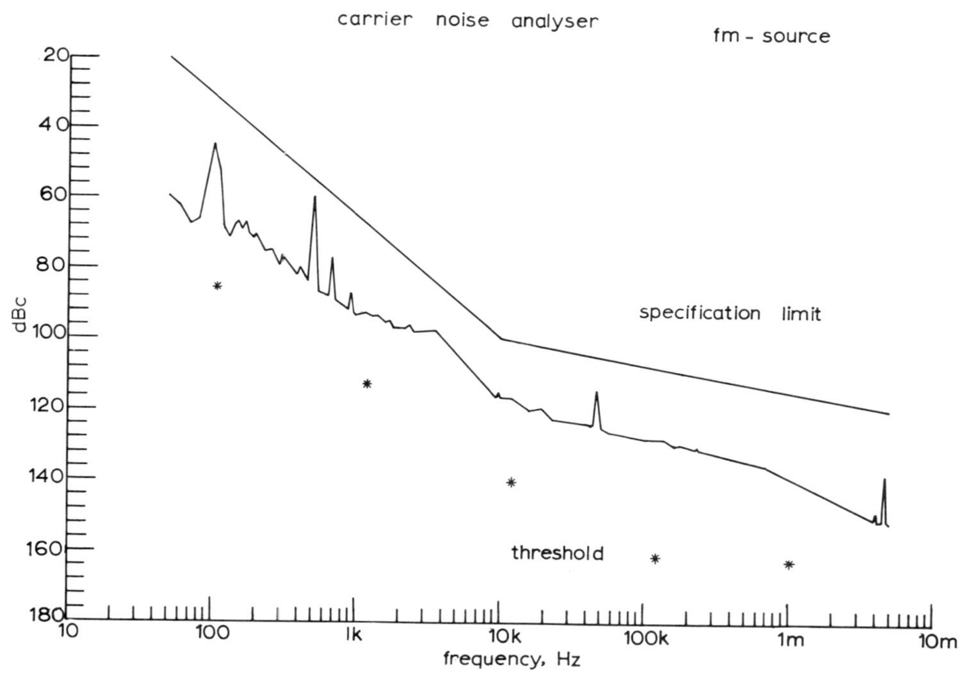


Figure 13

CARRIER NOISE ANALYSER TEST RESULTS

DEVICE UNDER TEST

CLASS AMPLIFIER*/OSCILLATOR*
TYPE No. ..............................
SERIAL No. ............................
FREQUENCY .............................
POWER OUTPUT ..........................
TEMPERATURE ...........................

REMARKS

* DELETE AS APPROPRIATE

NOISE ANALYSER

TYPE .....................................
REF. .....................................
LOCATION OF TEST .........................

MIXER DRIVE:-

SIGNAL ARM mW
REFERENCE ARM mW

MIXER DIODES

TYPE:- ...................................
SERIAL No:- ..............................

CAVITY

TYPE:- ...................................
SERIAL No:- ..............................
LOADED Q FACTOR:- ........................

OPERATOR

..........................................

DATE .....................................

OSCILLATOR*/AMPLIFIER* SIDEBAND NOISE LEVELS. A.M.*/F.M.* MEASUREMENTS

| SIDEBAND FREQUENCY | MEASUREMENT BANDWIDTH | CALIBRATION | | | TOTAL NOISE | SYSTEM NOISE | SOURCE NOISE | | SYSTEM THRESHOLD |
|---|---|---|---|---|---|---|---|---|---|
| | | SIDEBAND | DEVIATION | AMP.O/P. | | | AMP. O/P. | SIDEBAND | |
| Hz | Hz | $dB_c$ | Hz | mV | µV | µV | µV | $dB_c$/Hz | $dB_c$/Hz |
| 225 | 10 | | | | | | | | |
| 475 | 10 | | | | | | | | |
| 975 | 10 | | | | | | | | |
| 1975 | 10 | | | | | | | | |
| 5 k | 100 | | | | | | | | |
| 10 k | 100 | | | | | | | | |
| 20 k | 100 | | | | | | | | |
| 50 k | 100 | | | | | | | | |
| 100 k | 100 | | | | | | | | |
| 200 k | 100 | | | | | | | | |
| 500 k | 100 | | | | | | | | |
| 1 M | 1 k | | | | | | | | |
| 2 M | 1 k | | | | | | | | |
| 5 M | 1 k | | | | | | | | |
| 10 M | 1 k | | | | | | | | |
| 20 M | 1 k | | | | | | | | |
| 50 M | 1 k | | | | | | | | |
| 100 M | 1 k | | | | | | | | |

Figure 14

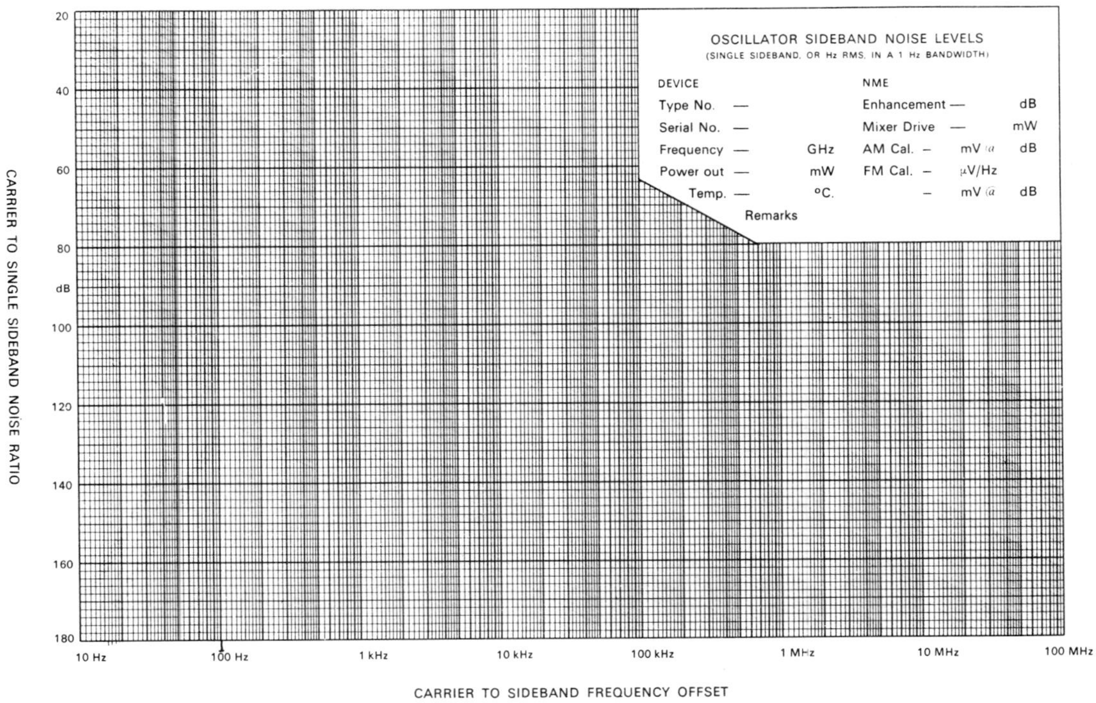


Figure 15

## (b) PULSED OSCILLATORS

B. R. Mazonas

Microwave Associates Ltd., UK

### INTRODUCTION

One of the most important measurements made on any oscillator is the determination of the frequency of oscillation. This measurement varies in difficulty depending on the magnitude of the actual frequency of the oscillator. In general the accurate measurement of low frequencies is achieved easier and with less expense than the measurement of high frequencies. The problems encountered in the measurement of high frequencies such as those in the microwave region are a result of the small periods of oscillation (typically 100 picoseconds) which impose severe restrictions on the response time and resolution of the measuring circuit. The difficulty in the measurement of frequency also depends on the conditions of operation of the oscillator. If the oscillator is operating under continuous wave (CW) conditions then the frequency is easily determined using a CW frequency counter or wavemeter. However, if the oscillator is operating under pulsed conditions then the measurement of the frequency can be relatively difficult especially if the pulse width is small ($< 10\mu s$). The frequency of a pulsed oscillator can also vary with time during a pulse. This variation of frequency is known as chirp and in many applications of pulsed oscillators such as in radar systems it is necessary to measure the chirp. Methods used in the measurement of the frequencies of pulsed oscillators are described in the following sections. These methods are also used in the measurement of chirp and in some instances they can be adapted to measure 'post tuning drift'. An example of post tuning drift is the drift in frequency of a varactor tuned oscillator after the application of a different varactor bias voltage.

### TECHNIQUES USED IN THE MEASUREMENT OF THE FREQUENCIES OF PULSED MICROWAVE OSCILLATORS

#### Pulsed Frequency Counters

The frequencies of pulsed oscillators can be measured using a pulsed frequency counter in the circuit configuration shown in figure 16. The counter indicates the average oscillation frequency in a time interval defined by a control pulse which triggers the counting procedure. The control pulse width can be reduced to less than 100 nanoseconds and by adjusting the control pulse position relative to the RF pulse it is possible to measure the oscillation frequencies over the entire pulse. This technique is not only directly applicable to the measurement of chirp but also to post tuning drift. The main disadvantage of this method is the expense of the pulsed frequency counter.

#### Frequency Down Conversion

This technique can be used to measure the frequency and/or chirp characteristic of a pulsed microwave oscillator. The required circuitry is illustrated in figure 17. The pulsed RF signal is passed through a waveguide isolator and variable attenuator to the input port of a magic T hybrid. The second input port of the hybrid is connected to a CW local oscillator. The pulsed RF signal ($f_p$ (t)) is mixed with the local oscillator signal ($f_{cw}$) at the crystal detector. The mixed signal is passed to an oscilloscope where the signal at the frequency ($f_p$ (t) - $f_{cw}$) is displayed on a screen. Some typical waveforms which are observed on the oscilloscope for the same RF signal but different local oscillator frequencies are shown in figure 18. The pulsed RF signal frequencies are determined as a function of time by the simple addition of the local oscillator frequency $f_{cw}$ and the frequencies $f_p$ (t) - $f_{cw}$ which can be calculated from the waveforms displayed on the oscilloscope screen. Hence, the accuracy of this method of frequency/chirp measurement depends on the accurate determination of $f_{cw}$, which is constant, and $f_p$ (t) - $f_{cw}$ which can vary with time during the pulse. The local oscillator frequency $f_{cw}$ can be determined accurately using a CW frequency counter or absorption wavemeter. However, $f_p$ (t) - $f_{cw}$ is calculated from readings taken from an oscilloscope screen which makes it difficult to maintain a high accuracy. The acquisition of accurate results can be made easier by photographing the waveform on the oscilloscope screen and extracting the freqencies $f_p$ - $f_{cw}$ from the photograph. Using different local oscillator frequencies and plotting the frequency versus time graphs enables accurate results to be obtained by drawing a line of least error. A chirp characteristic obtained by this method using six different local oscillator frequencies is illustrated in figure 19. It is seen that the maximum error of any one measurement is 0.5MHz. In practise it is observed that the above accuracy can be achieved for chirps of up to 25MHz when using X-band RF pulses of typically 1 microsecond duration.

#### The Absorption Wavemeter

The absorption wavemeter can be used to measure chirp and post tuning drift. The circuitry required for these measurements is shown schematically in figure 20. The RF pulses are passed through the wavemeter where the signals at a frequency corresponding to the wavemeters resonant frequency are absorbed. If the RF pulses contain chirp then the absorption of these particular signals results in a 'dip' in the magnitude of the crystal detected pulses. The position of the 'dip' indicates where the frequency of oscillation equals the resonant frequency of the wavemeter. This resonant frequency is determined from a calibration chart which converts resonant cavity length to frequency. Hence the frequencies of oscillation over the entire RF pulse can be determined by adjusting the length of the resonant cavity of the wavemeter and noting the positions of the 'dips' in the magnitude of the crystal detected pulses.

The chirp characteristic of a pulsed microwave oscillator measured by the wavemeter technique and by the frequency down conversion method is shown in figure 21. The total magnitude of chirp $|\Delta f|$ over the pulse measured by the wavemeter is 41 MHz whereas 33 MHz was measured by frequency down conversion. In practice it is found that the measurement of chirp by frequency down conversion is reasonably accurate. Thus it appears that the measurement of chirp by a wavemeter can result in significant errors. These errors result mainly from the finite Q-factor of the wavemeter which enables it to absorb a band of frequencies. This effect not only makes the calibration of the wavemeter inaccurate but also makes it difficult to determine the exact position of the 'dips' in the magnitude of the crystal detected pulses.

## Filter Frequency Discriminators

This technique of frequency measurement depends on the frequency discriminating properties of a waveguide single sided filter (Rhodes (4)). The required circuitry is shown in figure 22. In this section the use of a filter frequency discriminator for the measurement of chirp is described, however, the technique can also be adapted to the measurement of post tuning drift. The initial step when using this method of chirp measurement is to measure the central frequency of oscillation ($f_c$) of the pulse. This can be achieved using a wavemeter or spectrum analyser. The filter is then tuned so that the 'principle falling edge' of its attenuation versus frequency characteristic contains the center frequency ($f_c$). A typical filter attenuation versus frequency characteristic showing the principal falling edge and the center frequency $f_c$ is illustrated in figure 23. The filter characteristic is measured by connecting a sweep oscillator to the input of the filter and displaying the transmitted signal on a CRO screen. The output power from the sweep oscillator is levelled so as to be constant over the frequency range of interest. Thus the filter attenuation characteristic can be calculated from a measurement of the input power of the filter and measurements of the transmitted power. After calibration the filter is placed in the discriminating circuit as shown in figure 22. Since the frequency of oscillation changes during a pulse then the filter attenuation will also vary. Before measuring the filter attenuation over a pulse it is necessary to consider any amplitude changes during the unfiltered pulse. Sometimes these amplitude changes are small and can be neglected but in general they must be taken into account. Typical measurements of the filter attenuation during a pulse are illustrated in figure 24. These measurements are converted to a chirp characteristic by reference to the filters predetermined attenuation versus frequency characteristic.

In practice it is found that the chirp characteristics of an oscillator measured by the filter discriminating technique agree very favourably with results obtained by frequency down conversion. This indicates that the technique gives reasonable accuracy, however, when the two methods are compared it is found that the filter discriminating method is more time consuming and presents greater difficulties in obtaining accurate results. The major disadvantage of the filter discriminating technique is that if the central frequency of oscillation ($f_c$) of the RF pulse changes significantly so that it is no longer contained within the principle falling edge of the attenuation versus frequency characteristic of the filter then it is necessary to recalibrate the filter. Thus the measuring system is relatively narrowband. Changes in the magnitude of chirp do not affect the system accuracy providing none of the frequencies contained within the RF pulse are located outside the falling edge of the attenuation versus frequency characteristic.

## Delay Line Frequency Discriminators

This method of frequency measurement can be applied to the measurement of chirp and post tuning drift. The particular delay line frequency discriminator to be described was developed from polar discriminators commonly used for instantaneous frequency measurement in military electronic counter measure systems (Lipsky (5)). The delay line frequency discriminator uses a hybrid magic T to mix the RF signal with a delayed version of itself thus enabling the frequency information to be converted to amplitude changes. The circuitry required for the measurement of chirp using this technique is shown schematically in figure 25 which includes a detailed circuit diagram of the discriminator inside the dashed rectangle. The theory of the discriminator is explained by considering its response to an input signal $V\sin 2\pi ft$.

Initially phase shifter A is set to $90^{\circ}$ and phase shifter B is set to $45^{\circ}$. Therefore the signal entering port 1 of the magic T can be expressed as:

$$a = \frac{k_1 V}{2} \sin 2\pi ft \times \mathrm{EXP}\left[-j\left(\frac{\pi}{2} + kf\right)\right] \quad \ldots\ldots 9$$

where $k_1$ = the attenuation of phase shifter A and other relevant associated circuitry and $k = (2\pi l/c)$ where $l$ = the length of the delay line and $c$ = the velocity of light in free space. The signal entering port 2 of the magic T is given by

$$b = \frac{k_2 V}{2} \sin 2\pi ft \times \mathrm{EXP}\left[-\frac{\pi}{4} j\right] \quad \ldots\ldots 10$$

where $k_2$ = the attenuation of phase shifter B and other relevant associated circuitry.

The signals entering the magic T are added vectorially as shown in figures 26a and 26b resulting in the square law detected outputs D1 and D2 respectively. These outputs are given by

$$D1 = |a|^2 + |b|^2 + 2|a| \times |b| \sin\left(\frac{\pi}{4} - k^f\right) \quad \ldots\ldots 11$$

$$D2 = |a|^2 + |b|^2 - 2|a| \times |b| \sin\left(\frac{\pi}{4} - k^f\right) \quad \ldots\ldots 12$$

Consequently the output (A1) of the subtractor is given by

$$A1 = D1 - D2 = 4|a| \times |b| \sin\left(\frac{\pi}{4} - Kf\right) \quad \ldots\ldots 13$$

The discriminator output given by equation 13 is of limited use since it not only depends on frequency but also on $|a|$ and $|b|$ which depend on amplitude changes during the pulse. Consequently if there are significant amplitude changes during the pulse it becomes difficult to extract the required frequency information from the discriminator. The effect of these amplitude changes can be eliminated by using the above discriminator output (A1) together with the output when phase shifter A is set to 45° and phase shifter B is set to 90°. By using similar theory to that described above the new phase shifter settings result in a discriminator output

$$A2 = 4|c| \times |d| \cos\left(\frac{\pi}{4} - kf\right) \quad \ldots\ldots 14$$

Assuming that the attenuation of phase shifter A (set to 90°) equals the attenuation of phase shifter B (set to 90°) and the attenuation of phase shifter A (set to 45°) equals the attenuation of phase shifter B (set to 45°) then $|a| = |c|$ and $|b| = |d|$. (The phase shifters used were chosen such that these conditions were satisfied). Using this assumption gives the ratio of the discriminator outputs A1 to A2 as

$$B = \frac{A1}{A2} = \tan\left(\frac{\pi}{4} - kf\right) \quad \ldots\ldots 15$$

Differentiating equation 15 with respect to frequency and letting $C = \tan^{-1} B$ gives

$$\frac{dC}{df} = -k \quad \ldots\ldots 16$$

$$\text{Hence,} \quad \delta f = \frac{-1}{k}\,\delta C \quad \ldots\ldots 17$$

where $\delta C$ is the change in $\tan^{-1} B$ resulting from a change of frequency $\delta f$.

Hence the change in frequency between two different points contained within an RF pulse can be calculated using equations 15 and 17 and measurements of A1 and A2 at each of the points. Taking several of these measurements enables the frequency changes over the entire pulse to be calculated. These results and the measurement of any one particular frequency during the pulse, which can be obtained using a wavemeter, give the chirp characteristic.

A chirp characteristic measured by the delay line frequency discriminator described above is illustrated in figure 27 where the same characteristic measured by frequency down conversion is also shown. It is seen from these results that the chirp characteristics agree to within an accuracy of ±0.5MHz. The major disadvantage of this frequency measurement technique is the narrowband nature of the measurement circuit.

This narrowband effect results in the technique only exhibiting reasonable accuracies for the measurement of chirps up to typically 30MHz in an X-band oscillator.

## The RF Spectra of Outputs from Pulsed Oscillators

If the frequency of a pulsed oscillator is constant during the pulses then the frequency spectrum will be a pure SINC function. In this situation the frequency of the oscillator is indicated by the amplitude peak of the spectrum. However, the accurance of chirp can affect the spectrum of the output pulses in many different ways. The extent to which the spectrum is affected depends on the magnitude of the chirp, the shape of the chirp characteristic and amplitude changes during the pulse.

A discrete spectrum synthesis of an RF pulse train containing chirp is possible using the ideas forwarded by McRae (6). This synthesis simulates theoretically the RF spectral envelope of a pulse train from frequency and amplitude information derived experimentally from the pulses. The synthesis involves dividing an RF pulse into elements of constant oscillation frequency and determining the spectrum of each element by Fourier transform methods. These individual spectra are then added together to give the spectrum of the entire pulse train V(w) which can be expressed as

$$V(\omega) = 2\pi H\left(\frac{\tau^1}{T_p}\right) \sum_{m=-\frac{(M-1)}{2}}^{\frac{(M-1)}{2}} \left\{ \sum_{n=-\infty}^{n=+\infty} \operatorname{sinc}\left(n\omega_b \frac{\tau^1}{2}\right) e^{-jnm\omega_b \tau^1} \times \delta\left[\omega + \omega_o - m\omega_d - n\omega_b\right] + \sum_{n=-\infty}^{n=+\infty} \operatorname{sinc}\left(n\omega_b \frac{\tau^1}{2}\right) e^{-jnm\omega_b \tau^1} \times \delta\left[\omega - \omega_o + m\omega_d - n\omega_b\right] \right\} \quad \ldots\ldots 18$$

where
- $H$ = the RF pulse amplitude
- $\tau^1$ = time duration of each individual pulse element
- $T_p$ = RF pulse period
- $M$ = Number of elements into which the RF pulse is devided
- $\omega_o$ = Center frequency of pulse
- $\omega_b$ = $2\pi/T_p$
- $\omega$ = $2\pi \times$ frequency

The RF frequency corresponding to the mth element is assumed to be $(\omega_o - m\omega_d)$.

Equation 18 is solved using a computer which determines the envelope of the RF pulse spectrum. The spectra of RF pulses containing non-linear frequency modulation can be simulated using the above method by assuming unequal frequency increments $\omega_d(t)$. Similarly, amplitude modulation effects can be introduced by weighting the amplitude H at each time interval.

The circuitry required for the measurement of the RF spectra of pulsed oscillators is shown in figure 28. The measured spectrum of a pulsed oscillator displaying 11.5MHz of linear chirp is compared to the theoretically determined spectrum in figure 29a. It is seen that the magnitude of chirp corresponds approximately to the width of the 'plateau' which features on both spectra. In general this effect provides a quick, if not accurate, method of determining the magnitude of chirp during a pulse. However, if the chirp characteristic is non-linear, as in figure 29b , or the RF pulse contains significant amplitude modulation, as in figure 29c , then it is not a simple exercise to obtain chirp information from the RF spectrum.

## SUMMARY OF THE METHODS USED IN THE MEASUREMENT OF THE FREQUENCIES OF PULSED MICROWAVE OSCILLATORS

Frequency measurement techniques using frequency down conversion, filter frequency discriminators and delay line frequency discriminators are capable of measuring X-Band frequencies during pulses of up to 10 microseconds duration to within an accuracy of typically 0.5MHz. However, the extensive calibration and the precision necessary in acquiring accurate results from the narrowband filter and delay line frequency discriminators makes these particular methods difficult and time consuming. Alternatively the measurement of frequencies by down conversion is relatively wideband and only requires the calibration of the CW local oscillator which is simply accomplished by using a CW frequency counter. The measurement of frequencies using a pulsed frequency counter is reasonably accurate but the capital outlay for the counter is most restrictive. The simplest methods of frequency measurement which have been presented involve the use of a wavemeter or spectrum analyser. Although these methods are not as accurate as the frequency down conversion technique they are wideband and require little calibration and results can be acquired easily and quickly. However, if the RF pulse contains significant amplitude modulation or severely non-linear frequency modulation then the extraction of frequency information from the RF spectrum can be very difficult.

## REFERENCES

4 Rhodes, J.D., 1975, "Explicit design formulas for waveguide single sided filters," Trans. IEEE, Vol. MTT-23, No. 8, 681-688.

5 Lipsky, S.E., 1970, "Measure new threat frequencies instantly," Microwaves, December, 49-56.

6. Mc.Rae, C.J., 1974, "RF spectra of pulsed transferred electron microwave oscillators," Proceedings of the eleventh New Zealand National Electronics and Geophysics Convention, University of Auckland, Stream Al, 1 - 11.

## QUESTIONS

1. Estimate the frequency versus time characteristic of the oscillator whose down converted pulsed RF signal is as shown in figure 30 . The frequency at the center of the pulse has been measured to be 10 GHz. The pulse duration is one microsecond.

2. The RF pulses from the output of an oscillator are passed through a filter. Estimate the frequency versus time characteristic of the oscillator if the filter attenuation versus time characteristic for the duartion of the pulses is as shown in figure 24 . It is to be assumed that the filter attenuation versus frequency characteristic is as shown in figure 23 . The pulse duration is one microsecond and the pulses contain no significant amplitude modulation.

3. The relative filter attenuation versus time chacteristic for one microsecond RF pulses is as shown in figure 31b The filter attenuation versus frequency characteristic is as shown in figure 31a. Estimate the frequency versus time characteristic of the pulses assuming that there is no significant amplitude modulation and the frequency at the center of the pulse is 10 GHz. From the frequency versus time characteristic estimate the total magnitude of chirp over the pulse. If the amplitude of the RF pulse increases by 3dB during the pulse then estimate the new magnitude of chirp over the pulse assuming its filter attenuation versus time characteristic remains unchanged.

4. The output pulses of an X-band oscillator are passed through a delay line discriminator as shown in figure 25 . The outputs of the subtractor A1(A set 90° and B set to 45°) and A2 (A set to 45° and B set to 90°) are given in table (1).

TABLE 1 - Outputs of delay line frequency discriminator

| Time After Start of Pulse (nanoseconds) | A1 (mV) | A2 (mV) |
|---|---|---|
| 0 | 20 | 36 |
| 100 | 21 | 34.5 |
| 200 | 22 | 33 |
| 300 | 22.5 | 31.5 |
| 400 | 23 | 30.3 |
| 500 | 24 | 29 |
| 600 | 25 | 27.5 |
| 700 | 25.5 | 26 |
| 800 | 26.2 | 25 |
| 900 | 27 | 23.5 |
| 1000 | 28 | 22 |

Given that the length of the delay line is 100 centimetres and the frequency after 500 nanoseconds is 10GHz calculate the frequency versus time characteristic of the RF pulse.

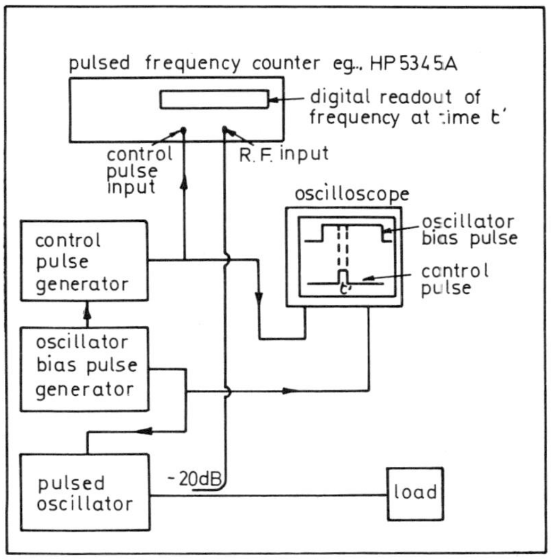


Figure 16 Frequency measurement using a pulsed frequency counter

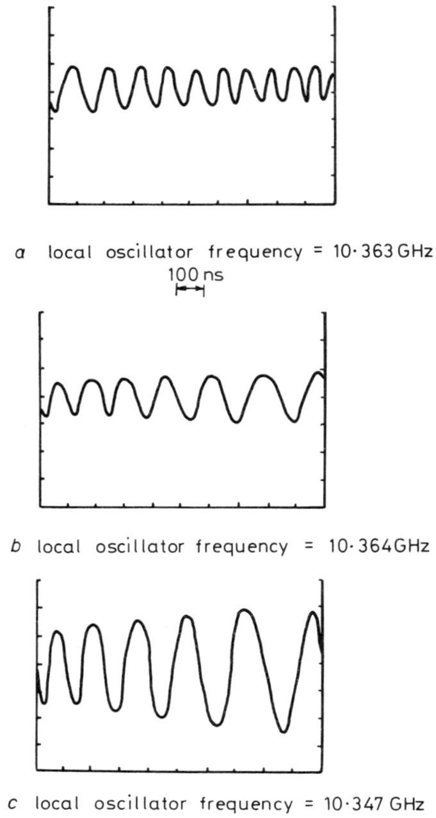


Figure 18 Displayed waveforms of down converted RF pulses

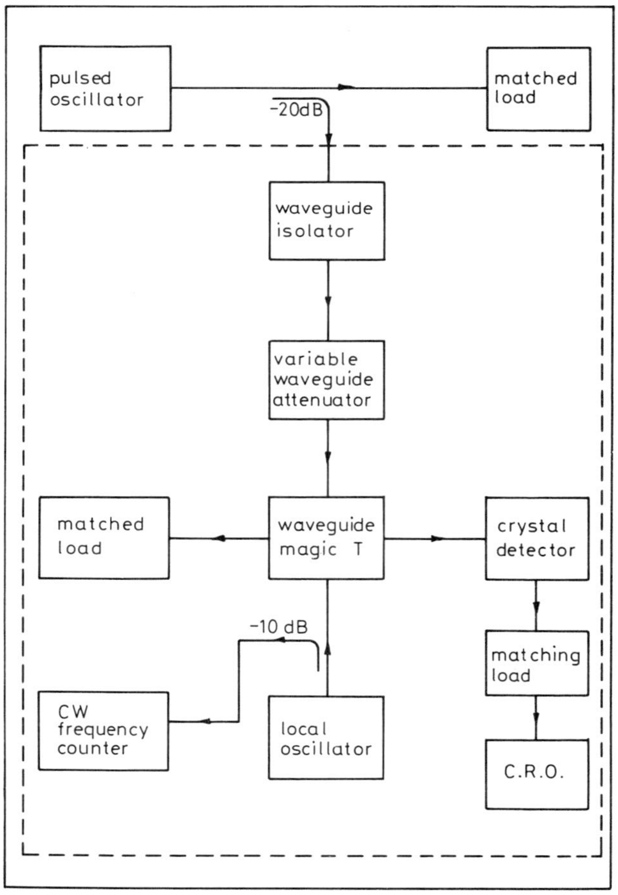


Figure 17 Frequency measurement by frequency down conversion

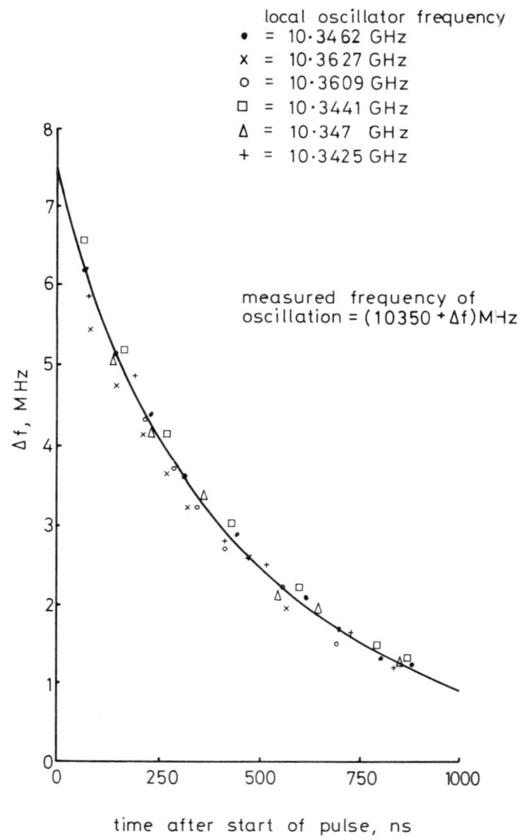


Figure 19 Frequency versus time characteristic measured by frequency down conversion

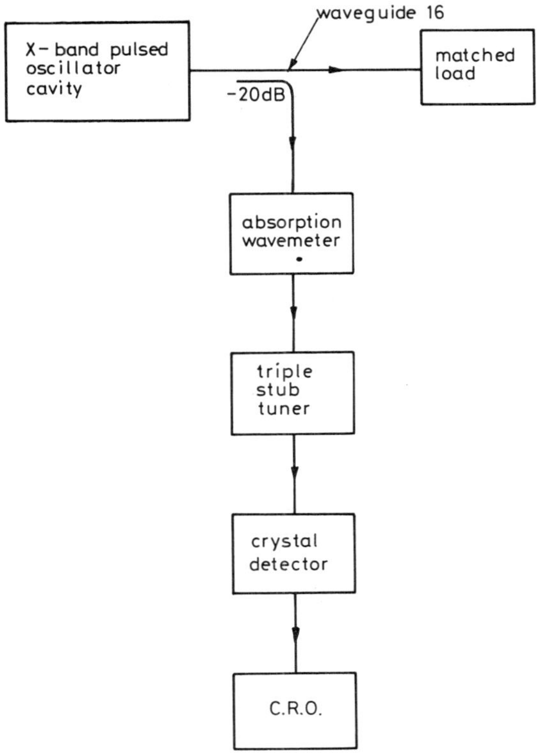


Figure 20 Frequency measurement using an absorption wavemeter

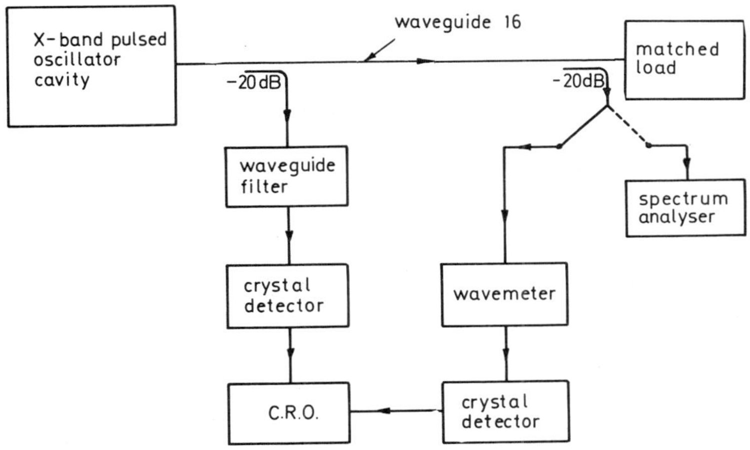


Figure 22 Frequency measurement using a filter frequency discriminator

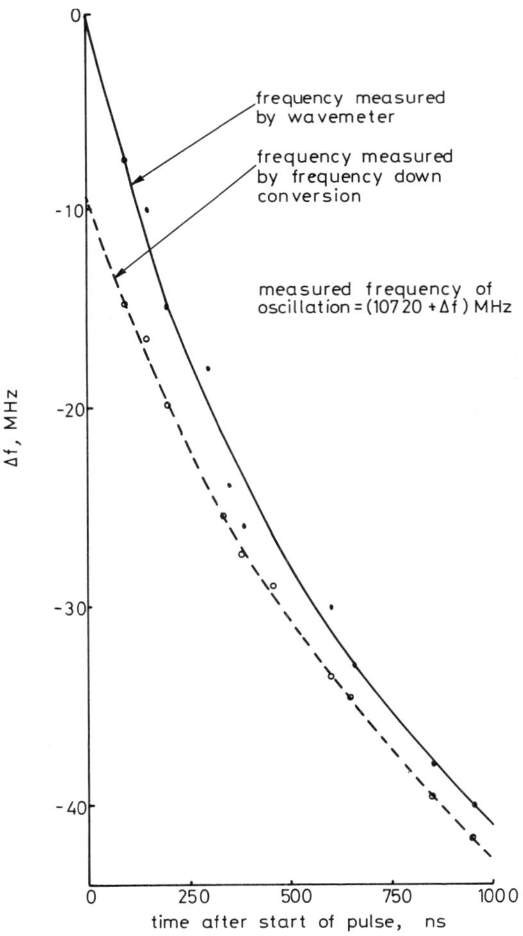


Figure 21 Frequency versus time characteristic measured by absorption wavemeter

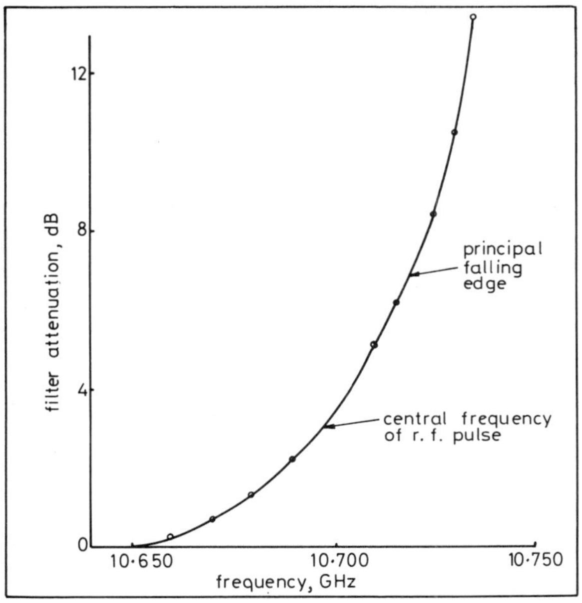


Figure 23 Attenuation versus frequency characteristic of filter

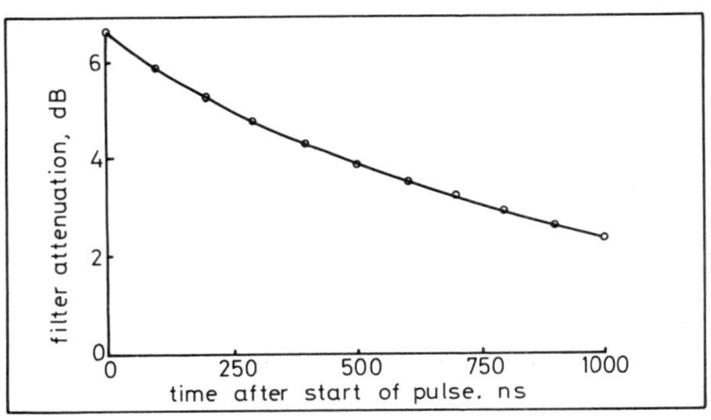


Figure 24 Filter attenuation of RF pulse

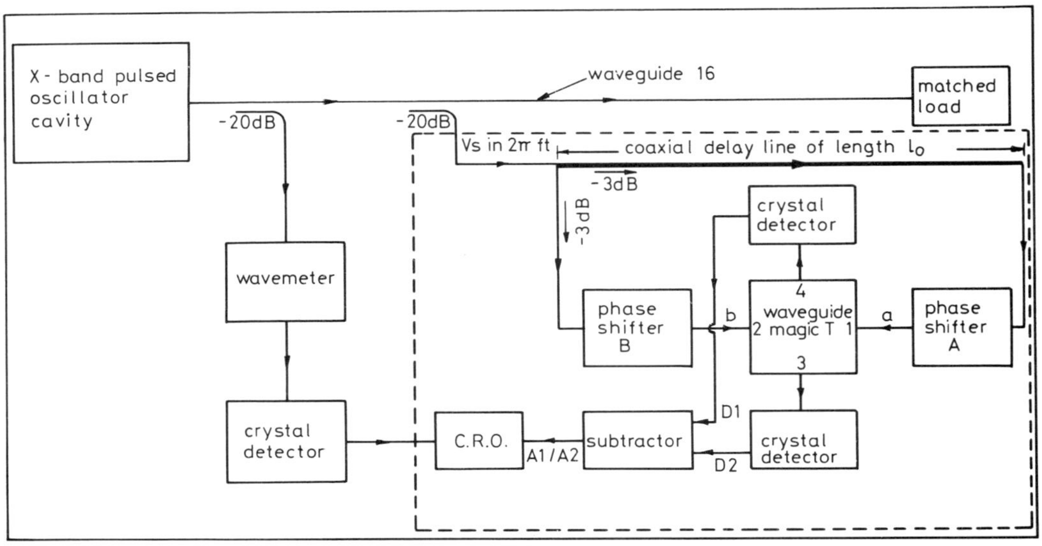


Figure 25 Frequency measurement using a delay line frequency discriminator

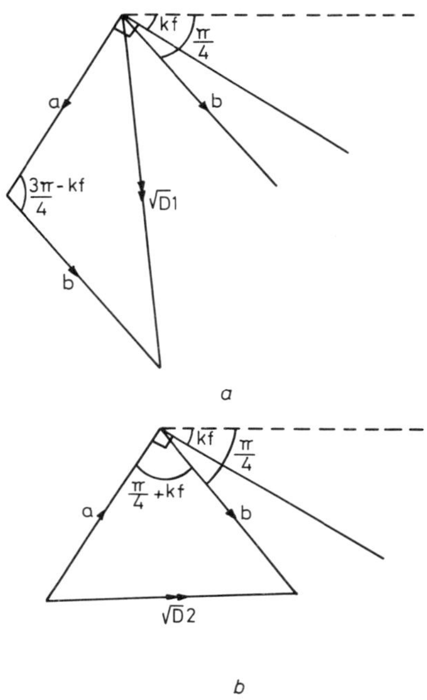


Figure 26 Addition of signals inside the hybrid magic T

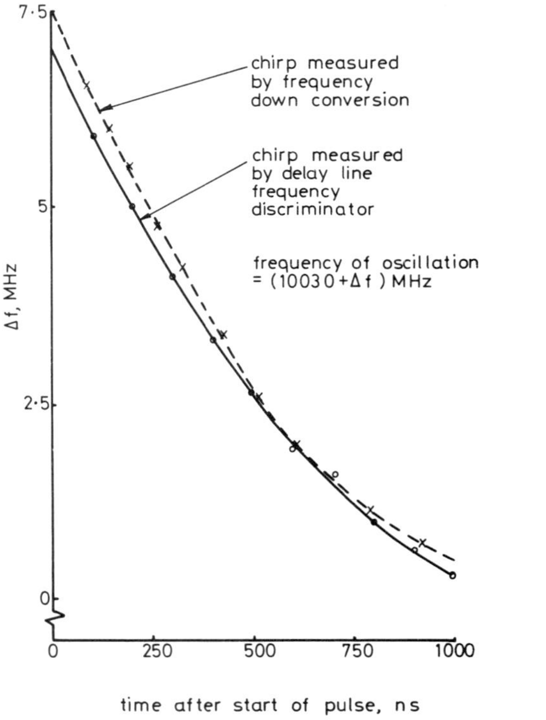


Figure 27 Frequency versus time characteristic measured by delay line discriminator

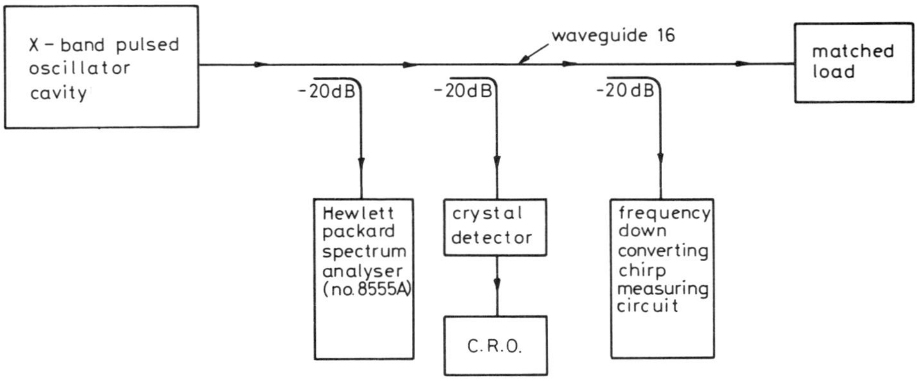


Figure 28 Circuit for measurement of frequency, amplitude changes and spectra of RF pulses

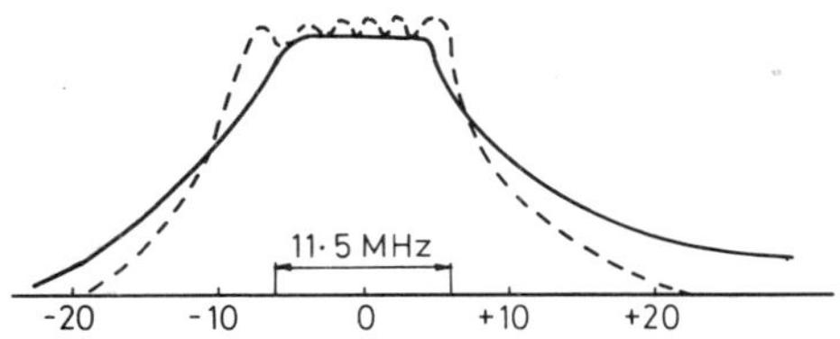


frequency = (Δf+9864)MHz

*a* pulse contains 11·5MHz of linear chirp

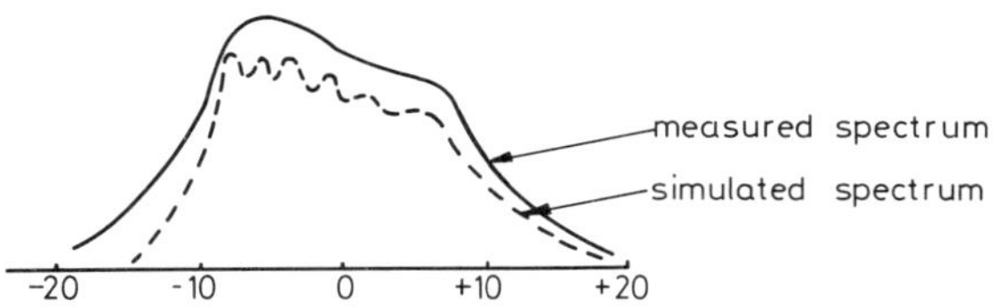


frequency = (10587+Δf) MHz

*b* pulse contains 17MHz of non-linear chirp

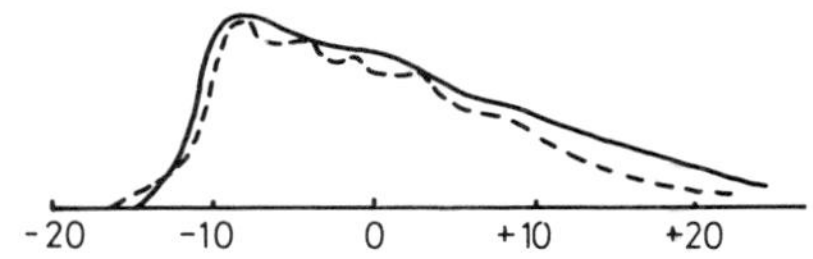


frequency = (9761+Δf) MHz

*c* pulse contains significant amplitude modulation and linear chirp

Figure 29 Measured and computer simulated spectra of RF pulses

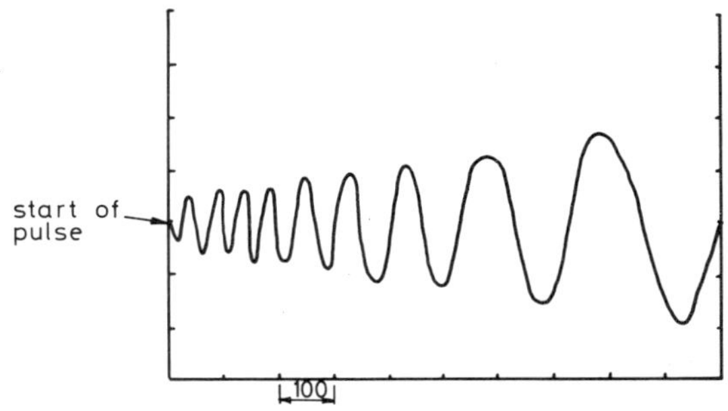


Figure 30 Down converted RF pulse signal (local oscillator frequency = 9.993GHz)

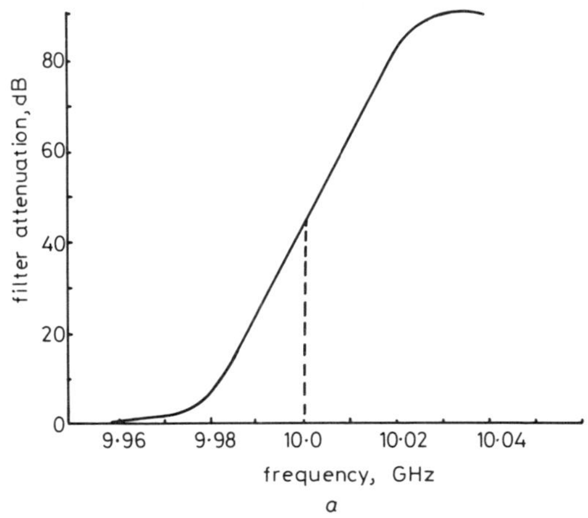


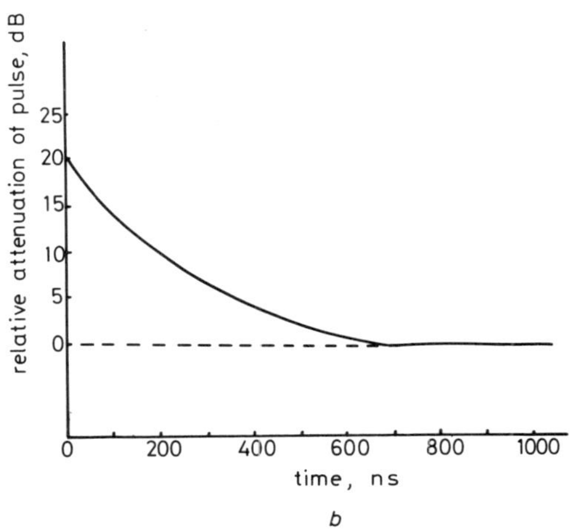


Figure 31 (a) Filter attenuation characteristic (b) Relative filter attenuation during pulse

# CHAPTER 14
# Mixers and detectors

Dr. T. Oxley

# MIXERS AND DETECTORS - Chapter 14

T.H. Oxley

AEI Semiconductors Ltd., Lincoln.

## INTRODUCTION

These notes briefly review theoretical and design parameters of semiconductor devices and circuits applied as mixers and detectors in microwave systems.

## SUPERHETERODYNE AND VIDEO DETECTION SYSTEMS

Receivers operating in the microwave region are usually characterized by four main features: sensitivity, bandwidth, dynamic range and power handling. Sensitivity describes their ability to detect the weakest possible signal, which may be comparable to or even weaker than the noise in the system. Considerable gain is then required to bring this noise level, and signals which are comparable with it, up to a useable level. The receiver must also be able to deal with pulses covering an extremely wide dynamic range and may be required to accept a wide band of radio frequency signals. Receiver power handling describes the ability of the receiver to withstand electrical overloads without degradation in the sensitivity performance.

The most common types of receiver used in practical microwave systems are based on the superheterodyne and video-detector principles of operation.

Sensitivities of about -100 dBm and -50 dBm for superheterodyne and video-crystal type receivers may be achieved respectively assuming a comparable bandwidth (about 1 MHz). Thus the sensitivity of the videc-crystal receiver is low compared with that of a superheterodyne receiver. However, this is not prohibitive in many cases, and the advantages of simplicity, small size, and wide bandwidths make this an attractive system for applications where high sensitivity is not demanded.

### Superheterodyne Receiver

A schematic diagram of a simple heterodyne receiver is shown in figure 1(a).

Heterodyne reception provides conversion of the carrier frequency of the signal to a new value. This is accomplished by means of a local oscillator and a nonlinear rectifying element. The local oscillator and the signal are coupled into the nonlinear device and generate (among other frequencies) the frequency difference between the signal and local oscillator frequencies, normally termed the intermediate frequency. The local oscillator may have either a lower or higher frequency than the signal, since it is the difference which is of interest. The amplitude of this difference (or intermediate) frequency is proportional to the signal amplitude and is substantially independent of the amplitude of the local oscillator voltage. If the signal is an amplitude-modulated wave, the difference frequency output will consist of a carrier at intermediate frequency plus sideband which reproduce the original modulation of the signal. The circuit which contains the nonlinear rectifying element, the means for introducing the signal and local oscillator, and provides the output terminals for the intermediate frequency is usually termed the 'mixer' or 'down converter' and the process is referred to as 'mixing' or 'frequency conversion'. The rectifying nonlinear device is usually a diode rectifier known as a 'mixer diode'. The intermediate frequency voltage is amplified and the complete circuit is referred to as a 'superheterodyne receiver'.

Application of the local oscillator and signal voltage to a mixer diode rectifier (shown by a diode I-V characteristic of the form I = f(V)) is indicated in figure 1 (b). The local oscillator voltage is adjusted to a higher level than that of the signal in order to bias the signal voltage to the optimum region of the rectifying element. A local oscillator power at about the milliwatt level is normal. Under these conditions a standing mean current is produced by rectification of the local oscillator voltage, and is normally termed the 'mixer diode' rectified current.

The nonlinear element used in frequency conversion may be any type of rectifying device, in so far as its characteristics and parasitic components allow efficient rectification at the frequency of interest.

### Video-Crystal Receiver

A schematic diagram of a simple video-crystal receiver is shown in figure 2(a). This system provides direct demodulation of the radio frequency input signal, the signal being detected by rectification at a nonlinear element. The nonlinear rectifying device is usually a diode rectifier, normally termed a 'video-crystal detector' or 'low level detector'. Such a detector operates in the square law region of its nonlinear characteristic, i.e. the rectified output voltage of the detector is directly proportional to the radio frequency (signal) input power (square of the input voltage). The video output voltage of the detector is amplified, and the complete circuit is usually referred to as a 'crystal video' or 'video-detector' receiver.

Application of the radio frequency (signal) voltage to the diode rectifier (shown by a diode I-V characteric of a form I = f(V) is indicated in figure 2(b). The signal voltage is applied at the optimum region of the nonlinear element: this can necessitate the use of a forward d.c. bias voltage.

The nonlinear element used as a low level detector may be any type of rectifying device, in so far as its characteristics and parasitic components allow efficient rectification at the frequency of interest. In general, almost any rectifier will function as a square-law detector when the applied signal is sufficiently small.

## THE NONLINEAR RECTIFYING ELEMENT

Rectification may be defined as an operation on an a.c. voltage to produce a unidirectional component. This component arises from the fact that the average resistance to current flow is less in one direction than in the other. In addition to the d.c. component in the rectifier output there are also present harmonics of the input signal which arise because of the nonlinear character of the rectifying element.

The nonlinear element normally considered is a rectifying semiconductor device. The static rectifier characteristics of this element are of interest on several counts. They are of importance in the theory of rectification, in that an adequate theory must predict quantitatively the features of the current-voltage characteristic and the relative amplitude of the harmonics of the input signal and magnitude of the d.c. component dependency on the shape of the current-voltage characteristic. For efficient rectification, a high reverse resistance and a low forward resistance together with a sharp curvature at the origin are obvious desirable features. However, it must be emphasized that a good static characteristic is not a sufficient condition for a good microwave performance. At sufficiently high frequencies the shunting effect of the capacitance of the barrier at the rectifying contact becomes important. Also, in microwave applications, other factors such as noise and impedance are significant.

Semiconductor diode rectifiers for microwave application are mainly based on the phenomenon of majority carrier injection across a Schottky-type barrier between a metal and semiconductor, such as in the conventional point-contact diode and the more recently introduced Schottky barrier diode. However, the diode properties resulting from the mechanism of quantum mechanical tunnelling of current carrier across a p-n junction, such as in the tunnel or backward diode, are also of interest for some special applications.

### Metal-Semiconductor Junction

The metal-semiconductor contact is described by the Schottky theory of rectification. Unlike the p-n junction diode, the rectifying contact is based on majority carrier conduction, and in normal operation exhibits virtually no storage of minority carriers. This results in more efficient rectification at high frequencies.

The operation of a metal-semiconductor diode can best be understood by referring to its appropriate electron energy diagrams shown in figure 3. These show the energies of free electrons in the metal in, for example, an n-type semiconductor under various conditions of bias. Briefly, at zero bias, there is a constant interchange of electrons between the two materials. The flow of electrons from the semiconductor to the metal results in a current $I_S$, normally termed the diode saturation current, with an equal and opposite current $-I_S$ flowing from the metal to the semiconductor to give zero current flow across the junction. ($I_S$ is dependent on the area of the junction, the diffusion potential, and temperature.)

Under forward bias conditions the energy of the electrons in the conduction band of the semiconductor is increased, permitting them to overcome the diffusion potential and to be injected into the metal. The application of the external bias $V_E$ results in a lowering of the diffusion barrier by an amount $V_A$, that is the height of the potential barrier is reduced as seen by the electrons in the semiconductor looking towards the metal. ($V_A = V_E - IR_S$, where $IR_S$ represents a voltage drop across the semiconductor junctions and $R_S$ is the diode series resistance.) This reduction in the barrier results in a current flow which is exponentially dependent on the applied voltage $V_A$. The current increases exponentially with voltage until with high voltage the semiconductor conduction band is flattened and all the electrons in the conduction band have sufficient energy to flow into the metal. Beyond this point the current is limited by the diode series resistance. The diffusion potential $V_B$ thus determines the voltage at which high forward conduction takes place. $V_B$ is mainly dependent on the metal and semiconductor in use, that is $\phi_B$ and $E_G$. For a perfect contact, $\phi_B = \phi_m - \chi_s$ where $\phi_m$ is the work function of the metal and $\chi_s$ is the electron affinity of the semiconductor. $\phi_B$ is thus strongly affected by the work function and surface state of the metal-semiconductor contact interface (1).

Silicon, germanium, and gallium arsenide are the most common semiconductor materials normally considered for microwave application, and it is possible to obtain some guidance to their relative diffusion potential by consideration of their energy gap properties. For example, the energy gaps are 0.68 eV for germanium, 1.08 eV for silicon, and 1.4 eV for gallium arsenide. These give rise to the diffusion potentials $V_B$ increasing with the energy gap. However, as already indicated, the work function of the metal contact will have some influence on the value of $V_B$, and this can be made use of for particular applications.

Under reverse bias conditions the number of electrons in the conduction band of the semiconductor having sufficient energy to overcome the barrier is decreased. In the ideal case the reverse current will be a constant, $-I_S$; in practice, however, surface potential and image-force effects result in higher than ideal reverse leakage currents. Breakdown voltage is the voltage limit when the current increases extremely rapidly: avalanche multiplication, as in p-n junctions, is usually responsible for this phenomenon.

Both diffusion and diode theories predicts an ideal I-V characteristic of the form shown in figure 3, illustrating the law

$$I = I_s(e^{\alpha V} - 1) \tag{1}$$

where

$$\alpha = \frac{q}{nkT} .$$

Thus

$$I = I_s\left[\exp\left(\frac{qV}{nkT}\right) - 1\right] . \tag{2}$$

Where $I_S$ is the saturation current

e is the electron charge $= 1.6 \times 10^{-19}$ (coulomb)

T is the absolute temperature ($^oK$)

K is Boltzmann's constant $= 1.38 \times 10^{-23}$ (Joule/$^oK$)

V is the voltage across the diode junction (Volts)

n is termed the ideality factor and should equal 1.0 for an ideal characteristic.

That is $\frac{e}{nkT} = 40\ (\text{Volt})^{-1}$.

It should be noted that V is not the voltage $V_A$ across the rectifier, but is less than $V_A$ by the voltage drop across the series resistance of the rectifier. That is,

$$V = V_A - IR_S.$$

This is, however, normally neglected for small currents.

The saturation current is related to the barrier height $\phi_B$ by Richardson's equation

$$I_S = \alpha A^* T^2 \exp\left(-\frac{q\phi_B}{kT}\right) \qquad (3)$$

where α is the diode area and A* is Richardson's constant which is related to the effective mass of the semiconductor (1):

$$A^* = 9.6\ \text{A cm}^{-2}\text{K}^{-2} \text{ for silicon}$$

and

$$A^* = 4.4\ \text{A cm}^{-2}\text{K}^{-2} \text{ for gallium arsenide.}$$

Tunnel or Backward Diode

Since electrons can be considered to have wave properties, they have the ability to penetrate potential barriers. This effect, known as quantum mechanical tunnelling, is exhibited by the tunnel diode or Esaki diode, a p-n junction diode. The most familiar example is the tunnel diode where impurity concentrations on both sides of a p-n junction are such that an increase in tunnelling current is followed by a decrease, producing a negative resistance. The diode itself is made of heavily doped semiconductor material and is an abrupt junction with an extremely narrow depletion layer. Much of the importance of the tunnel diode arises from its negative resistance characteristic, but in the case of the backward diode the tunnelling is reduced from the high levels of a conventional diode to such levels that a negative resistance is almost nonexistent. The tunnel diode, although capable of microwave mixing and detecting, is difficult to stabilize and the negative conductance is a circuit problem. A good compromise is a backward diode which acts as a highly efficient rectifier over a limited voltage range but in the reverse direction, hence the term 'backward diode'. This type of characteristic is normally used for microwave mixer and detector applications.

The expression for the tunnelling current is given by (1)

$$J \propto \exp\left[\frac{-Am^{1/2}E_G^{3/2}}{F}\right] \qquad (4)$$

where A is a numerical constant, and F is the average electric field across the space-charge region. F is inversely proportional to the width of the space-charge region of the p-n junction. A wide space-charge region means low tunnelling current, a narrow one means high tunnelling current. For a backward diode, it is necessary to reduce the tunnelling current from the high levels of a tunnel diode to such levels that a negative resistance is almost non-existent. The expression for the width of the space-charge region of an abrupt junction is given by (1).

$$W = \left[\frac{2\varepsilon(N_D+N_A)}{qN_DN_A}(V_B + V_A)\right]^{1/2} \qquad (5)$$

Thus, if in fabrication one doping level is kept constant at some high level, the space charge can be adjusted by decreasing the other impurity level.

The consequences of forming a junction on a highly doped semiconductor may best be illustrated by simple energy band diagrams. These are shown in figure 4 for different conditions of bias.

Briefly, at zero bias conduction-band electrons from the n-type side can tunnel into the p-type valence band, and the valence-band electrons from the p-type side can tunnel into the conduction band. These two current components exactly balance each other at zero bias, so the net current is zero.

When a reverse bias is applied, all energy levels on the p-type side are increased in relation to those of the n-type side. As there are then many available states on the n-side facing the electrons in the p-type valence band, a large current of electrons can easily flow from the p-type to the n-type side.

For small forward-bias voltages, the energy level of the p-type side is decreased in relation to the n-side. In this the n-type conduction-band electrons are opposite unoccupied states in the p-type valence band, while the p-type valence-band electrons are opposite the forbidden energy gap of the n-type side, and thus the current flows in the forward direction.

When the forward voltage is increased further, some conduction-band electrons on the n-type side are opposite the forbidden-energy gap of the p-type side, and the conduction-to-valence-band is reduced. The forward current then decreases as the forward voltage increases. This current voltage relationship accounts for the negative-resistance region of the tunnel diode. Finally, for a forward-bias voltage equal to the valley voltage, tunnelling completely ceases. At voltages greater than the valley voltage, the height of the barrier is reduced to a level which permits conventional current to flow over the barrier.

The I-V characteristic of a tunnel diode is difficult to represent in a single equation. However, the predicted I-V characteristic of a backward diode is of the form (1).

$$I = C_1 V(S-qV)^2\{\exp[-C_2(E_G-qV)^{1/2}]\} \qquad (6)$$

where $C_1 = Aq/kT$ (A is a numerical constant), S = penetration of Fermi level, $C_2$ = constant determined by semiconductor carrier properties.

Diode-Rectifier Characteristics

The basic microwave frequency characteristics of a diode rectifier are its input r.f. admittance and sensitivity. Sensitivity is related to the diode rectification efficiency and is normally expressed in terms of receiver overall noise figure for mixer applications or detectable signal level for video detector applications.

Diodes for these applications may be described electrically by the simple equivalent circuit shown in figure 5 where $C_B$ is the barrier (junction or contact) capacitance, $R_B$ is the barrier resistance, and $R_s$ is the series (spreading) resistance. Stray parasitic components introduced by insertion of the diode into the microwave circuit, such as device encapsulation, are represented by the series inductance $L_s$ and parallel capacitance C.

For a mixer diode which normally operates under large signal conditions (that is, with local oscillator powers of about a milliwatt), the rectification efficiency $\eta$ is the ratio of resistance in the forward direction to the impedance in the reverse direction. That is,

$$\text{Forward resistance} = R_s \quad (R_B \ll 1/\omega C_B \text{ and } R_s)$$

$$\text{Reverse impedance} = R_s - \frac{j}{\omega C_B} \quad (R_B \text{ large})$$

$$\eta = 1 - \frac{1}{\omega C_B R_s} \tag{7}$$

This analysis is sometimes used to express the frequency capability of the mixer diode rectifying contact expressed in terms of cut-off frequency, which is given by

$$f_{co} = \frac{1}{2\pi R_s C_B} \tag{8}$$

For a detector diode, signal levels are normally low (that is, in the microwatt region) and the concept of front to back ratio for the mixer applications has little significance. The equivalent circuit becomes a current generator shunted by the barrier admittance, and the diode operates as a square law device. That is:

$$P_{in} \triangleq V_{in}^2(1+\omega^2 C_B^2 R_B R_s) \quad \text{for } R_B \gg R_s \tag{9}$$

and

$$I \propto \frac{P_{in}}{1+\omega^2 C_B^2 R_B R_s} \tag{10}$$

where I is proportional to $V_{in}^2$ for square law operation.

Thus for high rectification efficiency, $C_B$ and $R_s$ must be minimized for mixer and detector considerations, with detector sensitivity depending on the square of $C_B$ in contrast to linear dependence for the mixer.

From the equivalent circuit of figure 5, the r.f. admittance of the diode rectifier may be given by:

$$\frac{1}{Z} = \frac{1}{[(1/R_B)+j\omega C_B]^{-1}+j\omega L_s+R_s} + j\omega C \tag{11}$$

Obviously the equivalent circuit has two resonant frequencies, the series and the parallel. If the ohmic resistances $R_B$ and $R_s$ are neglected, then in the case of the series resonance, the admittance of the tuned circuit is infinite and the resonant frequency is

$$\omega_{\text{series res}} = \frac{1}{(L_s C_B)^{1/2}} \tag{12}$$

In the case of the parallel resonance, the admittance of the tuned circuit is zero and the resonant frequency is:

$$\omega_{\text{parallel res}} = \omega_{\text{series res}}[1+C_B/C]^{1/2} \tag{13}$$

Thus the values of the frequencies corresponding to the series and parallel resonance are determined by the values of $L_s$, $C_B$ and C. By adjusting the values of these parameters it is possible to adjust the resonant frequencies of the diode rectifier, and hence the device r.f. admittance for a given design frequency.

The barrier resistance $R_B$ and the barrier capacitance $C_B$ are both functions of the current through the diode. $R_B$ is the reciprocal of the slope resistance of the I-V characteristic at the operating point. The relationship between $R_B$ and I can be shown by rearranging equation (1). That is:

$$V = \frac{1}{\alpha} \ln \frac{1 + I_s}{I_s}$$

and the 1/slope resistance

$$R_B = \frac{dV}{dI} = \frac{1}{\alpha(I + I_s)} \tag{14}$$

or

$$R_B = \frac{1}{\alpha I} \text{ for } I \gg I_s. \tag{15}$$

The relationship between the barrier capacitance and applied voltage can be obtained for the depletion layer capacitance expression for a step junction by

$$C_{B(V)} = \frac{C_{B(0)}}{(1 - V/V_B)^{1/2}} \tag{16}$$

where $C_{B(0)}$ is the zero bias capacitance and $V_B$ is the diffusion potential.

Interpretation of this expression may therefore be used to determine the effective diffusion potential for a step junction, from capacitance - voltage plots (1).

## MIXER PARAMETERS

### Receiver Overall Noise Figure

All noise sources are usually compared to the available noise power from a resistor which is said to have an available noise power given as kTB. This concept is useful in that it enables a figure for noise power to be quoted that is independent of the resistor value.

$$kTB = 4 \times 10^{-21} \text{ watts per cycle bandwidth}$$
$$= 4 \times 10^{-15} \text{ watts for 1 MHz bandwidth.}$$

In the case of conventional amplifiers these have an inherent noise made up of shot noise, Johnson's noise, etc., and in order to set a measure to the value of this noise power, a receiver is said to possess a 'noise figure' which is defined as:

$$F = \frac{\text{Signal/noise at the amplifier network input}}{\text{Signal/noise at the amplifier network output}}$$

$$= \frac{S_1/N_1}{S_0/N_0} \quad \text{where } N_1 = kT_0B. \tag{17}$$

Using the expression it may be shown that the overall noise figure performance of a superheterodyne receiver may thus be given by (1)

$$F_0 = L_c(F_{if} + N_r - 1) \tag{18}$$

assuming that any local oscillator noise is adequately suppressed. Where $L_c$ is the conversion loss of the mixer diode, defined as the ratio of the available input power from the generator at signal frequency to the available output power at intermediate frequency, $N_r$ is the noise temperature ratio of the mixer diode, defined as the ratio of available output noise power from the mixer to that of a resistor equal to the output resistance of the mixer, at room temperature, and $F_{if}$ is the noise figure of the following i.f. amplifier.

Hence $L_c$ and $N_r$ are the mixer diode parameters determining the value of the overall noise figure, and these should be minimized.

Frequency changing or mixing may broadly be defined as the conversion of a signal from one frequency to another by combining it with a local oscillator in a nonlinear device. This mixing process will result, however, in a complex frequency spectrum.

Frequencies of major interest in the mixing process are indicated in figure 6(a) and are summarized below:

| | |
|---|---|
| Local oscillator | l.o. |
| Local oscillator harmonics | n(l.o.) |
| Intermediate | i.f. |
| Signal | l.o.+i.f. |
| Image | l.o.-i.f. |
| Harmonic sidebands | n(l.o.)±i.f. |

The image frequency is of special importance on two counts. It is generated as a result of the direct beat between the signal and the l.o. second harmonic and also as a result of the beat between the i.f. and l.o., and the power consumed at the image is taken from the available signal power with resultant lowering of the conversion efficiency. Termination of this frequency and the phase relationship to the signal is thus an important consideration of mixer design.

The broadband or matched input circuit is favoured for most microwave receiver applications and the noise figure performance of mixer diodes is usually measured and specified under these conditions, that is when the same load admittance is presented to the image and signal channels. The theoretical limit for the diode conversion loss set by the image-matched condition is 3 dB (that is half the signal power is converted to image power and absorbed in the image termination). Thus the theoretical limit for the overall receiver performance assuming $N_r = 1.0$ is given by

$$F_0 = 2F_{if}$$

e.g. 4.5 dB for 1.5 dB i.f. amplifier.

With reference to figure 6(b), the simple mixer can be represented by a network with signal, image, and i.f. terminals. Isolation between these terminals is dependent on the conversion loss of the mixer. Thus for broadband (image matched) operation, matched terminations of all these ports is a desirable feature for reducing unwanted interaction effects.

In addition, it is obvious from figure 6(a) that the received signal must be separated from the l.o. frequency by an amount equal to the i.f., but it may be above or below the l.o. frequency.

Thus the receiver can convert two different signal frequencies to i.f.: one above and one below the l.o. frequency. The desired frequency is the signal, and the other is the image.

Double-sideband (DSB) noise figure thus signifies the operation of a mixer when it is receiving usable signals in both the signal and image bands, and single-sideband (SSB) when it is receiving a useful signal in the signal band only. Thus for a double-sideband receiver, equation (18) becomes

$$F_0 = \frac{L_c}{2}(N_r + F_{if} - 1)$$

assuming equal conversion efficiency for both channels. (It should be noted that it is general practice to use a broadband noise source, with available power at both signal and image, for the measurement of receiver overall noise figure.

Again assuming the same conversion efficiency for both image and signal channels, it is necessary to add 3 dB to the double-sideband conditions to convert it to the single-sideband case).

The local oscillator can, however, provide a further source of noise in the superheterodyne system. This appears as an r.f. output component, which enables the noise to pass through the mixer-amplifier system. This, stated as 'excess local oscillator noise', $N_0$ is the ratio of the available noise output from the oscillator to the available noise output from a resistor kTB, expressed as the actual noise output in watts/cycle/mW.

When considered in expression (18) the noise is converted to noise power at i.f., and the i.f. noise power

$$t_0 = \frac{N_0 \text{ watts/MHz/mW}}{L_c kTB} \tag{19}$$

and

$$F_0 = L_c(F_{if} + N_r - 1 + t_0). \tag{20}$$

If the noise figure of the oscillator is quoted in terms of effective increase of $N_r$, it is necessary to quote $L_c$, i.f. and the bandwidth, and the l.o. power.

## Diode Conversion Loss

It has already been shown in figure 5 that a diode rectifier may be represented by the simple equivalent circuit of the voltage-dependent barrier resistance $R_B$, shunted by the voltage-dependent barrier capacitance $C_B$. this combination being in series with the fixed series (spreading) resistance $R_s$. The actual mixing is accomplished by the non-linear barrier resistance, the series resistance, and the barrier capacitance acting as parasitics which degrade the conversion loss. This neglects the presence of the series inductance $L_s$.

The conversion loss $L_c$ of the mixer diode may be represented by

$$L_c = \frac{P_s}{P_{if}} = \frac{\text{Signal power available at the input terminals}}{\text{i.f. power available at the output terminal}}$$

The theory shows that this is dependent on the low frequency conversion loss $L_o$, determined basically by the analysis of the I-V characteristic and represented by

$$L_0 = \frac{P_{in}}{P_{out}} = \frac{\text{Signal power into the barrier resistance}}{\text{i.f. power out of the barrier resistance}}$$

and the degrading effect of the diode parasitics $R_s$ and $C_B$ represented by

$$L_1 = \frac{P_s}{P_{in}} = \frac{\text{Signal power available at the input terminals}}{\text{Signal power absorbed in the barrier resistance}}$$

such that

$$L_c = L_0 L_1 \tag{21}$$

The theory of mixing (1) predicts that a resistive mixer may be represented by a three-port network with terminals at the signal, image, and intermediate frequencies. Using such a network it can be shown that the conversion loss is dependent on the termination at the image and i.f. ports as well as the signal port. Such a network may be represented by a matrix whose elements may be derived from the I-V characteristics.

For example, assuming zero reverse current, the forward I-V characteristic can be represented by an equation of the form

$$\frac{I}{I_0} = \left(\frac{V}{V_0}\right)^{\bar{\chi}} \tag{22}$$

and the theory shows that the minimum low frequency conversion loss $L_o$ that can be realized under various conditions of image termination is a function of the diode exponent $\bar{\chi}$. The value of $\bar{\chi}$ is the slope of the log I versus log V of the diode forward I-V characteristic, i.e.

$$\bar{\chi} = \frac{d \log I}{d \log V} \tag{23}$$

Theoretical diode conversion loss as a function of the diode exponent $\bar{\chi}$ is shown in figure 7 for different image terminations. That is, image open circuit (o/c), short circuit (s/c), and matched conditions.

With reference to the equivalent circuit of figure 5, it can be shown that an angular frequency $\omega$

$$L_1 = \frac{R_s + R_B/[1+(\omega C_B R_B)]^2}{R/[1+(\omega C_B R_B)]^2} \tag{24}$$

Thus, when simplified, the conversion loss $L_c$ is given by

$$L_c = L_0 \left[ 1 + \frac{R_s}{R_B} + \omega^2 C_B^2 R_s R_B \right] \tag{25}$$

where $L_0$ = low frequency conversion loss determined by the I-V characteristic, $R_B$ barrier resistance, $C_B$ barrier capacitance, and $R_s$ = series resistance.

If the rectifier is represented by $R_B$ only, the conversion loss decreases and approaches a value of 3 dB (image matched) as the local oscillator power (and therefore diode rectified current) is increased. However, the effect of $C_B$ on the conversion loss is most pronounced at low values of local oscillator drive; as this is decreased the value of $R_B$ increases, more current is shunted by $C_B$ and therefore the conversion loss increases. The effect of $R_s$ on conversion loss, however, is most pronounced at high levels of local oscillator drive; as the l.o. drive is increased, the value of $R_B$ is decreased, more power is dissipated in $R_s$. and again the conversion loss increases. Obviously there is an optimum drive condition for maximum voltage across $R_B$ and maximum power at $R_B$, and minimum conversion loss will occur when $R_B$ is equal to $1/\omega C_B$, or

$$L_c = L_0(1 + 2\omega R_s C_B). \tag{26}$$

Thus at any frequency at which the barrier capacitance is significant, the product of the parasitics (series resistance and barrier capacitance) must be minimized for optimum conversion loss and thus satisfactory mixer performance. That is,

$$L_c \alpha C_B R_s. \tag{27}$$

It should be noted that this is consistent with the rectification efficiency and the cut-off frequency expressions of equations (7) and (8) respectively. A reasonable correlation between the conversion loss $L_c$ and rectification efficiency is also observed in practice.

## Diode Noise Temperature Ratio

The contributions to the diode noise temperature ratio are:

(a) Thermal noise of the series resistance $R_s$. This is due to the random motion of electrons in resistive elements. A short-circuit noise generator feeding the resistance R is given by the mean square current fluctuations

$$\bar{I}^2 = 4\frac{kTB}{R} \tag{28}$$

(b) Shot noise of the rectifying barrier $R_B$. This manifests itself as random fluctuations in the mean electron current, arising from the discreteness of electronic charge and the random fluctuations in the number of electrons constituting the current. The mean square fluctuation is shown by the theory to be

$$\bar{I}^2 = 2qIB \tag{29}$$

where I = mean current ($I + 2I_s$) for a metal-semiconductor device).

(c) Flicker noise. This manifests itself as noise in excess of shot or thermal at low frequencies of the form

$$\bar{i}^2 \propto \frac{I^m}{f^n} \qquad (30)$$

where m and n are about 2 and 1 respectively.

This noise is commonly termed flicker, 1/f, or excess. The form is shown in figure 8. At intermediate frequencies in the MHz range, the major contribution to the diode noise temperature ratio is usually shot noise, but at lower frequencies flicker noise predominates. The frequency at which the flicker noise starts to prevail is often termed the noise corner.

For mixer operation, the noise temperature ratio $N_r$ is the ratio of the total noise measured at the output terminals of the mixer, with the diode operating in a specified circuit under local oscillation excitation, to an equivalent resistance of the mixer, at room temperature. The mixer equivalent resistance is the resistance measured at the i.f. output terminals of the mixer due to l.o. excitation. Local oscillator excitation also implies a possible noise contribution due to the reverse part of the I-V characteristic. In practice, the mixer noise temperature ratio is usually observed to be greater for dynamic conditions (with l.o. excitation) than static conditions (equivalent d.c. bias), with a higher noise corner frequency (due to flicker noise) for the former situation.

The mechanism of dynamic mixer diode noise temperature ratio, particularly flicker noise, is not completely understood. It is known that it is related to the metal-semiconductor interface, junction formation, the rectifying contact area, ohmic contact, and finally the reverse current leakage. A large junction area, low reverse current leakage, and low diode rectified current are all desirable features for minimizing the diode noise temperature ratio.

## R.F. Admittance

The r.f. impedance (admittance) of the mixer is of importance in the mixer design. Not only can r.f. impedance mismatch result in signal loss due to reflection, but this can effect the mixer i.f. impedance, particularly in the case of low conversion loss diodes, see figure 6.

The diode parameters effecting r.f. admittance are given by equations (11, 12 and 13). A Smith chart interpretation is shown in figure 9.

It is desirable to adjust the diode parameters to provide an r.f. impedance near that of the system transmission line impedance, if power losses associated with matching circuits to provide large impedance transformation are to be minimized, and wide operation bandwidths obtained. In practice a compromise is usually adopted with the diode parameters for sensitivity and r.f. admittance. Operation near parallel resonance and d.c. admittance is usually avoided, with operation nearer but below series resonance preferred. The application of high l.o. power (thus reduced $R_B$) is general practice, particularly to meet the modern requirement of 50 ohm transmission line systems.

## I.F. Impedance $Z_{if}$

The mixer i.f. impedance is the impedance at the i.f. terminals of the mixer under operating conditions, and is of prime importance for the design of the i.f. amplifier input circuit, for ensuring a mixer - i.f. amplifier interface to give optimum receiver overall noise figure performance. To achieve this situation the design of the i.f. amplifier input circuit should be such that the optimum i.f. amplifier noise figure is obtained for a noise source resistance equal to the mixer i.f. output impedance, $Z_{if}$. It is important to note that this may not necessaryily be the same as the i.f. amplifier input impedance which gives maximum power transfer (1).

For conventional broadband (image matched) operation, however, a power match at the i.f. terminals is a desirable feature in order to reduce the effect of i.f. reflections complicating the signal and image impedance characteristics, figure 6(b).

The i.f. impedance is related to the diode high level signal a.c. resistance provided by the l.o. excitation. Thus the value is clearly a function of local oscillation power, diode rectified current, and any applied d.c. bias. However, as for conversion loss, the theory of mixing also shows that the i.f. impedance is dependent on the terminations presented at the image and signal ports, and the value of conversion loss. A high i.f. impedance is predicted for the reactive o/c image, low impedance for the reactive s/c image, and an intermediate value approaching the a.c. value when all the frequencies are matched (1).

## D.C. Bias

The main advantage of forward d.c. bias is that it enables operation with reduced local oscillator drive, or operation at reduced impedance levels (r.f. and i.f.). A possible disadvantage for the low local oscillator drive is reduced dynamic range. d.c. forward current will, however, increase the diode noise temperature ratio and this will provide a limit for this d.c. component.

The advantage of reverse d.c. bias, providing that the reverse I-V characteristic has low current leakage over the range of interest, is high local oscillator power operation, with the possibility of increased dynamic range.

The application of forward bias required a voltage supply, and different polarity lines may be required when diodes of opposite polarity are used (e.g. the balanced mixer (see Mixer Circuits). If the bias is derived from a low source impedance, then this value can be optimized so that the negative voltage developed by the diode rectified current will compensate for the applied positive d.c. bias, resulting in a variable bias with local oscillator drive and consequent performance fairly independent of local oscillator drive.

The reverse bias situation can be derived directly from the negative voltage developed across the load resistance, and thus by the choice of load resistance and rectified current.

## Dynamic Range

The dynamic range of any receiver is usually defined as the power difference between the minimum detectable signal and the maximum input signal that can be accepted before compression or undesirable intermodulation effects take place. In the case of the superheterodyne receiver, the minimum detectable signal is determined by the overall noise figure and the maximum input signal is normally taken to be the point where a specified compression (normally 1 dB) occurs in the r.f. to i.f. overall gain response (figure 9). Assuming that the i.f. amplifier does not limit the system, this implies compression in the mixer conversion loss response.

For example assuming the minimum detectable signal (MDS) sensitivity of a receiver is expressed as the signal power which is 3 dB greater than the system noise power, then

$$\text{MDS} = 2(KT_0BF_0)\ \text{(Mixer Parameters)} \tag{31}$$

$$\begin{aligned} \text{Thermal noise} &= kT_0B \\ &= 4 \times 10^{-18}\,\text{mW/Hz} \\ &= -174\ \text{dBm} \\ \text{System noise} &= -174\ \text{dBm}+10\log B+F_0\,\text{dB} \\ \text{MDS} &= -174\ \text{dBm}+3\ \text{dB}+10\log B+F_0\,\text{dB} \\ &= -171\ \text{dBm}+10\log B+F_0\,\text{dB} \end{aligned}$$

where B is the receiver bandwidth in Hz and $F_0$ is the receiver overall noise figure. The output signal level which produces receiver compression is related to the overall r.f. to i.f. gain G of the receiver, and dynamic range DR is the difference between the two input levels $P_{comp.}$ and $P_{MDS}$, such that

$$\text{DR} = (\text{output } P_{comp}\text{dBm} - G\text{dB}) - P_{MDS}\text{dBm (dB)}$$

Thus for a receiver with

$$B = 10\ \text{MHz}$$
$$F_o = 10\ \text{dB}$$
$$G = 20\ \text{dB}$$

and

(output level for 1 dB compression)$P_{comp}$ = +15 dBm,

$$\begin{aligned} \text{DR} &= [(+15 - 20) - (-171 + 70 + 10)]\ \text{dB} \\ &= 86\ \text{dB}. \end{aligned}$$

The individual mixer compression response may be obtained by plotting the conversion loss $L_c$ against the input or output power.

A further method of expressing the upper limit of dynamic range for a receiver is the intercept point technique. This is a measure of the third-order intermodulation products generated by a second input signal arriving at the mixer signal port along with the desired signal. Due to the inherent nonlinearities of the receiver (mixer - i.f. amplifier), it will generate new signals which are called intermodulation products. The intermodulation products of chief concern are the 2nd and 3rd order products, with the 3rd order being the most important because it usually falls within the pass band of the receiver. If the two signals f1 and f2 of equal amplitude are plotted in log-log form (i.f. output power as a function of r.f. input power) these are on a 2.1 and 3.1 slope respectively in relation to the 1.1 slope of the fundamental i.f. signal. The frequency term corresponding to the third order is $(2fS_2 - fS_1)\pm fL$ where fS represents the signal inputs and fL the l.o. frequency.

For example, figure 10 demonstrates the intercept point technique to determine the distortion-free dynamic range. The linear section of the curves are extended and intersect at the intercept point. It is generally accepted that the upper dynamic range limit determined by the intercept point technique is about 10 dB greater than that determined by the 1 dB compression point (1).

In practice (neglecting the i.f. amplifier contribution), for high dynamic range the mixer diode should be operated in its linear rectification region and the input signal level should be maintained below the applied local oscillator power. Desirable features are mixer diodes with low series resistance, large linear range I-V characteristic, low reverse current leakage, and high reverse breakdown voltage, with the application of high levels of local oscillator power. The high local oscillator power situation indicates an advantage for the use of mixer diodes which exhibit large value of diffusion potential $V_B$ or the use of reverse bias.

## Burnout

Microwave receiver reliability is largely dependent on the burnout properties of the diode effecting the frequency changing or detection of microwave pulses. The most common forms of microwave systems are the transmit-receive pulse radar, c.w. or pulsed c.w. doppler radar, and wideband video systems encountered with interrogating transmitters.

Burnout is a term given to the overload effect which results in significant deterioration in diode performance. The most frequently encountered type of mixer electrical overload in radar systems arises in the transmitting pulse by the function of a transmit-receive (t.r.) cell or gas switch. Power leakage through the cell results in the diode being subjected to a voltage/time pulse consisting of a spike followed by a flat response, as shown in figure 11.

The physical cause of diode burnout is complicated, but it is usually accepted that the major cause may be attributed to the high temperatures produced at the junction leading to diffusion or melting (1). This mechanism is in good agreement with practical experience, so the major points of thermal hypotheses are considered very briefly.

The theory is mainly based on the assumption that burnout is a function of temperature effects. Two aspects of major importance emerge. Namely, in the case of the microwave pulse (or c.w.) power conditions, the pulse width is longer than the thermal time constant* of the diode (this is normally considered to be about $10^{-8}$s) and, assuming a circular junction, the junction temperature may be given by

$$T \propto \frac{P}{4r(k_s + k_m)} \tag{32}$$

*The 'thermal time constant' or 'thermal relaxation time' is defined as the time required to reach a steady state temperature if the temperature were to continue to rise at its initial value.

where P is the peak pulse power, r is the junction contact radius and $k_s$ and $k_m$ are the thermal conductivities of the semiconductor and the junction metal respectively.

Thus, to reduce T for a given P, the inference of increasing r or the thermal conductivity of the junction is fairly straightforward.

However, in the case of a t.r. cell spike leakage, the spike width (normally considered to be about $4 \times 10^{-9}$s) is shorter than the diode thermal time constant, the time thus being too short for heat to appreciably diffuse away from the locality where it was generated, and assuming a circular junction the temperature may be shown to be given by

$$T \propto \frac{1}{C} \frac{Pt}{4\pi r^2 d} \qquad (33)$$

where Pt becomes the energy dissipated, r is the junction radius, d is the distance from the junction edge, and C is the thermal capacity.

The significance is: (i) temperature is proportional to the energy dissipated; (ii) temperature is inversely proportional to $r^2$ or junction area; (iii) there are extreme temperature edge effects. It may also be shown that heat is generated within the semiconductor, not the junction metal.

Burnout is thus a function of the type of junction (e.g. alloyed or metal-semiconductor), the metallization system, and for a given junction it is dependent on the junction radius, which should be as large as is compatible with the sensitivity and the r.f. admittance considerations. Burnout due to reverse voltage breakdown is only considered to be of significance under pulse power conditions, where the design of high reverse voltage breakdown devices is obviously desirable.

In practice, diode r.f. burnout is a complex function of the device r.f. impedance and hence absorbed power, r.f. circuit bandwidth, and (as already indicated) nature of the power, whether c.w., pulsed, or t.r. cell leakage are involved. Therefore r.f. burnout ratings in terms of maximum c.w. power, peak power, or spike energy tend to be rather ambiguous unless the r.f. circuit conditions are carefully defined. Although attempts have been made to define diode burnout ratings by using simulated t.r. cell spikes by coaxial line d.c. spikes (1), these unfortunately have not yet established a good correlation between r.f. and d.c. performance. Hence actual r.f. burnout tests give the most reliable information, but it is essential that assessment tests are carefully defined under specified controlled r.f. circuit conditions to ensure useful comparative data.

'Temporary deterioration' is a term given to an effect related to mixer burnout, which takes the form of a temporary increase in the receiver noise figure performance in a radar system using t.r. cell protection against overloads by its own transmitter. The deterioration in noise figure only occurs while the transmitter is operating, and the noise figure recovers to its initial value after the power is removed. In practice this may occur with some mixer diodes at a t.r. cell spike leakage level below that which produces permanent burnout damage, and lasts for about half a minute after the power is removed, before recovery.

## DETECTOR PARAMETERS

Detectors function on the basis of direct rectification of a microwave signal and operate in two distinct regions: low level and high level.

In the low level situation, the detector operates at small-signal levels in the square law region. That is the rectified output of the diode is proportional to the input power, and the output voltage (or current) is proportional to the square of the input voltage (or current). Such operation is commonly known as square law, video, low level, and small-signal detection.

The detector for the high level case operates at large-signal levels where there is a linear relationship between the rectified output current and the input voltage. This operation is commonly termed linear, high level, large-signal, and peak detection.

### Video Receiver Sensitivity

A noise figure cannot be defined (as for a superheterodyne receiver) for the sensitivity of a crystal video receiver. Instead a signal level is found such that the output signal power is just equal to or a specified level above the output noise power. The sensitivity is determined by a number of factors: detector diode rectification efficiency, video impedance, and noise properties, together with the video amplifier noise figure and bandwidth.

A widely used performance criterion for video detectors is 'tangential sensitivity' which indicates the ability of the detector to detect a signal against a noise background, and includes the noise properties of the detector and video amplifier. This is defined as the amount of signal power, below a one milliwatt reference level, required to produce an output pulse whose amplitude is sufficient to raise the noise fluctuations by an amount equal to the average noise level. Tangential sensitivity corresponds to a signal-to-noise ratio of about 2.5, and is thus approximately 4 dB above the signal which corresponds to a signal-to-noise ratio of about 1.0.

The tangential sensitivity performance in watts may be derived from (1)

$$S_t = \frac{1}{\beta Z_v} [4YkTB_v(F_v + N_r - 1)]^{1/2} \qquad (34)$$

where β is the short circuit current sensitivity, $Z_v$ is the video impedance, $B_v$ is the video amplifier bandwidth in Hz, $F_v$ is the video amplifier noise figure, $N_r$ is the noise temperature ratio of the detector diode, and Y is an experimentally determined factor, relating the edge to edge height of a band of Gaussian noise to its r.m.s. value when displayed on an oscilloscope:

The term 'figure of merit' is sometimes used to express quantitatively the quality of a video receiver, in terms of the detector diode short circuit current sensitivity (rectification efficiency) and video impedance, by

$$M = \beta Z_v^{1/2} \qquad (35)$$

It is important to note, however, that this only expresses the quality of a video crystal detector operating under zero bias conditions. The expression does not include the noise properties of the diode, i.e. the shot and flicker noise introduced by bias current, and does not therefore represent the true character of devices under conditions of forward bias.

Tangential sensitivity is therefore dependent on the amplifier noise figure and the amplifier bandwidth (usually varying as the square root of the bandwidth). The detector diode parameters of interest are the video impedance, current sensitivity, and noise temperature ratio.

Assuming a diode noise temperature ratio of 1.0 and negligible noise contribution from the video amplifier, than an approximate relationship between 'figure of merit' and tangential sensitivity for a 1 MHz video bandwidth may be derived from the equation (34):-

$$S_t = 51 + 10\log\frac{M}{60} - \text{dBm} \qquad (36)$$

which implies that a tangential sensitivity of -51 dBm corresponds to a figure of merit of 60.

Video Impedance and Current Sensitivity

The diode video impedance is defined as the dynamic impedance of the diode under operating conditions and is the reciprocal of slope resistance of the I-V characteristic at the operating point. That is:

$$Z_v = \frac{dV}{dI} \qquad (37)$$

or

$$Z_v = R_B \ (R_B \gg R_s).$$

The rectification efficiency of the diode is normally given in terms of current sensitivity or voltage sensitivity. The current sensitivity $\beta$ is expressed as the short circuit rectified current per unit power. That is:

$$\beta = \frac{I_{sc}}{P_{in}} \ \text{(normally expressed in } \mu A/\mu W) \qquad (38)$$

and the voltage sensitivity r is expressed as the voltage developed across a specified load resistance per unit power. For example,

$$r = \frac{V_{oc}}{P_{in}} \ \text{(normally expressed in } mV/\mu W) \qquad (39)$$

where $V_{oc}$ is the open circuit voltage. The two are related by

$$I_{sc} = \frac{V_{oc}}{Z_v} \qquad (39)$$

The theoretical low frequency current sensitivity $\beta_o$ (neglecting the effect of $C_B$ and $R_s$) can thus be determined by the device short circuit rectified current $I_{sc}$ and the r.f. power $P_{in}$ absorbed in the device in terms of voltage applied across the device V.

By analysis of an I-V characteristic of the form given in equation (1) it may be shown that (1)

$$\beta_0 = \frac{\alpha}{2} \qquad (41)$$

With reference to equation (2), this clearly indicates that ideally a low frequency current sensitivity of about 20 μA/μW should be obtainable assuming $q/nkT = 40\ V^{-1}$.

At microwave frequencies, with reference to figure 5, the voltage across the barrier $V_B$ is degraded by the effect of the diode parasitics $R_s$ and $C_B$, and it may be shown that (1)

$$\frac{I_{sc}}{P_{in}} = \frac{\frac{1}{4}V_B^2\alpha/(R_B+R_s)}{\frac{1}{2}V_B^2(1/R_B+R_s/R_B+\omega^2C_B^2R_B^2)}$$

and

$$\beta = \frac{\alpha}{2(1+R_s/R_B)^2}\,\frac{1}{(1+\omega^2C_B^2R_sR_B^2)/(R_s+R_B)}$$

Assuming $R_B$ $R_s$, this reduces to

$$\beta = \frac{\alpha}{2}\,\frac{1}{1+\omega^2R_B^2R_sR_B} \qquad (42)$$

or

$$\beta \propto \frac{\beta_0}{1+\omega^2C_B^2R_BR_s} \qquad (43)$$

This is the same form as the rectification efficiency given by equation (10).

Similarly, with the backward diode, the theoretical current sensitivity may be determined from an analysis of the current-voltage characteristic of the form

$$I = f(V)$$

and shows that

$$\beta_0 = \left[\ C_2\Delta E^{-1\ \ 2} - \frac{2}{S}\right] \qquad (44)$$

where S is the total penetration of the Fermi level into the conduction and valence bands of n- and p-type respectively, and $C_2$ is a constant determined by the semiconductor properties.

By submitting the appropriate semiconductor constants in equation (44) it is possible to determine the potential theoretical current sensitivities for semiconductor materials suitable for fabricating backward diodes. For example, with germanium, the first and second terms in equation (44) are 4.2 and 16 respectively, giving a theoretical current sensitivity of 11.8 μA/μW. For gallium arsenide the two terms are 1 and 5 respectively, giving 4 μA/μW.

Curve fitting to the graphical form of figure 4 also gives a potential $\beta_0$ of about 12.0 for a germanium backward diode(1). $\beta_0$ can then be related to the microwave current sensitivity by

$$\beta \propto \frac{\beta_0}{1+\omega^2C_B^2R_BR_s}$$

Similar considerations thus apply to detectors and mixer diodes, in that for high efficiency, $C_B$ and $R_s$ must be minimized. Detector sensitivity, however, depends on the square of $C_B$

in contrast to linear dependency for the mixer.

D.C. Bias

For detector application, the best type of I-V characteristic is one with maximum curvature at the operating point. The noise contributions to the diode noise temperature ratio $N_r$ are, however, still the thermal, shot, and flicker as already described for mixers in the section in Diode Noise Temperature Ratio. Although the detector situation has not the added complexity of local oscillator excitation, it is obvious that it is desirable for the operating point to occur at zero bias (in which case $N_r = 1.0$), as the use of d.c. bias current may introduce additional noise into the system resulting in an increase in the value of $N_r$.

Performance advantages can, however, be derived by the application of forward d.c. bias to detector diodes with large zero bias values of barrier resistance $R_B$ normally associated with the metal-semiconductor device. These arise directly from the reduction of $R_B$ with bias current.

The relationship may be seen by rearranging equation (1), such that

$$V = \frac{1}{\alpha} \ln \frac{I+I_s}{I_s}$$

and the slope resistance

$$Z_v = \frac{dV}{dI} = \frac{1}{\alpha(I+I_s)} \quad (45)$$

$$= \frac{1}{\alpha I} \quad (I \quad I_s) \quad (46)$$

$$= \frac{26}{I} \text{ ohms} \quad (47)$$

where I is in mA (ideal Schottky barrier diode), or

$$Z_v = \frac{1}{\alpha I_s} \quad (I = 0, \text{ i.e. zero bias}). \quad (48)$$

The low frequency current sensitivity $\beta_0$ is generally independent of forward d.c. bias, but $\beta$ is dependent on the frequency, with the shunting effect of $C_B$ becoming more severe with large values of $R_B$. This situation can be improved by the use of forward d.c. bias, as already indicated; however, the diode noise temperature ratio $N_r$ increase with bias will set an upper limit to the bias current. The optimum bias will depend on the reactance of $C_B$ (thus the frequency), $R_B$, and the diode $N_r$. The video amplifier bandwidth, especially the lower cut-off frequency, will also have some influence on the diode $N_r$, due to the onset of flicker noise.

In addition the application of forward d.c. bias has other beneficial effects in that it reduces the r.f. and output (video) impedance of the diode. The lower r.f. impedance results in an easier r.f. match to the r.f. source, preferably 50 ohms for the modern equipment, and increases the r.f. bandwidth response. The lower output impedance allows the use of a lower input impedance amplifier and results in an increased video bandwidth.

A constant current bias arrangement, i.e. high impedance source, is commonly used for detector diodes, but this can result in a rapid change in r.f. impedance with signal power, due to the effective change in $R_B$. A constant voltage bias source (low impedance) is thus preferred when it is required to maintain a match to the r.f. source at signal power levels above tangential.

Dynamic Range

The dynamic range (DR) of the video receiver is given as the difference between a specified input signal level, expressed as the minimum detectable-signal (MDS), and the input signal level to give a specified deviation from square law response, such as the 1 dB input compression point in the voltage output.

That is,

$$DR = P_{cp} \text{ dBm} - P_{MDS} \text{ dBm} \quad (dB) \quad (48)$$

MDS is defined as the smallest signal which can be observed and is assumed to correspond to a signal-to-noise ratio of unity. Tangential sensitivity corresponds to a signal-to-noise ratio of approximately 2.5. For example,

$$DR = -10 \text{ dBm} - (-54 \text{ dBm})$$
$$= 44 \text{ dBm}$$

for $P_{cp} = -10$ dBm and $P_{MDS}$ and $P_{s_t}$ equal -54 dBm and -50 dBm respectively.

Obviously the compression point seen at the detector diode output will depend on the gain of the video amplifier. The square law characteristic of the diode, neglecting any contributions from the video amplifier, may be determined by plotting the diode open circuit voltage $V_{OC}$ as a function of input power $P_{in}$, the upper limit being defined as the input power where a specified (normally 1 dB) departure occurs from square law.

Burnout

The most common forms of electrical overlaids encountered by a detector diode are c.w. and pulse power. Thus, assuming that burnout is a temperature effect, as discussed in the mixer section in burnout, then equation (3) is of interest. Unfortunately, the requirements for a large area for burnout consideration conflict directly with the sensitivity depending on the square of the barrier capacitance for a detector diode. In practice a compromise situation is accepted leading to similar burnout ratings to those of the mixer diode.

## DIODE REALISATION

### The Point-Contact Diode

Point-contact technology has extensively been employed to achieve the required small-area junction for microwave mixer and detector applications. Not only has the choice of whisker material to be consistent with the achievement of the junction rectification properties, but compatibility with the formation of the junction is also essential.

A model of the conventional form of point contact, assumed to be circular, in contact with a bulk semiconductor material, is shown in figure 12. The contact area is small compared with the thickness of the bulk semiconductor layer, resulting in the concept of spreading resistance, given by

$$R_s = \frac{\rho}{4r} \qquad (50)$$

$$= \frac{1}{4rNe\mu}$$

That is

$$R_s \propto \frac{1}{rN\mu} \qquad (51)$$

where $\rho$ is the resistivity, N the majority carrier density, $\mu$ the carrier mobility, and r is the radius of contact.

The barrier capacitance is obtained from the Schottky theory of metal-semiconductor contacts,

$$C_B = \pi r^2 \left( \frac{\varepsilon\varepsilon_0 Ne}{2\phi} \right)^{1/2}$$

and

$$C_B \propto r^2 N^{1\ 2} \varepsilon^{1/2} \qquad (52)$$

Thus, by combining the product $C_B R_s$, it may be shown that for mixer operation

$$L_c \propto C_B R_s \propto \frac{r\varepsilon^{1/2}}{N^{1/2}\mu} \qquad (53)$$

This expression indicates that the contact radius should be made as small as possible. In practice this is determined by design considerations compatible with satisfactory mechanical stability and resistance to electrical overloads. In general the equation provides a figure of merit for the semiconductor material: the higher the value of mobility the better the material for reduction of $C_B R_S$ and thus the higher cut-off frequency capability. Semiconductor material carrier concentrations of about $5 \times 10^{18}$ carriers/$cm^3$ for p-type silicon and about $1 \times 10^{18}$ carriers/$cm^3$ for n-type germanium are generally used in practice.

In device fabrication the whisker material should have the desirable mechanical and electrical properties to allow formation of the metal-semiconductor interface to give the required rectifying characteristics. In general, silicon point-contact diodes uses p-type semiconductor material with a tungsten or a tungsten based wire, and germanium point-contact diodes use n-type semiconductor material with a titanium wire. Unfavourable barrier potentials associated with n-type silicon point-contact diodes have not allowed advantage of the higher mobility n-type compared with p-type silicon to be made use of in practice. Formation of the metal-semiconductor interface is an important aspect of point contact fabrication, the aim being to obtain a clean intimate contact free of high current density spot locations: the contact is mechanically formed in the case of silicon, and electroformed in the case of germanium.

In the case of gallium arsenide a phosphor bronze whisker with an electroformed contact has generally been used, but such devices have not been fabricated in production quantities as in the case of silicon and germanium.

## THE SCHOTTKY BARRIER DIODE

The metal-semiconductor junction of the Schottky barrier device is formed with an evaporated contact rather than a whisker. A model is shown in figure 13.

This approach maintains the required clean interface between the metal and semiconductor, with resultant low internal noise generation, the advantages of diode design for elimination of possible instability of the whisker pressure contact and compatibility with integrated circuits. The junction are and consequent barrier capacitance are normally larger than associated with point-contact technology (using bulk semiconductor material); however, by use of epitaxial semiconductor material, the series resistance is reduced to maintain the desired product for $C_B R_S$.

The contact area is thus large compared with the thickness of the epitaxial layer, and the series resistance is given by

$$R_s = \frac{wP}{\pi r^2} \qquad (54)$$

that is

$$R_s \propto \frac{w}{r^2 N\mu} \qquad (55)$$

where w is the thickness of the epitaxial layer. Hence from equation (52) it may be shown that for mixer operations

$$L_c \propto C_B R_s \propto \frac{w\varepsilon^{1/2}}{N^{1/2}\mu} \qquad (56)$$

This expression indicates that the epitaxial layer should be made as thin as possible. A theoretical lower limit is set by the need to accommodate the depletion layer associated with the potential barrier, but practical limits may be greater than this. Provided therefore that an epitaxial layer of the requisite thickness can be achieved and a suitable junction can be formed, the expression suggests further freedom in device design than with point-contact technology (1). Further, the elimination of a forming procedure to produce an intimate metal-semiconductor interface allows the practical use of higher mobility semiconductor materials, such as n-type compared with p-type silicon, and germanium arsenide. Although theoretically superior to silicon (i.e. higher mobility), germanium is not normally used for Schottky barrier devices, due to the lack of epitaxial technology with this material.

As the Schottky barrier junction is non-penetrating it is possible to prepare devices on the thin epitaxial layers, i.e. about 0.2 µm thick. With material of this type, the junction can be made on the surface of the high resistivity layer and the breakdown voltage and other parameters associated with that doping level retained, whilst the substrate being of low resistivity reduces the bulk resistance of the device. Epitaxial layer carrier concentration about $10^{17}$ carriers/$cm^3$ are usual.

Three approaches are generally used for connection to the Schottky barrier contact metal.

(a) The so called multi-dot approach as shown in figure 13, in which a matrix of holes is etched through an oxide layer on the semiconductor surface, with metal contacts then being deposited in the etched out areas to form the rectifying elements, care being taken to create a depression at each contact by ensuring that the metal layer is thinner than that of the oxide, and finally bringing a pointed wire into contact with the oxide surface until it locates a depression to establish electrical contact with the rectifying contact metal. Contact diameters as small as about 2 μm diameter can generally be achieved with this technique.

(b) The bonded wire approach shown in figure 13, in which a gold overlay structure is produced and a gold wire is thermocompression bonded to the overlay metal. A technique with this structure is to produce several rectifying contacts of different diameters on a single chip, the appropriate capacitance being chosen depending on the application. Contact diameters are normally in the region of about 5 to 20 μm.

(c) The use of beam lead technology described in Section Diode Encapsulation. Contact diameters are normally in the range of about 3 to 10 μm.

An important aspect of the Schottky barrier diode is that it gives the design freedom for the application of different metals to form the rectifying contact. With reference to the metal-semiconductor junction, the work function of the metal contact (although affected by surface state and image forces) influences the barrier height $\phi_B$ of the metal-semiconductor contact and thus the potential barrier $V_B$. $V_B$ is important in mixer operation in that it determines the local oscillator power necessary to bias the diode into its nonlinear region; this is of interest in some system applications, particularly where the available local oscillator power is limited. In addition, if $V_B$ can be reduced sufficiently by lowering of the barrier, then this should result in the capability for zero bias detector operation.

An indication of the range of potential barriers that may be achieved in practice is very briefly summarized in Table 1 by giving very approximate values of $V_B$ for different semiconductor-metal combinations.

In general, diodes exhibiting values of $V_B$ which allow optimum mixer performance to be obtained at local oscillator powers less than 1.0 mW are termed low barrier height devices, and devices which require local oscillator powers in excess of 1.0 mW are known as high barrier height devices. As seen from Table 1, the low barrier height devices are normally produced using silicon with titanium as the barrier metal.

Devices with sufficiently low barrier potential to allow zero bias detector operation are termed 'zero bias Schottky barrier detectors', i.e. $V_B \precsim 0.25$. The barrier height lowering to achieve this form of operation is normally achieved by high temperature processing to produce some form of alloying (1), and there is some doubt as to whether this type of device is a true metal-semiconductor device conforming to the Schottky theory of rectification.

It may be shown however (1) that for this form of device, $Z_V$ is temperature dependent and will decrease with temperatures higher than room temperature and increase with temperatures lower than room temperature, with obvious implications on device sensitivity.

In practice, this temperature dependency may restrict the use of the zero bias Schottky barrier detector to application where a wide operating temperature is not required, such as test equipment. The device may not be suitable for military environments unless temperature compensation circuit techniques are employed.

It should be noted, however, that these temperature dependencies would not apply to the situationfor a d.c. biased detector or a local oscillator biased mixer (1).

## The Backward Diode

As already indicated, the backward diode (1) is similar in concept to the tunnel diode, and equations for calculations of the barrier capacitance $R_B$ and seriest resistance $R_s$ are of the form

$$C_B = \pi r^2 \left( \frac{\varepsilon q}{2V_B} \frac{N_D N_A}{N_D - N_A} \right)^{1/2} \qquad (57)$$

and

$$R_s = \frac{1}{4rNe\mu} \qquad (58)$$

where r is the radius of the contact area.

Carrier concentrations of about $10^{19}/cm^3$ are required and a high mobility is desirable to maximize $R_s$. Ge, GaAs and GaAn all have acceptable properties. At present however, microwave backward diodes are normally fabricated using n-type Ge with a gallium or aluminium doped junction contact.

Alloyed junctions and pulse bonded techniques are often employed for their fabrication. Although etched mesa techniques have been used with limited success, a sufficiently low capacitance is difficult to achieve with such a structure and the use of a pulse bonded method with its associated small contact area is an attractive method for obtaining the required capacitance (1). However, with improvements in planar technology it is now possible to consider planar structures in which the junction is defined by an oxide without subsequent etching or processing. Models of the pulse bonded and planar alloyed structure are shown in figure 14.

## Diode Characteristics

I-V characteristics of different diode types, with mixer and detector operation are shown in figures 15, 16 and 17.

The merits of different diode types designed for about 10 GHz frequency, their theoretically predicted performance potential and X-band microwave characteristics are summarized in Tables 2 and 3.

It should be noted that:-

(a) The point contact metal-semiconductor contacts do not exhibit ideal Schottky barrier I-V characteristics, and ideality factors n greater than 1.5 are normally observed.

The value of the forward I-V characteristic $\bar{\chi}$ is usually in the range of 5 to 9, giving rise to a low frequency mixer conversion loss $L_c$

(image matched) of about 4.2 dB.

An ideality factor n of about 2.0 for point-contact diodes predicts a low frequency sensitivity of about 10 μA/μW and this suggests an ideal tangential sensitivity performance of about -58 dBm for a 1 kΩ video impedance. However, this is not observed in practice and values nearer -50 dBm are usual. The difference is mainly considered to be due to the forward bias current requirement to reduce the video impedance to a usable level, resulting in an increase in the barrier capacitance and the diode noise temperature ratio.

(b) The Schottky barrier evaporated contact approach maintains the required clean interface between the metal and semiconductor with resultant low internal noise generation and I-V characteristic which approach ideal, i.e. ideality factor of about 1.05.

The value of the forward I-V characteristic exponent $\bar{\chi}$ can be as high as 22, with predicted low frequency mixer conversion loss of about 3.3 dB (image matched).

The higher energy gap and resultant higher diffusion potential of GaAs normally implies the need for increased local oscillator power requirements compared with silicon.

Flicker noise as in the case of the point-contact diode, can still predominate at low intermediate frequencies, and values of about 10 dB may be observed at about 10 kHz.

Due to the very high zero bias impedance levels, acceptable detector sensitivity depends critically on forward bias.

Similarly to the point-contact diode, although the low frequency and the microwave tangential sensitivity values are improved by about 2 dB compared with the point-contact case, the theoretically predicted tangential sensitivity of about 60 dBm ($Z_V$ = 1 kΩ) is not realized at microwave frequencies.

As a result of the design flexibility provided by the Schottky barrier diode design approach, that is, choice of contact metal and contact diameter yet still maintaining acceptable sensitivity performance, resistance to electrical overloads is easier to design for than in the point-contact case.

(c) Due to the inherent I-V characteristic of the backward diode the value of the I-V forward exponent $\bar{\chi}$ is normally about 2, with consequent low frequency mixer image matched conversion loss of about 5.5 dB. Lower local oscillator power and better flicker noise properties are obtained with the backward diode compared with the metal-semiconductor devices.

Considering detector operation, the I-V characteristic of the Ge backward diode gives rise to a potential low frequency current sensitivity of about 12 μA/μW. Although this value is less than theoretically predicted for metal-semiconductor devices, this is not degraded by a great extent in practice. Optimum operation takes place at zero bias (thus $N_r$ = 1.0) with a low value of video impedance.

## DIODE ENCAPSULATION

As already indicated in Section The Nonlinear Rectifying Element, the diode encapsulation can play a significant role in the diode operational characteristics, not only as regards the frequency response, but also as regards the associated mount to introduce it into the microwave circuit.

Some forms of diode encapsulation, based on point contact technology, for waveguide and coaxial applications are shown in figure 18.

At the early stages of the advent of the Schottky barrier diode, efforts were made to produce retrofit devices for the point-contact structures, for use in pre-tuned mounts and thus application in existing equipments. From equation (11) the inductance associated with the whisker wire plays an important role in the r.f. admittance of the device and this it is desirable to retain this length of wire or to introduce the appropriate inductance, together with providing the correct value of barrier capacitance $C_B$. The lower value of series resistance $R_S$ associated with the Schottky barrier could however present a difficult problem in designing a direct r.f. admittance point-contact diode retrofit. The multi-dot Schottky barrier design approach (shown in figure 13(a)) obviously offered a solution to obtaining the desired equivalent inductance, and this technique has generally been adopted by diode manufacturers.

More advanced diode structures for use in planar transmission lines, such as microstrip, for application to Microwave Integrated Circuits (MICs), are shown in figure 19.

Direct bonding of the Schottky barrier chip into the microstrip transmission line can produce problems, as the diode r.f. performance cannot be assessed to allow pre-selection before mounting. Further, the length of the connecting wire from the rectifying contact, and hence the inductive parasitic due to it, cannot be controlled sufficiently to assume repeatable r.f. characteristics. An alternative method of introducing planar devices into MICs which is now finding wide acceptance is the use of the LID (leadless inverted device) ceramic channel carrier for the active chip (1). This structure is illustrated in figure 19(a). The configuration allows full r.f. characterization before permanent bonding into the circuit. There is, however, an inherent inductance of about 1 nH associated with the structure, which normally leads to a series resonant frequency of about 10 GHz and the restriction to obtaining broadband operation in 50 ohms systems above about 12 GHz.

An obvious requirement for a high operating frequency is a reduced inductance structure, and the beam-lead technology inherently lends itself to this design objective (1). Briefly, beam-lead techniques imply the formation of two co-planar beams, one contacting the rectifying junction and the other providing an ohmic contact. A beam-lead structure is shown diagrammatically in figure 19(b).

The equivalent circuit for this type of device is given in figure 20. One of the chief design requirements of the beam lead structure is minimization and control of the stray capacitance $C_S$ introduced by the junction contacting beam overlaying the dielectric layer.

This is effectively in parallel with the junction capacitance and series resistance with undesirable implications on the device sensitivity performance and the series resonant frequency. Design aspects affecting this parameter are the thickness of the dielectric layer and the beam geometry, i.e. overlaying beam area, the latter having to be compatible with the strength of the beam attachment.

Gallium arsenide beam lead diodes have been designed with cut-off frequencies greater than 1000 GHz, and with self-resonant frequencies suitable for frequency operation as high as 100 GHz.

## MIXER CIRCUITS

### Single-Ended Mixer

The single-ended mixer illustrated in the block diagram, figure 21, is the simplest form of mixer circuit. It consists essentially of r.f. and i.f. terminals, with both the signal and local oscillator frequencies being applied at the r.f. port. The circuit should provide:

(a) An i.f. filter that will not allow r.f. power to couple into the i.f. load circuit.

(b) A d.c. low resistance path for the diode rectified current.

(c) Any r.f. matching necessary to ensure that all the r.f. signal power is absorbed in the diode.

(d) Appropriate harmonic terminations, the l.o. second harmonic being the one of prime importance.

The main disadvantages presented by the single-ended mixer are injection of the local oscillator power and elimination of local oscillator noise sidebands.

The local oscillator coupling mechanism should have negligible effect on the signal admittance, and it should not introduce significant signal power loss. The most common method adopted is the use of a directive coupler arranged in the r.f. feed to the mixer (shown in figure 21). For example, a 10 dB coupler would imply the requirement for at least 11 mW available l.o. power to deliver 1 mW to the mixer terminals, and a signal loss of 0.4 dB.

Amplitude and frequency modulation of the local oscillator signal will produce noise sideband with a spectrum of the form shown in figure 22. It is obviously desirable to suppress this noise power if it is available at the signal and image frequencies as it will cause a deterioration in the receiver overall noise figure performance as indicated by equation 20. A common approach for the elimination of this noise contribution is the use of a filter. The value for the loaded Q of the cavity will depend on the local oscillator and intermediate frequencies, noise output of the l.o. at a specified power level, and the conversion loss of the mixer.

### Balanced Mixer

A balanced mixer consists of two separate mixers of the single ended form coupled together so that the l.o. noise products are balanced out.

Such a mixer is shown schematically in figure 23. The four-port 3 dB coupler may provide a 180° phase shift (e.g. a rat race coupler, a magic tee, etc.) or a 90° phase shift (e.g. branch arm, proximity, slot-hybrid etc.). The local oscillator and the signal are fed into adjacent ports and the single ended mixers are situated on the other two.

As the balanced mixer consists essentially of two diodes fed by a four-port 3 dB coupler, the overall performance is dependent on the characteristics of the coupler in addition to the performance of the mixer diodes. For example, the 3 dB coupler may be designed to provide 90° or 180° phase difference between outputs, and both types will provide a.m. noise suppression of the l.o. noise sidebands. However, each type of coupler possesses certain advantages from the point of view of port v.s.w.r. and isolation between ports. Briefly, the isolation is less dependent on the match of the coupled arms in the 180° case, but at the expense of v.s.w.r. and vice versa in the 90° case.

In summary, the single-ended mixer has the advantage of simplicity and wide r.f. bandwidth, whereas the balanced mixer provides the advantage of cancellation of sideband a.m. noise generated by the l.o., isolation of the l.o. from the antenna, and the possibility of lower l.o. power requirement, but is bandwidth limited by the coupler characteristics.

### Ring Mixers

The microwave ring mixer design is based on the well known switching ring modulator which has found general application at audio frequencies (1). The microwave form is often referred to as a Double Balanced Mixer (DBM) but this should not be confused with the double balanced mixer circuit described in the next section.

The ring mixer is essentially a four-diode mixer connected as shown in figure 24. At low frequencies conventional centre tapped toroid transformers may be used for applying the l.o. and r.f. signals and extracting the i.f. These, however, obviously apply restrictions at higher frequencies and microwave designs tend to use broadband balanced microstrip lines (balun transformers) to apply the l.o. and signal, with fine wire microwave chokes and miniature decoupling capacitors to isolate the i.f. from the signal and l.o.

In operation, the mixer can be visualized as a switch activated by the l.o. Unlike the balanced mixer, in which the signal is switched during half of the l.o. signal, the r.f. voltage is alternatively connected to the ring mixer output during both halves of the cycle. This results in a considerable difference in the output waveform of the two types of mixer, with no even harmonics of the l.o. or r.f. signals appearing at the output of the ring mixer, and a considerable reduction of the odd order intermodulation products (e.g. 2 l.o.± signal and 2 signal ± l.o.).

The ring mixer will thus in general provide similar properties, i.e. cancellation of l.o. a.m. noise and isolation of the l.o. from the antenna, as in the balanced mixer, with the potential merits of a compact design and bandwidth properties not restricted by the use of a 3 dB coupler. Characteristics are inherent high isolation between all ports, i.e. signal, i.f. and l.o.; suppression of even harmonics of the r.f. and l.o. signals, i.e. high rejection of even-mode harmonics and reduction in the total number of possible intermodulation products; high signal handling capability (due to the high levels of l.o. power that can be accommodated in a four-diode system), which can result in high dynamic range; and finally low i.f. impedance (resulting from four diodes in parallel). The r.f. matching considerations are similar to those of the single-ended mixer.

The sensitivity, i.e. overall noise figure performance, of the ring form of mixer is mainly characterized by the mixer diodes as discussed in the section on Mixer Parameters and the small signal conversion matrix analysis for a single ended mixer is generally applicable.

However, the practical operation of the ring mixer, as in the balanced mixer, does depend on circuit and diode balance, in this case transformer symmetry and matched quad diodes. The high l.o. power requirement to drive four diodes can be a disadvantage for some system applications, and the ring configuration does not allow monitoring of the diode rectified currents or the application of d.c. bias. Finally, the circuit complexity may impose a lower frequency design limitation than the conventional balanced mixer.

## Image Suppression and Image Recovery Mixers

Image suppression may be defined as the suppression of r.f. input signals at the image frequency, whilst image recovery (sometimes termed image enhancement) is defined as recovery of the image power generated by the mixer with re-conversion to intermediate frequency power leading to improved receiver overall noise figure. An image recovery circuit inherently provides image suppression, but not necessarily vice versa.

As already discussed in the section on Mixer Parameters, the theory of mixing shows that the conversion loss of a mixer diode is dependent on the image termination. Briefly, in the more conventional image-matched situation the image power generated in the nonlinear element is dissipated in the input load and can present about 3 dB loss in signal power. If some of this power can be recovered, however, then the performance of the mixer will be improved.

The mixer diode nonlinear element requirements for low conversion loss reactively image terminated mixers have already been indicated in Section Mixer Parameters. Briefly, these may be summarized as a high value for the diode forward exponent $\overline{\chi}$ and a high ratio of cut-off frequency to operating frequency. This clearly identifies the desirability for the use of gallium arsenide Schottky barrier diodes.

The section on Mixer Parameters has already noted that the diode i.f. impedance also depends on the image terminations, and thus the diode requirements which result in low conversion loss imply a greater variation in this parameter. The magnitude of the variations in the i.f. impedance can be as great as 4 to 1 from the image short-circuit to open-circuit case for a low loss diode, the former case resulting in the low impedance.

This situation must obviously be accommodated in the design of the i.f. amplifier input circuit to provide the correct source impedance for optimum receiver overall noise figure.

The simplest way to achieve and understand the operation of this form of circuit is to consider a mixer with a filter in the r.f. input line which will pass signal but will suppress image frequency input signals. The filter, however, will also present a reactive impedance to the image power generated by the mixer and will return this power to the mixer for possible conversion to intermediate frequency power, provided that it is in the correct voltage phase. If an isolator is placed between the filter and the mixer, then the diode image power will be absorbed and the mixer will operate under image matched conditions, but the circuit will have image suppression properties.

An alternative to the filter approach, which lends itself to the design aim of image suppression or recovery, is the phasing circuit technique. The most common form is based on the well known single side-band circuit which makes use of the fact that the two i.f. signals produced by signals on either side of the local oscillator are $180^{o}$ out of phase. By a proper phasing arrangement, the image can be 'out-phased' or rejected while the desired signals are unaffected. The circuit normally consists of two mixers and, in the case of image recovery, it recovers the image power generated within the circuit (1). This type of circuit does not incur the signal losses and bandwidth restrictions of the filter circuit, but it does present greater complexity in controlling the voltage phasing for image recovery.

The basic design concept is shown schematically in figure 25. Briefly, (with reference to figure 25), the design aim for image recovery is such that the image power generated in mixer 1 is converted to i.f. by mixer 2, and the image power generated in mixer 2 is converted to i.f. by mixer 1, with the resultant generated i.f. signals being in phase with those generated by the primary signal. The design is thus very dependent on the control of the phase relationship between the signals and thus critical control, with minimization, of the transmission line lengths; this requirement normally leads to the adoption of microwave integrated circuit techniques for the r.f. circuitry. Alternative approaches may be used for combining the intermediate frequency outputs of the two mixers with the appropriate $90^{o}$ phase shift, namely, positioning of the $90^{o}$ i.f. combiner directly following the two mixers feeding a single i.f. amplifier, or the use of a separate i.f. amplifier for each mixer before the i.f. power combiner.

In practice, for image recovery, the mixers are normally operated under image s/c conditions. This provides a low i.f. impedance suitable for operation with a single i.f. amplifier system. In addition, it is beneficial to provide very short line lengths connecting each mixer at the signal power splitter and this tends to promote the use of very compact mixer designs, such as the ring mixer structure.

The circuit of figure 25 may easily be converted to image suppression only by the use of a signal power splitter which provides isolation between the mixers and a load for the dissipation of the image power generated by both mixers, such that each mixer operates under image matched conditions.

ACKNOWLEDGEMENTS

Part of this work has been carried out with the support of the Procurement Executive, Ministry of Defence, sponsored by DCVD.

The work described in these notes was carried out at GEC Hirst Research Centre to whom acknowledgement is given for permission to publish.

REFERENCE

1. T.H. Oxley: The Application of Variable Impedance Devices in Mixers and Detectors. Chapter 4 - Wiley Series in Solid State Devices and Circuits.

TABLE 1 - Metal-Semiconductor Potential Barriers

| Metal | $V_B$ (Volts) | |
|---|---|---|
| | Silicon | Gallium Arsenide |
| Au | 0.8 | 0.9 |
| Ti | 0.4 | 0.8 |
| Ni | 0.5 | 0.9 |
| Rh | - | 0.9 |
| Pt | 0.8 | |
| Pd | 0.8 | |
| Au-Ge | | $\sim$0.25 |
| Au-Ag | | $\sim$0.25 |
| Cr-Au | $\sim$0.25 | |

TABLE 2 - Mixer Diode Characteristics

| | Point Contact Si | Schottky Barrier Si | Schottky Barrier GaAs | Back-diode Ge |
|---|---|---|---|---|
| **Low frequency characteristics** | | | | |
| Diffusion Potential $V_B$(V) | 0.3 | 0.5 | 0.8 | Conduction at zero bias |
| Reverse Voltage (10 μA)$V_{RB}$(V) | 2.0 | >6 0 | >6.0 | 0.5 |
| Series Resistance $R_s$ (ohms) | 20 | 10 | 5 | 2 |
| Junction Capacitance $C_j$ (pF) | 0.2 | 0.15 | 0.15 | 0.3 |
| I-V Ideality Factor n | 1.5 | 1.05 | 1.05 | - |
| Fwd I-V Exponent $\chi$ | 7.0 | 22.0 | 22.0 | 2.0 |
| **Ideal predicted performance** | | | | |
| Conversion loss $L_o$ | | | | |
| Image matched (dB) | 4.3 | 3.3 | 3.3 | 5.5 |
| Image s/c (dB) | 3.6 | 1.8 | 1.8 | 5.5 |
| Image o/c (dB) | 2.5 | 1.3 | 1.3 | 5.5 |
| Overall noise figure $F_o$ ($F_{if}$ = 1.5 dB, $N_r$ = 1.0) | | | | |
| Image matched (d) | 5.8 | 4.8 | 4.8 | 7.0 |
| Image s/c (dB) | 5.1 | 3.3 | 3.3 | 7.0 |
| Image o/c (dB) | 4.0 | 2.8 | 2.8 | 7.0 |
| **X-band characteristics** | | | | |
| Overall noise figure $F_o$ | | | | |
| ($F_{if}$ = 1.5 dB 45 MHz i.f. (dB) | 7.5 | 6.5 | 6.0 | 8.0 |
| 10 kHz i.f. (dB) | 28.0 | 18.0 | 18.0 | 10.0 |
| Conversion loss $L_c$ (dB) (image matched) | 5.0 | 5.0 | 4.5 | 5.5 |
| l.o. power range (mW) | 0.4-2 | 1-10 | 2-20 | 0.1-0.5 |
| Input level for 1 dB (4 dBm) departure from $L_c$ linearity | -18.0 | -5.0 to 0 | 0 to + 10 | -30.0 |
| **Burnout** | | | | |
| c.w. Power (mW) | 200 | 500 | 500 | 50 |
| Peak pulse power (1.0 μs, 1000 p.p.s.) (W) | 1.0 | 0.5 to | 2.0 | 0.1 |
| t.r. cell spike leakage e/s | 0.2 | 0.2 to | 0.5 | 0.05 |

TABLE 3 - Detector Diode Characteristics

| | Point Contact Si | Schottky Barrier Si | Schottky Barrier GaAs | Back-diode Ge |
|---|---|---|---|---|
| Low frequency characteristics | | | | |
| Diffusion Potential $V_B$(V) | 0.3 | 0.5 | 0.8 | Conduction at zero bias |
| Reverse Voltage (10 μA)$V_{RB}$(V) | 2.0 | >6.0 | >6.0 | 0.5 |
| Series Resistance $R_S$(ohms) | 20 | 10 | 5 | 2 |
| Junction Capacitance $C_j$ (pF) | 0.2 | 0.15 | 0.15 | 0.3 |
| I-V Ideality Factor n | 1.5 | 1.05 | 1.05 | - |
| Fwd I-V Exponent $\overline{\chi}$ | 7.0 | 22.0 | 22.0 | 2.0 |
| Ideal predicted performance | | | | |
| Current sensitivity $\beta_o$ (μA/μW) | 10 | 18 | 18 | 12 |
| Tangential sensitivity $T_S$ (dBm) | -57 | -59 | -59 | -55 |
| $N_r$ = 1.0, $F_v$ = 0dB<br>$Z_v$ = 1 kΩ for S/B<br>$Z_v$ = 300 Ω for B/D | | | | |
| X-band characteristics | | | | |
| Tangential sensitivity $T_S$ (dBm)<br>$F_v$ = 2 dB<br>$Z_v$ = 1 kΩ for S/B<br>$Z_v$ = 300 Ω for B/D | 50 | 52 | 52 | 54 |
| Input level for 1 dB (dBm) Departure from square law | -18 | -15 | -15 | -20 |

PROBLEMS

1.

(a) Give the expression for the overall noise figure of a superheterodyne receiver.

(b) Define the parameters involved.

(c) Given a diode conversion loss of 5 dB, calculate the receiver overall noise figure for a diode noise temperature ratio of 1.5 times, when the mixer is operating in conjunction with a following i.f. amplifier of 2 dB noise figure.

2. Define detector sensitivity (MDS and tangential). What are the important diode parameters?

3. Sketch the approximate I-V characteristics of typical silicon and gallium arsenide Schottky barrier diodes. Compare these with the silicon point contact and backward diode. Briefly indicate their application merits.

4. Why is DC bias sometimes applied to a mixer or detector diode? Discuss this for both mixer and detector applications.

5. Derive the expression for the 'figure of merit' of the semiconductor material for mixer applications. Indicate which semiconductor material has the best potential for high frequency operation.

6. Define the important diode parameters which influence the microwave performance of a mixer diode? A mixer diode has the following junction parameters and characteristics:-

Barrier capacitance $C_B$. $C_B$ = 0.2 pF

Series resistance $R_S$. $R_S$ = 10 Ω

Low frequency conversion loss $L_o$.

$L_o$ = 3.5 dB

Noise temperature ratio $N_r$.

$N_r$ = 1.0 at 60 MHz i.f. and 40.0 times at 10 kHz i.f.

Assume that the local oscillator drive is adjusted for minimum conversion loss and calculate the overall noise figure performance of a receiver operation at 10 and 20 GHz, where $F_{i.f.}$ = 1.5 dB, for

(a) an i.f. of 60 MHz
(b) an i.f. of 10 kHz

7. Discuss the technological approaches which may be applied to the fabrication of Schottky barrier diodes. What are their application merits?

8.

(a) How is the simple equivalent circuit for a mixer diode modified by the application of beam lead technology.

(b) A conventional circuit mounting of a mixer diode, with the junction properties of Question 6, introduces a series inductance of 1 nH. Application of planar technology reduces the series inductance to 0.1 nH, however, this introduces a stray overlay capacitance $C_S$ of 0.1 pF.

Assume the simple case that $C_S$ is in parallel with $C_B$ and calculate:-

(i) the series resonant frequency for the conventional and planar diode circuit mounting,

(ii) the effect on noise figure performace, by the planar approach, for the receiver of Question 6, operating at 10 and 20 GHz in conjunction with a 1.5 dB 60 MHz i.f. amplifier.

9. What is a balanced mixer? Discuss its merits compared with a single-ended mixer.

10.
(a) What is meant by the terms 'image frequency'? What approaches are taken to discriminate against the image frequency. Indicate the difference between image suppression and image recovery (enhancement).

(b) For the measurement of overall receiver noise figure, what factor should be added to the available noise source power if:-

(i) a broadband noise source is used with power available at the image and signal channels

(ii) a simple single section filter with a $Q_L$ of 130 is positioned between the broadband noise source and the mixer. The signal frequency is 9.375 GHz. Calculate the image contributions for an i.f. of 30 MHz and 80 MHz.

(c) What special precautions need to be taken if an R.F. amplifier stage is considered desirable to improve the sensitivity of a superheterodyne receiver.

11. Define receiver dynamic range. Express this for both the superheterodyne and video receiver.

Calculate the dynamic range of a superheterodyne receiver giving the following parameters:

Receiver bandwidth B = 10 MHz
Overall noise figure $F_O$ = 10 dB
Overall gain G = 20 dB
Output level for 1 dB compression = +15 dBm.

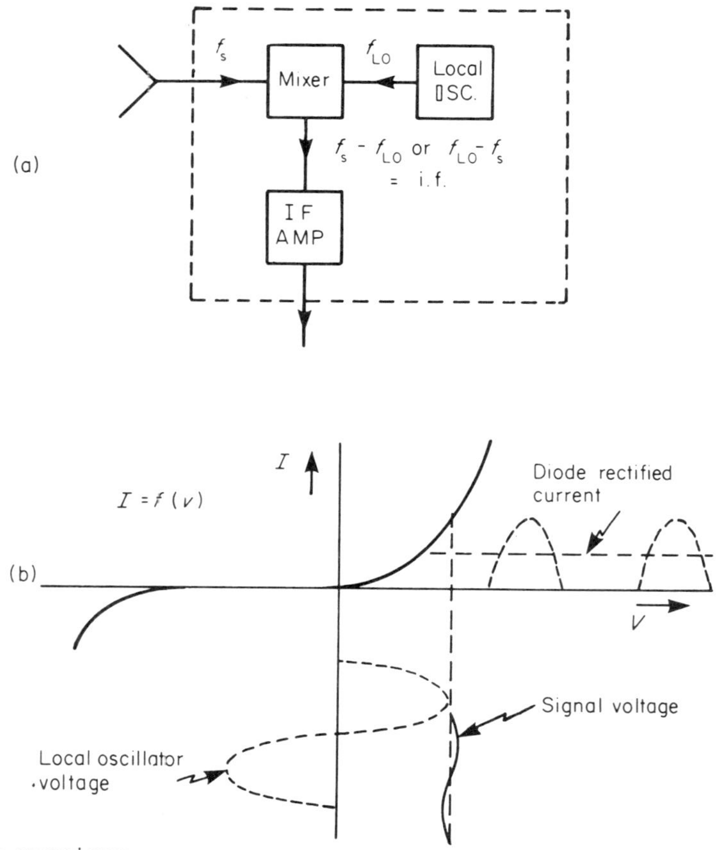


Figure 1 Superheterodyne receiver
(a) Block diagram (b) Application of local oscillator and signal voltages to diode rectifying element

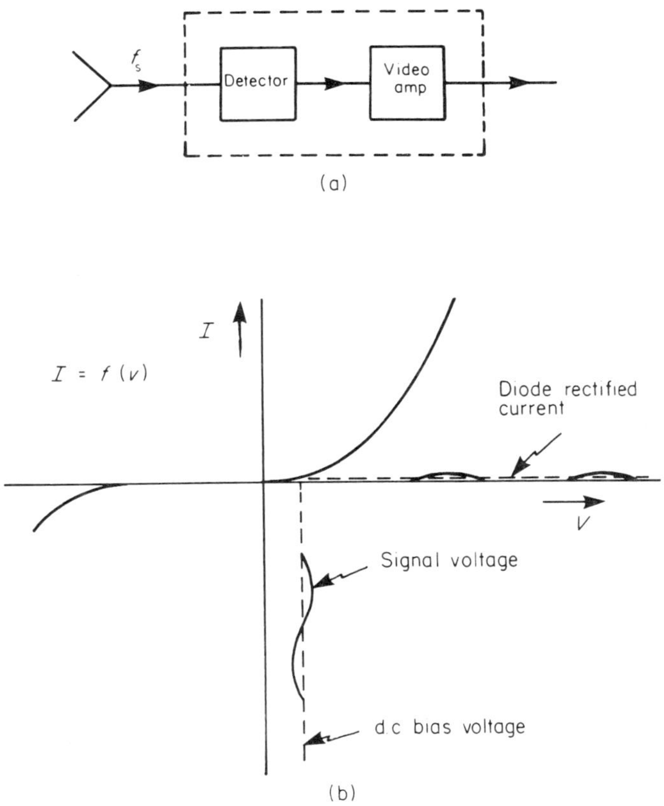


Figure 2 Video crystal receiver
(a) Block diagram (b) Application of signal voltage to diode rectifying element

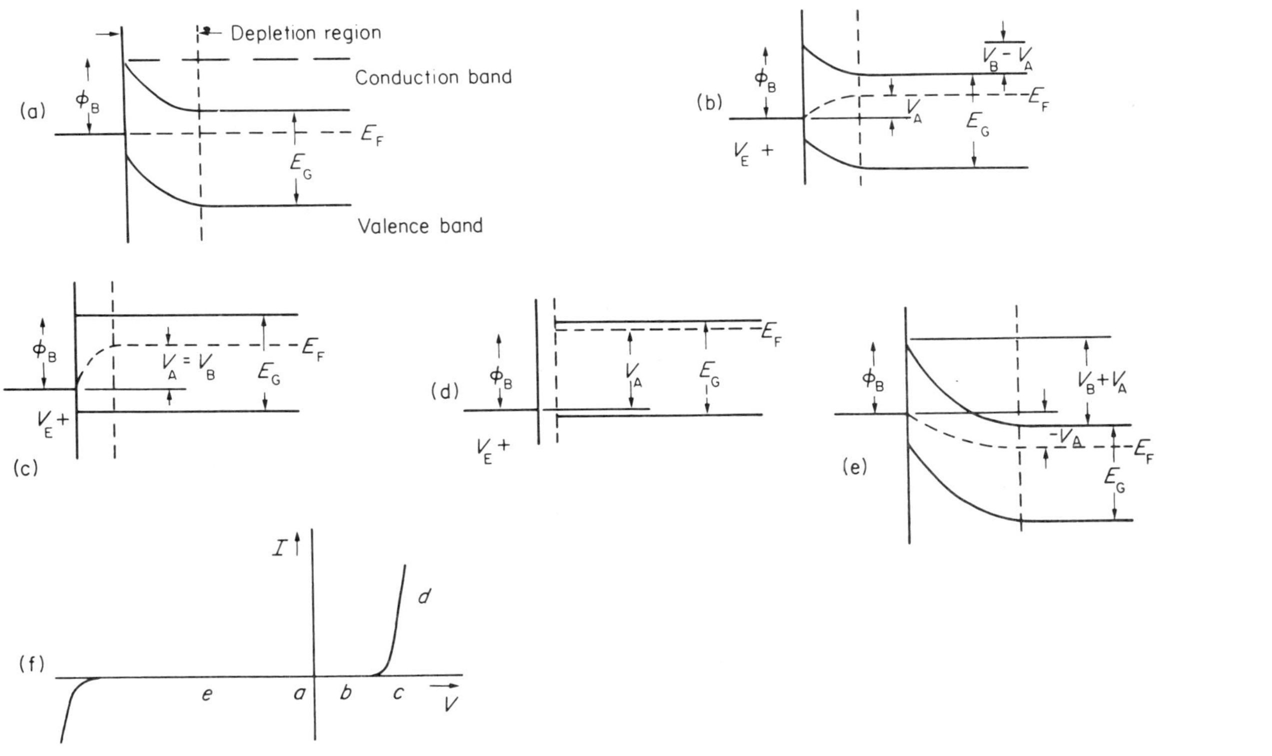


Figure 3 Metal-semiconductor energy band diagrams
(a) Zero bias (b) Forward bias (c) Forward bias (d) Forward bias (e) Reverse bias (f) I-V characteristic

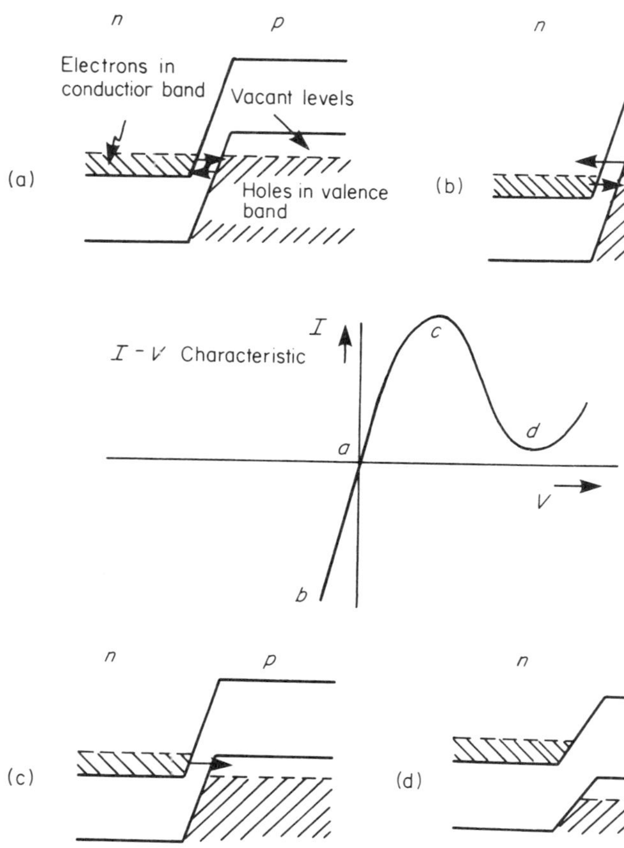


Figure 4 Tunnel diode energy band diagrams

(a) Zero bias (b) Reverse bias (c) Forward bias at peak (d) Forward bias at valley

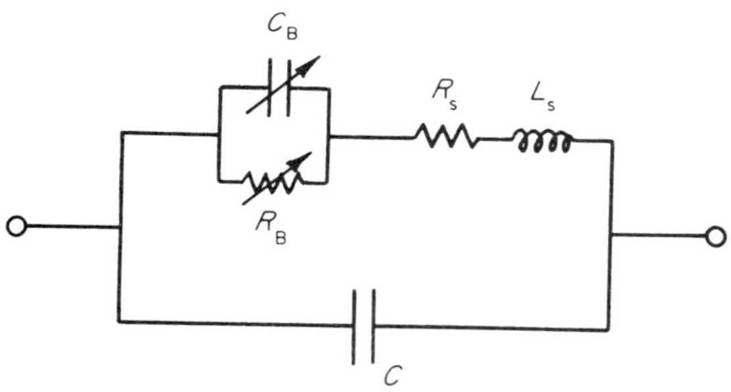


Figure 5 Simple equivalent circuits for diodes

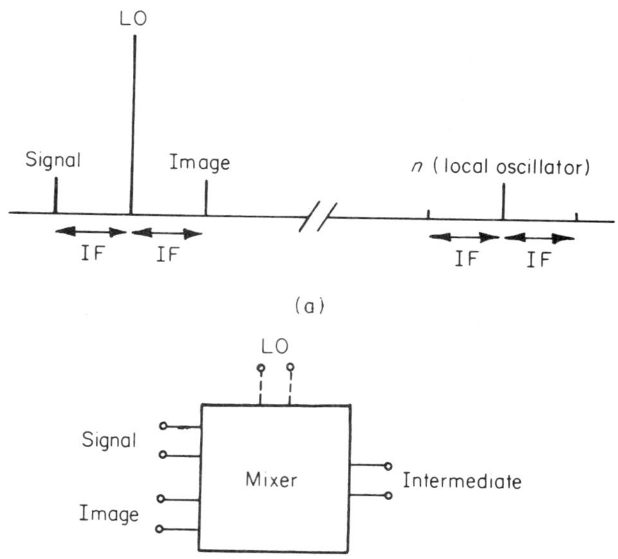


Figure 6

(a) Mixer spectral diagram
(b) Mixer network representation

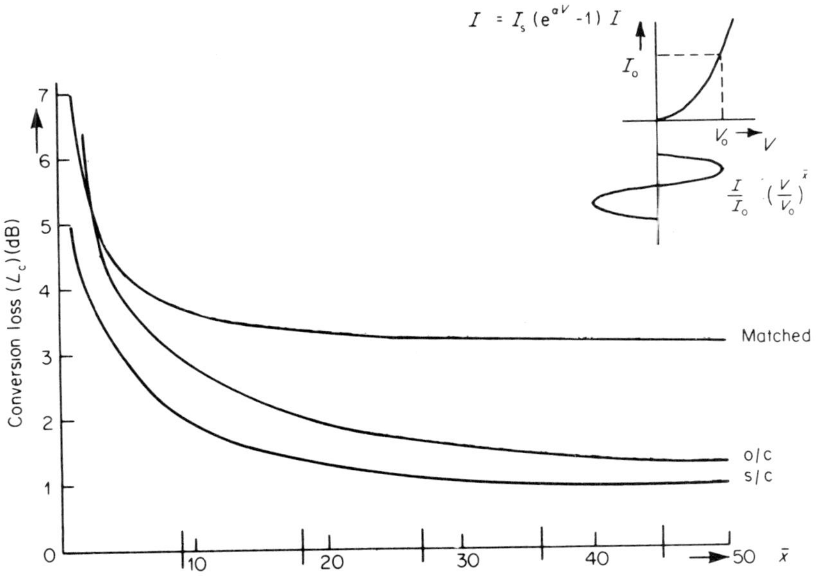


Figure 7 Low frequency mixer conversion loss as a function of diode exponent x

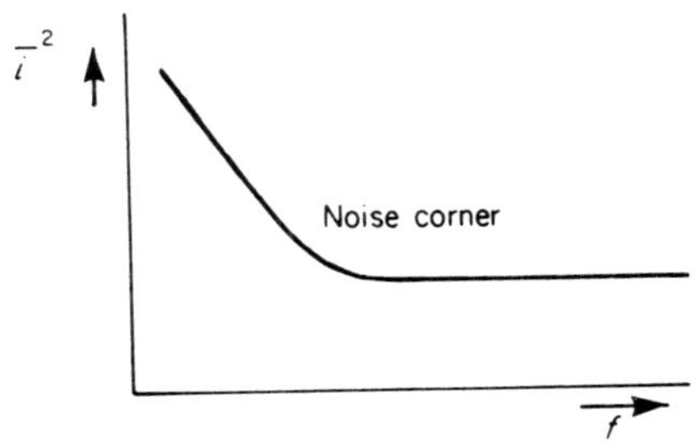


Figure 8 Flicker noise spectrum

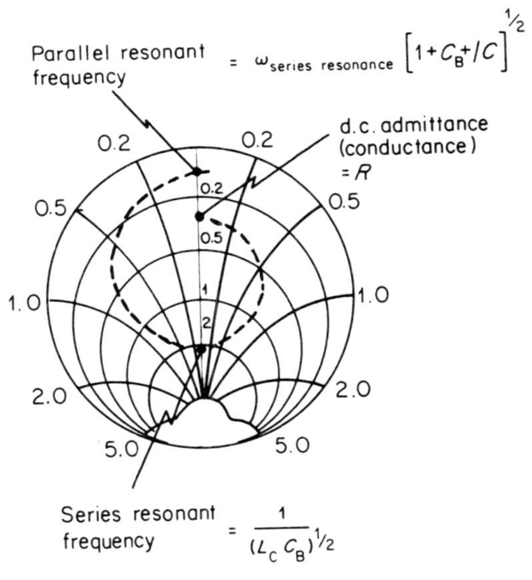


Figure 9 Smith chart interpretation of a diode r.f. admittance with respect to 0.02 mhos

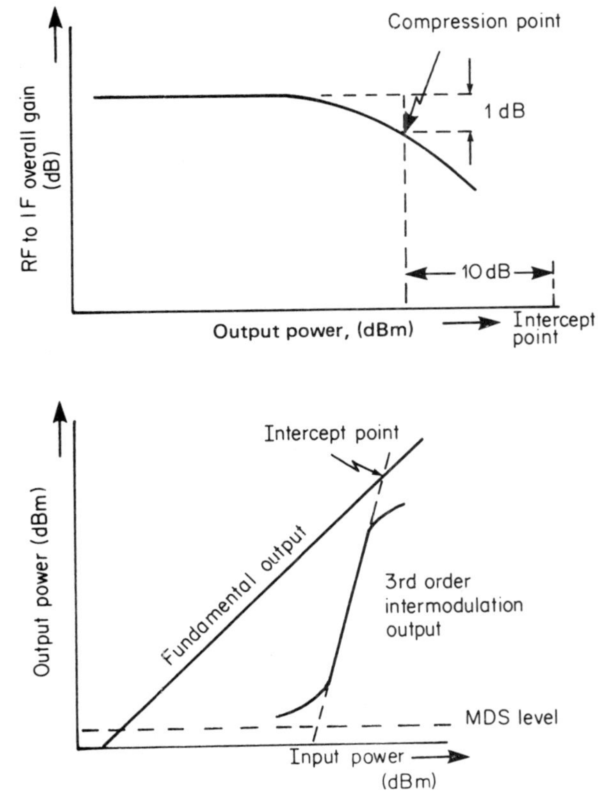


Figure 10 Dynamic range compression and intercept points

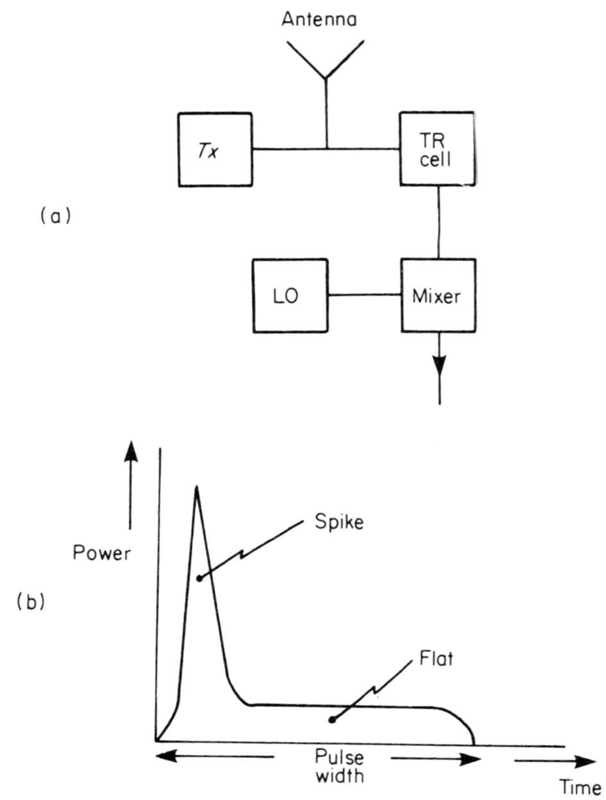


Figure 11 T.R. cell spike leakage

(a) Simple pulse radar system
(b) Leakage through t.r. cell

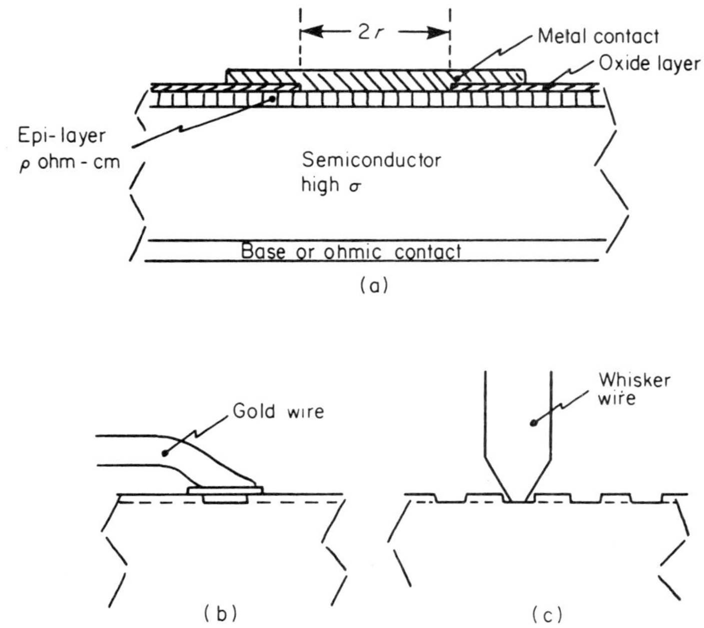


Figure 13 Schottky barrier diode technology

(a) Basic structure
(b) Thermocompression bonded contact
(c) Multi-dot wire pressure contact

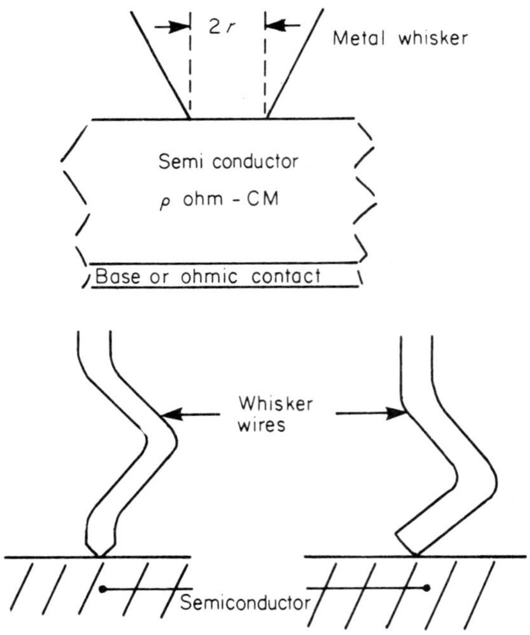


Figure 12 The point contact structure

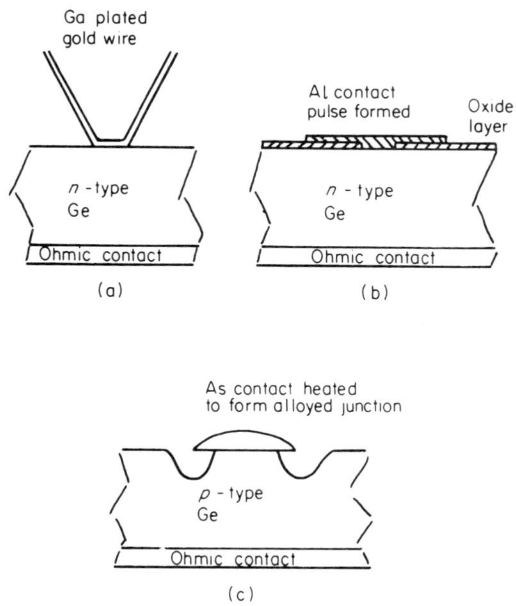


Figure 14 Backward diode structures

(a) Pulse bonded
(b) Planer
(c) Etched mesa

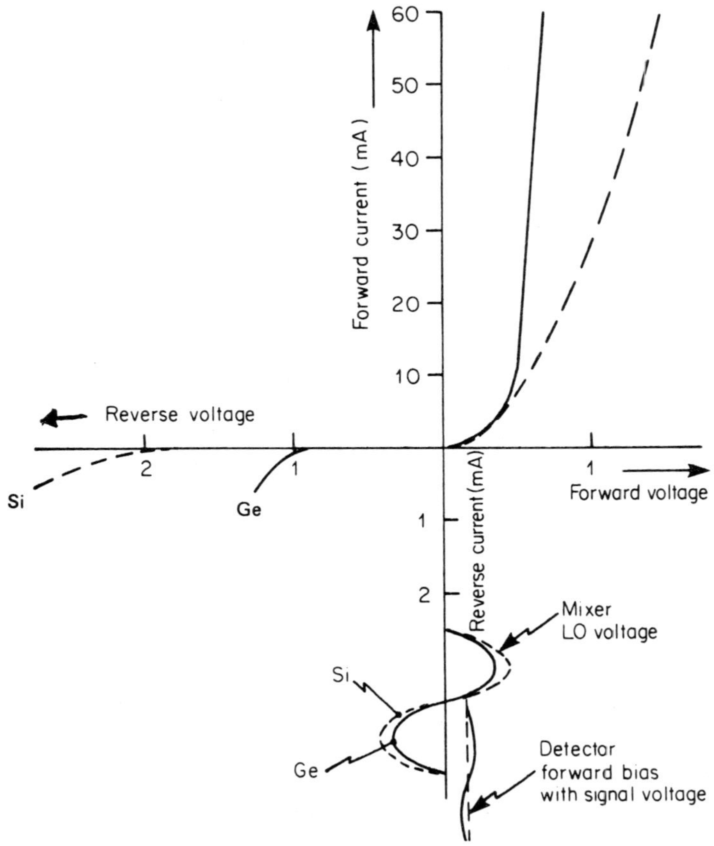


Figure 15 Point-contact I-V characteristics

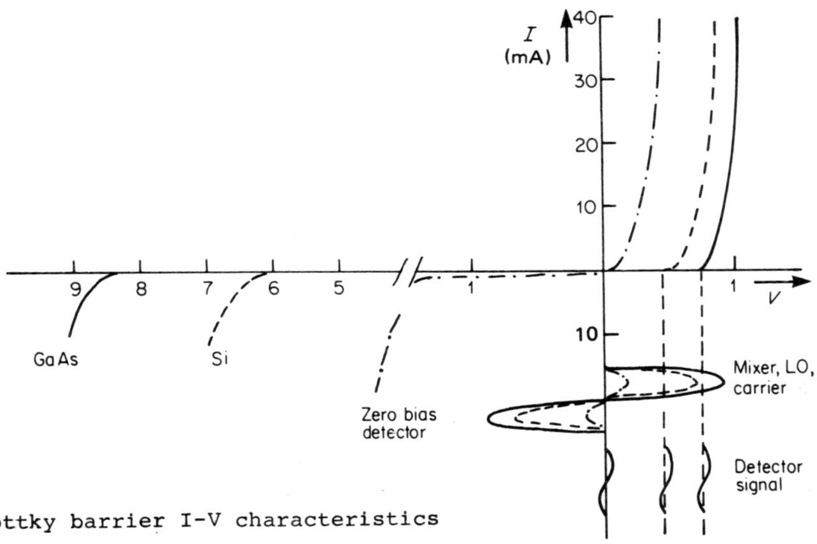


Figure 16 Schottky barrier I-V characteristics

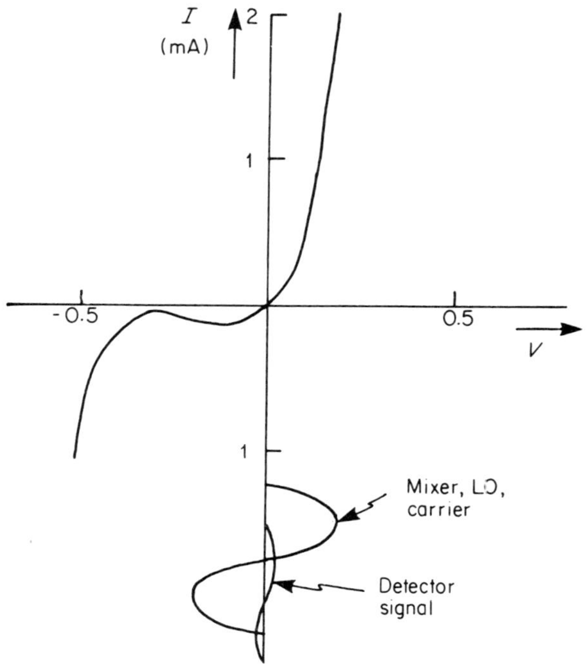


Figure 17 Backward diode I-V characteristics

(a)

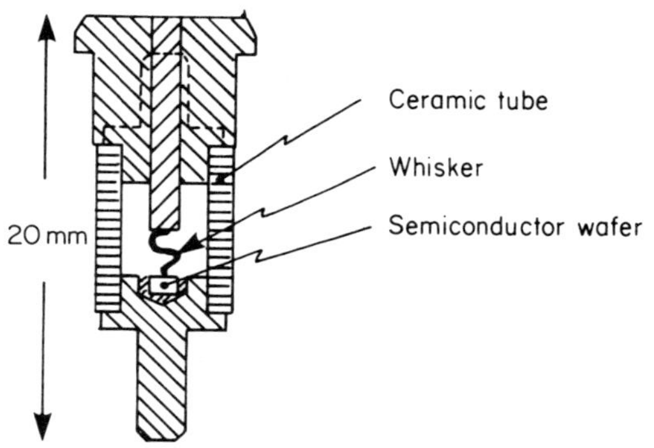


Figure 18 Forms of diode encapsulation

(a) Capsule

(b)

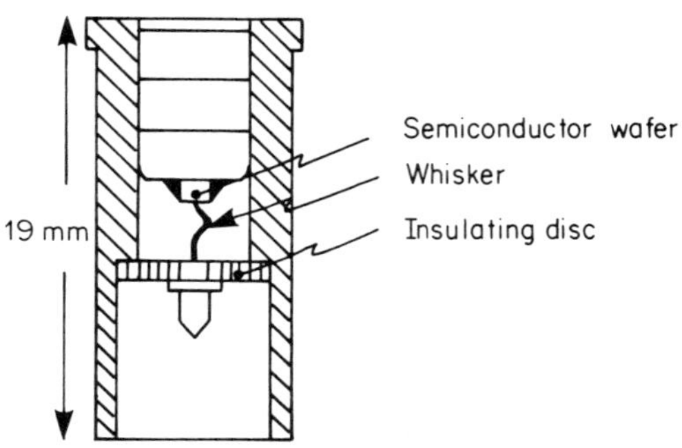


(b) Coaxial

(c)

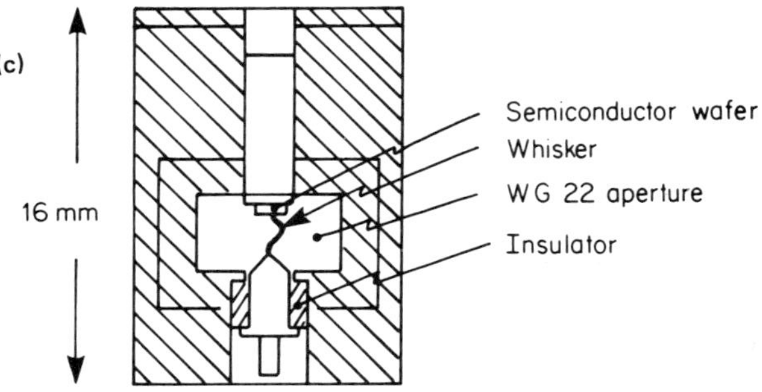


(c) Plug-in waveguide

(d)

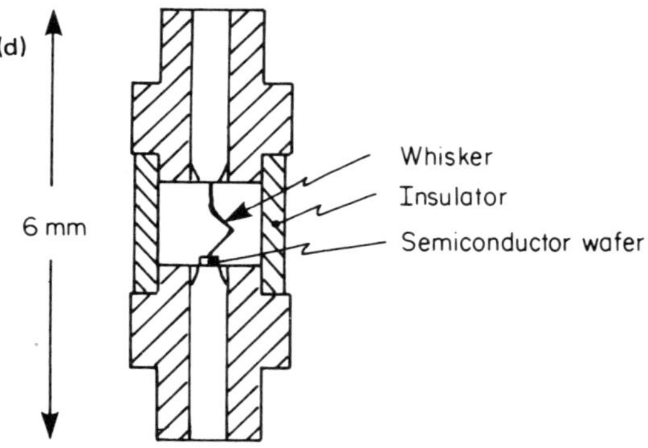


(d) Miniature double-ended capsule

(a)

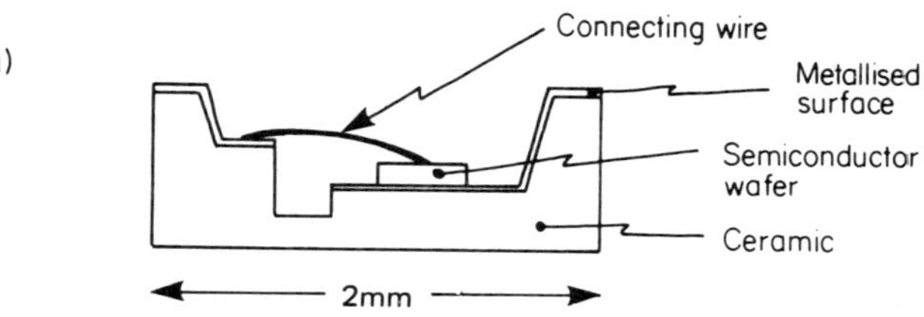


(b)

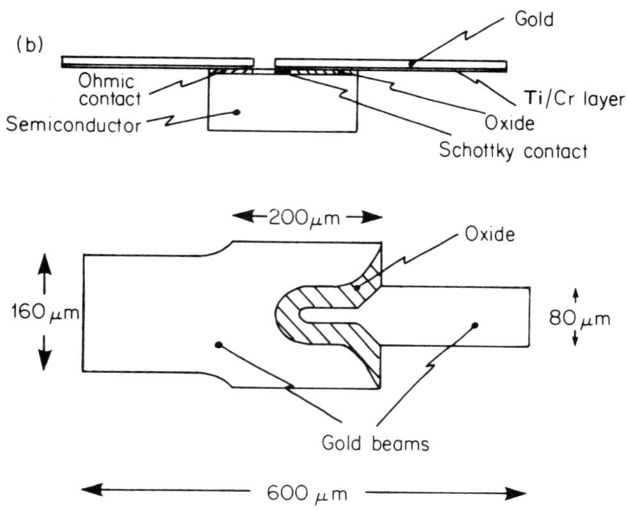


Figure 19 MIC diode structures

(a) Leadless inverted device (LID)
(b) Beam-lead

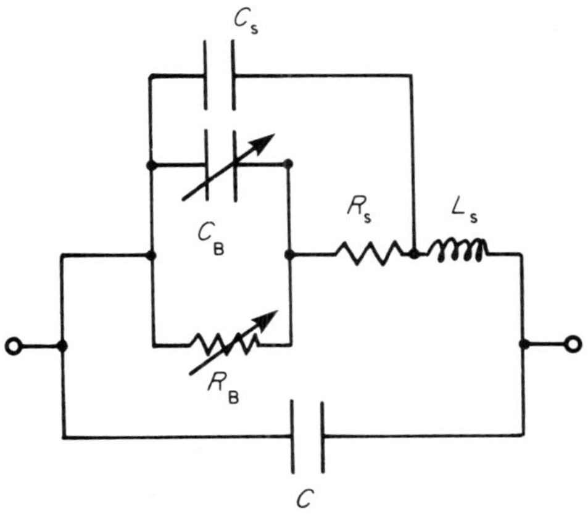


Figure 20 Simple equivalent circuit for beam-lead structure

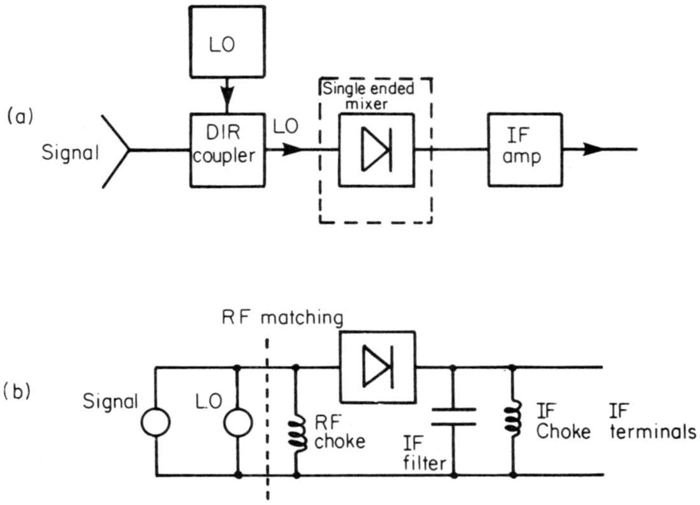


Figure 21 Single ended mixer
(a) Application circuit
(b) Basic mixer circuit

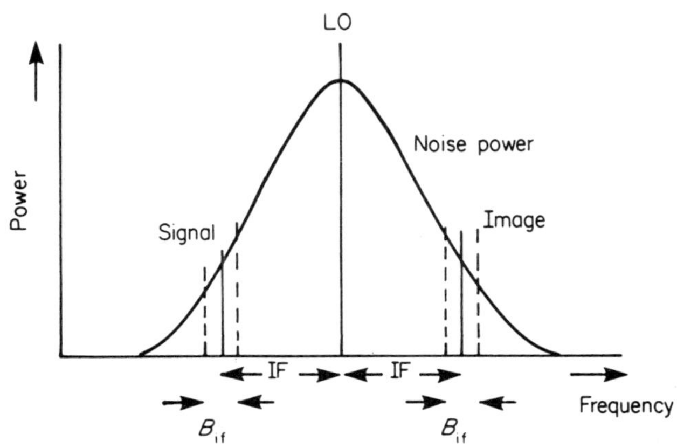


Figure 22 Local oscillator noise power spectrum

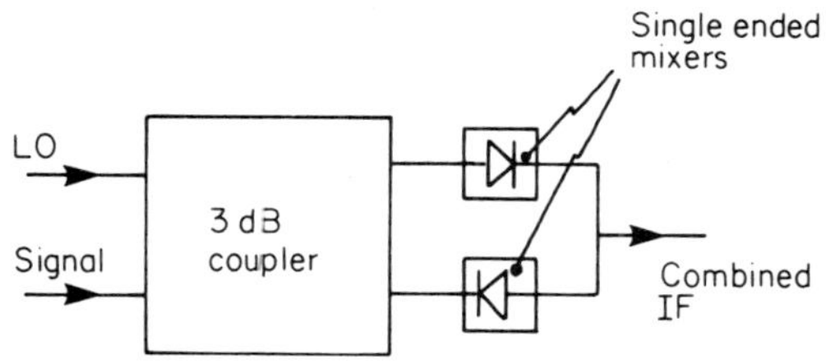


Figure 23 Balanced mixer

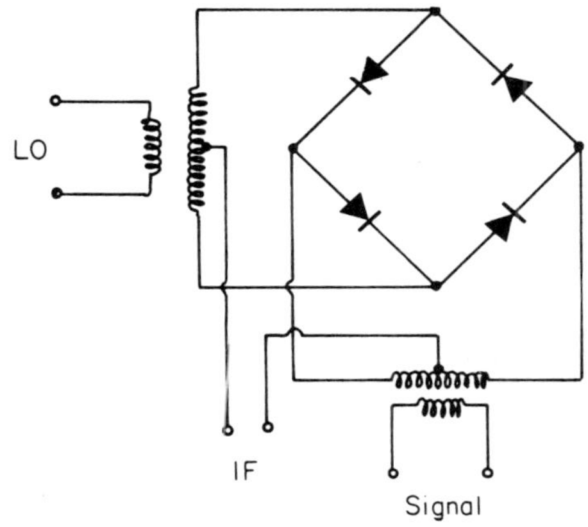


Figure 24 Ring mixer connection

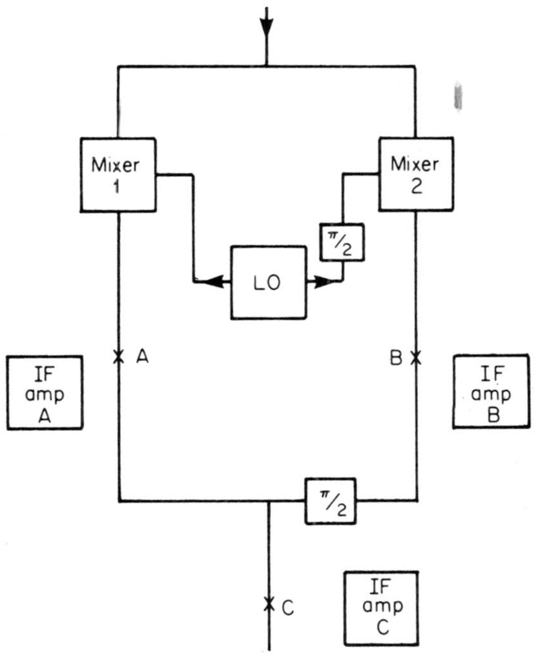


Figure 25 Phasing type double balance mixer circuit. Single i.f. amplifier at C or two matched i.f. amplifier at A and B

INDEX

INDEX

INDEX